Waste Management and the Environment III

WITPRESS

WIT Press publishes leading books in Science and Technology.
Visit our website for the current list of titles.
www.witpress.com

WITeLibrary

Home of the Transactions of the Wessex Institute.
Papers presented at WASTE III are archived in the WIT elibrary in volume 92 of
WIT Transactions on Ecology and the Environment (ISSN 1743-3541).
The WIT electronic-library provides the international scientific community with immediate
and permanent access to individual papers presented at WIT conferences.
http://library.witpress.com

THIRD INTERNATIONAL CONFERENCE ON
WASTE MANAGEMENT AND THE ENVIRONMENT

WASTE MANAGEMENT III

CONFERENCE CHAIRMEN

V. Popov
Wessex Institute of Technology, UK

A. Kungolos
University of Thessaly, Greece

C. A. Brebbia
Wessex Institute of Technology, UK

H. Itoh
University of Nagoya, Japan

INTERNATIONAL SCIENTIFIC ADVISORY COMMITTEE

J-H. Abraham
S. Bethanis
M. Cardinaletti
M. Gehring
D. Kaliampakos
M. Kuras
A. Mavropoulos
A. Moutsatou
K. Okada
J. A. Pascual
A. Purandare
J. Tharakan
A. Zouboulis

Organised by
Wessex Institute of Technology, UK

Sponsored by
WIT Transactions on Ecology and the Environment

Transactions Editor

Carlos Brebbia
Wessex Institute of Technology
Ashurst Lodge, Ashurst
Southampton SO40 7AA, UK

Email: carlos@wessex.ac.uk

WIT Transactions on Ecology and the Environment

Editorial Board

A H-D Cheng
University of Mississippi
USA

C-L Chiu
University of Pittsburgh
USA

A Cieslak
Technical University of
Lodz
Poland

W Czyczula
Krakow University of
Technology
Poland

M da Conceicao Cunha
University of Coimbra
Portugal

M Davis
Temple University
USA

A B de Almeida
Instituto Superior Tecnico
Portugal

K Dorow
Pacific Northwest National
Laboratory
USA

C Dowlen
South Bank University
UK

R Duffell
University of Hertfordshire
UK

J P du Plessis
University of Stellenbosch
South Africa

A Ebel
University of Cologne
Germany

D Elms
University of Canterbury
New Zealand

D M Elsom
Oxford Brookes University
UK

D Emmanouloudis
Technological Educational
Institute of Kavala
Greece

J W Everett
Rowan University
USA

R A Falconer
Cardiff University
UK

D M Fraser
University of Cape Town
South Africa

G Gambolati
Universita di Padova
Italy

N Georgantzis
Universitat Jaume I
Spain

F Gomez
Universidad Politecnica
de Valencia
Spain

W E Grant
Texas A & M University
USA

A H Hendrickx
Free University of
Brussels
Belgium

I Hideaki
Nagoya University
Japan

W Hutchinson
Edith Cowan University
Australia

K L Katsifarakis
Aristotle University of
Thessaloniki
Greece

B A Kazimee
Washington State
University
USA

D Koga
Saga University
Japan

B S Larsen
Technical University of
Denmark
Denmark

D Lewis
Mississippi State
University
USA

K G Goulias
Pennsylvania State
University
USA

C Hanke
Danish Technical University
Denmark

S Heslop
University of Bristol
UK

W F Huebner
Southwest Research Institute
USA

D Kaliampakos
National Technical
University of Athens
Greece

H Kawashima
The University of Tokyo
Japan

D Kirkland
Nicholas Grimshaw &
Partners Ltd
UK

J G Kretzschmar
VITO
Belgium

A Lebedev
Moscow State University
Russia

K-C Lin
University of New
Brunswick
Canada

J W S Longhurst
University of the West of
England
UK

U Mander
University of Tartu
Estonia

J D M Marsh
Griffith University
Australia

K McManis
University of New
Orleans
USA

M B Neace
Mercer University
USA

R O'Neill
Oak Ridge National
Laboratory
USA

J Park
Seoul National University
Korea

B C Patten
University of Georgia
USA

V Popov
Wessex Institute of
Technology
UK

M R I Purvis
University of Portsmouth
UK

T Lyons
Murdoch University
Australia

N Marchettini
University of Siena
Italy

J F Martin-Duque
Universidad Complutense
Spain

C A Mitchell
The University of Sydney
Australia

R Olsen
Camp Dresser & McKee Inc.
USA

K Onishi
Ibaraki University
Japan

G Passerini
Universita delle Marche
Italy

M F Platzer
Naval Postgraduate School
USA

H Power
University of Nottingham
UK

Y A Pykh
Russian Academy of
Sciences
Russia

A D Rey
McGill University
Canada

R Rosset
Laboratoire d'Aerologie
France

S G Saad
American University in
Cairo
Egypt

J J Sharp
Memorial University of
Newfoundland
Canada

I V Stangeeva
St Petersburg University
Russia

T Tirabassi
Institute FISBAT-CNR
Italy

J-L Uso
Universitat Jaume I
Spain

A Viguri
Universitat Jaume I
Spain

G Walters
University of Exeter
UK

A C Rodrigues
Universidade Nova de
Lisboa
Portugal

J L Rubio
Centro de Investigaciones
sobre Desertificacion
Spain

R San Jose
Technical University of
Madrid
Spain

H Sozer
Illinois Institute of
Technology
USA

E Tiezzi
University of Siena
Italy

S G Tushinski
Moscow State University
Russia

R van Duin
Delft University of
Technology
Netherlands

Y Villacampa Esteve
Universidad de Alicante
Spain

Waste Management and the Environment III

Editors

V. Popov
Wessex Institute of Technology, UK

A. Kungolos
University of Thessaly, Greece

C. A. Brebbia
Wessex Institute of Technology, UK

H. Itoh
University of Nagoya, Japan

 **WIT**PRESS Southampton, Boston

V. Popov
Wessex Institute of Technology, UK

A. Kungolos
University of Thessaly, Greece

C. A. Brebbia
Wessex Institute of Technology, UK

H. Itoh
University of Nagoya, Japan

Published by

WIT Press

Ashurst Lodge, Ashurst, Southampton, SO40 7AA, UK
Tel: 44 (0) 238 029 3223; Fax: 44 (0) 238 029 2853
E-Mail: witpress@witpress.com
http://www.witpress.com

For USA, Canada and Mexico

WIT Press

25 Bridge Street, Billerica, MA 01821, USA
Tel: 978 667 5841; Fax: 978 667 7582
E-Mail: infousa@witpress.com
http://www.witpress.com

British Library Cataloguing-in-Publication Data

A Catalogue record for this book is available
from the British Library

ISBN: 1-84564-173-6
ISSN: 1746-448X (print)
ISSN: 1743-3541 (online)

*The texts of the papers in this volume were set
individually by the authors or under their supervision.
Only minor corrections to the text may have been carried
out by the publisher.*

Preface

This volume of the Transactions of Wessex Institute contains the papers presented at the 3rd International Conference on Waste Management and the Environment that was held in Malta in June 2006. The conference offered the opportunity for professionals involved in the waste management sector, industrial sector, governmental and non-governmental organizations as well as other interested parties to be involved in discussions on key issues and challenges in waste management and to exchange experiences and views on the current technologies and strategies applied in different parts of the world.

Waste management is one of the key issues of modern society and requires constant advancement in the adopted practices and technologies in order to keep up with the new industrial technologies and commodities introduced to the market every year, which can have varying impacts on the environment. Sustainable strategies in waste management are focused on reducing the amount of generated waste and recycling of the waste, the creation of which cannot be avoided with current technologies.

The proceedings of the conference address a wide range of the current waste management issues, which are conveniently arranged into the following sessions: Advanced waste treatment technology; Air pollution control; Biological treatment of waste; Clean technologies; Community involvement and education; Construction and demolition waste; Costs and benefits of waste management options; Hazardous waste, disposal and incineration; Landfills, design, construction and monitoring; Methodologies and practices; Resources recovery; Soil and groundwater cleanup; Waste incineration and gasification; Waste management, strategies and planning; Waste pre-treatment, separation and transformation; Waste reduction and recycling; Water and wastewater treatment.

The Editors are grateful to all the authors for their excellent contributions and in particular to the members of the International Scient.fic Advisory Committee for the review of the abstracts and the papers and their help on ensuring high quality standards for the conference.

The Editors
Malta, 2006

Contents

Section 4: Water and wastewater treatment

Section 15: Waste management

Section 16: Construction and demolition waste

Section 17: Costs and benefits of waste management options

Section 1
Advanced waste treatment technology

Recovery of rare metals from spent lithium ion cells by hydrothermal treatment and its technology assessment

H. Itoh, H. Miyanaga, M. Kamiya & R. Sasai
*Division of Environmental Research, EcoTopia Science Institute,
Nagoya University, Japan.*

Abstract

A novel cobalt recovering process from a cobalt-based cathode electrode in a lithium secondary cell was developed using a hydrothermal treatment combined with a pyrometallurgical technique. A cobalt-based cathode electrode was prepared by casting the mixture of $LiCoO_2$, poly(vinylidene fluoride) and conductive carbon black onto an Al foil, and this model electrode was used as a test sample. A hydrothermal treatment of the cathode sample was carried out using pure water as a solvent in the temperature range of 423–473 K for the duration of 0–40 h. The hydrothermal treatment at 473 K for more than 15 h led the cathode sample to the disintegration into powdery particles. By the hydrothermal treatment under the optimum condition at 473 K for 20 h, more than 99.9 mass% of Co and 98 mass% of Al in the cathode sample was reclaimed as a form of spinel type $Co(Co_xAl_{2-x})O_4$ (0<x≤2), and most of Li and F could be dissolved into the solution. Subsequently, metallic cobalt was successfully recovered from the spinel compound and carbon mixture by the pyrometallurgical treatment with additives for slag formation under a reducing condition at 1623 K for 5 h.

Keywords: spent lithium ion cell, hydrothermal treatment, cobalt recovery, pyrometallurgical treatment.

1 Introduction

In the past one decade, lithium ion cell has been spreading its occupation in the world cell market because of its properties superior to other conventional secondary cells [1]. Due to its incomparable advantages, the lithium ion cell has been exploited for many electric appliances or even in a medical field [2]. Furthermore, the lithium ion cell contains less toxic metals comparing to the

WIT Transactions on Ecology and the Environment, Vol 92, © 2006 WIT Press
www.witpress.com, ISSN 1743-3541 (on-line)
doi:10.2495/WM060011

nickel-cadmium cell. Taking these overall advantages into consideration, it is quite explicit that the lithium ion cell will be equipped continuously in a number of new electric applications.

It has no doubt that a tremendous number of the spent lithium ion cells will be disposed when they reach the end of their practical life [3]. These wastes should be appropriately treated in order to recover cobalt in the spent cells since cobalt is classified as one of rare metals. The consumption of cobalt always leads to the market price appreciation due to little reserves of cobalt-containing ore in the earth's crust. In addition, a large amount of cobalt is demanded not only in lithium ion cells, but also in various sorts of industrial materials such as hard alloys, paints or magnets. Therefore, recovering cobalt from spent lithium ion cells is quite beneficial to solve the pressing problem on stable cobalt-supply.

Up to the present time, most of practical recycle process for the spent lithium ion cells is based on a pyrometallurgical method. This process mainly consisted of two heat-treatment stages, *i.e.*, the gasification removal of lithium and fluorine in the spent cell, and the reduction and condensation of cobalt from the residue. It is obvious that this process needs less chemical reagents, but a large amount of energy is required for the whole heat-treatment processes. Additionally, there is a serious problem in the first heat-treatment stage that a furnace wall has a severe damage by corrosive gasses.

Recently, a cobalt reclamation process has been investigated by a hydrometallurgical technique of the cathode materials in the spent lithium cell. For example, Zhang et al. [4] attempted to recover cobalt from the cathode by three steps, *i.e.*, (1) the cobalt dissolution step from the cathode using hydrochloric acid, (2) the extraction step from the leached liquid into kerosene and (3) the inverse extraction step with sulfuric acid to form cobalt sulfate. Nan et al. proposed another hydrometallurgical process, which was based on the dissolution-deposition and extraction methods [5]. These reports imply that cobalt-recovering process has an advantage from the standpoint of smaller energy consumption compared with the pyrometallurgical method; however, this process requires an appreciable amount of toxic chemicals or complex multistage-treatments.

In order to reduce the environmental loads arising from both the pyrometallurgical and hydrometallurgical treatments, we propose a novel cobalt-recovering process based on a hydrothermal technique, which is incorporated into the current pyrometallurgical treatment by replacing the heat-treatment process for the gasification of lithium and fluorine. In this paper, the disintegration process of the cobalt-based cathode electrode by hydrothermal treatment in pure water was investigated in detail and the feasibility of cobalt reclamation was also examined.

2 Experimental

2.1 Preparation of LiCoO$_2$ powder

Commercially available Li$_2$CO$_3$ and Co$_3$O$_4$ powders were used as starting materials for the preparation of LiCoO$_2$ powder. LiCoO$_2$ was synthesized through the solid-state reaction technique reported by R. Gupta and A. Manthiram [6]. The Li_2CO_3 and Co_3O_4 were first mixed by rotary ball-milling with ethanol, and then dried at 333 K for 1 h. An excess amount of about 3 mol% Li was added to the mixture to compensate for any loss of lithium that may occur during the firing. Then, the powder was precalcined in an O_2–N_2 stream (mole ratio of 20: 80) of for 5 h at 773 K. After the pulverization by ball-milling, the mixture was calcined again in the stream of O_2–N_2 for 24 h at 1123 K to form $LiCoO_2$. Finally, the agglomerated $LiCoO_2$ was further pulverized by ball-milling to obtain powdery $LiCoO_2$ for subsequent cathode preparation.

2.2 Preparation of cathode electrode sample

Commercially available conductive carbon black, poly(vinylidene fluoride) (abbreviated as PVDF) for a supporting binder, an aluminum foil for a current collector (purity 99.5%, thickness: 10 μm) and the synthesized $LiCoO_2$ powder were used for the components of a model cathode electrode sample.

The conductive carbon black, PVDF and $LiCoO_2$ powders were, respectively mixed at the ratio of 10: 5: 85 mass% by ball-milling. Then, N, N-dimethyl-formamide was added to the mixture to form viscous slurry, and the resulting slurry was cast onto one side of the aluminum foil (size: 2×2 cm). The sample was cold-isostatically pressed at 200 MPa for 1 min, and then vacuum-dried at 393 K. About 450 mg of the mixture was cast onto the Al foil to ensure an enough amount of the treated sample for subsequent analyses.

2.3 Hydrothermal treatment

Hydrothermal treatment was conducted using a stainless steel vessel lined with poly(tetrafluoroethylene) (PTFE) in an inner volume of 60 mL. A cathode electrode sample was placed into the vessel and then immersed with 30 mL of deionized and distilled water. The experiment was performed in the subcritical temperature range of 423–473 K for duration up to 40 h. At any experiment, the heating and isothermal step was regulated by applying an external heating, but the cooling process was spontaneously done in ambiance. After the hydrothermal treatment, the solid component was collected by filtration using a membrane filter (pore size: 0.45 μm) under reduced pressure.

2.4 Pyrometallurgical treatment

Cobalt aluminates ($Co(Co_xAl_{2-x})O_4$, x=0) were prepared from γ-Al_2O_3 (particle size < 20 nm) and Co_3O_4 by referring to the paper reported by Casasdo and Rasines [7]. The synthesized $CoAl_2O_4$ was mixed with additives, such as SiO_2, $CaCO_3$, α-Al_2O_3 and carbon black. The mixing molar ratio of $CaCO_3$: SiO_2: α-Al_2O_3: $CoAl_2O_4$ = 1: 0.8: 0.2: 0.02 was applied for slag formation. The powders were mixed by ball-milling, uniaxially pressed into a pellet with diameter of 15 mm, and then cold-isostatically pressed at 200 MPa for 1 min. The green compact was charged into a graphite crucible, muffled with carbon in mass quantities, and thermally treated in an electric furnace in Ar flow at 1623 K for

5 h. This temperature was determined so as not only to reduce the $CoAl_2O_4$ but also to generate the liquid slag phase in the CaO-SiO_2-Al_2O_3 system by referring to the phase diagram in the ternary system of CaO, SiO_2 and Al_2O_3 [8], and the Richardson-Jeffes diagram for the each oxide [9].

3 Results and discussion

3.1 Hydrothermal treatment of the cathode electrode

Figure 1 shows the photographs of (a) as-prepared cathode and the hydrothermally treated specimens in H_2O at 473 K for (b) 5 (c) 15 and (d) 40 h. As-prepared cathode was collapsed depending on the treatment time; the cast mixture of $LiCoO_2$, carbon black and PVDF was peeled off from the Al foil, and then was disintegrated into powdery particles; the Al foil became thinner and was broken apart with duration of the treatment. The fractured Al foil can be seen in the disintegrated mixture at the treatment time of 5 h (see Fig. 1(b); bright gray fragments are Al foil), but the foil disappeared completely when the treatment time was more than 15 h (see Figs. 1(c), (d)).

Figure 1:   The appearance variation of the cathode electrodes though the hydrothermal treatments; (a) an untreated cathode and hydrothermally treated cathode for (b) 5, (c) 15 and (d) 40 h at 473 K in H_2O.

3.1.1 Analyses of the solution after hydrothermal treatment

Relationship between the pH values of (■) calculated and (▲) measured of the solution after the hydrothermal treatment at 473 K for various times is shown in Figure 2. When the sample was immersed in water, the pH value of the solution increased immediately. Subsequently, the pH value increased gradually with increasing hydrothermal treatment time, and reached about 12. The measured pH value almost corresponded to that of the calculated one according to the Li^+ dissolution reaction [10]:

$$LiCoO_2 + 5H_2O \rightleftharpoons [Li(H_2O)_4]^+ + HCoO_2 \qquad (1)$$

Thus, the increasing behavior of pH value was mainly caused by the ion-exchange and hydration reaction between $LiCoO_2$ and H_2O accompanying Li^+ dissolution. Figure 3 shows the dissolution of lithium (Li^+), fluorine (F^-), aluminum (Al^{3+}) and cobalt (Co^{2+} and Co^{3+}) ions into the solution during the hydrothermal treatment. Fluorine ion was detected by the anionic chromatography, indicating that the cleavage of C-F bond in PVDF progressed on time and led to the formation of F^-. A similar dissolution behavior between Li^+ and F^- was observed; *i.e.*, both dissolution amounts gradually increased, and then reached over 97 mass% at the treatment time of 20 h. This trend resulted from Li^+ dissolution accompanied by the increase in pH value, because the hydrolytic cleavage of PVDF became more susceptible in higher concentrated alkali solution. The dissolution of cobalt ion was less than the detection limit by ICP-AES under the whole treatment conditions. The increase of pH value is also expected to have an effect on the Al foil corrosion because metal Al can dissolve into a high pH solution. As mentioned previously, the Al foil was fractured with increasing treatment time, and finally disappeared when the treatment time reached 15 h (see Fig. 1). However, the dissolution of Al^{3+} initially increased up to approximately 40 mass% until the treatment time of 10 h, but successively decreased to approximately 2 mass% at the treatment time more than 20 h. The reason for decreasing Al^{3+} dissolution will be deduced to the formation of $Co(Co_xAl_{2-x})O_4$.

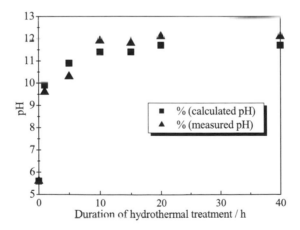

Figure 2:   Relation between (■) the calculated and (▲) the measured pH value of the solution after the hydrothermal treatments in H_2O at 473 K for various treatment times.

With regard to the Al^{3+} dissolution, an electrolysis effect was considered to play an important role as another effect except for the pH factor. In order to examine this effect, the Al foil coated with or without PVDF was treated in various solutions at 473 K for 40 h. The Al^{3+} dissolution result analyzed by ICP-AES is listed in Table 1. NaCl was chosen as an electrolyte instead of LiOH to avoid the Al^{3+} dissolution by the pH effect. These data showed that the Al foil was able to dissolve only when the foil with PVDF was treated in NaCl electrolytic solution (sample No. 1; the dissolution was nearly 76 mass%). On the contrary, the Al foil without PVDF could not dissolve both in H_2O and NaCl solutions (sample No. 3 and 4), and the foil with PVDF also could not dissolve in H_2O (sample No. 2). Consequently, it was clarified that the electrolysis effect must be involved in the Al^{3+} dissolution process during the hydrothermal treatment under the electrolyte (e.g., Li^+ and OH^-) present conditions.

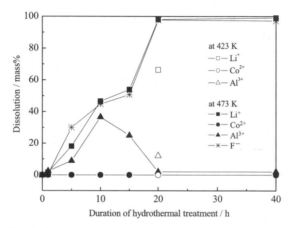

Figure 3: Dissolution of lithium, cobalt, aluminum and fluorine ions into the solution after the hydrothermal treatments of the cathode electrode at 423 and 473 K for various treatment times.

Table 1: Al^{3+} dissolution from Al foil under hydrothermal condition.

Sample No.	Sample *1)	Solution	Dissolution (mass%)
1	Al foil with PVDF	1M NaCl aq.	79.1
2	Al foil with PVDF	H_2O	0.1
3	Al foil	H_2O	Not detected
4	Al foil	1M NaCl aq.	Not detected

*1) Size of Al foil was 1×1 cm for each sample.

3.1.2 Analyses of the solid residues obtained after hydrothermal treatment

Figure 4 shows XRD profiles of (a) the as-prepared cathode, and hydrothermally treated cathodes with 30 mL of deionized H_2O for (b) 1, (c) 5, (d) 10, (e) 15, (f) 20 and (g) 40 h. XRD profiles in Figs. 4(a) and (b) were acquired from the cathode surface cast by the mixture of $LiCoO_2$, conductive carbon black and PVDF. Only peaks derived from the layered rock salt structure of $LiCoO_2$ were detected on the as-prepared cathode. After the hydrothermal treatments for 1, 5, 10 and 15 h, new peaks of hydrogen cobaltate ($HCoO_2$) and/or spinel-type cobalt aluminates ($Co(Co_xAl_{2-x})O_4$ ($0<x\leq2$)) were identified in addition to $LiCoO_2$

peaks, and an unknown peak at around $17°$ (assigned as +) appeared also for the samples treated for 5 and 10 h. The unknown peak was derived from a reactant between the carbon black and $LiCoO_2$ because the peak at around $17°$ appeared when a green compact of carbon black and $LiCoO_2$ mixture was treated under the above hydrothermal condition. Metallic Al peaks were observed in the process of the treatment for 5 and 10 h (Figs. 4(c) and (d)) due to the disintegration of the cathode sample, and the Al peaks entirely disappeared when the treatment time was 15 h and more (Figs. 4(e), (f) and (g)), which were in good agreement with the result of appearance change shown in Figure 1. Only a spinel-type $Co(Co_xAl_{2-x})O_4$ $(0<x\leq2)$ was formed when the treatment time was 20 and 40 h, as seen in Figs. 4(f) and (g). Taking the ICP-AES and XRD results into consideration, it was implied that all cobalt existed in the solid phase, i.e., $LiCoO_2$, $HCoO_2$ or $Co(Co_xAl_{2-x})O_4$, depending on the treatment time. It was found that a spinel-type Co_3O_4 with no Al formed from a green compact of $LiCoO_2$ and carbon black mixture by a hydrothermal treatment. The diffraction peaks of $(Co(Co_xAl_{2-x})O_4)$ and Co_3O_4 could not be distinguished each other because their peaks generally overlapped.

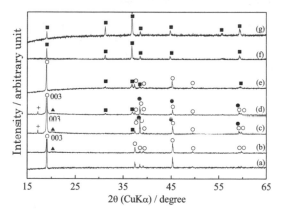

Figure 4:   XRD profiles of (a) as-prepared cathode, and hydrothermally treated cathodes in H_2O at 473 K for (b) 1, (c) 5, (d) 10, (e) 15, (f) 20 and (g) 40 h. Symbols are (■) $Co(Co_xAl_{2-x})O_4$ $(0<x\leq2)$, (□) Co_3O_4, (●) Al, (○) $LiCoO_2$, (▲) $HCoO_2$ and (+) unknown peak.

Figure 5 shows representative (a) secondary electron image (SEI), and EDX element maps corresponding to (b) Co, (c) O, (d) Al and (e) C on the hydrothermally treated cathode. The EDX maps clearly verified that particles consisting of Co, Al and O existed in the treated cathode, while particles consisted of carbon were originated from carbon black and PVDF. Although the $Co(Co_xAl_{2-x})O_4$ particles were confirmed by the EDX analysis, the Co_3O_4 particles were not observed. The hydrothermally treated cathode was believed to contain Co_3O_4 particles, but the amount of Co_3O_4 seemed too small to be identified by EDX analysis.

Figure 5: Representative (a) secondary electron image (SEI) and element distributions of (b) Co, (c) O, (d) Al, and (e) C on hydrothermally treated cathode at 473 K for 40 h.

3.1.3 Optimum condition of hydrothermal treatment

The hydrothermal treatment at 423 K for 20 h was carried out in order to lower the treatment temperature, but the Li^+ dissolution was found insufficient (see Figure 3). Therefore, it was found that the treatment temperature at 473 K or more was required for the separation of Li from Co and Al. Taking the results of ICP-AES, IC, XRD and XPS analyses into consideration, the hydrothermal treatment at 473 K for 20 h was determined as an optimum condition for the separation of Li and F from the cathode electrode, when more than 99.9 mass% of Co and 98 mass% of Al were efficiently reclaimed as a form of solid-state $Co(Co_xAl_{2-x})O_4$ ($0<x\leq2$), and most of Li and F could be dissolved into the solution. XPS analysis exhibited that no fluorine was detected on the solid residue after the hydrothermal treatment over 20 h, corresponding to the result of F^- dissolution shown in Fig. 3. Residual carbon analysis by CHN coder revealed that the recovered spinel compound contained nearly 12 mass% of carbon, which originated from the conductive carbon black and PVDF. This carbon contamination has no problem in the subsequent pyrometallurgical treatment because the carbon can be used as a reductant.

3.2 Recovery of metal cobalt by pyrometallurgical method

Figure 6 shows XRD profiles of the mixture of SiO_2, $CaCO_3$, $\alpha-Al_2O_3$, $CoAl_2O_4$ and carbon black before (a) and after (b) a pyrometallurgical treatment at 1623 K for 5 h in Ar flow. The peaks of SiO_2, $CaCO_3$, $\alpha-Al_2O_3$ and $CoAl_2O_4$, which were the components of the starting mixture, were identified in Fig. 6(a). On the other hand, the peaks of calcium silicate, larnite, gehlenite and metallic cobalt were identified with no peaks of the starting components, as seen in Fig. 6(b). Hence, by adding the SiO_2, $CaCO_3$ and α-Al_2O_3 to the $CoAl_2O_4$ and carbon black mixture, we succeeded in incorporating Al into the slag phase and obtaining the metallic Co without formation of any alloy. The effect of such additives is explained by the difference in Gibbs free energy change for the formation of each oxide from metallic Co and Al. As generally known, the more negative a free-energy change for oxidation is, the easier the formation of an oxide is. The free energy change for the formation of aluminum oxide is more negative than that for the formation of cobalt oxide according to the Richardson-Jeffes diagrams [9]. In addition, α-Al_2O_3 can form a liquid slag phase with the additives at 1623 K in terms of the equilibrium phase diagram of the SiO_2-CaO-Al_2O_3 system [8]. Thereupon, Al was transferred into the slag phase and Co was separated as a pure metal.

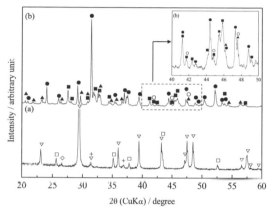

Figure 6: XRD profiles of the mixture (a) before and (b) after a pyrometallurgical treatment at 1623 K for 5 h. Symbols are ($\Diamond$) SiO_2, ($\square$) $CaCO_3$, ($\square$) α-Al_2O_3, (+) $CoAl_2O_4$, ($\bullet$) $Ca_2Al_2SiO_7$ (gehlenite), ($\blacktriangle$) Ca_2SiO_4 (calcium silicate), ($\blacksquare$) Ca_2SiO_4 (larnite) and ($\circ$) Co.

Considering the results of the hydrothermal and the pyrometallurgical treatments, we propose a novel cobalt recovering process. This new process consisted of two stages; the first is the hydrothermal treatment for the dissolution of Li and F into the solution and the second is the pyrometallurgical treatment for the reduction and condensation of cobalt from the spinel compound. It is quite obvious that the hydrothermal treatment can operate at temperatures lower than the present pyrometallurgical process for the gasification of Li and F with no corrosive gas generation and with no utilization of toxic or expensive solvents such as acid or alkali solution. Thus, a cost for equipment maintenance of damaged furnace wall is expected to be reduced by preventing the corrosive gas generation. Additionally, the energy consumption in hydrothermal treatment can be compensated by the utilization of waste heat generated from the subsequent pyrometallurgical treatment. In the second pyrometallurgical step, the present-operating furnace is probably utilized without any modification, which means that an investment for new equipments is expected to be reduced. Therefore, taking all advantages into account, this new process can be recommended as an environmental-friendly and low-cost process.

4   Summary

(1) Metallic cobalt was successfully reclaimed from a cathode electrode by a hydrothermal treatment in pure water at 473 K for 20 h and subsequent pyrometallurgical treatment at 1623 K for 5 h under Ar flow.
(2) More than 99.9 mass% of Co and 98 mass% of Al in the cathode sample was converted to solid state spinel $Co(Co_xAl_{2-x})O_4$ $(0<x\leq2)$ and most of Li and F could be dissolved in the solution after the hydrothermal treatment under the optimum condition at 473 K for 20 h.
(3) The hydrothermal treatment for the separation of Li and F in the cathode electrode could lower the treatment temperature from 1273 K to 473 K with no corrosive gas generation and with no utilization of toxic and expensive solvents.

References

[1]   Rdyh, C.J. & Svard, B., Impact on global metal flows arising from the use of portable rechargeable batteries. *The Science of the Total Environment*, **302**, pp. 167-184, 2003.
[2]   Castillo, S., Ansart, F., Laberty-Robert, C. & Portal, J., Advances in the recovering of spent lithium battery compounds. *J. Power Sources,* **112**, pp. 247-254, 200.
[3]   Lain, M.J., Recycling of lithium ion cells and batteries. *J. Power Sources*, **97-98**, pp. 736-738, 2001.
[4]   Zhang, P., Yokoyama, T., Itabashi, O., Suzuki, T.M. & Inoue, K., Hydrometallurgical process for recovery of metal values from spent lithium-ion secondary batteries. *Hydrometallurgy,* **47**, pp. 259-271, 1998.
[5]   Nan, J., Han, D. & Zuo, X., Recovery of metal values from spent lithium-ion batteries with chemical deposition and solvent extraction. *J. Power Sources*, **152**, pp. 278-284, 2005.
[6]   Gupta, R. & Manthiram, A., Chemical extraction of lithium from layered $LiCoO_2$. *J. Solid State Chemistry* **121**, pp. 483-491, 1996.
[7]   Casasdo, P.G. & Rasines, I., The series of spinels $Co_{3-S}Al_SO_4$ $(0 < S < 2)$: study of Co_2AlO_4. *J. Solid State Chemistry*, **52**, pp. 187-193, 1984.
[8]   Levin, E.M., McMurdie, H.F. & Hall, F.P., in: *Phase Diagrams for Ceramists*, American Ceramic Society: Columbus, Ohio, pp. 313-326, 1956.
[9]   Gaskell, D.R., in: *Introduction to Metallurgical Thermodynamics*, Second Edition, Hemisphere Publishing Corporation: UAS, 1981.
[10]   Watanabe, T., Uono, H., Song, S-W., Han, K-S. & Yoshimura, M., Direct fabrication of lithium cobalt oxide films on various substrates in flowing aqueous solutions at 150 °C. *J. Solid State Chem.*, **162**, pp. 364-370, 2001.

Current developments and future directions in nuclear waste immobilisation

E. R. Maddrell [1] & N. B. Milestone [2]
[1]Nexia Solutions, Sellafield, UK
[2]Department of Engineering Materials, University of Sheffield, UK

Abstract

Current development work towards novel wasteforms required for the on-going decommissioning of the UK's nuclear facilities is described. A discussion of possible management options for future spent nuclear fuel arisings is given.
Keywords: nuclear, waste, wasteform, cement, glass, ceramic, spent fuel.

1   Introduction

Two principal technologies are currently employed for the immobilisation of radioactive waste in the United Kingdom: vitrification for high level waste and cementation for various intermediate level waste streams.

Cementation is used in a number of plants on the Sellafield site for the immobilisation of secondary waste streams arising from fuel reprocessing operations such as: metal swarf from the decladding of Magnox reactor fuel; PWR and AGR oxide fuel assembly hulls and ends; ferric hydroxide and other flocs; and barium carbonate slurries. Composite cement systems are used based on binary blends of Ordinary Portland Cement [OPC] with high replacement levels of either Blast Furnace Slag [BFS] for swarf, fuel hulls and slurries, or Pulverised Fly Ash [PFA] for flocs. For fuel assembly components the cement is mixed externally, poured into the drum and allowed to permeate around the metal swarf, whilst for flocs and slurries an internal mixing operation is required to produce a homogeneous blend of liquid waste and cement. Examples of typical immobilised product are shown figure 1. The main objective of cementation is the physical consolidation and stabilisation of the wastes until such a time that they can be moved to safe disposal.

WIT Transactions on Ecology and the Environment, Vol 92, © 2006 WIT Press
www.witpress.com, ISSN 1743-3541 (on-line)
doi:10.2495/WM060021

Figure 1:   Externally mixed Magnox swarf (left) and internally mixed EARP floc (right).

Vitrification is employed to immobilise the primary highly active [HA] waste stream from fuel reprocessing. This stream is a nitric acid solution containing the radioactive fission products and minor actinides, and inactive fuel additives [e.g. Al, Mg, Fe] that remain after uranium and plutonium have been removed from the dissolved spent fuel by solvent extraction. The HA waste is evaporated to dryness and calcined in a rotary calciner to convert it to oxides. Thereafter, it is incorporated into a mixed alkali borosilicate glass formulation by melting at $1050 - 1100°C$. Unlike the cementitious wasteforms, the glass is designed to exhibit long term durability.

Whilst these are mature technologies, the progression of clean-up and decommissioning of the UK's nuclear sites is leading to waste streams that are not amenable to immobilisation by these processes. Current work on the development of alternative low and high temperature wasteforms will be reviewed in this paper.

In addition to existing wastes, the maintenance of the UK's nuclear generating capacity, and its possible replacement with new-build reactors, will lead to an additional inventory of spent fuel for which no long term management strategy has been identified. A range of spent fuel management options which the authors believe should be considered in this context will be discussed.

2   Current developments

2.1 Low temperature wasteforms

Following are brief descriptions of developments in a range of low temperature wasteforms. A more complete discussion of these is given by Milestone [1].

2.1.1 Modified cement formulations

Whilst the success of OPC/BFS and OPC/PFA cement formulations is undisputed, further improvements are possible. The formulations being used were chosen to control temperature rise but the amount of hydration is limited due to the low amount of cement present. To produce a cement grout that has the necessary fluidity and workability requires the addition of a significant excess of water relative to the optimum amount required for the hydration reactions that occur during cement setting. This residual water is held in pores which can be detrimental to wasteform quality. The use of superplasticisers to improve fluidity is discouraged by the UK Nuclear Industry Radioactive waste EXecutive [NIREX] because of uncertainties over their organic degradation products which may have the ability to complex and mobilise actinide ions. Improved fluidity and workability for a given water to solids ratio could be achieved by using ternary OPC/BFS/PFA cement blends, potentially allowing the water to solids ratio to be reduced from as high as 0.42 to 0.32. This is still higher than the theoretical amount of 24% by weight of the cement needed for full hydration. Ternary cements could lead to an improved immobilised product for swarf, hulls and ends. Cement fluidity is crucial to ensure that it can flow evenly through a convoluted network of metallic residues. As decommissioning proceeds, demolition rubble will become a major waste and more fluid cement systems will better enable the rubble to be stabilised for transport to a disposal site. Decontamination prior to demolition will mean that the rubble is low level waste.

2.1.2 Calcium sulfo-aluminate cements

One problematic waste stream at Sellafield is the Cs and Sr loaded clinoptilolite ion exchange compound. The OPC based cements have an internal pore solution pH > 12 and this will attack the zeolite causing the destruction of its framework with concomitant release of Cs. One solution to this reaction requires utilising a completely different cement based on the calcium sulfo-aluminate system. This cement system is extensively used for construction in China and its use for radioactive waste is being examined both there and in the UK. Its formulation is such that the hydration products are free of $Ca(OH)_2$ which reduces the pH in the internal pore solution to 10-11 and preserves the integrity of the clinoptilolite. Moreover, the binder formed, ettringite, $3CaO.Al_2O_3.3CaSO_4.32H_2O$ binds large amounts of water making it unavailable for any ongoing reactions such as corrosion, and it can be extensively substituted making it ideal for immobilisation of a number of toxic species. A further application of this system is for the encapsulation of metallic aluminium wastes which evolve hydrogen in contact with the highly alkaline pore solution of traditional OPC based cements.

2.1.3 Geopolymers

An emerging class of low temperature wasteform binder is generically referred to as geopolymers. These are amorphous, 3-dimensional polymeric aluminosilicate networks formed from alkali silicate solution activated glasses such as PFA. A typical atomic structure of a geopolymer is shown in figure 2. In addition to the PFA, the starting materials are 5-8 M sodium hydroxide and

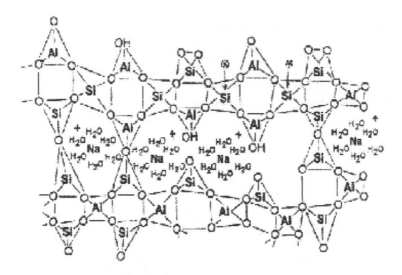

Figure 2:   Atomic structure of a geopolymer alkali aluminosilicate binder.

4-5 M sodium silicate. Clearly this system is highly alkaline, however, in contrast to OPC based systems this does not preclude their use for the immobilisation of zeolites. Rather, the zeolite can be used in place of the PFA component and the Cs remains bound within the aluminosilicate network. Residual water which is not required for the binding phases is contained within pores in the final wasteform and can be lost on heating. Because the binder is not a hydrate it will function at high temperature.

2.2 High temperature wasteforms

High temperature wasteforms are preferred to the low temperature alternatives when long term wasteform performance criteria are more stringent. Currently the principal requirement for new high temperature wasteforms is for the immobilisation of wastes and residues containing plutonium and other actinides. These range from wastes that contain plutonium in excess of the level that defines plutonium contaminated material, through to separated plutonium that is surplus to national requirements. Current work is aimed at developing a generic process capable of producing a range of wasteforms tailored to specific waste streams. The process is described in more detail by Scales et al. [2].

2.2.1 Glass ceramic wasteforms

The initial driver for the development of a glass ceramic wasteform was a range of highly heterogeneous wastes and residues arising from previous plutonium processing work on the Sellafield site. It is noted that the term glass ceramic is used for convenience to describe a wasteform containing both glass and crystalline phases, and not in the strict Materials Science definition of the term. The principle behind the wasteform is that the miscellaneous and highly variable components of the wastes are immobilised in an alumino-borosilicate glass matrix whilst the plutonium partitions into a crystalline zirconolite phase [ideally $CaZrTi_2O_7$]. Zirconolite is a naturally occurring mineral which possesses considerable chemical flexibility, thus making it an ideal wasteform phase. In nature, it contains significant quantities of uranium and thorium which have been retained for hundreds of millions of years, despite having experienced aggressive weathering conditions [3]. Consequently, zirconolite has been identified as a key phase for the immobilisation of plutonium and other man made actinides. A typical microstructure for this wasteform, imaged using back scattered electrons in the scanning electron microscope, is shown in figure 3. The light phase is zirconolite, the mid grey phase is calcium fluoride, which occurs in some of the residues, and the background is the glass matrix. The wasteform is highly durable and also demonstrates excellent resistance to the retrieval of plutonium.

Figure 3:   Microstructure of glass ceramic wasteform. Light phase is zirconolite; mid grey phase is calcium fluoride; dark background is glass matrix.

One of the main requirements for processing the residues into the wasteform is to minimise, and preferably avoid, the generation of any secondary wastes. A key means by which this is achieved is through the use of a wholly dry process. The proposed process starts with calcination of the residue and coarse, medium and fine size reduction stages. The fine size reduction will be carried out using an attrition mill and at this stage the residue will be blended with the precursor materials. The output powder will then be granulated to reduce dust and allow it to flow freely through to the consolidation stage.

Consolidation will be achieved through the use of hot isostatic pressing [HIP]. This has been chosen for a number of reasons, including improved flexibility and reduction of volatilisation compared to conventional sintering methods. The HIP cans are made to a novel dumbbell design which collapse during HIPping to form regular right cylinders for subsequent ease of handling and storage.

2.2.2 Ceramics for plutonium immobilisation

As the development of the above process progressed, it was recognised that there was a wider inventory of actinide wastes at Sellafield and it would be advantageous for the process to be able to handle these. A good example of this is the existence of a quantity of separated PuO_2 that, during storage over a number of decades, has become contaminated with chlorine from PVC. Purification of this material such that it would be an acceptable feed to the Sellafield MOx Plant would be uneconomic and hence the material has been designated for immobilisation. The choice of host phase for the plutonium is clearly of great importance.

One candidate phase is the zirconolite used in the glass ceramic. An alternative, developed by the US Department of Energy, [4] is a wasteform based on a related, pyrochlore structured, phase with an approximate target composition of $(Ca_{0.89}Gd_{0.11})(Pu_{0.22}U_{0.45}Hf_{0.22}Gd_{0.11})Ti_2O_7$. This wasteform was referred to as pyrochlore but the closest mineral analogue is actually betafite, ideally $CaUTi_2O_7$. On the basis of data from natural analogues, zirconolite has been shown to exhibit superior long term resistance to alteration than betafite. It has been demonstrated that both of these phases can be produced using the dry processing line developed for the glass ceramic wasteform.

An important factor behind the definition of the US DoE wasteform was the use of multiple barriers to prevent criticality: Gd and Hf both act as neutron poisons, whilst the addition of depleted uranium dilutes the fissile U-235 that the Pu-239 will have all decayed to within 250,000 years. This criticality control philosophy was based on the assumption that no credit could be afforded to the long term durability of the wasteform. The approach towards criticality control that will be taken within the UK is currently under discussion.

It has been suggested that a weakness of these titanate mixed oxide wasteforms is that they become amorphous as alpha decay damage accumulates, and this will lead to a deterioration in their leach resistance. Consequently, there has been much attention in recent years directed at zirconia based wasteforms such as zirconate pyrochlores $[Gd_2Zr_2O_7]$ [5] and cubic zirconia solid solutions $[(Zr,Y,Pu)O_{2-x}]$ which remain crystalline even after extensive radiation damage. It has recently been demonstrated, however, [6] that amorphisation of titanate wasteforms does not lead to a significant increase in leach rates and hence the importance of resistance to radiation damage may not be as great as claimed. Zirconia based wasteforms also have less chemical flexibility than the titanates and require significantly higher consolidation temperatures, making them unsuitable for the process under development. We are also evaluating the potential of other phases for plutonium immobilisation such as britholite $[Ca_2Gd_8(SiO_4)_6O_2]$ and kosnarite $[NaZr_2(PO_4)_3]$.

2.2.3 Ceramics for MOx residues.

Residues from the Sellafield MOx Plant are predominantly a solid solution of PuO_2 in UO_2. From the brief discussion of wasteforms for separated plutonium above, it can be seen that the US DoE pyrochlore formulation is clearly a favoured wasteform in that it facilitates much higher waste loadings than zirconolite. Again, the process initially described is amenable to the immobilisation of MOx residues.

3   Future directions – to immobilise or to isolate?

The maintenance of the UK's current nuclear generating base is the subject of significant debate and Government review [7]. It is axiomatic that management options for future arisings of spent fuel are included in this discussion. Whilst the existing strategy for spent fuel management in the UK is based on reprocessing, worldwide there is a growing move towards a once-through fuel cycle followed by direct disposal of the spent fuel. The difference between these two options exemplifies the two competing philosophies at the heart of nuclear waste management: to immobilise or to isolate? Immobilisation requires the conversion of the waste into a wasteform from which the hazardous nuclides cannot be leached. Isolation relies on a combination of engineered and geological barriers that retard the movement of nuclides back to the biosphere – central to this philosophy is that the return of a nuclide to the biosphere is controlled by its solubility in groundwater and that this is independent of the wasteform.

Management of spent fuel is not limited to these two options and the following is a brief discussion of a wider range of technologies that might be considered with a qualitative description of their relative advantages and disadvantages. The options are arranged in increasing order of the complexity of processing involved. A key factor, upon which public consensus should be sought, is whether the environmental impact of disposal over future millennia should satisfy criteria defined today as being an acceptable hazard, or whether society should strive for the best immobilisation and disposal combination possible. In the discussion, an effluent is taken to be a by-product that is discharged to the environment whilst a waste is disposed of to a repository.

3.1 Direct disposal

This is the embodiment of the isolation option and is the simplest concept in that, after a period of cooling, spent fuel is put into an overpack and sent to a repository. Economically, fuel processing costs are avoided, which is attractive, although this benefit will be eroded by the cost of the exotic alloys used for overpacking and other engineered barriers. Although the fuel is not processed, it is claimed that the UO_2 matrix of the fuel is an effective wasteform. Nevertheless, long lived fission products such as Caesium-135 (half-life 2.3 million years) and Iodine-129 (half-life 15.7 million years) are not effectively immobilised by UO_2 [8] and have high solubilities in groundwater. Many environmental models of direct disposal show that a major dose contributor is I-129, [9] and that this dose occurs within the first half life of I-129; hence the 'decay' of I-129 is due to it being flushed from the repository system into the sea. Given that in the existing reprocessing scenario I-129 is discharged directly to sea – in accordance with IAEA recommendations – it might be argued that this dose to future societies due to I-129 is incurred needlessly.

3.2 Spent fuel conditioning

If reprocessing of spent fuel to recover plutonium for future use is not necessary, it might be argued that direct disposal is a satisfactory end point. However, spent fuel is not an optimised wasteform. A fuel processing cycle can be developed in which uranium only is extracted from the dissolved spent fuel and the remainder, comprising fission products and all transuranic elements, is immobilised in an appropriate wasteform. The high plutonium content of this waste stream would require the use of a wasteform such as the titanate ceramic Synroc, and this would be demonstrably more proliferation resistant than the original spent fuel. The use of titanate ceramics as a wasteform challenges the above notion that wasteform durability is unimportant to the long term environmental impact of waste disposal. McGlinn [10] has confirmed that elemental concentrations in leachates from titanate ceramics are below their solubility limit, which indicates that superior wasteforms can reduce the environmental impact of disposal. In this scenario the I-129 would be discharged directly to sea, however, we are also evaluating wasteforms for the immobilisation of iodine.

3.3 Reprocessing using PUREX technology

A continuation of the UK's current reprocessing strategy requires that plutonium is viewed as a future asset. There are a number of enhancements that could be made such as single cycle solvent extraction, minimisation of effluents, the reduction of low and intermediate level waste and the use of improved wasteforms such as titanate ceramics.

3.4 Novel non-aqueous reprocessing technology

One argument against existing PUREX reprocessing technology is that aqueous processes inevitably lead to effluent discharges and this has led to interest in non-aqueous technologies to avoid effluents. A prominent example of a non-aqueous fuel processing technology is commonly referred to as pyroprocessing. This involves electrochemical separations of the spent fuel using a molten KCl-LiCl electrolyte. Whilst this may be successful in eliminating effluents, the resulting waste stream is not readily amenable to immobilisation, leading to low waste loadings and increased volumes of a less than optimum wasteform.

This raises the question of whether today's society would favour an effluent free technology combined with a poor wasteform, potentially leading to a greater long term environmental impact; or a safe level of effluent – rapidly dispersed in the oceans – combined with a wasteform of maximum durability.

3.5 Enhanced separation and selective immobilisation

This technology further supports the position that the durability of the wasteform is important in the disposal scenario. Moreover, it extends this to the view that certain components of the high level waste stream are best immobilised using different wasteforms. Enhanced separation cycles are used and the waste streams are processed into the preferred wasteforms. Whilst this is viewed as further reducing the environmental impact in the disposal scenario, this benefit must be assessed against increased operational hazards such as dose to plant operators.

3.6 Partitioning and transmutation

To many people, this represents the holy grail of nuclear waste management. Rather than dispose of a radioactive waste, the key long lived nuclides are separated by chemical means and made into targets for irradiation. In principle, irradiation converts long lived radioactive nuclides into stable ones that can be safely disposed of. Short lived nuclides are simply allowed to decay before disposal as inactive waste. In practice there are some fundamental obstacles, for example, the nuclear physics can mean that transmutation half-lives are long and multiple irradiations are required before a particular nuclide can be destroyed. Also, additional complications arise for certain elements in the spent fuel. One example is that the destruction of the Cs-135 isotope is accompanied by conversion of inactive Cs-133 - from which it is inseparable - to further Cs-135.

4 Conclusions

The development of new waste immobilisation technologies in support of the decommissioning of the UK's nuclear sites has been described. Options for the management of future arisings of spent nuclear fuel, for which no long term strategy has been defined, are discussed. It is recommended that a more detailed, quantitative assessment of these is conducted.

Acknowledgements

The authors acknowledge the Nuclear Decommissioning Authority and British Nuclear Group for funding much of the work described in this paper. The views expressed by the authors do not necessarily reflect UK Government policy.

References

[1]   Milestone, N.B., Reactions in cement encapsulated nuclear wastes: need for toolbox of different cement types. *Advances in Applied Ceramics,* **105(1),** pp. 13-20, 2006.

[2]   Scales, C.R., Maddrell, E.R., Gawthorpe, N., Day, R.A., Begg, B.D., Moricca, S.S. and Stewart, M.W.A., A flexible process for the immobilisation of plutonium containing wastes. *Proceedings of Global 2005,* Paper No. 249.

[3]   Lumpkin, G.R., Day, R.A., McGlinn, P.J., Payne, T.E., Giere, R. and Williams, C.T., Investigation of the long-term performance of betafite and zirconolite in hydrothermal veins from Adamello, Italy. *Scientific Basis*

for Nuclear Waste Management XXII, eds. D.J. Wronkiewicz & J.H. Lee, MRS Symposium Proceedings **556.** pp. 793-800, 1999.

[4]   Ebbinghaus, B.B., Armantrout, G.A., Gray, L., Herman, C.C., Shaw, H.F. and Van Konyenburg, R.A., Plutonium immobilization project baseline formulation. LLNL report UCRL-ID-133089, 2000.

[5]   Ewing, R.C., Weber, W.J. and Lian, J., Nuclear waste disposal – pyrochlore ($A_2B_2O_7$): Nuclear waste form for the immobilization of plutonium and "minor" actinides. J. App. Phys., **95(11),** pp. 5949-5971, 2004.

[6]   Strachan D.M., Scheele, R.D., Icenhower, J.P., Buck, E.C., Kozelisky, A.E., Sell, R.L., Elovich, R.J. and Buchmiller W.C., Radiation damage effects in candidate ceramics for plutonium immobilization: final report. PNNL-14588, Pacific Northwest National Laboratory, Richland, WA. 2004.

[7]   Our energy challenge; securing clean affordable energy for the long-term. UK DTI Energy review consultation document. Jan 2006. http://www.dti.gov.uk/energy/review/energy_review_consultation.pdf

[8]   Poinssot, C., Jegou, C., Toulhoat, P., Piron, J.-P. and Gras, J.-M., A new approach to the RN source term for spent nuclear fuel under geological disposal conditions. *Scientific Basis for Nuclear Waste Management XXIV,* eds. K.P. Hart & G.R. Lumpkin, MRS Symposium Proceedings **663.** pp. 469-76, 2001.

[9]   Wilson, M.L., Swift, P.N., McNeish, J.A. and Sevougian, S.D., Total system performance assessment for the Yucca Mountain site. *Scientific Basis for Nuclear Waste Management XXV,* eds. B.P. McGrail & G.A. Cragnolino, MRS Symposium Proceedings **713.** pp. 153-64, 2002.

[10]   McGlinn, P.J., Personal communication, 31st Jan. 2006, Australian Nuclear Science and Technology Organisation.

Physical-chemical characterization of a galvanic sludge and its inertization by vitrification using container glass

F. Andreola, L. Barbieri, M. Cannio, I. Lancellotti, C. Siligardi & E. Soragni
Dipartimento di Ingegneria dei Materiali e dell'Ambiente, University of Modena and Reggio Emilia, Modena, Italy

Abstract

Several industrial processes produce large amounts of heavy metals-rich wastes, which could be considered as "trash-can raw materials". The incorporation in ceramic systems can be regarded as a key process to permanently incorporate hazardous heavy metals in stable matrixes. In particular the aim of this work is to prepare and evaluate environmental risk assessment of coloured glass and glass-ceramic with the addition of chromium(III) galvanic sludge having a high content of Cr_2O_3 (15.91 wt%). Trivalent chromium compounds generally have low toxicity while hexavalent chromium is recognized by the International Agency for Research on Cancer and by the US Toxicology Program as a pulmonary carcinogen. The sludge has been characterized by ICP –AES chemical analysis, powder XRD diffraction, DTA, SEM, leaching test after different thermal treatments ranging from 400°C to 1200°C. Batch compositions were prepared by mixing this sludge with glass containers. The glass container composition is rich in SiO_2 (69.89 wt%), Na_2O (12.32 wt%) and CaO (11.03 wt%), while the sludge has a high amount of CaO (42.90 wt%) and Cr_2O_3 (15.91 wt%). The vitrification was carried out at 1450°C in an electrical melting furnace for 2 h followed by quenching in water or on graphite mould. Chromium incorporation mechanisms, vitrification processability, effect of initial Cr oxidation state, and product performance were investigated. In particular toxic characterization by leaching procedure and chemical durability studies of the glasses and glass-ceramics were used to evaluate the leaching of heavy metals (in particular of Cr). The results indicate that all the glasses obtained were inert and the heavy metals were immobilized.

Keywords: chromium electroplating sludge, thermal behaviour; vitrification, chemical glass durability.

WIT Transactions on Ecology and the Environment, Vol 92, © 2006 WIT Press
www.witpress.com, ISSN 1743-3541 (on-line)
doi:10.2495/WM060031

1 Introduction

Chromium has been used in industry for various applications over a century, including steel production, plating, anodising of aluminium, leather tanning, wood preservation, water-cooling, etc. The use of chromium (VI) containing solutions (CrO_3 250g/l and H_2SO_4 2,5 g/l) as deposition metal is the unique technique employed in the plating industry [1], despite of its high toxicity [2]. After electroplating is completed, the plated parts are rinsed with water. Eventually this rinse water becomes highly contaminated with plating solution and must be replaced. This fact presents a serious environmental problem since the water is highly concentrated in chromate ions. The hazardousness of this galvanic waste is not only due to Cr(VI) presence but also to the high concentration of heavy metals, as Ni, Ti, etc. This chromium rich solution needs a chemical treatment based on a two-stage process: the first stage reduces hexavalent chromium Cr(VI) to a harmless $Cr_2(SO_4)_3$ by using chemical agents as sodium bisulfate ($NaHSO_3$), sodium or calcium meta-bisulfite($Na_2S_2O_5$). In the second stage trivalent chromium and other metals are precipitated by the addition of calcium or sodium hydroxide in their hydroxide form, which can be easily separated and disposed in landfills. Nowadays the cost of landfill disposal and the decrease in the number of disposal sites have led to considerate stabilization of waste into a glassy matrix or ceramic materials an accepted treatment process [3, 4]. In the last 30 years it has been reported many examples about recycling of industrial waste in cement, glass, and metallurgy industry [5], but few works can be found in literature regarding recycling of galvanic chromium sludge [6] into glassy matrixes or ceramics compounds [7].

The aim of this work is to characterize the chromium galvanic sludge and to study the vitrification process as an efficient method to immobilize the hazardous components of this galvanic waste. In particular it is investigated the stabilization process evaluating the chemical resistance of the obtained glasses and glass-ceramics compounds.

2 Experimental

The electroplating sludge employed in this investigation has been collected from a local plant, while the glass is from container waste. These wastes have been characterized by ICP -AES technique (Varian, Liberty 200). Galvanic sludge previously milled for 40 min in a porcelain jar was added in the percentages of 0.5, 1.5, 3, 6 wt% corresponding to Cr_2O_3 content of 0.12, 0.25, 0.5 and 1 wt% to the container glass composition and the dry mixtures were ground in a ball-mill for 20 min. The mixture was fired in mullite crucibles at 1450°C for 1 h in electrical melting furnace. The melts have been quenched in water to obtained the frits or poured into a graphite mould and quenched in air to obtain a bar shape glass. The glass shape were heated at 750°C with 1 h of soaking time to eliminate the internal stresses due to the fast cooling rate in air. Frits can facilitate and reduce the time required to ground in the preparation of glass ceramics. The container glass and frits in the above reported composition have been humidified with 6 wt% of water and pressed in order to obtain cylindrical samples. Subsequently these pellets were fired at different temperatures from 600 to 1000°C obtaining glass ceramics.

The mineralogical analysis of the galvanic sludge and glass-ceramics has been carried out by an X-ray (CuKα) powder diffractometer, XRD (Philips PW 3710, Holland). The XRD patterns were collected in the 5-60° (2θ) range at room temperature. The thermal behaviour of chromium galvanic waste has been characterized by means of differential thermal analysis (DTA and TG) (Netzsch DSC 404). Scanning electron microscopy (SEM) (PHILIPS XL 40) has been performed on representative samples of sludge and glass-ceramics to observe phase distribution and microstructural features. Furthermore the assessment of the chemical durability in water, acid (HCl 6M boiling, 3h) and alkaline (NaCO$_3$ 0.5 M and NaOH 1 M. boiling,1h) conditions was carried out on representative specimens of the manufactured glasses following ISO/R 719, ISO/R 695 and DIN 12116 standards [8].

3   Results and discussion

3.1   Thermal behaviour of chromium electroplating sludge

The chemical analysis of galvanic waste indicates the following composition in wt%: 0.81 SiO$_2$, 0.8 Al$_2$O$_3$, 42.50 CaO, 1 MgO, 6.1 Na$_2$O, 0.5 K$_2$O, 1.3 Fe$_2$O$_3$, 0.01 TiO$_2$, 0.2 NiO, 0.5 CuO, 0.2 MnO, 15.91 Cr$_2$O$_3$, 0.02 CoO, 0.02 CdO and the L.OI is 30% Figure 1 illustrates the TG and DTA curves of the galvanic sludge during the heating process.

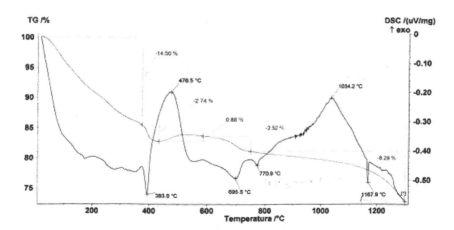

Figure 1:   DTA and TG curves of as-received galvanic sludge.

The DTA curve shows exothermic peaks at 476.5 and 1034.2°C, while endothermic peaks appeared at 393, 695.5, 770.9 and 1167.9°C. The total weight loss is 28%. This result is in close agreement with that obtained from calculations based on chemical analysis and calcination at 1200°C (about 30%) It is presumed that the weight loss is related to CO_2, H_2O. The exothermic peak at 393°C is probably related to the release of cristallization water; the peak at 770.9°C corresponds to the $CaCO_3$ decomposition and 1034.2°C could be due to a crystallization. The endothermic peak at 475°C represents the organic compound combustion and the one at 1167.9°C could be a melting reaction. X-ray diffraction was performed in the as received waste and in samples previously submitted to different isothermal treatments from 400 to 1200°C with 1h of soaking time. Fig. 2 resumes the XRD data showing $CaSO_4$, Na_2SO_4, $CaCrO_4$ and $CaCO_3$ as main constituents, labelled a, b, c, and d respectively.

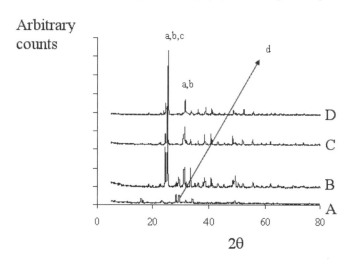

Figure 2:   XRD spectra of galvanic sludge dried at 105°C (A) and thermally treated at 500°C (B), 800°C (C) and 1200°C (D) respectively

It is important to observe, in agreement with the DTA results, that the $CaCO_3$ peaks disappear at 800°C. The elementary analyses performed on the dried samples and on those obtained at 700 and 800°C show a decrease in C, confirming the $CaCO_3$ decomposition, as Table 1 reports.

Table 1:   Elementary analysis of galvanic sludge dried at 105°C and thermally treated at different temperatures from 700°C up to 1200°C.

Sample	N %	C %	H %	S %
105°C	0.00	1.76	1.91	10.99
700°C	0.00	0.64	0.00	9.59
800°C	0.00	0.00	0.00	9.50
1000°C	0.00	0.00	0.00	6.53
1200°C	0.00	0.00	0.00	6.39

—— 5 µm.   —— 5 µm

Figure 3:   SEM micrographs of dried galvanic sludge and thermally treated at 1100°C.

3.1.1 Glasses and glass ceramic characterization

Table 2 reports the chemical composition from theoretical calculations and the ICP chemical analysis of the V1 and V2 glasses. It is evident from these results that a very low crucible refractory contamination occurs. In fact, it can be observed an increase in the amount of Al_2O_3. It is important to remark that no significant variation of the Cr_2O_3 percentage is recorded. This fact indicates that vitrification could be considered a good alternative to stabilize wastes with a high content of heavy metals.

Table 2:   Chemical composition from theoretical calculations of V1, V2, V3 and V4 and the ICP chemical analysis of the V1 and V2 glasses.

Oxide	V1 Theoret. wt%	V2 Theoret. wt%	V3 Theoret. wt%	V4 Theoret. wt%	ICP V1 wt%	ICP V2 wt%
SiO_2	69.57	69.21	68.52	67.18	68	68.5
Al_2O_3	2.64	2.62	2.6	2.55	6.2	7.1
CaO	11.28	11.56	12.1	13.50	10.57	11.50
MgO	2.24	2.24	2.23	2.20	1.92	2.22
Na_2O	12.27	12.24	12.2	12.10	11.61	12.30
K_2O	1.48	1.47	1.46	1.45	1.51	1.42
Fe_2O_3	0.12	0.12	0.13	0.15	0.19	0.13
BaO	0.13	0.13	0.13	0.13	0.12	0.11
ZrO	0.13	0.14	0.14	0.13	0.13	0.13
Cr_2O_3	**0.12**	**0.25**	**0.49**	**0.96**	**0.11**	**0.22**

The water chemical durability of the obtained glasses remains constant with respect to container glass up to 3% in waste showing a medium resistance [8].

These results are in agreement with the increase of pH and conductivity observed with the increasing waste percentage. The damage of the glass network lead to a release of OH^- and therefore to an increase in pH, while an increase of the mobile ions Na^+ and K^+, which are easily leached, causes an increase in conductivity. The examined glasses are slightly attacked in basic and acid medium. The weight loss of the glass containing 6 wt% of galvanic waste, expressed as weight loss per surface unit, is equal to 174 mg/100 cm^2 and 2.69 mg/100 cm^2 in alkali and acid respectively [8].

In order to verify the possibility to obtain a semi-crystalline technologically interesting such as glass-ceramics different thermal treatments have been performed on the glassy samples. XRD analysis shows the presence of crystalline phases dispersed in the glassy matrix, in particular wollastonite is the main crystalline phase (Fig.4).

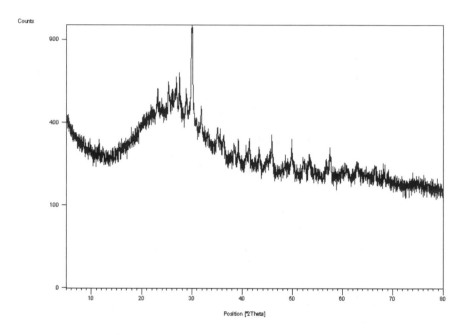

Figure 4:   XRD spectrum of a sample fired at 900°C showing wollastonite as main crystalline phase.

Figure 5 shows a SEM micrograph of a sample of these glass-ceramics compounds showing a microstructure where needle grains of a crystalline phase, that could be wollastonite, are embedded in an amorphous glassy matrix. These needle like grains are presented in the glass-ceramic microstructure from 750°C (Fig.5). The porosity of these glass-ceramic results low, which could be an indication of good mechanical properties.

4   Conclusions

The results of this investigation highlight the possibility of incorporating galvanic waste as economical raw material in glassy products. Adequate thermal treatments could lead to the production of glass-ceramic products. Therefore vitrification and devitrification could represent adequate techniques for inertization of industrial sludge with heavy metals content and alternatives to the disposal in landfills. In future more studies will be performed both on glasses and glass ceramics to evaluate the influence of an increase in the amount of galvanic sludge on the chemical and mechanical properties of the products and to better analyse the sintering process.

[1500 x]

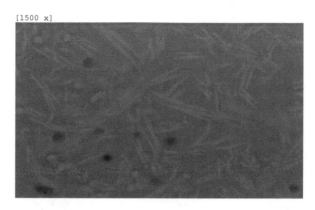

Figure 5:   SEM micrograph of a sample fired at 750°C showing needle-like grains of wollastonite crystalline phase.

Acknowledgement

We are grateful to MIUR for financial support.

References

[1]   Fink C. G., U.S. Pat. 1581188, 1926.
[2]   Katz S.A., Salem H., The Biological and Environmental Chemistry of Chromium, Verlag Chemie, Weinheim, 1994.
[3]   Andreola F., Barbieri L., Corradi A., Lancellotti I., Falcone R., Hreglish S., Glass-ceramics obtained by the recycling of end life cathode ray tubes glasses, *Waste Management* 25, 2005, pp.183-189.
[4]   Leonelli C.; Boccaccini D. N; Veronesi P.; Barbieri L.;. Lancellotti I; Andreola F.; Microwave thermal inertization of asbestos containing waste and recycling in traditional ceramics, *Journal of Haz. Mat.*, in press.
[5]   Sophia A.C., Swaminathan K. Assessment of the mechanical stability and chemical leachability of immobilized electroplating waste, *Chemosfere* 58, 2005, 75-82.
[6]   Cheng N., Wei Y.L., Hsu L.H., Lee J.F., XAS study of chromium in thermally cured mixture of clay and Cr-containing plating sludge, J. Electron Spectroscopy and Related Phenomena 144-147, 2005, 821-823.

[7]   Manufacture of chromium-containing aventurine glass.   (Mukai, Kenichi, Japan). Jpn. Kokai Tokyo Koho, 1985, Patent written in Japanese.   Application: JP 83-121669 19830706.

[8]   Engineered Materials Handbook, Ceramics and Glasses prepared under the direction of the ASM International Handbook Committee, Volume 4, (1991), *Chemical durability*, pp.856.

Weathering of metallurgical slag heaps: multi-experimental approach of the chemical behaviours of lead and zinc

N. Seignez[1], D. Bulteel[2], D. Damidot[2], A. Gauthier[1]
& J. L. Potdevin[1]
[1] UMR 8110 PBDS, University of Lille I, France
[2] Civil Engineering Department, Ecole des Mines de Douai, France

Abstract

The production of lead and zinc in the Lead Blast Furnaces of metallurgical factories produces a huge amount of slags. These slags are a Pb-Zn-rich vitreous material with a CaO-SiO_2-FeO-rich glass matrix containing metallic lead droplets and Fe oxides. The resulting slag heap is exposed to weathering and can be a source of chemical pollution. In this study, different experiments allow the study of the behaviour of Zn and Pb during the weathering of slags. First, a Teflon leaching column (ascending flux) equipped with samplers in order to collect the fluid inside and at the outlet of the column was used. The second experiment was a pure water flow on a polished section of slag. The third system was an outdoor experiment where the slag was exposed to natural weathering comparable to the heap. ICP-AES and ESEM-EDS have been used to follow the evolution of the water chemistry and the slag compositions and textures during the experiments. Two episodes of Pb and Zn releases were noticed at the column outlet. A first brief episode might be caused by some colloids or slag dust releases. The second episode essentially results from glass and metallic lead dissolution. Hydrocerusite locally appeared in the column. Investigations carried out on the polished sections show complex alteration patterns. They suggest that the behaviour of the glass and of minor phases is closely linked to the local structure and composition at the slag surface. Outdoor experiments show slower kinetics than in the other experiments, but the alteration type is similar despite some differences.
Keywords: metallurgical slag, iron-silica-lime glass, waste leaching, lead and zinc releases.

WIT Transactions on Ecology and the Environment, Vol 92, © 2006 WIT Press
www.witpress.com, ISSN 1743-3541 (on-line)
doi:10.2495/WM060041

1 Introduction

In the industrial basin of the Nord-Pas-de-Calais (France), blast and smelting furnaces have been used to produce lead and zinc during the last century. The Noyelles-Godault factory has generated a huge amount of waste. In its Lead Blast Furnace production unit, the production of 1 ton of metal generated approximately 1 ton of metallurgical slags [1]. These slags contain several percents of Pb and Zn [1–3] and others pollutants as Cr, Cd, As [2]. They were not reused as raw materials. Four million tons of such primary smelting slags were then stored in a slag heap spread on 1 km^2 next to the factory [2].

The pollutant releasing abilities of various lead-zinc metallurgical slags have already been discussed in numerous previous works. Indeed, contaminant mobility has been tested in different environments in order to evaluate the ways slags can be reused as raw materials or just to evaluate the stabilization abilities of their storage conditions. The studies concerning the leachability in concrete-like conditions [3, 4], when used as road materials [5], in pure or acidic water [2, 4, 6, 7] or in contact with organic solutions [8] are more or less common. More specific studies on glass leaching abilities have also shown that the glass composition [9–12], the local environment of iron [13, 14] and the pH and temperature [10] are very important parameters concerning the leaching resistivity in lime-silica glasses. Secondary phase formations [2, 6, 7, 8, 15] and sorption phenomena on glass or crystalline phases [16–20] can also play an important role on the pollutant mobility/immobility.

Slag heaps are generally not protected from rainfalls. Slags are thus exposed to weathering conditions. The most accessible part of the heap for the water may be governed by flow conditions. Yet, most of previous studies were made in batch conditions. To our knowledge, no study using the present Lead Blast Furnace slags in open flow system using pure water has been performed.

This work proposes to follow Zn and Pb mobility in two original laboratory leaching experiments in pure water flow conditions and to compare the laboratory results with slag alteration under natural weathering conditions in draining and non-draining conditions.

2 Materials and methods

2.1 Slag characterization

Slags are generally smaller than 2 mm. The size distribution is lognormal with a 500 μm dominant class [1]. The phase arrangements in the slags have already been described at micrometric scale by Deneele [2] estimated the waste is composed at 80% (in volume) of a $FeO-SiO_2-CaO$ glass with few percents of Pb and Zn. Slags also contain 19% of crystallized iron oxides and 1% of plurimicrometric droplets of metallic lead. Automorphous spinels are spread into the iron-lime-silica glass matrix. This glass also contains few mass percents of others elements like Al, Mg, Na and K. Spinels have a chromium rich centre ($MgCr_2O_4$) while the composition evolves to franklinite ($ZnFe_2O_4$) or sometimes in magnetite ($Fe^{2+}Fe^{3+}_2O_4$) to the borders [2]. Plurimicrometric and dendritically shaped oxides have also been described as Zn substituted wüstite ($Fe_{0.85-x}Zn_xO$) [1, 2]. Metallic lead droplets which sizes range between 1 μm and 400 μm [1, 2] have also been identified.

2.2 Column test

The LBF slags used here come from Noyelles-Godault (France). A significant volume of 300 kg of slags has been sampled at the end of the metallurgical process during 1 month (frequency: 2 or 3 samplings for a week) [2]. The diameters (d) of the slag used for this test range between 2 and 2.5 mm. The slag surfaces have been cleaned from most of dust by three successive ultra-sonic baths in alcohol.

The column is a 220 mm high PTFE cylinder with an internal diameter (D) of 25 mm. The D/d ratio of 10 and the column height must create a homogeneous flux. Teflon is used to avoid sorption phenomena on the column wall. At three levels of the column, openings filled with septa allow to do in-situ leachate samplings with specific needles [15]. In front of these openings, PTFE cylinders called samplers are placed (Figure 1). Hydrodynamic tests using tracer have shown, they do not create any notable disturbance on the ascending flux. At the column inlet, a Teflon piece is placed to homogenize the flux. The column contains about 150 g of slags. The column is connected to a pump (Figure 2) and a pure water reservoir (pH = 5.6). The pump provides a constant flow at 60 ml.h^{-1}. At the column outlet, a fraction collector driven by a computer allows to sample the leachate with a high frequency (for example, 1 sample by hour).

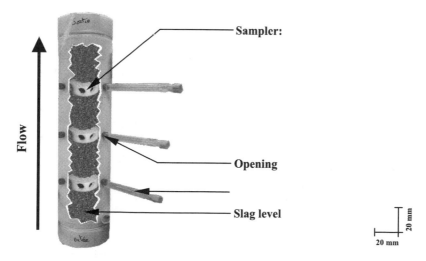

Figure 1:   Photomontage showing the column interior and exterior.

A Thermo ICP-AES has been used to estimate the concentration of the following elements: Al, As, Ba, Ca, Cd, Cr, Cu, Fe, K, Mg, Mn, Na, Ni, Pb, Si, Ti, Zn. The detection level is approximately about 20 ppb for the majority of the analysed elements. Cu detection level is a little higher: 50 ppb.

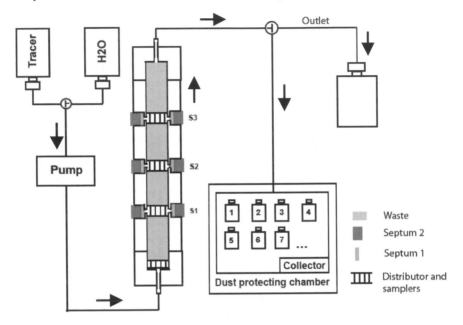

Figure 2:   Scheme of the experimental device.

2.3 Polished section test

This leaching experiment is very close to the one described before. The column is replaced by a polished section container. The container is supplied with pure water connected to a peristaltic pump which provides a constant flow of $60 ml.h^{-1}$. The sections are 60 mm long and 20 mm wide. To make these polished sections, slag grains are included in an epoxy resin and are polished in order to expose slag surfaces (40% of the total upper face). There are approximately 3 mm between the section surface and the upper side of the container. Then, the Darcy velocity in the container is about $1m.h^{-1}$. The sections were observed by ESEM 15 and 26 days after the beginning of the experiment. The chemical analyses have been obtained at 26 days with ESEM-EDS.

2.4 Outdoor weathering

In this experiment, draining and non-draining conditions which may occur in the heap are simply reproduced with the help of specific jars. In each sort of experiments polished sections were exposed to climatic conditions. The evolution of the slag aspect was followed by ESEM.

2.5 Electron microscopy

The morphology and composition of slag sections before and after alteration was studied using a FEI Quanta 200 Environmental Scanning Electron Microscope (ESEM). The tungsten electron source is used at 20 kVolt. The spatial resolution of its X-ray Energy Dispersive System (EDS) probe is about 1 μm^3. The accuracy of the analyses can reach 0.5% (in weight) in high vacuum. In order to be reused, the polished sections were not carbon coated as usual and were inserted in the ESEM chamber in low vacuum mode. The relatively low pressure in ESEM chamber (≤ 0.45 Torr) reduces the analysis accuracy. Then, only the global trend has been considered.

3 Results and discussion

3.1 Column test

3.1.1 Leachate chemistry

During the first hours of experiment, the pH value at the column outlet is enhanced to 7 and increases to 7.8. The pH (Figure 3 A) rapidly decreases to reach a value near to 6 (*i.e.* near pure water pH). Concerning ICP-AES analyses, only four elements (Ca, Si, Pb and Zn) were detected in relatively high amount in the leachates. No iron was detected. In the first hours, Pb, Zn and Ca releases (Figure 3 B, C, D) may be linked to the pH variations.

Indeed, the concentrations of Pb and Ca in the first samples were the highest measured during the test. This can be connected to dust cleaning at the slag surfaces. Indeed, tests have shown that dust is very difficult to keep out before the experiments. After the pH stabilization at 6, Ca, Pb and Zn progressively reached an approximately constant releasing level at 0.5, 0.15 and 0.15 ppm respectively. Si releasing slowly increases during the experiments (Figure 3 E) to reach 0.45 ppm at $t_{1400\,h}$.

The Ca and Si loss indicate the glass matrix alteration was persistent throughout the experiment. Considering the electron microscopy observations, Ca and Si may not be fixed in secondary phases. The Ca releasing rate is around 30 $\mu g.h^{-1}$ in the second part of the experiment and the Si releasing rate ranges between 10 and 30 $\mu g.h^{-1}$. The Pb and Zn releasing rate is around 10 $\mu g.h^{-1}$ in the second part of the experiments

3.1.2 Waste observations at t_{end}

ESEM-EDS analyses show leaching evidences on the glass and, particularly, near the column inlet. Some glass fields are highly fractured (Figure 4 A), however it may be due to desiccation of hydrolysed glass in the ESEM chamber. Traces of glass alteration are less important when slags are situated further from the column inlet. EDS analyses have shown the further from the inlet the glass is, the less important the Si and Ca loss is. Then, glass alteration essentially occurs at the bottom of the column (Table. 1) Leachate compositions are closely linked to this phenomenon. Indeed, the in-situ samplings show that the leachate chemistry does not really change between the three samplers and the outlet.

Conversely, superficial lead droplets often react in the entire column. Punctually, big lead droplets are not totally dissolved and plurimicrometric carbonates have been formed near (Figure 4 B). The iron oxides seem to be relatively stable. Thus, Zn may not be released from the dendritical wüstite spread in the glass. Its source may be only the glass.

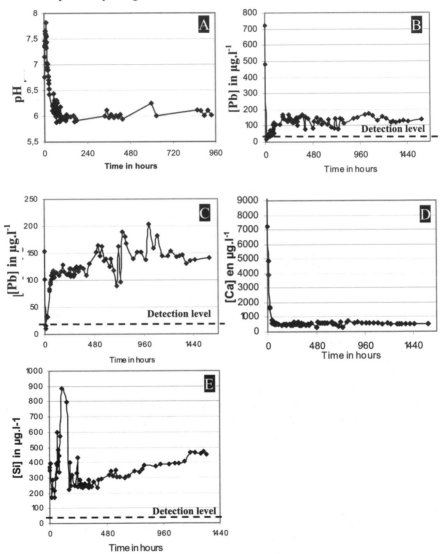

Figure 3: Graphs showing the pH evolution at the column outlet (A). Graphs (B), (C), (D) and (E) respectively show Pb, Zn, Ca and Si releasing during the experiment.

3.2 Polished section tests

ESEM showed the same type of glass alteration (Figure 4 C) and lead carbonate formations as in the column test. Laminar flux encountered in this test enhanced micro-canyon formation (by glass alteration) less observed in column test. Glass can be kept fresh in some saved regions. Nevertheless, Ca and Si losses correspond to altered glass regions similar to these observed in the column (Table 1). Highly fractured and lightening zones often correspond to regions downstream dissoluting big lead droplets (Figure 4 D).

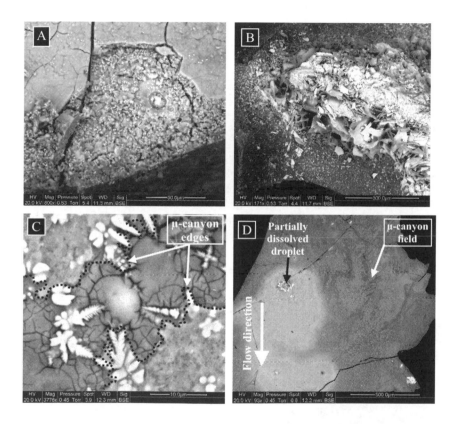

Figure 4:   SEM images of waste surfaces altered in the different experiments: (A) and (B) in column, (C) and (D) in the polished section experiments.

Some ESEM images might be consistent with [16,18] results showing lead sorption are supposed to happen in conditions similar to such influenced zones. Nevertheless no such evidence has been found with EDS punctual analyses. In altered zones, preliminary EDS analyses treatment may indicate that Ca is generally highly released conversely to Si, which releasing rate may be assumed to be lower. In these zones, there is a relative Fe enrichment.

Table 1: Statistics of EDS analyses of fresh and altered glass (% in weight).

		SiO_2	CaO	FeO
Fresh glass (**167 analyses**)	mean	34.35	28.92	36.72
	min	24.54	24.61	26.30
	max	40.13	34.98	47.55
First slag level of the column (**24 analyses**)	mean	13.78	3.12	83.10
	min	7.70	0.87	44.10
	max	36.02	36.02	91.44
Glass leached in container (**21 analyses**)	mean	30.41	11.06	58.53
	min	10.39	4.63	45.01
	max	42.37	23.04	84.99

3.3 Outdoor weathering

The results of this experiment are highly comparable to the polished section ones. During the glass alteration, there also is a Ca loss. However, the Si loss is more difficult to observe. Glass is also highly fractured and sometime lightens (Figure 5 A). Some micro-canyons can be observed in the altered glass. Lead carbonates are smaller (Figure 5 B) and apparently amorphous lead deposits which have not been observed in the laboratory experiments are here visible. Suspicion of sorption phenomena might be rarely observed on iron oxides near lead droplets. Thus, slags altered in weathering conditions may have a globally similar behaviour than in laboratory experiments.

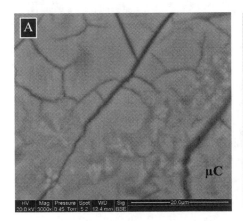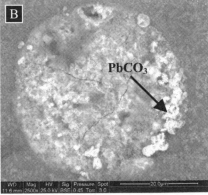

Figure 5: (A) altered glass matrix with fractures and micro-canyons (μC) both induced by leaching, (B) altered lead droplets and lead carbonates.

4 Conclusion

The laboratory experiments proposed here are useful to study the behaviour of Pb and Zn during metallurgical slag weathering. Indeed, they globally conduct to a comparable waste evolution than observed when altered in climatic conditions. It appears that when in contact with pure water in open flow conditions, slag releases relatively high and constant amounts of pollutants (Pb, Zn). It is linked to combined dissolution effects affecting glass and lead droplets. The advance of the glass alteration may be doped by big droplet dissolution. Pollutant immobilization may occur by lead carbonate formation but lead sorption on glass might be a minor mechanism. Nevertheless, complementary investigations (FIB, TEM-EDS, XANES...) will be necessary to better describe pollutant releases and glass leaching mechanisms.

Acknowledgements

This work is supported by the Nord-Pas-de-Calais region Council and the French Agency of Environment (ADEME). We would like to thank every Engineer who has helped us to use the ESEM (S. Gouy and P. Recourt) and the ICP-AES (B. Malet, J.-F. Barthe).

References

[1]   Sobanska, S., Ledésert, B., Deneele, D., Laboudigue, A., Alteration in soils of slag particles resulting from lead smelting. *Comptes Rendus de l'Académie des Science de Paris, Earth and Planetary Sciences*, **331**, pp. 271-278, 2000.

[2]   Deneele, D., Caractérisation, simulations experimentales et thermodynamiques de l'altération de déchets vitreux. *Unpublished University PhD Thesis*, University of Lille1, France, pp. 187, 2002.

[3]   Gervais, C., Evaluation environnementale des perspectives de valorisation en BTP de scories de premières fusion de plomb et de zinc. *Unpublished PhD Thesis*, Institut National des Sciences Appliquées de Lyon, France, pp. 218, 1999

[4]   Ettler, V., Mihaljevic, M., Touray, J.C., Piantone, P., Leaching of polished sections: an integrated approach for studying the liberation of heavy metals from lead-zinc metallurgical slags. *Bulletin de la Société Géologique de France,* **173(2)** pp.161-169, 2002.

[5]   Barna, R., Moszkowicz, P., Gervais, C., Leaching assessment of road materials containing primary lead and zinc slags. *Waste Management,* **24**, pp. 945-955, 2004.

[6]   Ettler, V., piantone, P., Touray, J.C., Mineralogical control on inorganic contaminant mobility in leachate from lead-zinc metallurgical slag: experimental approach and long-term assessment. *Mineralogical Magazine,* **67(6)**, pp. 1269-1283, 2003.

[7]   Ettler, V., Komarkova, M., Jehlicka, J., Coufal, P., Hradil, D., Machovic, V., Delorme, F., Leaching of lead metallurgical slag in citric solutions – implications for disposal and weathering in soil environments. *Chemosphere,* **57**, pp. 567-577, 2004.

[8]   Ettler, V., Jehlicka, J., Masek, V., Hruska, J., The leaching behaviour of lead metallurgical slag in high-molecular-weight (HMW) organic solutions. *Mineralogical Magazine*, **69(5)**, pp. 737-747, 2005.

[9]   Perret, D., Crovisier, J.L., Stille, P., Shields, G., Mäder, U., Advocat, T., Schenk, K., Chardonnens, M., Thermodynamic stability of waste glasses compared to leaching behaviour. *Applied Geochemistry*, **18**, pp. 1165-1184, 2003.

[10]   Wolff-Boenisch, D., Gislason, S.R., Oelkers, E.H., Putnis, C. V., The dissolution rates of natural glasses as a function of their composition at pH 4 and 10.6, and temperature from 25 to 74°C. *Geochimica et Cosmochimica Acta*, **68(23)**, pp. 4843-4858, 2004.

[11]   Pisciella, P., Crisucci, S., Kamaranov, A., Pelino, M., Chemical durability of glasses obtained by vitrification of industrial wastes. *Waste Management* **21**, pp. 1-9, 2001.

[12]   Wang, P.W., Zhang, L., Structural role of lead silicate glasses derived from XPS spectra. *Journal of Non-Crystalline Solids*, **194** pp. 129-134, 1996.

[13]   Karamarov, A., Pelino, M., Crystallization phenomena in iron-rich glasses. *Journal of Non-Crystalline Solids*, **281** pp; 139-151, 2001.

[14]   Rossano, S., Ramos, A., Delaye, J.M., Creux, S., Filipponi, A., Brouder, Ch., Calas, G., EXAFS and Molecular Dynamics combined study od $CaO-FeO-2SiO_2$ glass. New insight into site significance in silicate glasses. *Europhysics Letters*, **49(5)**, pp. 597-602, 2000

[15]   Seignez, N., Bulteel, D., Damidot, D., Gauthier, A., Potdevin, J.L., une nouvelle approche méthodologique pour caractériser l'altération d'un déchet industriel. *Proc. of the first International Conference On Engineering for Waste Treatment*, p 221, 2005

[16]   Dimitrova, S.V., Metal sorption on Blast-Furnace slag. *Water Research*, **30(1)**, pp. 228-232, 1996.

[17]   Dimitrova, S.V., Effect of the heat treatment on the morphology and sorption ability to metal ions of metallurgical slag. *Journal of Materials Science*, **36**, pp. 2639-2643, 2001.

[18]   Dimitrova, S.V., Use of granular columns for lead removal. *Water Research*, **36,** pp. 4001-4008, 2002

[19]   Benjamin, M.M.,. Leckie, J.O., Multiple-site adsorption of Cd, Cu, Zn, and Pb on amorphous iron oxyhydroxide. *Journal of Colloid and Interface Science*, **79(1)**, pp. 209-221, 1981.

[20]   Farley, K.J., Dzombak, D.A., Morel, F. M, A surface precipitation model for the sorption of cations on metal oxides. *Journal of Colloid and Interface Science*, **106(1)**, pp. 226-242, 1985.

Anaerobic codigestion of biowastes generated in Castilla-La Mancha (Spain): batch studies

L. Rodríguez, J. Villaseñor, V. Sánchez, F. J. Fernández & I. M. Buendía
Department of Chemical Engineering. ITQUIMA, University of Castilla-La Mancha, Spain

Abstract

The combined anaerobic treatment of waste activated sludge (WAS) with olive oil mill wastes (OMW) and waste activated sludge with wine vinasses (VN), was investigated in batch reactors. The codigestion was studied by mixing the WAS with different ratios of OMW or VN. The criteria used for judging the success of the codigestion was the COD reduction, total methane production and methane yield. The results indicate that COD removal is better when biowastes are mixed. Total methane production in codigestion was higher than that from anaerobic digestion of raw wastes. A maximum was reached when WAS:OMW and WAS:VN were mixed in ratios of 6:94 and 25:75, respectively. When the ratio of OMW was increased in the mixture, the total methane production decreased. This effect could be explained because of the inhibition caused by the polyphenol contained in this waste.
Keywords: anaerobic digestion, codigestion, olive oil mill waste, wine vinasses waste activated sludge.

1   Introduction

Agro-industries play a significant role in and represent a considerable share of the Spanish economy, mainly in central Spain in regions such as Castilla-La Mancha. Every year 1.4-1.8 million tons of olive oil are produced in Mediterranean countries, resulting in 30 million m^3 waste [1]. Amongst other wastes, the main agro-industrial wastes generated in Castilla-La Mancha are olive oil mill wastes and wine vinasses. The food and drink industry of Castilla-La Mancha represents the 28.25% of the regional industry. These industries

WIT Transactions on Ecology and the Environment, Vol 92, © 2006 WIT Press
www.witpress.com, ISSN 1743-3541 (on-line)
doi:10.2495/WM060051

produce a high volume of wastes. In 2003, for instance, 5 million tons of wastes were generated by this kind of industries in Spain [2]. Nevertheless, a fraction of these agro-wastes can be considered a by-product and can be useful to other industries. This has many environmental and economic advantages and allows the natural resources contained in the waste to be recovered.

The biowaste mentioned above present certain problems, such as a high COD content, disproportionate C/N/P ratio, high generation rate, seasonal production and expensive management and storage. In view of these problems, anaerobic digestion seems to be a reasonable solution whose performance could be increased by mixing different biowastes (anaerobic codigestion), which leads to better conditions for the growth of microorganisms.

The co-digestion is expected to be cheaper than the treatment of individual waste, mainly due to the lower cost per volume treated, and because the moisture and rheological conditions can be adjusted. In addition, by mixing the wastes, the inhibitory and toxic effects of certain compounds can be minimised. Furthermore, global methane production can be higher than the amount generated by the addition of the methane generated by the digestion of the individual wastes [3-5].

In this context, the aim of this work was to study the feasibility of anaerobic codigestion of WAS, OMW, VN from Castilla-La Mancha (Spain) in batch reactors.

2 Materials and methods

2.1 Substrates and inoculum

Activated waste sludge was picked up from the wastewater treatment plant situated in Ciudad Real (Spain).

Olive oil mill waste came from *Aceites Pina* located in Villarta de San Juan (Ciudad Real, Spain).

Wine vinasses came from *Bodegas Cruz-Vega,* in Madridejos (Toledo, Spain).

The inoculum used in the batch experiments was taken from the anaerobic digester of the Ciudad Real wastewater treatment plant.

2.2 Batch experiments

The co-digestion was studied in 1*l* batch digesters with a working volume of 250 mL and a temperature of 35°C. The inoculum was concentrated until it reached a volatile suspended solid content of 10000 mg/L. The experiments were duplicated and the results were evaluated by comparing the rate of methane production in the headspace of the reactors (750 mL). During the experiments, the pressure in the gas phase was measured with barometric sensors incorporated at the top of the reactors. In addition, control experiments were carried out to determine the endogenous biogas production.

Mixtures of WAS, OMW and ultra pure water, like mixtures of WAS, VN and ultra pure water, at different mixed ratios from 0 to 100%, were performed to evaluate the potential degradation efficiency and inhibition. The length of each experiment was about 1 month.

2.3 Analytical methods

Using the methods described previously in literature, the methane and carbon dioxide contents of biogas and VFA in the bulk liquid were measured by GC [6]. Volatile solids (VS), total solids (TS), COD and pH were determined using *Standard Methods* [7]. Total nitrogen, total phosphorus and dissolved nitrogen were also determined using *Standard Methods* [7].

3 Results and discussion

3.1 Waste characterisation

The characterisation of the waste used in the experiments was carried out following a standard procedure [8]. The results of the characterisation are presented in Table 1. The results indicate that the codigestion of the wastes is feasible and could be the best option for degrading these wastes.

For example, the pH of WAS was higher than that of OMW, and WAS would compensate the OMW pH deficiency, allowing the growth of the methanogenic bacteria. The C/N/P ratio in the mixtures was more suitable for anaerobic treatment than the individual C/N/P ratio.

Table 1: Characterisation of the wastes.

	WAS	VN		OMW
TS (g/L)	21	20	TS (%)	46.42
VS (g/L)	18	14	VS (%)	44.28
COD (g/L)	32	22	COD (g/L)	685
pH	6.89	6.86	pH	5.62
Total N (mg-N/L)	80	9.7	Total N (mg-N/g)*	0.35
Total P (mg-P/L)	3.6	1.2	Total P (mg-P/g)*	0.003
Ammonia (mg-N/L)	16.3	1.2	Ammonia (mg-N/g)*	0.22

*gram of wet waste.

3.2 Batch experiments

Several batch experiments were carried out to study the anaerobic treatment of the biowaste mentioned above and to evaluate the influence of the mixture ratios.

Figures 1 and 2 show the cumulative biogas production of each feedstock separately WAS, OMW and VN and three mixtures vs. reaction time in the batch reactors.

Regarding both Figures 1 and 2, cumulative biogas production was higher when WAS was mixed with OMW and with VN. These results indicated that codigestion increased the anaerobic biodegradability.

As a summary of the results, the COD reduction (%) and cumulative biogas production expressed as total methane production (mg COD$_{methane}$/L) are showed in Table 2.

These results indicate that COD removal was also better when biowastes were mixed.

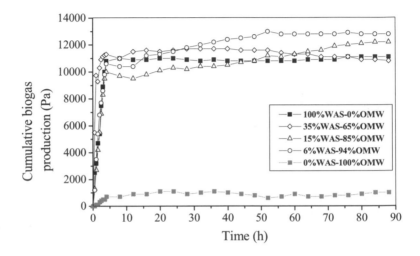

Figure 1:   Codigestion of WAS and OMW. Cumulative Biogas Production.

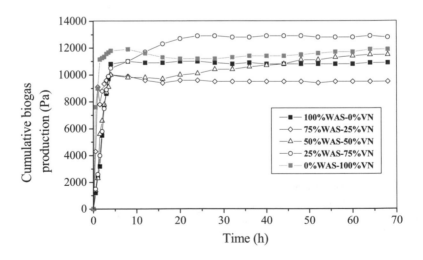

Figure 2:   Codigestion of WAS and VN. Cumulative Biogas Production.

Table 2: COD reduction (%) and Total methane production (mg COD$_{methane}$/L).

WAS:OMW	COD reduction	Methane production	WAS:VN	COD reduction	Methane production
100:0	58.25	947.20	100:0	58.25	947.20
35:65	56.08	998.40	75:25	46.88	870.40
15:85	57.06	1126.40	50:50	56.61	1100.80
6:94	78.53	1152.00	25:75	59.83	1152.00
0:100	66.56	102.40	0:100	53.91	1100.80

3.2.1 WAS:OMW codigestion

Regarding Table 2, the most suitable mixture to be anaerobically codigested, in terms of COD reduction, is 6:94 (WAS:OMW). This value is not in accordance with the theoretical mass balance, which indicated that mixtures with higher amounts of WAS were more efficient than OMW. This can be explained, as not all fractions of C/N/P were available to the methanogenic bacteria. In the case of the OMW, some of the nutrients contained are in the olive stone and this cannot be digested by anaerobic biomass within such a short time. When the amount of OMW is increased, total methane production decreased due to inhibitory effects of polyphenols on methanogenic bacteria [9].

3.2.2 WAS:VN codigestion

The methane production obtained at the end of the experiment for the WAS and VN were 947.20 and 1100.80 mg COD$_{methane}$/L, respectively, indicating that these wastes were of a similar biodegradability. When WAS and VN were codigested in a ratio of 25:75, a maximum in the COD reduction and methane production was reached. However, the values of these parameters were very similar when single wastes were digested.

4 Conclusions

The results and evaluations of this study may be summarised as follows:
- In the batch reactors, it was found that both COD removal and total methane production increased when biowastes were mixed.
- A maximum was reached when WAS:OMW and WAS:VN were mixed in ratios of 6:94 and 25:75, respectively.
- It was observed that by increasing the amount of OMW, the total methane production decreased due to inhibitory effects of polyphenols.

References

[1]   Angelidaki, I., Ahring, B.K., Deng, H. & Schmidt, J.E., Anaerobic digestion of olive oil mill effluents together with swine manure in UASB reactors. *Water Science and Technology*, **45 (10)**, pp. 213-218, 2002.

[2]   National Institute of Statistics, Encuesta industrial de empresa, Spain, 2003.
[3]   Angelidaki, I. & Ahring, B.K., Codigestion of olive mill wastewaters with manure, household waste or sewage sludge. *Biodegradation*, **8 (4)**, pp.221-226, 1997.
[4]   Marques, I., Anaerobic digestion treatment of olive mill wastewaters for effluent re-use in irrigation. *Desalination*, **137**, pp. 233-239, 2000.
[5]   Monnet, F., An introduction to anaerobic digestion of organic wastes. *Remade Scotland*, pp. 1-48, 2003.
[6]   Angelidaki, I., Petersen, S.P. & Anhring, B.K., Effects of lipids on thermophilic anaerobic digestion and reduction of lipid inhibition upon addition of bentonite. *Applied Microbiology Technology*, **33**, pp. 469-472, 1990.
[7]   APHA-AWWA-WPCF, *Standard Methods for the Examination of Waste and Wastewater*, American Public Health Association, Washington, DC, 1998.
[8]   Federal Compost Quality Assurance Organization (FCQAO), *Methods book for the analysis of Compost*, Bundesgütegemeinschaft Kompost e.V., Germany, 1994.
[9]   Fountoulskid, M.S., Dokianakis, S.N., Kornaros, M.E., Aggelis, G.G., Lyberatos, G., Removal of phenolic in olive mill wastewaters using the white-rot fungus *Pleurotus ostreatus*. *Water Research*, **36 (19)**, pp. 4735-4744, 2002.

Section 2
Clean technologies

Production of synthetic zeolites from lignite - calcareous Greek fly ashes and their potential for metals and metalloids retention

A. Moutsatsou[1] & V. Protonotarios[2]

[1]National Technical University of Athens, Laboratory of Inorganic and Analytical Chemistry, Department of Chemical Engineering, Greece
[2]National Technical University of Athens, Laboratory of Inorganic and Analytical Chemistry, Department of Chemical Engineering, Greece

Abstract

The current state of the art focuses on the production of synthetic zeolites from lignite - calcareous fly ashes. The synthetic minerals are produced by the hydrothermal treatment of Greek fly ashes originating from Ptolemais and Megalopolis Power Plants in Greece. The products are examined in terms of their mineralogical composition, for the formation of known zeolites, morphology, thermal behavior, cation exchange capacity, pH, porosity and specific surface area. Zeolitic products, along with untreated fly ash, are investigated for their potential of retaining heavy metals from liquid wastes. The mechanisms of metal retention by zeolites in artificial solutions including lead, zinc, copper, arsenic and cadmium were examined. Solutions including the same pollutants, mobilized from a contaminated soil, utilizing selected solvents were produced. Results revealed the capability of zeolitic materials and fly ashes to retain metals in significant concentrations. Furthermore, a number of different retention mechanisms are developed, including ion exchange, surface adsorption and precipitation of metals.
Keywords: fly ash, zeolites, metals, metalloids, retention mechanisms.

1 Introduction

Fly Ash (FA) is a by product derived from the combustion of coal or lignite, mainly in electric power stations. More than 11.5 million tons of lignite fly ash is

WIT Transactions on Ecology and the Environment, Vol 92, © 2006 WIT Press
www.witpress.com, ISSN 1743-3541 (on-line)
doi:10.2495/WM060061

produced annually in Greece [1]. Throughout EU, a small percentage (less than 30%) of the produced FA is used for lignite quarries reclamation, in cement industry as cement additive, for road construction, in structural materials and as a soil additive [2]. Nevertheless, the largest amount of the waste product is directly discharged into ponds and landfills. Possible application for the effective consumption of FA could be the formation of zeolites. The particular process is similar to the formation of natural zeolites from volcanic deposits or other high Si-Al materials [3–5]. The main process from which zeolites can be naturally formed is through the influence of hot groundwater on the glass fraction of volcanic ash. In the laboratory, the activation solution is usually NaOH or KOH in different molarities and according to atmospheric or water vapor pressure, at a temperature 80-200°C and for 3 to 96 h different types of zeolites can be synthesized [3–9].

Metals and metalloids such as Pb, As, Cu, Zn, Fe and Mn are well known as a serious environmental hazard [10]. Intensive industrial activity has resulted in the accumulation of high concentrations of heavy metals and toxic elements in the environment, causing health, ecological and socio-economic problems [11]. New technologies in the area of environmental relief from metal pollution include the utilization of industrial by-products (fly ashes and slag) presenting high adsorption capacity with respect to metals [12] resulting to environmental and economical, benefits.

The current state of art focuses on the following topics:

1. Formation of zeolites from Greek FAs
2. Investigation of zeolitic materials for a number of critical attributes in the direction of optimum metal and metalloid retention
3. Application of the zeolitic materials and the untreated FAs for metals and metalloids retention, under acidic conditions, utilizing both artificial and real polluted solutions.

2 Materials and methods

2.1 Zeolites synthesis

Ptolemais FA (PFA) and Megalopolis FA (MFA) underwent an alkaline hydrothermal treatment at 90°C, using NaOH 1M as an activation solution, in a 1L stainless steel reactor. The incubation time was set at 24 h, stirring at 150 rpm. Then, the mixture was filtered with a 0.45 μm paper, leached with water until no NaOH was detected and the collected solid residue was dried at 40°C for 24 h. The final solid product, along with the FAs, was subjected to XRD analysis and SEM investigation for identification of known zeolites. The pH, Cation Exchange Capacity (CEC), Specific Gravity and Specific Surface Area (SSA) of the FAs and the zeolitic products were also determined. Two different solution/FA ratios were examined for each of PFA and MFA. In Table 1 the four synthesized materials are presented.

Table 1:   Alkali activation of Greek FA$_s$ under 90°C, NaOH 1M.

FA Origine	FA/Solution Ratio (g L^{-1})	2.1.1 Product Code
Ptolemais	50	ZP$_{50}$
Ptolemais	100	ZP$_{100}$
Megalopolis	50	ZM$_{50}$
Megalopolis	100	ZM$_{100}$

2.2 Metals and metalloids retention from artificially polluted sample

Synthetic materials were examined for their potential of retaining pollutants from artificially polluted solutions. In particular, a hydrochloric acid base solution with concentrations of Pb, As, Cd, Cu, Zn, Fe and Mn was prepared. The pollutants concentrations were selected so as to be similar to that derived from the effect of HCl to contaminated soil samples (see 2.3). Metal and metalloids concentrations are illustrated in Table 2. Solutions were mixed in a solid to liquid ratio of 20 g L^{-1} with the zeolitic materials and the FAs.

2.3 Metals and metalloids retention from contaminated soil

Soil sample was collected from a former mining – metallurgical site in the city of Lavrion, Greece. Pollution is typical of a heavy mining – metallurgical industry and consists mainly of slag, sulphur compounds and oxidized phases, with an extremely complicated mineralogy. After mobilization of metal and metalloid content by 1 M HCl, further adsorption of the pollutants on synthesized zeolites and FAs was examined. The chemical composition of the HCl extraction solution, concerning the selected pollutants, is shown in Table 2. For retention experiments, a solid to liquid ratio of 20 g L^{-1} was used.

Table 2:   Concentrations of artificial and soil solutions.

Element	Concentration in artificial sample (ppm)	Concentration in soil sample (ppm)
Fe	2000	2500
Cu	75	75
Zn	1000	1375
Mn	100	100
Pb	800	844
As	150	176

3   Results and discussion

3.1 Zeolites synthesis

FA/NaOH = **50 g L^{-1}** for PFA, XRD pattern and SEM graphs, confirm an essential dissolution of the predominant fly ash phases (alumino-silicate glass, quartz) and the formation of NaP1 (tabular crystals) and NaP Zeolites ("blade" crystals) (Table 3). The semi-quantitative estimation of zeolite content is 30-35%. Results are similar for MFA and s/l =**50 g L⁻¹**. In case of MFA, the zeolitization yield is higher, fluctuating between 40 and 45%. This fact was rather expected, due to the lower calcium and the higher silica content of the particular FA [5, 10].

Table 3:   Zeolites identified from XRD patterns.

Zeolite	ZP_{50}	ZP_{100}	ZM_{50}	ZM_{100}
NaP1	+		+	
NaP	+	+	+	+
Herschelite				
Tobermorite				+
Hydroxy-cancrinite				+
Hydroxy-sodalite				+

When s/l increases further to **100 g L⁻¹** for **PFA** NaP zeolite is detected (15-20%) while formation of NaP1 is insignificant. Due to the increase of FA with respect to NaOH, a significant dissolution of the initial FA phases is not achieved. SiO_2 and $CaCO_3$ remain almost intact and no significant crystallization is taking place. In case of **MFA,** in addition to NaP, Tobermorite, Hydroxy-cancrinite and Hydroxy-sodalite are detected and their semi-quantitative percentage is estimated to 10-15%.

The **pH** is slightly higher for the greatest s/l ratio, for both FAs (Table 4). After the zeolitization process, the initial **CEC** of PFA and MFA (0.030 and 0.023 meq g⁻¹) increase significantly, reaching a maximum for ZP_{50} and ZM_{50} (0.870 and 1.164 meq g⁻¹) (Table 4). However, CEC decreases with further increase of s/l (100 g L⁻¹) for MFA. That is rather normal, since for the highest FA/NaOH ratio, the pure zeolite content of the products is lower. For s/l up to 50g L⁻¹, experimental products from MFA present higher CEC than the respective from PFA, but for higher FA/NaOH ratio, ZP_{100} excel ZM_{100} with respect to their CEC. It should be mentioned that, the morphology of ZP_{100} and ZM_{100}, seems promising for probable use of the aforementioned materials as adsorbents.

Hydrothermal treatment of FA may have conflicting impacts on **SG** of the final products (Table 4). As long as the NaOH solution penetrates the cenospheres, allows the escape of the trapped air and thus increasing the SG [12]. Nevertheless, as the zeolitization process evolves, the subsequent crystallization of the experimental products yields to a larger pore volume, thus the SG decreases. In the present case, SG decreases as PFA/NaOH ratio increases, while, in case of MFA, SG presents a minimum for ZM_{100}.

The **SSA** of PFA increases up to 400% for s/l = 50 g L^{-1} and then slightly decreases with further rise of PFA/NaOH ratio (Table 4). Results are even better for MFA, since SSA increases, with respect to the raw material, amounts to 800%. Nevertheless, there is a significant recession of SSA value for ZM$_{100}$. From the aforementioned results it is obvious that the alkali activation of the FA$_s$ results in decreasing the particle size of the raw materials. The latter phenomenon is probably attributed to the dissolution (etching) of the FA cenospheres and the subsequent exposure of their inner surface [13]. As it was expected, SSA best results are obtained for ZM$_{50}$ and ZP$_{50}$ for which the maximum zeolitization yields (i.e. the best dissolution of the cenospheres) are observed.

3.2 Metals and metalloids retention from artificially polluted sample

Retention yields seems to be very high as it is illustrated in Table 5 for PFA substrates, especially in case of Mn, Cu and Pb. Slightly lower retention yields are observed in case of As, while significantly lower retention is achieved for Fe and Zn. When MFA substrates are used, Mn, Cu and Pb present a very satisfactory retention, while Fe retention is lower than in case of PFA substrates. In all cases, the capability of untreated FAs for pollutants retention is ameliorated with respect to Fe, Zn, Pb and As, while zeolitic materials and FAs are retaining the total quantity of Cu and Mn present in the artificial solution.

Table 4:   Selected properties of FAs and Zeolitic materials.

	pH	CEC (meq g^{-1})	SG (g ml^{-1})	SSA (g m^2)
PFA	9.9	0.03	2.6	5.1
MFA	10.1	0.02	2.7	7.4
ZP$_{50}$	11.8	0.87	2.5	20.3
ZP$_{100}$	12.0	0.81	2.2	17.6
ZM$_{50}$	10.0	1.16	2.3	60.7
ZM$_{100}$	10.5	0.51	2.4	16.4

Table 5:   Percentage retention pollutants from artificially pollutant sample.

	Fe	Cu	Zn	Mn	Pb	As
ZP$_{50}$	64.8	100.0	65.4	100.0	81.2	84.1
ZP$_{100}$	67.1	100.0	71.2	100.0	79.4	80.2
PFA	62.5	100.0	58.4	100.0	70.3	71.7
ZM$_{50}$	47.0	100.0	57.3	100.0	82.2	80.5
ZM$_{100}$	51.4	100.0	59.1	100.0	80.1	82.1
MFA	47.5	100.0	45.2	100.0	77.4	74.3

3.3 Metals and metalloids retention from contaminated soil

In Table 6, the respective retention yields of the zeolitic materials, with respect to soil pollutants, are presented. As it was expected results are significantly lower than those obtained for the artificially polluted sample, with the exception of As retained on ZP_{50}. Pure zeolites are expected to retain metals mainly through an ion – exchange mechanism. It should be mentioned that, for purposes of effective retention of metals and metalloids, the non-complete transformation of FA to zeolites is preferable. This is due to the fact that the portion of FA that has not been converted to zeolite, may trigger additional retention mechanisms such as precipitation or surface adsorption. Another important advantage concerning the presence of unconverted FA is the capability of further stabilization/solidification of the pollutants on the substrates [13]. Results presented in Table 6 are quite satisfactory, bearing in mind the extreme pollution load of the washing solution and the great variety of metals and metalloids compounds present. It is obvious that the results between the different substrates, verify the triggering of a variety of retention mechanisms. The latter are depended, not only on the zeolitic material, but also on the pollutant – solvent – substrate interaction. Finally, with the exception of Zn and Pb, there is a significant amelioration of the retention capacity of the FAs, after their hydrothermal treatment.

Table 6: Percentage retention of soil pollutants on Zeolitic materials and Fas.

	Fe	Cu	Zn	Pb	As	Mn
ZP_{50}	45.3	25.7	19.6	27.7	90.3	68.8
ZP_{100}	44.6	30.5	23.7	18.8	28.5	32.5
PFA	5.2	6.1	22.7	18.3	8.2	7.4
ZM_{50}	50.8	42.8	27.2	25.7	24.4	40.2
ZM_{100}	42.4	33.2	25.3	24.2	22.7	42.6
MFA	4.8	15.2	18.2	21.2	7.1	12.7

4 Conclusions

Two Greek FA_s (Ptolemais FA and Megalopolis FA) were subjected to hydrothermal treatment for purposes of zeolites production. The best FA/NaOH ratio for both quantitative and monomineral synthesis is spotted at 50 g L^{-1} (ZP_{50} and ZM_{50}) for which NaP1 and NaP zeolites are formed in a 30-45%. Both Cation Exchange Capacity and Specific Surface Area of the experimental products are very improved with respect to the original FA_s and ZM_{50} presents the better results. Experimental products, along with untreated FAs were examined for their capacity in retaining heavy metals and metalloids from contaminated and artificially polluted samples, under acidic conditions. Results were excellent with respect to artificially polluted samples, with the zeolitic materials present improved retention yields compared to the untreated FAs.

Relatively low, but satisfactory retention capacity was illustrated for soil polluted sample, with the synthetic materials being considerably more effective with respect to metals and metalloids retention, than the raw FAs. The presence of a significant unconverted portion of FA in all zeolitic materials, indicate the capability of developing different retention mechanisms, including, except from the ion-exchange, precipitation and surface adsorption. The latter could be critical in view of further stabilization/solidification of the retained pollutants.

References

[1]   Moutsatsou A., Stamatakis M. Hatzitzotzia K., Protonotarios, V. (2005): The utilization of Greek Ca-rich and Ca-Si-rich fly ashes in the production of synthetic minerals. Fuel, Vol .85/5-6 pp. 657-663.

[2]   Querol X., Alastuey A., Fernadez-Turiel L. J., Lopez-Soler A. (1995): Synthesis of zeolites by alkaline activation of ferro-aluminus fly ash. Fuel, Vol. 74, pp. 1226-1231.

[3]   Steenbruggen G., Hollman G. G (1997). The synthesis of zeolites from fly ash and the properties of the zeolite products. Journal of Geochemical Exploration, Vol. 62, pp. 305-309.

[4]   Hollman G. G, Steenbruggen G., Jourkovicova J. V. (1999): A two step process for the synthesis of zeolites from coal fly ash. Fuel, Vol. 78, p. 1225-1230.

[5]   Querol X., Moreno N., Umana C. J., Alastuey A., Hernadez E. J., Lopez-Soler A., Plana F. (2002): Synthesis of zeolites from coal fly ash: an overview. Coal Geology, Vol. 50, p. 413-423.

[6]   Norihiro M., Yamamoto H., Shibata J. (2002): Mechanism of zeolite synthesis from coal fly ash by alkali hydrothermal reaction. International Journal of Mineral Processing, Vol. 64, pp. 1-17.

[7]   Shih W. H., Chang H. L. (1996). Conversion of fly ash into zeolites for ion exchange applications. Materials Letters, Vol. 28, pp. 263-268.

[8]   Albert R. B., Cheetham K. A., Stuart, J. A Adams C. J. (1998). Investigations on P zeolites: synthesis, characterization and structure of highly crystalline low-silica NaP. Microporous and Mesoporous Materials, Vol. 21, p. 133-142.

[9]   Tanaka S., Sakai Y., Hino R. (2002). Formation of Na-A and Na-X zeolites from waste solutions in conversion of coal fly ash to zeolites. Materials Research Bulletin, Vol. 37, p. 1873-1884.

[10]   Nael, C. N., Bricka, M. R., Chao, A. C., 1997. Evaluating acids and chelating agents for removing heavy metals from contaminated soils. Environ. Prog. 16, 274-280.

[11]   Maiz I., Esnaola V.M., Millan E. (1997). Evaluation of heavy metal availability in contaminated soils by a short sequential extraction procedure. The Science of the Total Environment, Vol. 206, pp. 107- 115.

[12]   A. Moutsatsou, A. Karathanasis & V. Protonotarios (2004): Remediation of soils polluted by industrial activities utilizing hydrothermally treated calcareous fly ashes. 2nd International Conference on Waste Management

and the Environment, Section 6: Environmental Effects and Remediation, pp. 283-291, Rhodes, Greece.

[13] A. Moutsatsou M. Gregou, M. Liokalou, D. Matsas and Protonotarios V (2003): Mobilization of metals and metalloids from contaminated soil and readsorption on synthesized zeolites and zeolite-fly ash mixtures. 8th International Conference on Environmental Science and Technology, Lemnos, Greece, Vol. A, pp. 642-649.

Toward Cleaner Production technologies in surface treatment of metals

A. Nakonieczny & M. Kieszkowski
Institute of Precision Mechanics, Warsaw, Poland

Abstract

Cleaner Production technologies are an efficient tool to solve the environmental problems of the surface treatment of metals with a number of process-integrated and end-of-pipe techniques and to minimise emissions to the waters and to air, as well as to reduce the generation of hazardous wastes. This paper presents a short characteristic of the main environmental problems of surface treatment of metals with special consideration of hazardous wastes generated in the production of the corrosion-protective and decorative coatings. Key environmental issues are given. General principles of Cleaner Production technologies for surface treatment processes are discussed. Some practical methods of reducing water consumption and different techniques of the direct recovery of process solutions are shown. Practical examples of Cleaner Production methods implemented into Polish plating shops for zinc, nickel, chromium and tin plating are presented with a short description of the obtained efficiency of recovery and economy of water consumption.
Keywords: surface treatment of metals, environmental protection, Cleaner Productions, IPPC Directive, rinsing technology, water consumption, emissions minimisation, recovery of process solutions.

1 Introduction

Cleaner Production [1] technologies are among the main elements of the Best Available Techniques, which have to be implemented in plating shops to obtain the new integrated permit, according to European Directive 96/61/EC, so called IPPC Directive [2] (IPPC = Integrated Pollution Prevention and Control).

The strategy of Cleaner Production and Best Available Techniques are both based on the application of low-waste and waste-free technologies. Also the

WIT Transactions on Ecology and the Environment, Vol 92, © 2006 WIT Press
www.witpress.com, ISSN 1743-3541 (on-line)
doi:10.2495/WM060071

priorities of environmental protection measures of the Best Available Techniques are similar to Cleaner Production methods and are based usually on three types of activities [3]:

- elimination of sources of toxic emissions
- substitution of harmful substances by environmentally friendly ones
- minimisation of wastes which are difficult to eliminate.

2 Key environmental issues

Key environmental issues in the surface treatment industry are: raw materials, energy and water usage, emissions to water and to air, as well as generation of liquid and solid wastes. Therefore efforts to reduce environmental impacts of production activities in this sector are directed mainly at:

- efficient and economic usage of raw materials, energy and water
- minimisation of emissions by effective process control and waste water treatment
- minimisation of generated wastes
- prevention of environmental accidents with an effective protection of soils and groundwater with a view to future site decommissioning.

Technological and management measures to achieve better environmental protection are often complex and cover a variety of process-integrated and end-of-pipe techniques. As the majority of surface treatment processes are water-based, the water usage and its management with associated raw materials consumption, waste water treatment technology and waste generation are the most important issues. Key emissions to water are typically metals, which are used as soluble salts and, depending on the process, may include zinc, nickel, chromium, copper and other. Waste water may also contain many other species, including cyanides, surfactants, complexing agents (as e.g. EDTA) and different anion, like chlorides, sulphates, phosphates, nitrates etc.

Therefore, the production of metal platings for protective and decorative purposes generates waste water, solid wastes and smaller quantities of some volatile impurities:

⇒ waste water which may contain various hazardous compounds, such as metals, cyanides and some mineral acids, alkalis, mineral salts, oils, organic substances and other

⇒ solid wastes and concentrated liquids consisting mainly of neutralisation sludge and spent process solutions

⇒ volatile impurities, such as hydrogen chloride, sulphur dioxide, nitrogen oxides, aerosols, dust and other volatile substances which may be emitted to the atmosphere during technological processes.

3 General principles of Cleaner Production technologies

General principles and main elements of cleaner technologies for surface treatment processes are:

- reduction of any risk of environmental accidents resulting in unplanned emissions to the environment
- modifications of technological processes - substitution options for various harmful chemicals and processes, such as organic solvents (especially chlorinated hydrocarbons), EDTA and other strong chelating agents, cadmium, hexavalent chromium, nickel and others
- regeneration and other maintenance operations to extend the life of process solutions and to avoid or to reduce their dumping frequency, or to improve bath performance
- minimisation of drag-out of process solutions (specifically in rack and barrel lines)
- efficient and economic rinsing technology ("eco" rinse, multistage cascade rinsing, multistage static rinsing, minimisation of rinsing water consumption etc.)
- direct recovery of process baths (evaporation and other methods of rinsing water concentration, closed loop of materials)
- energy saving (electricity, bath heating, cooling, bath agitation)
- waste water (flow minimisation, identification and separation of problematic flows, waste water treatment technology)
- waste management
- air emissions.

4 Direct recovery of process solutions

Direct recovery of plating baths is based on the proper utilisation of effective and economic rinsing techniques. It consists of increasing drag-out recovery by returning the rinse water from the first rinse to the process solution. It is a good example of Cleaner Production technology with technical and economic advantages such as:

⇒ higher quality of plating
⇒ lower consumption of rinsing water
⇒ reduction of volume and load of generated waste water
⇒ reduction of waste water treatment cost
⇒ reduction of neutralisation sludge volume
⇒ improvement of final effluent quality.

In the simplest case, the direct recovery of plating baths is realised with a static, so called drag-out tank (economic rinse), which is commonly used after high temperature process baths. Rinse water from drag-out tank is used for the replenishment of volume losses of the process bath due to evaporation. Periodic water transfer from drag-out tank into process tank is made with a small pump. Practical recovery ratio obtained by this method is usually not higher than 50-60 %. Efficiency of the direct recovery can be significantly improved by the application of several drag-out tanks, arranged in series. Such a technique is successfully applied in the WK-1 automatic drag-out recovery system, developed at the Institute of Precision Mechanics in Warsaw. Schematic diagram of this system is shown in Fig. 1.

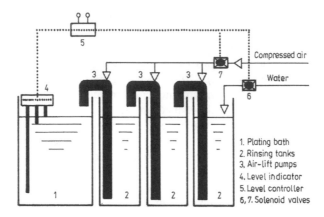

Figure 1:   WK-1 direct recovery system of plating baths.

Air-lift pumps made from PVC (3) are used for water transfer from one rinse tank (2) to another and from the first rinse tank to the process tank (1). Fresh water feed into the rinsing system and the operation of air-lift pumps are automatically controlled by solenoid valves (6) and (7). Opening and closing of both valves is controlled by an electronic level controller (5) coupled with conductometric probes of a level indicator (4) installed in the process tank (1). Demineralised water should be used to feed the WK-1 drag-out rinse tanks. Special construction of air-lift pumps makes possible the transfer of only the excess of water entering the rinse tank (as an overflow) and thus to maintain the same and constant water level in the rinse tanks. Normally the consumption of low pressure air to operate WK-1 air-lift pumps is in the range of 30-50 m^3/h. Depending on plating line output, evaporation losses of the bath, free room available at the plating shop etc., 3-4 or more drag-out tanks in series are normally installed. They can work in a closed loop system, i.e. practically with no effluent. Application of air-lift pumps of proper construction enables the creation of a multistage counter-flow (cascade) drag-out tank system, using individual rinse tanks without any design modifications, such as pipe fittings, overflows etc. It helps to simplify the implementation work and to reduce its cost. The WK-1 system has been installed in many plating shops in Poland and abroad, mainly for nickel, tin and copper plating baths and other hot bath recovery systems with an efficiency of 85-95 %. It has been also used for the recovery of process solutions operating at room temperature, such as zinc plating baths. In this case a small atmospheric evaporator is frequently installed to work in co-operation with the WK-1 system.

In case of chromium plating, where relatively small size process tanks are commonly in use, a good recovery rate is rather difficult to achieve, even with the WK-1 system. For this bath the IMPCHROME evaporator has been developed (Patents 149.478 PL and 149.479 PL). Partial evaporation of chromium bath enables the introduction of more rinse water into the plating tank and thus the improvement of the recovery efficiency.

The IMPCHROME evaporator comprises tubular modules made of PVC, in which evaporation occurs due to the action of low pressure air supplied from the central factory system or from an independent blower. The evaporator has no heating elements – the necessary heat energy is supplied by the hot bath. Because of the exothermic character of chromium plating process, the waste heat is used for evaporation. Therefore the chromium bath does not require any additional cooling system and the energy cost of the recovery process becomes very low. The consumption of low pressure air (0.02-0.04 MPa) is normally of the range of 200-300 m³/h. The plating bath is continuously circulated between the plating tank and IMPCHROME evaporator. The WK-1 system ensures the replenishment of plating bath with rinsing water transferred by automatically operated air-lift pumps. Schematic diagram of this system is shown in Fig. 2.

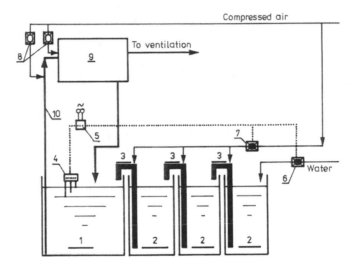

Figure 2:   Direct recovery system of chromium plating bath using IMPCHROME evaporator (1.Cr plating bath; 2.Rinsing tanks; 3.Air-lift pumps of WK-1 system; 4.Level indicator; 5.Level controller; 6,7.Solenoid valves; 8.Correction valves; 9.IMPCHROME evaporator; 10.Air-lift pumps for Cr-plating bath).

5   Practical examples of implemented techniques

Schematic of the modified semi-automatic barrel zinc plating line, installed at a plating shop in Poland, replacing a prior non-linear tank arrangement, is shown in Fig. 3. In this line, several cleaner technology elements were implemented, such as:

- 4-step counter-flow drag-out tank (8/9/10/11) for rinsing after five zinc plating tanks (7) equipped with an automatic WK-1 system for zinc bath recovery

- atmospheric evaporator located under the line, to assist and to improve recovery efficiency
- inter-connected 2-step counter-flow rinse tanks (2/3) and (5/6), where the same water is used twice – for rinsing after acid-dip (4) and finally after electrolytic degreasing (1)
- 2-step counter-flow rinse tank (13/14) after zinc passivation bath (12)
- flow-meters and water-counters in the rinse tanks
- micro-filtration unit for the purification and regeneration of degreasing bath.

A small RO unit for demineralised water supply is also installed.

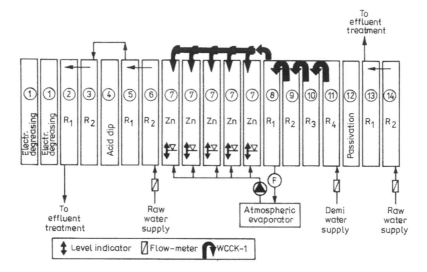

Figure 3:   Schematic diagram of the modified semi-automatic barrel zinc plating line equipped with cleaner technology elements.

With an average zinc plating output of 70.000 m²/year, the main economic effects observed after installation of the line were:

- drag-out recovery ratio of Zn-plating bath was 98 % (about 300 kg Zn/year), which means that amount of zinc drag-out from plating baths into waste water and then precipitated as useless sludge has been reduced by a minimum of 98 %
- average water consumption for rinsing purposes was reduced to 18 l/m². With the original water consumption of about 200 l/m² it means a reduction ratio of 91 %
- other savings included reduction of energy cost of the plating shop and waste water treatment system operation, reduction of anodes and plating salt consumption, reduction of chemicals consumption for waste water treatment processes, reduction of sludge de-watering and disposal costs, reduction of environmental charges etc.

In another case, similar modifications, with the application of Cleaner Production elements, were made for a semi-automatic rack nickel-chromium plating line. 4-step counter-flow drag-out tanks equipped with WK-1 bath recovery systems located after Ni-plating and Cr-plating baths were implemented together with an IMPCHROME evaporator for Cr-plating bath recovery. With an average Ni-Cr plating output of 28.000 m²/year, a 99 % drag-out recovery ratio of Ni-plating bath (about 330 kg Ni/year) and 97 % drag-out recovery ratio of Cr-plating bath (about 1250 kg CrO_3/year) were achieved. Water consumption was reduced to approx. 22 l/m².

In yet another case a small manual tin plating line was modified. 3-step counter-flow drag-out tank with the WK-1 bath recovery system was applied. With an average Sn-plating output of 6.000 m²/year, a 96 % drag-out recovery ratio of Sn-plating bath (about 32 kg Sn/year) was achieved. At the same time the rinsing water consumption dropped substantially.

Table 1 presents metal concentrations observed in the rinse tanks of the WK-1 recovery systems installed in the modified plating lines after 3 months of operation.

Table 1:   Average Zn, Ni, Cr and Sn concentrations in the rinse tanks of the WK-1 recovery systems applied to modified plating lines after 3 months of operation.

	zinc plating line [g/l Zn]	nickel-chromium plating line		tin plating line [g/l Sn]
		[g/l Ni]	[g/l Cr]	
plating bath	28,6	80,2	157	36,5
1° rinse tank	1,27	5,60	34,1	2,0
2° rinse tank	0,36	0,58	2,97	0,20
3° rinse tank	0,018	0,15	0,21	0,01
4° rinse tank	0,0036	0,006	0,007	-

As can be seen, a very low concentration of metals in the last rinse tanks of WK-1 systems was noted, which proves good efficiency of the recovery technique applied.

6   Conclusions

Various Cleaner Production elements of different kinds and different cost can be implemented in plating shops. Recovery of plating bath components and especially the direct recovery of plating baths play an important role in the cleaner technology applications. Obtained results in terms of the recovered material values (metals and other bath components), reduction of the operational costs, reduction of environmental pollution etc., as well as a significant reduction of rinsing water consumption are quite significant and enable minimisation of the environmental impact of plating production. At the same time the observed details of Cleaner Production show that plating quality level remains good and no trouble in plating bath maintenance occurs.

References

[1]   United Nations Environment Programme. Division of Technology, Industry and Economics. http://www.unep.org.
[2]   Council Directive 96/61/EC of 24 September 1996 concerning integrated pollution prevention and control. *Official Journal of European Communities,* L257, **39**, pp.15, 1996.
[3]   EIPPC Bureau. Reference Document on Best Available Techniques for the Surface Treatment of Metals and Plastics. Dated September 2005. http://eipccb.jrc.es/pages/Fmembers.htm.

Section 3
Waste reduction and recycling

Industrial treatment processes for the recycling of green foundry sands

S. Fiore & M. C. Zanetti
DITAG, Politecnico di Torino, Italy

Abstract

Three treatment processes for the reclamation of green moulding sands coming out from a cast iron foundry located in Northern Italy are considered in this study. A wet mechanical treatment, a dry mechanical treatment and a dry mechanical plus thermal treatment are compared to evaluate the efficiency of each process and to point out the best regeneration solution for the recycling of reclaimed sand in foundry operations. The inflow and the outflow samples of each process were characterized by means of particle-size analysis and the determination of silica, loss on ignition, acid request, oolitic and some metals contents.

A final evaluation of the three processes was performed taking into account the obtained quality of recovered silica sand and the economical aspects; the wet mechanical and the dry mechanical plus thermal treatments are the most effective for the recovery of green moulding sands coming out from Teksid foundry plant for cold-box core production.

Keywords: bentonite bonded moulding sand, green sand, foundry waste recycling.

1 Introduction

Exhaust sands represent a crucial issue in the management of foundry wastes: a ferrous foundry of secondary fusion produces an amount of wastes varying from 25% to 100% b.w. respect to the final product, and the 30-60% of these wastes is made of core and moulding sands (EPA [1]). Therefore the optimization of processes aimed to the recycling of exhaust foundry sands is particularly interesting.

WIT Transactions on Ecology and the Environment, Vol 92, © 2006 WIT Press
www.witpress.com, ISSN 1743-3541 (on-line)
doi:10.2495/WM060081

The Teksid cast iron foundry plant located in Crescentino (Vercelli, Northern Italy) and considered in this study actually produces about 700-800 t/d of casts for the automotive industry and about 750-800 t/d of wastes. The production process is made of the following phases: furnace charge preparation, mould and core manufacture, pattern making, melting and casting, shakeout, cleaning and finishing operations (steel shotting, painting and thermal treatment). Green sand moulding and hot/cold box processes are performed in Teksid plant respectively for mould and core manufacture.

Green sand moulding, accounting for about 85% of ferrous casts produced in the world (EPA [2]), employs a mixture of silica sand (80-95% b.w.), clay (3-10% b.w.), coal dust (2-10% b.w.) and water (3-4% b.w.) (Schleg [3]). The word "green" denotes the absence of any drying or baking phase (Luther [4]). Water activates the binding action of the clay on silica sand, and the coal dust burns off at contact with the molten metal, preventing its oxidation and increasing the refractory properties of the mould.

Cores are made of silica sand and organic binders, and should be harder and stronger than moulds, because they are required to resist the pressure of the molten metal that fills the mould (EPA [5]). The considered foundry plant employs two methods for core production: the hot-box process, in which silica sand is mixed with 1% b.w. phenolic resin and 0.8% b.w. of ammonium nitrate and urea as catalysts, and then heated at 230-260°C until solidification; and the cold-box process, in which silica sand is mixed with 1.3% b.w. phenolic resin, 0.9% b.w. isocyanic resin and 0.1% b.w. trimethylamine gas as catalyst, and no heating is required.

The wastes produced in Teksid foundry consist mainly of moulding sands and dusts from dust abatement plants on moulding lines (464 t/d), muds from dust abatement plants on moulding lines and furnaces (150 t/d), furnace and ladle slags (100 t/d), broken cores (50 t/d), powders and muds from dust abatement plant on furnaces and finishing operations (8 t/d), exhaust lime (6 t/d). The authors evaluated in another study (Zanetti et al. [6]) the reuse/recycle possibilities for all the mentioned residues.

Spent sands account for about the 60% of the total wastes generated by the considered plant and make a relevant amount of residues; the aim of this study is to evaluate a proper solution for their reclamation, taking into account both economic and environmental aspects.

The reclamation of moulding sands for the recycling in mould and core production is essentially based on the following aspects (EPA [2], Schleg [3]): right particle-size range (Teksid plant requirements consider a range between 0.1 and 0.4 mm), and elimination of bentonite, coal dust and metallic plus non-metallic impurities. The dimensional range and the coal dust and impurities elimination may be easily achieved by means of a sieving operation (metal fins and mould and core fragments cumulate preferentially in large fractions, while coal dust is particularly present in fine fractions). Instead bentonite is located as free particles in fine fractions, but also forms a hard shell (defined "oolitic") that surrounds the silica sand grains, which may be broken by means of mechanical treatments.

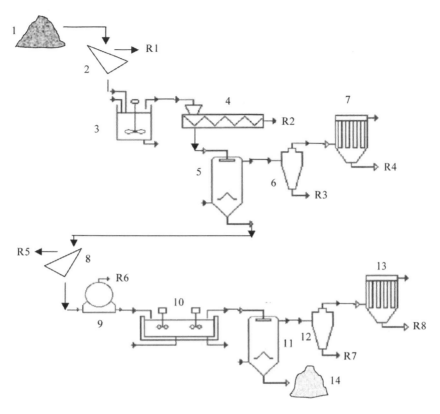

Legend: 1.feed, 2.screen 100 mm, R1. residues d>100 mm, 3. attrition cell, 4. screw conveyor, R2. fine particles (d<0.1 mm), 5,11. drier, 6,12. cyclone, R3,R7. cyclone coarse fraction, 7,13. bag filter, R4,R8. bag filter particles, 8. screen 0.7 mm, R5. residues d>0.7 mm, 9. rotating barrel magnetic separator, R6. magnetic fraction, 10. sulphuric acid lixiviation, 14. final product.

Figure 1:   Scheme of the wet mechanical reclamation treatment.

2 Treatment processes

Three reclamation processes, all performed on industrial scale on the spent moulding sands coming out from the Teksid plant, are compared in this study: a wet mechanical treatment, a dry mechanical treatment and a dry mechanical plus thermal treatment. The obtained clean sands from the three mentioned processes were chemically characterized.

2.1 Wet mechanical treatment

A wet mechanical treatment was tested on 500 t of green moulding sand coming out from the considered Teksid foundry in a plant, situated in Northern Italy, that usually performs the treatment of residues from granite mining for glass and ceramic production. The treatment cycle adopted is schematised in Figure 1.

The reclamation treatment produces 80% b.w. of clean sand, 15% b.w. of fine particles from the screw conveyor and 5% b.w. of magnetic products and fine particles from cyclone and bag filter. The authors proposed in another work (Zanetti and Fiore [9]) the recycle of the fine particles from the screw conveyor in mould manufacture and their reuse in the ceramic industry for tile production.

2.2 Dry mechanical treatment

A dry mechanical treatment was tested on 500 t of green moulding sand coming out from the considered Teksid foundry in a pilot plant that foresees a dry attrition phase. The reclamation treatment is based on a dry mechanical process: the dried sand grains are fed between metal blade drums that rotate at 1 rpm and a stone grinding wheel, which rotates at 40 rpm, thus obtaining the break of the oolitic shell and the rounding off of the silica particles. A scheme of the plant is shown in Figure 2.

The dry mechanical reclamation treatment produces about the 80% b.w. of clean sand.

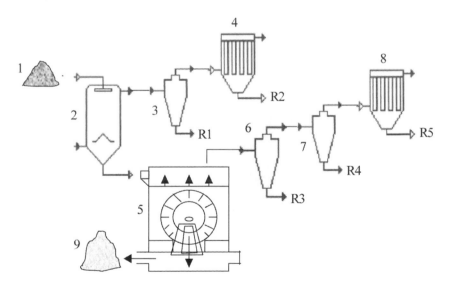

Legend: 1. feed, 2. drier, 3,6,7. cyclone, R1,R3,R4. cyclone coarse fraction, 4,8. bag filter, R2,R5. bag filter particles, 5. sand cleaner, 9. final product

Figure 2: Scheme of the dry mechanical reclamation treatment.

2.3 Dry mechanical plus thermal treatment

A dry mechanical plus thermal process is applied in a treatment plant situated in the considered Teksid foundry. The regeneration treatment is schematised in

Figure 3. After a sieving and a magnetic separation, a low intensity pneumatic mechanical scrubbing treatment is performed, then a thermal treatment at a temperature of 800 ÷ 900 °C, finally a high intensity pneumatic mechanical scrubbing treatment. The plant is designed to perform the reclamation of 600 t/d of green moulding sand for mould and core manufacture, with a theoretical efficiency of the 80%.

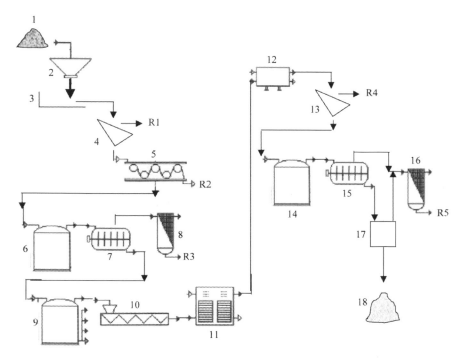

Legend: 1. feed, 2. shaking feeder, 3. shaking extractor, 4. screen, R1. iron residues (d>0.6mm), 5. magnetic belt separator, R2. magnetic fraction, 6. storage, 7. scrubber (high intensity), 8 and 16. dedusting (cyclone and bag filter), R3. not inert fine particles, 9. storage, 10.screw conveyor, 11. Hot Rec furnaces, 12. cooling, 13. sieve, R4. residues (d>0.4 mm), 14. storage, 15. scrubber (low density), R5. inert fine particles, 17. dedusting, 18. final product

Figure 3:   Scheme of the dry mechanical plus thermal reclamation treatment.

The produced residues are mainly of two kinds: not inert fine particles (15% of the feed) and inert fine particles (5% of the feed). The word "inert" refers to a minimum content of active clay that may be referred to the thermal resistance of the bentonite employed in green sand moulding. The authors proposed in another study (Zanetti et al. [6]) the reuse of not inert fine particles in mould manufacture, and the recycling of inert fine particles as raw material for the concrete industry.

3 Materials and methods

All the reagents employed are A.C.S. grade and the flasks and the glassware are A class. All the analyses were performed on dried samples, performing reference procedures (UNI [7]).

The particle-size analysis was realized on samples of about 5 kg reduced to a mass of about 400 g by means of a Jones splitter, using a Ro-Tap Tyler mechanical siever equipped with six Tyler mesh sieves ($2^{1/4}$ ratio) for 10 minutes. All the weighing operations were performed by means of a balance (0.01 g sensitivity). The inflow samples were at first wet sieved at a 0.025 mm and 0.038 mm dimension before the particle size analysis. The particle-size analysis below the 0.025 mm dimension was performed by means of the Andreasen apparatus.

The AFS grain fineness number shows the number of silica grains kept per 1 mm^2 of the sieve surface: a high value of this parameter means the existence of a high number of fine particles. This index is calculated after the particle-size distribution analysis, each elementary percentage being multiplied for a fixed number (Sundeen [8]). The sum of these values divided for the total weigh of the sample determines the grain fineness number.

The Acid Demand Value (at pH 5) is determined by means of the addition of 50 ml of distilled water and 50 ml of 0.1 N hydrochloric acid to a sample of a weight equal to 50 g. After 15 minutes of vigorous stirring, the sample is washed twice with 100 ml of distilled water, and the liquid phases are titrated, using a Orion 420 pH-meter, with a 0.1 N sodium hydroxide solution till a pH value equal to 5 is reached.

The oolitic content determination requires the addition of 200 ml of 6 N hydrochloric acid to a sample of a weight equal to 50 g that has previously been heated to a temperature of 900°C for 4 hours. The mixture is then boiled for 25 minutes. The sand is washed with distilled water until no more acidity is detected and then 250 ml of distilled water and 60 g of potassium hydroxide are added. The mixture is boiled for a further 25 minutes, the sand is washed again with distilled water until no more alkalinity is detected, and the residue is filtrated and dried. The difference (% b.w.) between the input sample and the treated one gives the oolitic content.

The Loss on Ignition (L.O.I.) was evaluated by roasting 2 g of dried sample at the temperature of 900°C for 3 hours. The weight difference (% b.w.) between the starting sample and the roasted one gives the searched value.

The metals (Na, K, Ca, Mg, Fe, Cr, Zn) contents were obtained through an acid digestion of 0.5 g sand with 6 ml of 32% hydrochloric acid and 2 ml of 65% nitric acid in a microwave oven. The digested samples were filtrated on a Whatman grade 44 filters and on the obtained solutions the metals contents determination was performed by means of an FAAS. For the silica determination the filtered solid residue underwent to a calcination phase (900 °C for 30 minutes) and a new acid digestion in the microwave oven with 6 ml of 40% fluoridric acid. The obtained solution is filtrated and the solid residue was calcinated at 900 °C for 30 minutes.

The Aluminium content was gathered through an acid digestion of 0.25 g sand with 2 ml of 96% sulphuric acid and 2.5 ml of 85% phosphoric acid in a microwave oven. 5 ml of fluoridric acid were added to the digested samples and a new digestion phase was performed. The digested samples were filtrated on a Whatman grade 44 filters and on the obtained solutions the Aluminium content determination was performed by means of FAAS.

4   Results and discussion

The efficiency of each one of the above-described treatments is evaluated to single out the different regeneration solutions for the recycling of the reclaimed sand in mould and core manufacture. The yields of the three reclamation treatments are comparable, as the amount of produced residues, for which the authors proposed some recycle/reuse possibilities in other studies (Zanetti et al. [6], Zanetti and Fiore [9]). The reclaimed sand obtained by means of the wet mechanical treatment has been tested by the authors, with positive results, for cold-box core manufacture and for glass production (Zanetti et al. [10]).

The inflow and the outflow samples of each process were characterized by means of a particle-size analysis and the determination of silica, loss on ignition, acid request, oolitic and some metals contents. The results of the performed analyses are compared in Table 1 with Teksid foundry requirements for reclaimed sand employed in cold-box core production.

Considering dimensional parameters, the dry mechanical treatment process produces a higher amount of fine particles, compared to the wet mechanical process. The fine particles amount produced by the dry mechanical plus thermal treatment is comparable with the value given by the dry mechanical process. The amounts b.w. of the particles having dimensions above 0.4 mm in the outflows are similar.

The Acid Demand Value (at pH 5) is the total acid quantity that the basic compounds contained in the recovered sand succeed in neutralizing. Alkaline substances (such as carbonates but also Na, K, Ca, Mg oxides) react with acidic catalysts added to the resin in core manufacture, thus reducing the efficiency of the resin-catalyst reaction; alkaline substances also reduce the refractoriness of the sand. The inflow Acid Demand Value is particularly high because of the significant content of fine particles, rich of bentonite (containing relevant amounts of alkaline substances and oxides). Wet mechanical and dry mechanical plus thermal treatments comply with foundry limits, otherwise the dry mechanical process produces a high residual Acid Demand Value, probably due to a not efficient bentonite removal. This is confirmed by the sodium and iron contents found in the outflow of the dry mechanical treatment, although the residual iron content is also caused by the absence of any magnetic separation phase.

The Loss on Ignition (L.O.I.) value, determined by organic substances (mainly coal dust and resins) and carbonates, represents a measure of the volume of gaseous products that will form at the contact of the molten metal with the mould and the core. Too high L.O.I. values are undesirable, because they may imply defects on casts surface. While all regeneration treatments respect foundry requirements, the dry mechanical plus thermal process is particularly effective in the abatement of the L.O.I. value, because of the high temperature achieved.

Table 1:   Characterization of the inflow and the outflows of the considered reclamation treatments evaluated for the regeneration of green moulding sands of Teksid foundry plant.

Parameter	Inflow	Outflows			Teksid Foundry requirements
		Wet mechanical treatment	Dry mechanical treatment	Dry mechanical + thermal treatment	
d<0.1 mm (% b.w.)	10	0.11	0.20	0.50*	0.30
AFS	50.38*	48.21	52.16	48.56*	44÷52
d>0.4 mm (% b.w.)	20	10	8.7	/	/
Acid Demand Value (ml)	48.0*	1.5	10.7	2.9*	0-6
Loss on Ignition (% b.w.)	3.02*	0.45	0.64	0.04*	2.00
Oolitic (% b.w.)	6.6	1.7	3.9	/	2.0
% Na	0.17	0.02	0.12	/	/
%K	0.04	/	0.02	/	/
% Ca	0.03	<0.01	<0.01	/	/
% Mg	0.32	/	0.03	/	/
% Fe	0.77	0.04	0.25	/	/
% Cr	<0.005	<0.005	0.010	/	/
% Zn	0.004	/	0.002	/	/
% Al	0.53	<0.005	0.13	/	/
% total oxides	2.88	0.13	0.87	/	1.00

*data obtained from an external laboratory.

The oolitic content indicates the ageing degree of the sand, because the clay used as binding agent after the melt creates a shell around the silica grains. This clay layer should not be very thick, or the casts may present some surface defects. The dry mechanical process doesn't comply with foundry limits, on the contrary the wet mechanical treatment, characterised by multiple attrition phases (initial attrition with water, then the action of the screw conveyor, finally the lixiviation with sulphuric acid) is particularly efficient in breaking the oolitic shell.

Total oxides must be kept below 1% b.w. to assure the refractoriness required in foundry operations. Both wet mechanical and dry mechanical processes comply with foundry requirements about total oxides contents; the higher value given by the dry mechanical treatment, mainly due to iron content, is caused by the absence of any magnetic separation phase.

Silica content of the reclaimed products has to be high, not less of the 90% b.w., to ensure the thermal resistance of moulds and cores; some tests performed by the authors showed that in all the three cases there aren't any problems in reaching that value, and that the best result was gathered by means of the wet mechanical treatment.

5 Conclusions

The results of the performed analyses show that the wet mechanical treatment and the dry mechanical plus thermal treatment are the most effective for the recovery of green moulding sands coming out from Teksid foundry plant for cold-box core production. In fact the oolitic content and acid demand values are better than the ones obtained from the dry mechanical treatment. However the wet mechanical treatment involves a noticeable sludge production and a treatment cost about equal to 0.02-0.03 €/kg. The involved cost of the dry mechanical plus thermal process is about equal to 0.03-0.04 €/kg. The dry mechanical process is economical (0.01 €/kg), but because of the high gathered acid demand value (about 10 ml) requires a change in the cold-box core making foundry operations for the compliance of the recovered foundry sand.

References

[1]   EPA United States Environmental Protection Agency, *Summary of factors affecting compliance by ferrous foundries*, vol.1, EPA-340/1-80-020, Washington, pp.34-49, 1981.

[2]   EPA United States Environmental Protection Agency, *Profile of the Metal Casting Industry*, EPA/310/R-97/004, pp.15-22, 77-84, 1997.

[3]   Schleg, F., Guide to casting and molding processes, *Engineered Casting Solutions* **2(3)**, pp.18-27, 2000.

[4]   Luther, N.B., Metalcasting and molding processes, *Casting Source Directory* **7(1)**, pp.29-35, 1997.

[5]   EPA United States Environmental Protection Agency, *Guides to Pollution Prevention, Metal Casting and Heat Treating* Industry, EPA/625/R-92/009, pp.5-12, 1992.

[6]   Zanetti, M.C., Fiore, S., Clerici, C., Recycling and reutilization of cast foundry wastes, *Proc. of XXII International Mineral Processing Congress*, eds. L. Lorenzen, D.J. Bradshaw, Document Transformation Technologies: Cape Town, ISBN: 0-958-46092-2, pp.1836-1844, 2003.

[7]   UNI, Italian National Standards Body, Foundry sand. Sampling and test methods, UNI 4628:1976.

[8]   Sundeen, S.P., Geological study of sand deposits in the State of Michigan, Phase II, final report – 1978, Open file report-OFR 78 04, Institute of Mineral Research Michigan Technological University, pp. 6-7. http://www.deq.state.mi.us/documents/deq-gsd-sanddune-OFR7804.pdf

[9]   Zanetti, M.C., Fiore, S., Foundry waste recycling in moulding operations and in the ceramic industry, *Waste Management & Research* **21**, pp. 235-242, 2003.
[10]   Zanetti, M.C., Fiore, S., Clerici, C., Reuse of foundry sands for core and glass production, *Journal of Solid Waste Technology and Management* **30(1)**, pp. 28-36, 2004.

Making waste minimisation a high impact activity in Southland, New Zealand

C. Dean
Invercargill City Council, FutureGenz Ltd, Invercargill, New Zealand

Abstract

Waste education in the Deep South of New Zealand is anything but boring!

For the past four years, schools and the general public have been enchanted, outraged, challenged and overjoyed by a steady stream of characters, storylines, clubs, events and challenges that have sprung forth from Southland's local waste educators. This lively approach to education is often so entertaining and engaging that the unsuspecting public barely even knows they're being educated. This is the pain-free method of education that we find works best.

Our educational messages are delivered through a wide variety of mediums and we regularly enlist the help of some 'larger than life' characters. These characters capture the imagination of our audiences and provide a 'face' that people can relate to easily.

The cornerstones of this successful approach to waste education are collaboration between agencies, offering a wide variety of opportunities for education, and providing comprehensive support for further exploration of the waste topic.

This paper describes the various aspects of our approach and discusses how we continue to make waste minimisation a high impact activity in Southland, New Zealand.

Keywords: waste minimisation, education, promotion, mediums, entertaining, collaboration, schools, community, Southland, New Zealand.

1   Introduction

Waste education in Southland is something of a success story, despite being afforded a relatively small budget in relation to many other regions' education programmes.

   WIT Transactions on Ecology and the Environment, Vol 92, © 2006 WIT Press
www.witpress.com, ISSN 1743-3541 (on-line)
doi:10.2495/WM060091

The key players in waste minimisation are the local councils, who provide waste services and support this with education delivered by a dedicated Waste Minimisation Officer, or an Environmental Education Officer. Over time, the councils have developed a wide range of opportunities for education, and each year fresh new approaches are added. These include class visits, stage shows, weekly newspaper columns, DVD movies and long-term programmes that develop a deeper understanding of the issues.

The use of 'characters' is another unique element within these educational experiences that helps bring the material to life and create a memorable experience for the target audience. Creating characters is similar to the strategy of 'branding' in marketing. Likeable characters provide an 'image' that the public can easily identify with and feel a positive association for in the future.

Staff from the individual councils also collaborate regularly on education and network with other likeminded agencies, such as the Department of Conservation, Southland Museum and Art Gallery, Southern Institute of Technology and relevant community groups. Working in this way creates powerful synergies and achieves far greater outcomes than if everyone worked in isolation.

2 Background

Located in the southernmost region of New Zealand, Southland is largely a farming province with a population of approximately 90,000 people (about 50,000 of these live in the main city of Invercargill). The Southland community is generally considered to be conservative and fairly resistant to change, including changes to the 'tried and true' methods of disposing of waste.

Nationally, the concept of waste minimisation was fairly slow to gather momentum, largely due to New Zealand's small population (4 million) and the fact that land for waste disposal has been historically easy to access. However, over the last decade there have been some changes. Landfills are now required to meet far more stringent environmental standards and the associated costs of landfilling have risen steadily, along with stronger public sentiment in support of waste minimisation. Consequently, waste minimisation and education have become a higher priority.

In terms of legislation, councils have an obligation to provide for efficient and effective waste management, and to consider disposal options based on the 5 R waste hierarchy - Reduce, Reuse, Recycle, Recover, Residual. However, waste minimisation is not compulsory and there are no mandatory waste reduction targets – only suggested targets in the National Waste Strategy. This set of circumstances has made waste minimisation a nice ideal, but certainly not imperative.

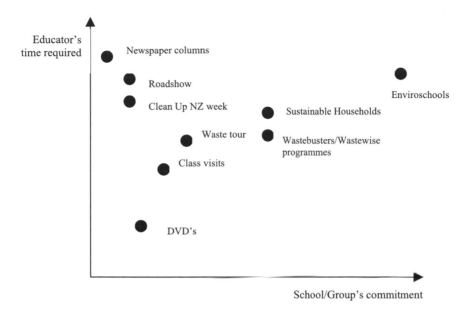

Figure 1:   Relative time commitment required for various education opportunities.

3   Discussion

3.1 Overview of approaches

Over a period of time, educators in Southland have built up an interesting repertoire of tools that enable schools and communities to explore the topic of waste to varying degrees. Many of these tools are listed below:

- Enviroschools programme
- Brucie/Lucy visits
- Educator visits
- Clean Up NZ week
- Reduce your Rubbish Roadshow
- Corporate Clean up Challenge
- Museum programmes/ screenings
- Wastebusters programme
- Wastewise worm farming
- Beat the Rubbish Blues DVD
- Trashtalk DVD
- Brucie's Buddies club
- Sustainable Households programme

- Lucy Grubb newspaper column
- Southland Wastebusters column
- Waste tour of facilities

Given that time is a significant restraint for educators and their audiences alike, this variety of programmes offers something that will suit most schedules. Figure 1 shows some of the relative time commitments.

3.2 Clean Up New Zealand week

Clean Up NZ week is a nationally co-ordinated campaign that encourages local areas to undertake their own 'clean up' activities for a week in September each year. Locally, Southland council's and other environmental agencies get together to plan a campaign that will inspire and motivate the community and provide an opportunity to take action. Planning for this week usually begins around May and staff time is fairly solidly committed to Clean Up Week activities throughout the month of September, followed by debriefing in October. Events include:

- Advertising campaigns
- Registering groups to undertake litter 'clean ups' in their community
- Providing free bags and gloves for those undertaking clean ups
- A Corporate Challenge for local businesses to 'clean up' an area
- A 'Roadshow' around Southland schools with colourful characters

3.3 Reduce Your Rubbish Roadshow

The Roadshow is generally held during Clean Up NZ week. Staging this event requires educators to develop a 25 minute 'show', source props and costumes, co-ordinate travelling to schools, handle equipment such as cameras, laptop and projector and play the characters in the show.

With three staff from the local councils, plus an additional helper, the travelling Roadshow visits twenty schools in one week, reaching approximately 2000 pupils with the 3 R's (Reduce, Reuse, Recycle) and 'clean up' messages. The main costs are staff time, travel expenses and some props and costumes, which are often picked up second-hand.

3.4 Characters

- Bruce C Gull (www.brucecgull.com)

Brucie (as he is known to his friends) is a character developed to assist with education about various environmental messages, including recycling and waste issues. With his mischievous antics and big hugs he is loved by children and enjoyed by adults everywhere he goes.

Brucie visits schools, attends many events and has his own club for children, which enables 'Brucies Buddies' to feel like they are helping Brucie to look after the environment. This concept has helped to mould a generation of youngsters who feel positive about environmental messages because of their early experience.

- Lucy Grubb

The Lucy Grubb character enables educators to demonstrate the *wrong* way to deal with rubbish. She is a naive old granny who thinks that these 'new-fangled' ideas about recycling are 'a load of old cods wallop' and advocates throwing rubbish out the window. Children know (helped along by Brucie) that Lucy is doing the wrong things, but they agree to help her learn the right ways when she asks for their help. This process helps to cement the children's own knowledge, or teach positive behaviours to those who are new to the concepts.

- Peti

Peti is the starring character in the recently produced 'Beat the Rubbish Blues' DVD that explores waste issues in a local context. Peti, a carefree student who learns about the 3 R's, is a character with wide appeal, particularly for students and Maori/Polynesian audiences.

- Can and Apple

Can and Apple are animated characters in another DVD movie called Trashtalk. In this ten minute movie, Can and Apple are on a quest to become something more than just litter – they want to reach the ultimate goal of being recycled. Their story appeals to children of all age groups and is suitable for pre-school children as an introduction to recycling.

3.5 Newspaper columns

The Invercargill City Council runs a weekly column in the free newspaper called 'The Southland Express', which is a popular publication with wide readership. The column is written from the perspective of Lucy Grubb, who is described in Section 3.4.

The local community group, Southland Community Wastebusters Trust, also have a weekly column in the region's principal paper, The Southland Times, which the Times print free of charge. Council staff collaborate with volunteers from the Trust to plan and write these columns, which is an ideal forum to promote topical issues.

3.6 DVDs

DVD's are always popular with children and this is an ideal medium for schools. The local councils have recently collaborated on two DVD's that provide an interesting overview of waste in the region, where it goes and the more sustainable alternatives of Reduce, Reuse and Recycle.

3.7 Ongoing education programmes

The council's educators provide programmes to schools and community groups. These can vary from a single visit to a comprehensive programme that requires a more significant commitment and therefore provides for a deeper understanding of the issues.

- Waste talks

Educators visit a classroom to give an overview of waste issues. Resources used can include DVD's, examples of recyclable material and samples of products made from recycled and biodegradable materials.

- Wastewise worm farming

A school commits to a series of three visits by an educator to learn about organic waste and worm farming as a way of recycling. The school then sets up a worm farm with the educator and they are periodically revisited by the educator to check progress.

- Wastebusters programme

Schools can choose to take part in one of four modules delivered by an educator – worm farming, composting, paper recycling or zero waste. The modules include videos, songs and practical or theoretical activities to support learning.

- Enviroschools (www.enviroschools.org.nz)

The Regional Council (Environment Southland) is the local provider for this nationally co-ordinated programme. Interested schools are required to make a three year commitment to the programme, which enables them to become an 'Enviroschool'. Enviroschools are supported in their journey by a trained facilitator and a comprehensive education kit which guides teachers through five topic areas. The programme also provides for professional development opportunities for teachers, attendance to the national 'hui' (get-together), a regional event each year and the option of entering into an awards scheme for achieving Enviroschool milestones.

- Sustainable Households (www.sustainablehouseholds.org.nz)

This programme is also nationally based and regionally co-ordinated. It is targeted at the wider community i.e. households, and enables individuals to learn about ways they can be more sustainable at home. A facilitator delivers several sessions covering topics such as waste, water, transport, shopping, organic gardening and so on. Sessions are generally held in the evening with people who have registered for the programme. A website supports these sessions and participants are able to access a wide range of additional information.

3.8 Collaboration

The common factor and essential ingredient for education in Southland has been collaboration. By working together we are able to stage major events and fund more programmes than we could as individual entities. We can also access a wider range of skills and experience when needed, which results in more successful outcomes.

Within the area of waste education, the key players are the local councils who employ either a Waste Minimisation Officer or an Environmental Education Officer. Other agencies that also have some common objectives are the Department of Conservation, the Southern Institute of Technology, the Southland Museum, the Southland Wastebusters Trust, the Invercargill Environment Centre, Venture Southland and other councils in neighbouring provinces. Southland's councils have liased and worked with all of these agencies to achieve common goals and to maximise efficiencies.

By collaborating in this way, the 'reduce your waste' message remains consistent and becomes somewhat omnipotent. This is a very powerful way to exert pressure on people to change their behaviour, and is ultimately more effective than councils just 'wagging their finger'.

4   Results

Generally speaking, more effort tends to go into developing and delivering educational opportunities than into evaluating the results in a quantitative way, which is an area where improvements could be made. However, some results are as follows:

- The 'Roadshow' visits 20 schools over one week each year, and directly reaches about 2,000 students, plus teachers and parents. It also engages the media and has regularly featured in the two local newspapers, local television and radio.

- Evaluation of the Roadshow in 2004 showed that students' knowledge of all the show's key environmental messages had improved after watching the show, and that learning had been retained for at least three weeks after the show (when post-surveying took place). Anecdotal evidence suggests that children still remembered and talked about the show many months later.

- Following a visit from the Roadshow, a number of schools went on to study waste as a major term topic (1/4 of the school year) in the following year.

- 55 guests (maximum capacity) attended the launch of the 'Beat The Rubbish Blues' DVD, including teachers, principals, councillors and community group representatives. Every school in Southland has been given a copy of the DVD.

- Participation in kerbside recycling in Invercargill has steadily increased since its introduction in 2003, and is now estimated to be around 70-80%.

- Brucie's Buddies club has membership of approximately 2,700 children and each year 350 children and parents attend his birthday party.

5 Conclusions

We consider our achievements to be considerable on what are relatively limited resources. Through collaboration, Southland educators and agencies are successful in keeping waste in the public eye at all times, through a variety of mediums, and public sentiment towards waste minimisation is positive. Some key conclusions from our experiences include:

- A variety of approaches keeps educational material fresh and reaches the widest possible range of people, from pre-schoolers to adult community groups, to individuals who need only pick up the newspaper to learn about waste minimisation.

- Creating a likeable character is similar to the marketing technique of 'branding' which helps to create positive associations for your message and provides a non-confrontational way to teach people.

- Collaboration with a wide range of groups maximises efficiencies, reduces costs, increases the skill base and maximises coverage of the message.

- A consistent message that comes from a range of sources, including community groups, is more likely to generate the social peer pressure that causes behaviour change.

Utilisation of two-stage waste incinerator bottom ash as a cement substitute in concrete

F. Kokalj & N. Samec

University of Mariboru, Faculty of Mechanical Engineering, Slovenia

Abstract

Waste incineration still seems to be an essential technology in the concept of integrated waste management. However, the desirable quantity of waste incineration residue needs to be as low as possible. Therefore, related optimization of two-stage waste incineration technology has been performed with the main goal of producing lower amounts of boiler ash, fly ash and flue gas-treatment residue, all classified as hazardous waste. Most of the combustion residue should be incinerator bottom ash. Tests were performed on light fraction of municipal solid waste in a two-stage pilot scale waste incineration plant.

The goal of this investigation is to present utilisation possibilities of waste light fraction incineration bottom ash as a cement substitute in concrete. The produced incinerator bottom ash, formed in the primary chamber of the two-stage incinerator, was analysed, tested and compared to other incinerator bottom ashes and cement. High resemblance of investigated bottom ash to cement was determined compared to other bottom ashes. Compressive and flexural strength and slump test were performed to characterise incinerator bottom ash and cement mixtures in concrete. It was found that after 28 days the flexural and compressive strengths of the binder linearly gradually decreases. The results show that it is possible to substitute 15 wt% of cement where low strength concrete is required.

Keywords: incinerator bottom ash, waste incineration, two-stage incinerator, concrete.

1 Introduction

Municipal waste – although classified as non-hazardous – still causes air, water and soil pollution during its decay when deposited in landfills without proper pre-treatment. Hence, today's modern landfills for municipal solid waste (MSW), built to prevent almost all pollution, have a complex design. This makes the dumping space for a volume unit of waste rather expensive even if a locally acceptable dumping site can be found in the first place.

The Slovenian population, just as in the rest of the European Union, produces approximately 500 kg of MSW per person each year. In order to meet modern standards for waste management, Slovenian society has accepted European philosophy and legislation in the field of waste management. Waste reduction, re-use and recycling are already successful in waste stream minimisation. In addition the composting of separately collected bio-degradable fraction has recently been under way. The remaining fraction of the waste stream still goes directly to landfills. National and local waste legislation sets high standards and there are some plans to build regional waste management centres involving smaller two-stage waste incinerators (TSWIs) especially for the energy utilisation of a higher calorific value waste fraction called light fraction.

TSWI was originally designed for industrial, medical and hazardous waste incineration since, in the past, legislation had set higher standards for the thermal treatment of hazardous waste compared to MSW. These incinerators had small capacity and were mostly batch fired. The main intention when installing a second combustion chamber was to improve the destruction of organic compounds.

In TSWI light fraction is reduced by about 90% of mass during combustion, leaving only 10% of incineration residues representing mostly IBA and minor amounts of fly ash, boiler ash and flue gas-treatment residue classified as hazardous waste. The IBA's composition and its amount of TSWI fed by light fraction differ drastically from the IBA of mass burn grate waste incinerators (MBGIs) fed by non separated MSW. It offers some additional possibilities for IBA utilisation from TSWI.

It is important to distinguish between the treatment of IBA for utilisation, and for disposal. When treatment occurs for the purpose of disposal, it is economically advantageous to limit the use and cost of the process treatment, e.g. additives, while maintaining compliance with the regulations. In the case of IBA utilisation, the processes' economics, including the composition of the final product, become very different, especially when the IBA component represents only a fraction of the final product [1]. The regulatory and economic circumstances relating to natural resources and civil engineering practices, prevailing in various countries, can reflect the attitude adopted towards the utilisation of IBA. The range of applications for which IBA from MBGI is used commercially usually includes: landfill cover (daily or final), road foundations, wind and sound barriers, lightweight concrete aggregate, structural fill, aggregate in asphalt, shore-line protection and marine reefs.

The utilisation of different kinds of ash from coal and MSW combustion for concrete production was recently reported in literature [2, 3, 4, 5, 6]. Because of typical composition (40 – 60 wt% of SiO_2 and 2 – 20 wt% of CaO), which is far from cement composition, most of the ash investigated in these studies has been used as filler substituting for a part of the natural aggregate. IBA from TSWI containing about 24 wt% of SiO_2 and about 40 wt% of CaO has quite similar composition to cement usually containing 17 – 24 wt% of SiO_2 and 55 – 65 wt% of CaO. It can be considered, with such composition as a good substitute for cement in concrete. Cement is a hydraulic binder and the fundamental ingredient of concrete. In general, it represents about 15% of a concrete's volume and around 45% of its cost. Concrete is one of the major materials in the construction industry and about 4.5 billion metric tons of concrete are cast each year worldwide [7]. In Slovenia this value is up to $1.5 \cdot 10^6$ tons ($6 \cdot 10^5$ m^3) [8]. Approximately 90% of this quantity is concrete with a characteristic compressive strength of 40 MPa or less.

The aim of the presented work has been the investigation of some significant concrete properties produced by the utilisation of IBA from the pilot scale TSWI. However, by incorporating waste materials into permanent compounds, their interaction with the environment is diminished and, at the same time, the required landfill space is radically reduced.

2   Description of pilot scale TSWI

Two-stage incineration technology shares the common idea of two divided chambers. The primary goal was to create even better and environmentally friendlier devices that will produce lower toxic emissions of particles and gases. This technology is today technically feasible and economically comparable with mass burn incinerators in MSW treatment. Figure 1 schematically shows the pilot scale TSWI used in this research work.

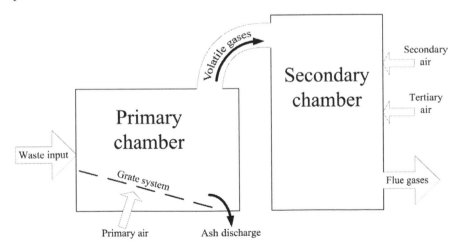

Figure 1:   Schematic presentation of pilot scale TSWI.

In general TSWI should provide better combustion leading to a lower release of volatile organic compounds and carbon monoxide. In addition, the low air flow in the primary chamber results in lower entrainment of particulate matter in the flue gases, which in turn reduces other particulate-borne pollutants such as heavy metals, dioxins and furans. In practice the two-stage combustion incinerator relies on semi-pyrolysis, based on sub-stoichiometric combustion of the waste in the primary chamber, and excess air in the secondary chamber, together assuring good combustion conditions, low emissions and lower consumption of added fuel, if necessary. The waste is combusted by the insufficiency of air providing incomplete combustion and, therefore, there are a high proportion of incomplete combustion products, which pass through to the second stage, where they finally burn out. In the primary chamber warming, drying and semi-pyrolitic gasification of the waste at temperatures from 800 to 1050 K take place, and in the secondary combustion chamber mixing of the volatile gases with air, ignition and complete combustion at temperatures up to 1550 K occurs. Relatively low entrainment of particulate matter in the first stage and a complete combustion condition in the second stage result in less – polluted raw gases. Consequently, the flue gas-treatment devices are less demanding, cheaper and produce less residues than standard MBGI. The quantities of total dust and some toxic pollutants in raw gas for different types of incinerators are given in Table 1 as an example. It can be seen that MBGI's produce many more pollutants than optimized TSWI. In the worst case scenario MBGI's produce more than two orders of magnitude more pollutants than TSWI, also causing higher amounts and more contaminated flue gas-treatment residuals.

Table 1: Emission's comparison of untreated flue gases from standard MBGI and optimised TSWI

Parameter/ Substance	WI Directive [15]	TSWI	MBGI [16]	MBGI [17]
Total dust	10	44	2000 – 15000	2000 – 6000
CO	50	7.5	50 – 600	20 – 500
TOC	10	1.6		
SO_2	50	10	200 – 800	300 – 600
NO_x	200	60	200 – 400	200 – 600
HCl	10	18	400 – 1500	500 – 1200
HF	1	0.28	2 – 20	1 – 10
PCDD/F	0.1	0.0025	-	1 – 10
NOTE: All concentrations are expressed in (mg/Nm^3), except PCDD/F which are expressed in $(ng\ TE/m^3)$				

It should be noted that the results for TSWI were achieved after a special optimisation process of the major operating conditions (i.e. amounts of air, temperatures, residence time, grate moving frequency, etc.). TSWI optimisation was divided into two parts. Firstly, the optimisation of combustion conditions in the primary chamber to produce the desired quality of bottom ash, and secondly the optimisation of conditions in the secondary chamber to achieve complete combustion of the volatile gases coming from the primary chamber.

In the primary chamber optimisation was the incineration residuals' oriented optimisation of the process parameters. The basic idea of substoichiometric conditions, relatively low temperatures and legislative demand for ash quality was retained at all times. The waste input stream for the presented investigation was the light fraction of MSW as mentioned previously in introduction. It is a waste fraction with a relatively high heating value of about 18 MJ/kg and mostly involves paper, chard board, wood, different sorts of plastics and textiles, representing approximately 50% of total MSW stream. Optimisation peak was achieved with the lowest possible temperature of 850 K and only 70% of the theoretically needed air for complete combustion in the primary chamber. Low temperatures and small amounts of oxygen did not allow most of the toxic heavy metals to leave the primary chamber and they remained in the bottom ash.

The composition of the IBA and the presence of toxic metals are directly dependent on the input waste's composition and also the incinerator's operating conditions. These factors have a direct influence on the IBA's quality and composition, also determining its utilisation possibilities. However, from a technical stand point more work could be done on operating condition optimisation, and incineration technology. The waste composition on the other hand depends strongly on legislative demands and population habits.

The second part for the optimisation of pilot scale TSWI operation was conducted with the help of a CFD programme package named CFX. The operating conditions and secondary and tertiary air flows of second combustion chamber were investigated and optimised, in order to achieve complete combustion conditions. The so-called 3T parameters (temperature, turbulence/mixing and residence time) were examined and optimised to achieve homogenous temperature filed (minimise regions of lower temperatures), increased mixing efficiency of air and volatile gases, and ensure proper residence time. Corrections were made to the design and operating conditions of secondary combustion chamber based on broad computer analyses. The results of a complete numerical combustion modelling optimisation of the secondary combustion chamber were tested on the existing waste incinerator and produced the expected good results. The raw gas emissions of this optimised combustion chamber are presented in Table 1. The emission results for the carbon monoxide prove that combustion is practically complete. More detailed results and discussion can be found in the literature [9].

3   Experimental set up

The study was performed with IBA from pilot scale TSWI only, operating under optimised conditions for the primary and secondary chambers. The binding abilities of the IBA from the primary chamber were investigated and an acceptable level of IBA in the total binder was determined and consistency, density and strength were studied.

The IBA's binding abilities were tested according to the standard EN 196-1 [10]. Specimen mixtures were prepared with 450 g of binder and 225 g of water, giving a water/binder (W/B) ratio of 0.5. The aggregate mass of 1350 g was quartz sand with grain sizes from 0 to 2 mm in accordance to EN 196-1 standard requirements. Pure cement designated as CEM I 42.5 R was used as binder for the preparation of a reference mixture. The standard EN 197-1 [11] prescribes Portland cement with a designation CEM I 42.5 R with at least 95 wt% of clinker minerals, having a compressive strength of at least 42.5 MPa after 28 days but not exceeding 62.5 MPa. R stands for the rapid development in strength during the early stage of hydration. Subsequent specimen mixtures were made with a binder which consisted of cement and IBA. The weight per cent of IBA in the total binder varied from 5 to 40 wt%. Originally the IBA taken from TSWI was in the form of flakes with sizes from 1 to 20 mm. It was ground using a ball mill and sieved through a 90 μm screen in order to obtain a suitable granular composition. The sizes and shapes of the IBA particles depend on the incineration process, as previously discussed. Literature [5] reported that the particles of IBA were rather globular in shape with sizes from 4 to 20 mm and other work [12] reported more than 37% of IBA was over 4.76 mm. The composition of IBA and CEM I 42.5 R was determined by X-ray diffraction (Table 2). Table 2 also presents the composition of some other IBA-s from different MBGI's. The presented results show that TSWI IBA is superior in terms of composition resemblance to cement when compared to other results from literature [6, 13].

Table 2: Chemical composition of cement CEM I 42.5 R and IBA from TSWI and other MBGI's

Oxide	Content (wt. %)			
	CEM I 42.5 R	IBA (TSWI)	MBGI IBA [6]	MBGI IBA[13]
SiO_2	22.3	24	41.13-56.99	41.2
Al_2O_3	5.83	14.8	9.2-11.35	12.7
Fe_2O_3	2.17	2.7	3.97-8.61	7.6
CaO	60.81	39	13.22-19.77	16.1
MgO	2.82	1.7	3.46-3.85	1.9
Na_2O	0.34	0.9	2.84-5.87	2.8
K_2O	0.72	0.2	1.35-1.57	1.0
SO_3	2.75	-	-	-

Nine specimens of size of 40 mm x 40 mm x 160 mm were cast from each mixture. They were kept in a mould for 24 hours. They were then stored in a 100% moist environment at 22 ± 2°C. Specimens were tested for flexural and compressive strength after 2, 7 and 28 days respectively.

The bulk density of the gravel with a granular composition was 1906 kg/m³. The referenced concrete mixture was designed to reach a compressive strength of 40 MPa. Fresh concrete was tested for its workability. Namely, the quality of casting and compacting depends on it and, thereby, workability has a significant impact on the strength and durability of the concrete. There is no widespread standard test available for workability, which is best defined as the amount of useful internal work necessary to produce full compaction [14]. Thus, the slump of a concrete cone is often measured and the term consistency is used to describe this property. After executing the standard slump test for the purpose of this experiment, six cubical specimens with edge lengths of 100 mm were cast from each mixture. Specimens were taken out of the mould after 24 hours and kept in a 100% moist environment at 22 ± 2°C. Compressive strength was determined after 7 and 28 days.

4 Results and discussion

The dependence of flexural and compressive strengths on the content of IBA is given in Figures 2 and 3. A linear decrease in both flexural f_f (Figure 2) and compressive f_c (Figure 3) strengths was observed. Hence, the following relationships were applied:

$$f_f = f_{f0} - k_f C \tag{1}$$

$$f_c = f_{c0} - k_c C \tag{2}$$

where C is the wt% of IBA in the binder, f_{f0} and f_{c0} denote the strength of the reference mix (without IBA), whereas the slope of the straight line denoted by k_f and k_c represents the reduction rate of strength. Values of all parameters after 2, 7 and 28 days are summarised in Table 3. Since the hydration of cementitious materials is a time – dependent process which can last up to several years, it was expected that the values of f_f and f_c would increase with elapsed time. Also, it is interesting to observe that the values of k_f and k_c decreased with time. This leads to the conclusion that IBA develops strength in the later stage of hydration compared to CEM I 42.5 R. A possible reason for this observed behaviour is that the temperature in the primary chamber is too low for the formation of alite ($3CaO \cdot SiO_2$). It is an essential compound for the strength of concrete in the early stage of the hydration process. Furthermore, the quantity of CaO is insufficient compared to cement, for the formation of all cementitious compounds.

Table 3:   Parameters defining the bottom ash (IBA) wt% in the binder having an influence on flexural and compressive strength. f_{f0} and f_{c0} denote flexural and compressive strength at 0 wt% of IBA. Curve slopes in fig 2 and 3 (eqn (1) and (2)) are denoted by k_f and k_c.

Time	Flexural		Compressive	
(days)	f_{f0} (MPa)	k_f (MPa/%IBA)	f_{c0} (MPa)	k_c (MPa/%IBA)
2	4.03	0.054	28.48	0.352
7	6.57	0.051	38.96	0.269
28	8.13	0.028	52.14	0.230

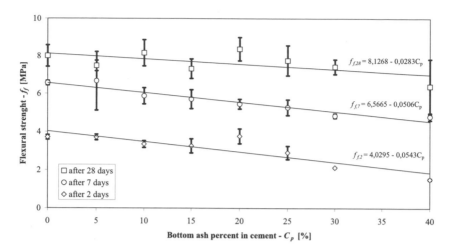

Figure 2:   Flexural strengths of cement after 2, 7 and 28 days regarding incinerator bottom ash portion

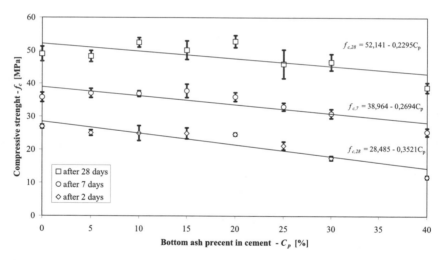

Figure 3:   Compressive strengths of cement after 2, 7 and 28 days regarding incinerator bottom ash portion

Standard EN 197-1 [11] sets two acceptance conditions for the compressive strength of cement CEM I 42.5 R:

- after 2 days it should be at least 20 MPa,
- after 28 days it should reach at least 42.5 MPa but should not exceed 62.5 MPa.

Results show that the early strengths are affected more when increasing the wt% of IBA in the binder. Thus the first above condition is more essential. Hence, the maximum content of IBA in the total binder calculated by eqn. (2), which still satisfies the second condition, is about 25 wt%. Taking into account a safety factor of 0.7, the acceptable level of IBA in the binder was established as 15 wt% and was further applied for the study of concrete.

The consistency was investigated using the slump test. The concrete with no IBA had a slump of 105 mm and the one with 15 wt% IBA had a slump of 55 mm. Interestingly the slump is significantly reduced when the binder contains 15 wt% of IBA. In general, lower slump should result in poorer workability, so one would expect lower density for such concrete mixtures. The results for the concrete mixtures' densities again display an interesting fact. The concrete density of cement is only 2310.7 kg/m^3 and 2324.8 kg/m^3 with 15 wt % of IBA. When the binder contains cement as well as IBA, the density of the concrete is unaltered although the slump is reduced.

The compressive strengths of concrete mixtures after 7 days are 31.6 MPa with no IBA content and 34.7 MPa with 15 wt % of IBA. After 28 days the compressive strength raises to 40.6 MPa (0 wt% IBA) and 43.4 MPa (15 wt% IBA). Although it was established that when IBA represents 15 wt% of the binder, flexural strength is reduced by approximately 10%, this was not generally observed in concrete specimens when the compressive strength was investigated. The compressive strengths of the mixtures with pure gravel aggregate even increased when IBA was used – surprisingly even after 7 days.

5   Conclusion

IBA composition and the presence of some hazardous components depend strongly on input waste composition and incinerator operating conditions. Furthermore, incineration technology (i.e. MBGI, TSWI, fluidised bed incinerator, etc.) can additionally influence IBA quality and composition, also determining its utilisation possibilities.

It was found that IBA, formed in the primary chamber of TSWI when it has been fed by light MSW fraction, involves the same significant components (i.e. SiO$_2$ and CaO) and comparable amounts as cement. Therefore, the feasibility of its application in concrete as a cement supplement was investigated and it was found that after 28 days the flexural and compressive strength of the binder's linearly gradually decreases. The presented results show that it is reasonable to use a binder containing IBA where a lower strength of concrete elements is required

Supposing that all MSW light fraction in Slovenia is incinerated with similar TSWI technology to that used in the experiment, then around $5 \cdot 10^4$ tons of IBA would be produced per year. About $4 \cdot 10^4$ tons of IBA or 80% of IBA could be re-used for the production of concrete with a characteristic compressive strength of 40 MPa or less by supplementing 15 wt% of cement with IBA. This would reduce the landfill space required by about $7 \cdot 10^4 \, m^3$.

References

[1]   Porteous, A.; Jones, G.; Frith, P.; Patel, N. Energy from Waste: A good practice guide, *CIWM*, 2003.

[2]   Kula, I.; Olgun, A.; Erdogan, Y.; Sevinc, V. Effects of colemanite waste, cool bottom ash, and fly ash on the properties of cement; *Cement & Concrete Research.* 2001, 31 (3), 491-494.

[3]   Targan, S.; Olgun, A.; Erdogan, Y.; Sevinc, V. Effects of supplementary cementing materials on the properties of cement and concrete; *Cement & Concrete Research.* 2002, 32 (10), 1551-1558.

[4]   Rémond, S.; Pimienta, P.; Bentz, D. P. Effects of the incorporation of municipal solid waste incineration fly ash in cement pastes and mortars I. Experimental study; *Cement & Concrete Research.* 2002, 32(2), 303-311.

[5]   Pera, J.; Coutaz, L.; Ambroise, J.; Chababbet, M. Use of incinerator bottom ash in concrete; *Cement & Concrete Research.* 1997, 27 (1), 1-5.

[6]   Filipponi, P.; Polettini, A.; Pomi, R.; Sirini, P. Physical and mechanical properties of cement-based products containing incineration bottom ash; *Waste Management.* 2003, 23, 145-156.

[7]   Su, N.; Miao, B. A new method for the mix design of medium strength flowing concrete with low cement content; *Cement & Concrete Research.* 2003, 25 (2), 215-222.

[8]   Statistical office of the Republic of Slovenia, on-line data, www.stat.si

[9]   Kokalj, F.; Samec, N.; Skerget. L. Combustion conditions and design control of a two stage pilot scale starved air incinerator by CFD. In: Popov, V., Itoh, H., Brebbia, C.A. & Kungolos, S. (eds.): *Waste management and the environment II*, pp. 25-34. WIT Press, Southampton, Boston. 2004.

[10]   EN 196-1. Methods of Testing Cement – part 1: Determination of Strength. European Standard; 1994.

[11]   EN 197-1. Cement – Part 1: Composition, Specifications and Conformity for Common Cements. European Standard; 2000.

[12]   Wang, K.-S.; Tsai, C.-C.; Lin, K.-L.; Chiang K.-Y. The recycling of MSW incinerator bottom ash by sintering; *Waste Management & Research.* 2003, 21, 318-329.

[13]   Bethanis, S.; Cheesman, C.R.; Sollars, C.J. Effect of sintering temperature on the properties and leaching of incinerator bottom ash; *Waste Management & Research.* 2004, 22, 255-264.

[14]   Glanvile, W. H.; Collins, A. R.; Matthews, D. D. *The grading of aggregates and workability of concrete.* Road Research Tech. Paper no. 5 HMSO, London, 1947.

[15]   Directive 2000/76/EC of the European Parliament and of the Council of 4 December 2000 on the Incineration of Waste.

[16]   Sattler, K.; Emberger, J. *Behandlung fester Abfaelle*, 4. ueberarb. Aufl., Vogel Verlag und Druck KG, Wuerzburg, 1995.

[17]   Kuerzinger, K.; Kleine – Moellhoff, P.; Morawa, J. *Moderne Verfahren zur Absorption.* Firmenschrift der Fa. Noel-KRC Umwelttechnik GmbH, Wuerzburg. 1994.

Development of a recycling technique of used glass by conversion to porous materials

K. Yanagisawa[1], N. Bao[1], M. Kariya[1], A. Onda[1], K. Kajiyoshi[1],
Z. Matamoras-Veloza[2] & J. C. Rendón-Angeles[3]
[1]Research Laboratory of Hydrothermal Chemistry, Kochi University, Japan
[2]Instituto Tecnológico de Saltillo, Facultad de Metal-Mecánica, México
[3]Centro de Investigación y Estudios Avanzados del IPN, Saltillo0, México

Abstract

The powders of used glass bottles were converted to porous plates by a newly developed technique, which involves two steps, hydrothermal treatment of glass powders at low temperatures and calcination at high temperatures in air for foaming. The ordinary method to prepare porous materials from glass needs vesicants such as calcium carbonate and silicon carbide that decompose at high temperatures to produce gas, and the gas is trapped in softened glass to form pores. The new technique can produce porous materials without any vesicants. Water incorporated into the glass structure by hydrothermal treatments acts as a vesicant. Water diffuses into the glass structure by hydrothermal treatments at low temperatures of around 200°C, and is released as vapor to form pores in the softened glass, when the hydrothermally treated glass is heated at high temperatures over 650°C. Thus, this process gives porous materials with a fine microstructure including closed pores at low temperatures in comparison with the ordinary method. Porous plates (45 x 45 x 3 cm³ in size) with a bulk density of 0.45g/cm³ were produced by hydrothermal treatment of the glass powder at 183°C in a large autoclave with an inside volume of 2.5 m³, followed by calcination at 800°C in a continuous furnace with 18 m in length.
Keywords: recycling of used glass, porous materials, hydrothermal treatment, foaming.

WIT Transactions on Ecology and the Environment, Vol 92, © 2006 WIT Press
www.witpress.com, ISSN 1743-3541 (on-line)
doi:10.2495/WM060111

1   Introduction

The material recycling is important for a sustainable development [1, 2]. In Japan, we produced 1,550,000 ton of glass bottles in 2004 [3]. After used, these glass bottles are separated into colors, transparent, brown and others, and recovered for recycling. Their recycled amount was 320,478 ton in 2004 [4]. The used glass bottles were mainly (69%) used as a raw material to produce new bottles, and the remainder (31%) was converted to other materials. The recycling of used glass is attractive for glass manufactures, but they need the used glass with similar composition controlled by severe color sorting and removing contaminants. It is easy to produce new bottles from separated transparent and brown glass bottles, but not from the bottles with mixed colors. Thus, the recycling of colored glass bottles has been received increasing interest. The construction industry has given successful recycling of used glass for heat insulation (fiber glass and light-weight aggregates), aggregates for concrete and asphalt, base and subbase filler materials, and cement constituent [5]. In this study, we developed a new technique to recycle the used glass bottles with mixed colors by conversion to porous plates.

The ordinary method to prepare porous materials from glass needs vesicants such as calcium carbonate and silicon carbide that decompose at high temperatures to produce gas, and the gas is trapped in softened glass to form pores. In this study, we developed a new method by using hydrothermal technique to convert the powder of used glass bottles to porous materials. Figure 1 schematically illustrates the procedures.

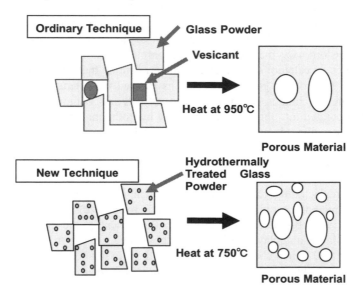

Figure 1:   The techniques for preparation of porous materials from glass powders.

The new method involves two steps, hydrothermal treatment of glass powders at low temperatures and calcination at high temperatures in air for foaming. Water incorporated into the glass structure by hydrothermal treatments acts as a vesicant. Water diffuses into glass structure by hydrothermal treatments, and is released as vapor to form pores in the softened glass, when the hydrothermally treated glass is heated at high temperatures. In this method, each glass particle after hydrothermal treatment can foam by heat treatment al low temperature in comparison with the ordinary method.

2 Preparation of porous materials

The starting glass powder was available from Toyo System Plant Co., Ltd., Japan. It was produced by milling of used glass bottles and sieved to be under 590 μm. The powder was placed in a stainless steel box (60 x 40 x 20 cm³) after mixed with water (20 mass%). The hydrothermal treatment was conducted in an autoclave (Tokai Concrete Ind. Co., LTD.) with inner volume of 2.5 m³. As shown in Figure 2, 9 boxes filled with the glass powder (320 kg, total glass powder) were hydrothermally treated at once at 183°C for 10 hours in saturated vapor.

Figure 2: Autoclave for hydrothermal treatment.

The hydrothermally treated glass was crashed, sieved again to get a powder to be under 590 μm, and heated in a continuous furnace (Toyo System Plant Co., Ltd., Japan) with width 1 m and length 20 m, as shown in Figure 3. The effective heating region was 18 m and was separated to three zones of which temperature can be independently controlled by oil burners. The hydrothermally treated glass powder was placed on stainless mesh belt and continuously transferred into the heating zones. In this study, the temperature of all heating zones was controlled to be 800°C and the transfer rate was selected to be 50 cm in a minute. Thus, the glass powder was heated at 800°C for 36 minutes. After heated in the continuous furnace, the foamed plate was immediately moved to a furnace at 450°C and kept for 4 hours.

Figure 3:   The continuous furnace for heat treatment.

3 Results and discussion

The change of crystalline phases by each procedure was observed by X-ray diffraction (Figure 4). The original glass powder was amorphous (Fig. 4(a)) but a few crystalline phases were formed after the hydrothermal treatment (Fig. 4(b)). They were considered to be hydrated crystalline phases, because they disappeared by heating at high temperatures for a short time. The amount of crystalline phases increased with the increase in hydrothermal reaction temperature and time.

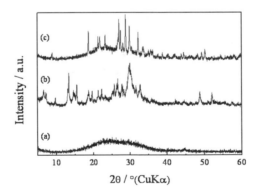

Figure 4:   XRD diffraction patterns of the original glass powder (a), hydrothermally treated glass powder at 183°C for 10 hours (b), and plate obtained after heat treatment at 800°C for 36 minutes (c).

It was confirmed by the other experiments that a glass block reacted with water by hydrothermal treatments to form a reaction layer consisting of the crystalline hydrated phases on its surface. The careful observation of the reaction layer showed that the reaction layer produced in water vapor had amorphous material between crystalline part and unreacted part. When the reaction layer formed by the hydrothermal treatment in water vapor was heated in air at 750°C for 30 minutes, pores were formed in the reaction layers. On the other hand, the reaction layers produced in a large amount of water by the hydrothermal treatment never gave pores. It is considered that the hydrothermal treatment in water vapor produces a reaction layer consisting of water diffused glass phase by ion exchange mechanism and the water diffused into the glass structure was released to produce pores when the glass starts to soften at high temperatures.

After heat treatment at 800°C for 36 minutes, the hydrothermally treated glass powders expanded and connected together to form a plate. Though the foaming was observed even at 650°C, the expansion was not enough and the mechanical strength of the plates was very small. In order to get high mechanical strength, heat treatment at high temperatures over 750°C was necessary. When the plate was immediately cooled after the heat treatment, cracks were formed during cooling. After the plate was transferred to the other furnace and kept at 450°C for 4 hours, the plate without cracks was successfully obtained. The plate was consisted of a few crystalline phases together with amorphous material (Fig. 4(c)). The amount of the crystalline phases increased with the increase in heating temperature and time.

The plate had high machinability, so that it was easily shaped into a rectangular parallelepiped. In this study, plates with 45 x 45 x 3 cm^3 in size were produced as shown in Figure 5. The average bulk density of the porous plates obtained in this was 0.45 g/cm^3. The bulk density depended on hydrothermal and heat treatment conditions. It was confirmed by the other experiments that the hydrothermal treatment at 200°C for 6 hours and heat treatment at 750°C for 30 minutes gave low density less than 0.3 g/cm^3.

Figure 5:   A shaped porous plate obtained in this study.

The porous plate obtained in this study has a fine microstructure as shown in Figure 6. The pores up to 500 μm were observed in the polished surface of the porous plate. The pore size also depended on the hydrothermal and heat treatment conditions. In general, higher hydrothermal and heat treatment temperature gave larger pores and larger pore diameter distribution. The plates included open and close pores, and floated on water.

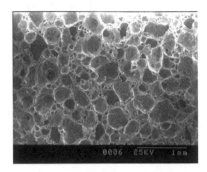

Figure 6:   Microstructure of the porous plate.

4   Conclusions

The porous plates were successfully produced from colored used glass bottles by hydrothermal treatment at 183°C for 10 hours, followed by heat treatment at 800°C for 36 minutes. The plates have following properties as average values; bulk density 0.45 g/cm^3, compressive strength 140 kg/cm^2, bending strength 55 kg/cm^2, thermal conductivity 0.18 W/mK, and line expansion coefficient 5.8x10^{-6}/K. Thus, these plates may be useful as a light weighted board for the void slab system, thermal insulators, sound and water absorbents, floats to form floating islands, and so on.

References

[1]   Kavouras, P., Komniou, Ph., Chrissafits, K., Kaimakamis, G., Kokkou, S., Paraskevopoulos, K., Karakostas, Th., Microstructural changes of processed vitrified solid waste products. *J. Eur. Ceram. Soc.*, 23, pp. 1305-1311, 2003.
[2]   Ogura, M., Astuti, I., Yoshikawa, T., Morita, K., Takahashi, H., Development of a technology for silicon production by recycling wasted optical fiber. *Ind. Eng. Chem. Res.*, 43, pp. 1890-1893, 2004.
[3]   Home page of The Japan Containers And Packing Recycling Association, www.jcpra.or.jp/data/index.html.
[4]   Home page of Glass Bottle Recycling Promoter Association, www.glass-recycle-as.gr.jp.
[5]   Sobolev, K., Recycling of Waste Glass in Eco-Cement. *Am. Ceram. Soc. Bull.*, (9), pp. 9501-9507, 2003.

Modelling of tribo-electrostatic separation for industrial by-products recycling

F. Cangialosi, F. Crapulli, G. Intini, L. Liberti & M. Notarnicola
Department of Environmental Engineering and Sustainable Development, Technical University of Bari, Taranto, Italy

Abstract

Tribo-electrostatic separation is a dry technology which allows the elimination of impurities from industrial wastes on the basis of their surface charging characteristics. In this paper results of investigations aiming to numerically simulate the tribo-electrostatic separation of fine particles for industrial by-products recycling are described. After charging in a pneumatic transport line, the mixture of the particles to be separated is injected in a separation chamber where a DC electric field is created using two parallel plate electrodes. The particle/gas flow inside the chamber was simulated using an extended commercial computational fluid dynamics (CFD) code. The three-dimensional turbulent flow was calculated. Based on the Lagrangian approach, the trajectories of the powder particles (<100 microns) were modelled considering electric and aerodynamic forces. Comparison of the simulation results with experiments carried out with a bench-scale separation unit are presented for the case of silica beads, whose electric characteristics resemble those of coal fly ashes. The effects of particle charge, electric field and injection velocity were investigated, revealing that a proper choice of separator geometry and flow parameters allows unwanted fine particles in industrial wastes like fly ash to be efficiently removed.
Keywords: powder wastes, tribo-electrostatic separation, computational fluid dynamics (CFD).

1 Introduction

For the sustainable development of our society, recycling of municipal and industrial wastes must continue to be pursued. Many industrial wastes (combustion fly ashes, residues of surface finishing process in soft metallurgy, commingled plastic waste from electric/electronic equipments, etc.) are in

 WIT Transactions on Ecology and the Environment, Vol 92, © 2006 WIT Press
www.witpress.com, ISSN 1743-3541 (on-line)
doi:10.2495/WM060121

powder or granulated form, having a grain size below 1 mm. Nearly all the valuable components in the industrial waste can be reused by subjecting the material to advanced treatment using effective methods to separate and purify each component. In the case of coal fly ash, more than 90% of the material can be re-used for cement and concrete production as long as the carbon content in the ash is below 5%, so that separation technologies are applied to meet the requirements in terms of product purity, whereas ABS, PS and PP were recovered from a plastic mixtures with recoveries above 73.0% and a grade of 92.1%, 84.9% and 90.0%, respectively [1, 2]. Interest in electrostatic separation is increasing because it appears to be the first dry method, which does not require wastewater treatment processes, to separate particles with alike density and electrical conductivity. Tribo-electrostatic separation is a technology developed in the field of the mining engineering which allows to separate materials on the basis of their surface charging characteristics: for dissimilar solids, which are initially uncharged, a transfer of a small electrical charge takes place from one material to the other as they make contact, where one becomes negative and the other positive. In the pneumatic tribo-electrostatic separation, particles to be separated are air-transported in pipelines where they collide with the inner wall of the tube and become positively and negatively charged; particles are then injected in a separation reactor where an electric field is applied to separate the particles. Although this technology works well in the separation of relatively coarse materials (diameter above 1 mm), much work has still to be done for making tribo-electrostatic separation effective in fine particles separation.

The main goal of the research activities here reported was to experimentally investigate the process parameters which mostly affect the separation process of powders and gain fundamental information for proper system design and operation by means of Computational Fluid Dynamics (CFD).

2 Experimental equipment

In order to elucidate the effect each parameter has on the system performance and avoid the arise of secondary effects related to particle-particle interactions, no mixtures were used in the separation tests: a single material (glass beads, average diameter 65 microns) was tested and the trajectories of the particles in the separation chambers at different voltages (negative and positive) and injection velocities were studied with CFD. A schematic of the bench-scale apparatus, having maximum throughput of 20 kg/h, employed for tribo-electrostatic separation tests is shown in Figure 1. A small amount of glass beads was placed in a vibratory feeder contained in a sealed tank. Falling from the feeder, the particles are transported through a 6.35 mm diameter stainless steel charging tube by air at a flow rate between 7 and 20 m/s controlled by a rotameter. The charger tube forms three loops that increase the number of collisions between particles and wall pipe, thus establishing a charge polarity on glass beads.

Increasing the length and turbulence within this charger section obviously enhances the chance of charging for particle-particle collisions. Injected with a diffuser through the flow-straighteners, the air-particle mixture enters the separation chamber (35 cm long) containing the electrodes made by two copper plates attached to its opposite side walls. Under the electric field generated by a high voltage power supply, the charged particles are deflected towards the electrodes according to their polarity. A flow splitter then convoys the two streams of particle-laden flows through two cyclones from which two product streams were collected and the mass of glass beads was measured. The injection velocity in all the experiments was set to a constant value and the velocities at the cyclones were adjusted in order to set a balanced flow between the product outlets, unless explicitly stated; the electric field ranged from -2 kV/cm to 2 kV/cm. The ratio between the injection velocity and the velocity measured at the cyclones was varied between 0.1 and 3, in order to study the effect of co-flow air entering from the flow-straighteners, whose primary role is to reduce the turbulence in the separation chamber.

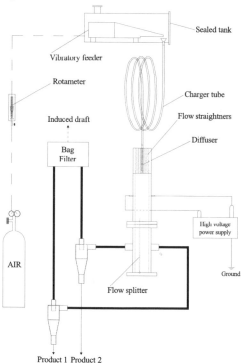

Figure 1:   Schematic of the experimental setup.

3   Numerical method

3.1   Gas flow

The experimental results (mass of glass beads recovered) were compared to those obtained with the use of CFX, a commercial CFD code. The separation chamber was discretized in one million of elements with an unstructured mesh and the Navier–Stokes equations with a standard k-ε model of turbulence were solved for the air flow distribution. The three-dimensional turbulent flow field was calculated in a computational domain with the size of 37.5 x 50 x 350 mm³ consisting of the separation chamber. A hybrid unstructured mesh of different elements (from 500,000 up to 1,500,000) is used in this study. Mass and momentum balance equations are integrated over the elements of the unstructured mesh [3]. The standard model k-ε was employed to simulate the turbulent motion with a sufficient accuracy and low CPU time cost. The following assumptions for the model were taken: (1) the fluid is incompressible, (2) the turbulent flow field is isotropic, (3) the boundary conditions of smooth wall and no slip were chosen. The two-phase flow enters the separator at 7 m/s through a slit 3 mm wide thus creating a plane jet. A honeycomb structure at the inlet of the separation chamber allows the air at atmospheric pressure to be sucked into the cell were a slight negative pressure is set by means of a vacuum pump. The simulations of the flow field were carried out investigating a range of velocity at the cyclones outlet ranging between 1.4 and 35 m/s.

3.2 Particle motion and electric field

Particle tracking inside the separation chamber allows trajectories of the charged particles to be followed from the diffuser to the cyclones outlet. The assumption of dilute flow was made during the simulation, so that influence of particle motion on gas flow and particle-particle collisions were neglected: the particle tracking routine employed was than based on the hypothesis of one-way coupling between gas and particles [4, 5]. Particles are injected in the separator with the same velocities as local air flow and with several angles φ from the direction of the z-axis. We assumed that φ is expressed by a normal distribution with average $\varphi_{av}=0$ and standard deviation σ_{φ}. The following steps could be undertaken to carry out the simulation of particle motion in the electric field generated by the parallel plate electrodes inside the separator:

1. solve the electric field neglecting the space charge ρ_E created by the charged particles using Laplace equation with the proper boundary conditions;
2. calculate the charged particle trajectories with the Lagrangian approach considering aerodynamic and electrical forces acting on the particles:

$$m_p \frac{dv_p}{dt} = \frac{1}{8}\pi\rho d^2 C_D |v_f - v_p|(v_f - v_p) + \frac{\pi d^3 \rho_f}{6}\frac{dv_f}{dt} + \frac{1}{6}\pi d^3 (\rho_p - \rho_f)g + F_E \quad (1)$$

where v_p and v_f are the particle and fluid velocity, respectively, C_D is the drag coefficient [5] and F_E is the electric force qE, q being the particle charge and E the electric field;

3. calculate the space charge distribution ρ_E inside the separator;
4. solve the electric filed by using the Poisson equation with the space charge obtained from the previous step;
5. repeat steps 2 to 4 until solution is convergent.

This numerical technique was originally proposed by Elmoursi [6]. In our case only steps 1 to 3 were considered as the space charge was considered sufficiently small to not affect the electric field distribution.

4   Results and discussion

4.1 Gas flow field

In order to evaluate the accuracy of the turbulence model in predicting the gas flow field, a grid-independence analysis was carried out with a number of grid elements ranging from 850,000 up to 1,500,000. Two control lines within the separator were chosen to compare the simulated mean gas velocity component along the z-axis (v_z), varying the number of elements, with experimental data obtained by other authors. The first line starts from the diffuser along the central axis of the separator (z-axis) to a distance 0.25 m downwards; the other line was set crossflow (x-axis) at 0.10 m downwards from the diffuser. In order to compare the results, the excess velocity (v_z-v_{coflow}) was made dimensionless by dividing it to the excess velocity along the centreline ($v_{z,centreline}$-v_{coflow}); the space scales were made dimensionless by dividing z and x to the diffuser width D and the distance from the center-line at which the velocity decreases to 50% ($r_{1/2}$), respectively.

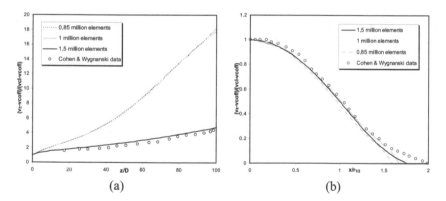

(a)   (b)

Figure 2:   Comparison between mean gas velocity v_z along the control lines in z-direction (a) and x-direction (b) with different number of elements and experimental data [7].

Figure 2 shows the results obtained from the grid-independence study: the data measured by Cohen and Wygnanski [7] are in good agreement with simulations of the gas flow field along the second control line (Figure 2(b)) even with the smallest numbers of elements considered in the simulations. Nevertheless, it was shown that 1,500,000 elements are required to model the process with sufficient accuracy along the first control line (Figure 2(a)). In all the simulations with particles, 1.5 million of elements were then used, which is the minimum number that allows one to obtain a good simulation with the standard k-ε model of turbulence.

4.2 Particles separation

The model validation was carried out by comparing the experimental results obtained with the bench-scale tribo-electrostatic separator previously described, with the outcome of simulations were an appropriate choice of calibration parameter was made: it was surmised that the standard deviation σ_φ of the particle injection angle φ distribution was the most important calibration parameter. During the first validation trials the flow inside the chamber needed to be balanced by regulating the suction velocities at the cyclones, as the flow splitter was not set in the middle of the separator but 1.2 cm close to the left plate. Glass beads in the size range 40-70 μm were employed in the experiments. The charge distribution of glass beads was experimentally and theoretically obtained in other works [8, 9]. The mass fraction of glass beads collected in the left cyclone (Figure 1) was compared to the number of particles leaving the separator form the same outlet divided to the total number of particles used in the simulation. The best agreement between measured (22%) and calculated (25%) mass fraction was obtained with the standard deviation σ_φ of particle injection angle set to 2.2°. The distribution of different injection angles was reproduced by considering five particle classes having injection velocity in the x-axis between -0.6 and 0.6 m/s. When the electric field was set to 2 kV/cm (a positive value indicates that the left plate is positively charged), the majority of particles were deflected towards the positive electrode, as indicated by other studies confirming that most of the glass particles become negatively charged during pneumatic transport [8]. The following table summarizes the results obtained in the calibration procedure.

Table 1: Comparison between experimental and simulation results of particle mass fraction deflected towards the left electrode.

Electric field (kV/cm)	Mass fraction (%)			
	Experimental		Theoretical	
	Left plate	Right plate	Left plate	Right plate
0	78	22	75	25
2	2	98	2	98

After the first calibration step, a voltage-dependence study was carried out to evaluate the influence of electric field strength on particle collection efficiency.

In order to assess whether the calibration parameter was robust enough to adequately describe the particles motion, the suction velocities were adjusted to establish an unbalanced flow in the separator: since the standard deviation σ_φ does not depend on the flow field, the value chosen in the previous experiments

(2.2°) should have given the same agreement between measured and simulated results. The electric field was varied between -2 kV/cm and 2 kV/cm, where a negative value corresponds to a negative charge polarity on the left plate and the yield is calculate as the mass collected at the right outlet over the entire mass fed to the system (Figure 3(b)). In Figure 4 the outcomes of this analysis are shown.

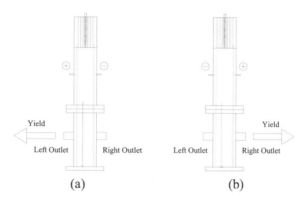

Figure 3: Schematic of the yield calculation for positive (a) and negative (b) electric field.

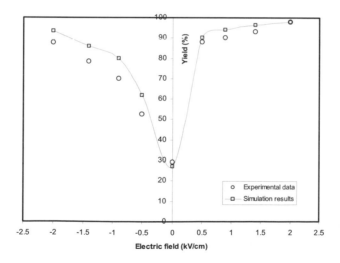

Figure 4: Comparison of yields between experimental tests and numerical simulations at different electric fields.

It must be pointed out that the good agreement between calculations and measured results without electric field remarks the adequacy of the calibration procedure of the standard deviation σ_φ, even with an unbalanced flow. When the electric field is increased until 2 kV/cm, the yield increases to 98% while in the case of negative fields, a mass fraction of 87% was collected in the right cyclone (E= -2 kV/cm). The asymmetry is the consequence of the fact that the flow splitter is set close to the left plate. Figure 5 shows the volume fraction distribution in the cross-flow direction (x-axis) of the five particle size classes without electric field (Figure 5(a)) and with a negative field of -0.9 kV/cm (Figure 5(b)).

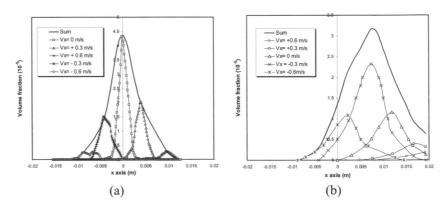

(a)   (b)

Figure 5:   Volume fraction distribution of particles in the separator without electric field (a) and with an electric field of -0.9 kV/cm (b).

Without electric field the particle distribution across the transversal section of the separator s not symmetric, due to the unbalanced flow field. With a negative electric field, the volume fraction distribution is significantly shifted towards the positive electrode (the right cyclone); some particles are however deflected towards the negative electrode because of either the initial horizontal velocity or the charge distribution which accounts for the fact that some particles can become positively charged in the transport line [9]. Besides the aspects related to particle movement due to electric field, another part of the investigation was focused on determining the effect of co-flow air in stabilizing the flow patterns in the separator. The lower the particle inertia, the more the effects of turbulence of the jet will be and, as a consequence, the particles are less likely to respond to electric force as compared to turbulent fluctuations. This occurrence could make tribo-electrostatic separation ineffective for small particles, unless turbulent eddies in the separator were dumped. The use of flow-straightners and the creation of a slight negative pressure in the separation chamber were experimentally found to have a great importance in the separation of low inertia particles by dumping the turbulent eddies [1, 8]. It was found that co-flow velocities slightly less than flow injection velocity cause only very small particles (<5 µm) to be entrapped in the eddies, while heavier particles follow the mean flow structures, as it happens when the ratio co-flow to flow velocity is greater than one. To simulate the condition without co-flow (Figure 6(a)), no slip end smooth wall conditions were assumed for the separator lateral boundaries, while atmospheric pressure was set at the two outlets [10].

Figure 6(b) shows the streamlines inside the separation chamber with co-flow 10% greater than flow, which points out the remarkable reduction in eddies due to entrainment [11].

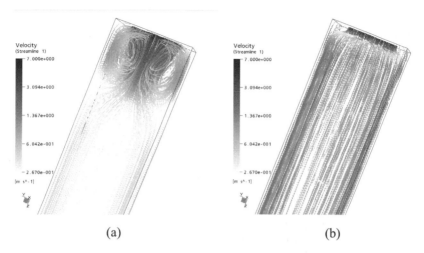

(a)   (b)

Figure 6:   Streamlines of particle jet in the separation cell (side view) without co-flow (a) and with ratio coflow-to-flow set to 1.1 (b).

The presence of additional co-flow air reduces the dissipation of the jet due to viscous forces, so that the average fluid and particle axial velocity diminishes more slowly allowing then the fine particles to be attracted by the plates, contributing to an overall enhancement of separation performances.

5 Conclusions

Some fundamental issues related to the motion of tribo-charged particles in an electrostatic separator were addressed. Separation performances of fine powders (glass beads) were analyzed experimentally and theoretically by the Lagrangian simulation method. Simulation results agreed very well with all the experiments when the calibration parameter, the standard deviation σ_φ of particle injection angle, was set to 2.2°. Once the model was calibrated, several simulations were carried out, with the aim of gain much fundamental understanding of particle motion in the separation chamber, which could not have been possible without computational fluid dynamics. First it was studied the evolution of volume fraction distribution of the particles in the separator due to the electric field and un-balanced flow. Then it was analyzed the importance of co-flow air related to mixing degree in the chamber (flow air to co-flow air ratio) and particles entrainment in turbulent eddies. The results show that in a free-fall separator, where no co-flow air is employed, particles re-circulate nearby the injection point, thus reducing the effect of electric force on the particle movement; on the other-hand, when the co-flow air is high enough to enhance the air entrainment in the separation chamber, the degree of mixing raises, the particle jet is more stable and particles are more easily deflected on the basis of their charge. Hence, this study confirms that tribo-electrostatic separation of fine particles deriving from industrial wastes is effective as long as the proper flow parameters and separator geometry are chosen: this occurrence opens new possibilities in the field of re-use of wastes made of fine particles by means of dry separation which has been halted to date by technological problems associated with free-fall separators.

References

[1]   F. Cangialosi. *Dry triboelectrostatic beneficiation of coal combustion fly ash*. Ph.D. dissertation, Department of Environmental Engineering and Sustainable Development, Technical University of Bari, 2005.

[2]   G. Dodbiba, A. Shibayama, T. Miyazaki, T. Fujita. Triboelectrostatic separation of ABS, PS and PP plastic mixture. *Materials Transactions*, **44**, 161-166, 2003.

[3]   S. Yuu, K. Ikeda, T. Umekage. Flow-field prediction and experimental verification of low Reynolds number gas-particle turbolent jets. *Colloids and surfaces A*, **109**, 13-27, 1996.

[4]   L.J.S. Bradbury & J. Riley. The spread of a turbolent plane jet issuing into a parallel moving air stream. *Journal of Fluid Mechanics*, **27**, 381-394, 1967.

[5]   G. Gousbet, A. Berlemont. Eulerian and Lagrangian approaches for predicting the behaviour of discrete particles in turbulent flows. *Progress in Energy and Combustion Science*, **25**, 133-159, 1998.

[6]   A.A. Elmoursi & G.S.P. Castle. Modelling of corona characteristics in a wire-duct precipitator using the charge simulation technique. *IEEE Trans. Ind. Appl.*, **23**, 95-102, 1987.

[7]   J. Cohen & I. Wygnanski. The evolution of instabilities in the axisymmetric jet. Part 1. The linear growth of disturbances near the nozzle. *Journal of Fluid Mechanics*, **176**, 191-219, 1987.

[8]   T. Li. *An experimental study of particle charge and charge exchange related to triboelectrostatic beneficiation*, Ph.D. dissertation, Department of Mechanical Engineering, University of Kentucky, Lexington, KY, 1999.

[9]   F. Cangialosi, L. Liberti, M. Notarnicola, J.M. Stencel. Monte-Carlo simulation of pneumatic tribo-charging in two-phase flow for high-inertia particles, accepted for publication on *Powder Technology*.

[10]   R.W. Schefer, V. Hartman, R.W. Dibble. *Conditional sampling of velocity in a turbolent non-premixed propane jet*. Sandia National Laboratories Report SAND87-8610, (1987).

[11]   T.B. Nickels & A.E. Perry. The turbolent co-flowing jet. *Journal of Fluid Mechanics*, **309**, 157-182, 1996.

Reuse of waste ashes formed at oil shale based power industry in Estonia

R. Kuusik, M. Uibu, A. Trikkel & T. Kaljuvee
Tallinn University of Technology, Estonia

Abstract

Estonian oil shale belongs to low-grade fossil fuels and by its combustion huge quantities of ash (currently about 5 million tons annually) are formed. Since 1959 pulverized firing (PF) has been used, and circulating fluidized bed combustion (CFBC) technology has recently been implemented – two units each of 215 MW capacity are in operation. In previous investigations of PF ashes the results have been used as a basis for several large-scale applications like production of construction materials and conditioning/neutralizing of soils; a developing of industrial applications for CFBC ashes is hindered by insufficient basic data. In the current investigation attention was focused on the reactivity of ashes towards SO_2 and CO_2 in heterogeneous gas – solid and gas – water – solid systems, being important for characterization of ashes as sorbents for the capture of acidic gases named from flue gases. The significant differences in chemical and phase composition as well as in surface properties of ashes have been shown. The SO_2-binding characteristics for CFBC ashes are higher than for PF ashes and they have more perspective as dry sorbents of sulphur dioxide. Also, under wet carbonization conditions, CFBC ashes can be carbonized more deeply as compared to PF ashes. As a result of carbonation, ashes could be environmentally friendly landfilled and abatement of CO_2 emissions will be achieved.
Keywords: Estonian oil shale, pulverized firing, circulating fluidized bed combustion, waste ashes, reactivity, sulphur dioxide, carbon dioxide.

1   Introduction

Estonian power supply is over 90% covered by oil shale fired thermal power plants. Local solid fossil fuel oil shale is characterized by low heating value

WIT Transactions on Ecology and the Environment, Vol 92, © 2006 WIT Press
www.witpress.com, ISSN 1743-3541 (on-line)
doi:10.2495/WM060131

(8.3-8.5 MJ/kg), moderate moisture content (10-13%), high content of minerals (60-65%) and by the unique and complicated chemical and mineralogical composition of last one. Mineral matter of the Estonian oil shale consists mainly of two components: carbonate matter and sandy-clay matter [1]. In the course of combustion, fuel inorganic part undergoes several chemical transformations, from decomposition processes to the formation of new secondary mineral compounds and phases at higher temperatures.

The mechanism of the processes that take place during pulverized firing (PF) of oil shale, including the formation of ash deposits at heat-transfer surfaces, is thoroughly examined by Ots [1], Ots *et al.* [2]. Investigations of fly ash composition and reactivity brought about developing and introducing ash-recycling processes like production of various building materials, cement with special properties and other products (Kikas [3]). Ash is also utilized as an expedient material in road construction as well as soil conditioner in agriculture (Kärblane [4]) and as a neutralizing additive in the production of mineral fertilizers (Veiderma *et al.* [5]). In addition, it could be used as a SO_2 and CO_2 sorbent (Trikkel [6], Kuusik *et al.* [7]).

Compared to PF, in the case of circulating fluidized-bed combustion (CFBC) the operating temperatures are considerably lower. Consequently, chemical and phase composition of CFBC waste ashes differ noticeably (Kuusik *et al.* [8]) and that is why the differences in chemical reactivity can also be expected.

The aim of the current study was to elucidate chemical reactivity of CFBC ashes compared to PF ones in the systems that are important in the formation of ash deposits in a boiler, at ash deposition at landfill, at flue gas desulphurization as well as at reducing CO_2 emissions.

2 Materials and methods

Ash samples used were collected from different points of the ash-separation systems of CFBC and PF boilers at the Estonian Thermal Power Plant. The CFBC ashes studied were bottom ash (CFBC/BA), intrex ash (CFBC/INT), economizer ash (CFBC/ECO), air preheater ash (CFBC/PHA), electrostatic precipitator ash from fields 1 and 4 (CFBC/ESPA 1 and 4) and mixture of ashes (CFBC/Mix) taken from a common ash silo where the ash from different units is collected before landfilling on ash fields. The PF ashes used were bottom ash (PF/BA), superheater ash (PF/SHA), economizer ash (PF/ECO), cyclone ash (PF/CA) and electrostatic precipitator ash from fields 1 and 3 (PF/ESPA 1 and 3). Chemical and phase composition as well as physical properties of these ashes are presented in Table 1 and discussed in more detail in paper of Kuusik *et al.* [8].

Reactivity of ashes towards SO_2 or CO_2 was tested in a heterogeneous gas-solid system using thermogravimetric equipment (Q-derivatograph, MOM) under isothermal conditions. In both cases the temperature of the isothermal experiment was 700°C, partial pressure of SO_2 or CO_2 was 190 and 144 mm Hg, respectively. The samples (100 ± 0.5 mg) were heated up to 700°C in air with heating rate of 10 K/min. The gas mixture was then fed into the furnace with the rate of 270 ml/min. The thickness of the sample layer in multiplate Pt crucibles was about 0.2–0.3 mm. To study the effect of grinding on SO_2- and CO_2-binding, the initial samples (except for CFBC/BA, which was slightly crushed to pass the 630 μm sieve) and the ground ones were used. Samples were ground in a one-ball vibration mill until the majority (approx. 85–100%) passed through the 45 μm sieve. In some experiments with CO_2, samples were heated up to 900°C to achieve full decomposition of carbonates and then cooled to 700°C to perform isothermal binding.

To characterize high-temperature binding of SO_2 and CO_2 by ashes, the following parameters were calculated:

SO_2- or CO_2-binding capacity (BC; weight of SO_2 or CO_2 bound by 100 mg of sample),

SO_2- or CO_2-binding rate (W; mg SO_2 or CO_2 per mg sample·min^{-1}),

SO_2- or CO_2-binding efficiency, (BE, %) showing the extent of utilization of CaO and MgO contained in the sample

The parameters were calculated on the basis of experimental and analyses data and the following summary binding reactions:

$$CaO(s) + CO_2(g) \rightarrow CaCO_3(s) \tag{1}$$
$$CaO(s) + SO_2(g) + \tfrac{1}{2}O_2(g) \rightarrow CaSO_4(s) \tag{2}$$
$$MgO(s) + SO_2(g) + \tfrac{1}{2}O_2(g) \rightarrow MgSO_4(s) \tag{3}$$

Table 1:   Chemical composition and physical properties of CFBC and PF ashes.

ASH	Content (%)							d_{mean} μm	SSA m²/g
	CaO_t	CaO_f	MgO_t	CO_2	S_t	$S_{sulfate}$	$S_{sulfide}$		
CFBC/BA	49.39	12.48	9.25	15.14	4.53	4.32	0.10	197	2.06
CFBC/INT	47.59	18.87	13.65	1.23	7.76	7.70	0.02	95	2.61
CFBC/ECO	32.84	10.40	9.50	5.48	2.32	2.22	0	27	6.89
CFBC/PHA	35.17	12.26	10.77	4.30	3.31	3.27	0.0013	32	5.40
CFBC/ESPA1	29.52	8.45	8.33	4.60	1.71	1.71	0	25	8.00
CFBC/ESPA4	28.88	2.82	9.35	3.80	2.23	2.21	0	23	7.92
CFBC/Mix	33.28	10.33	9.50	6.41	2.46	2.43	0.03	28	7.11
PF/BA	50.75	24.84	15.19	2.75	1.27	1.27	0	115	1.75
PF/SHA	54.71	23.08	7.81	0.96	1.98	1.93	0	105	0.50
PF/ECO	48.00	16.04	8.24	2.50	2.52	2.52	0.006	53	0.44
PF/CA	49.39	22.52	14.19	0.70	1.33	1.33	0	48	0.36
PF/ESPA1	36.08	13.56	11.26	1.16	2.74	2.74	0	24	0.61
PF/ESPA3	26.85	5.98	5.98	0.80	3.67	3.67	0	23	1.09

Carbonization of aqueous suspensions of ash with model gas, whose composition (10% CO_2 and 90% air) simulated CO_2 content of flue gases formed at oil shale combustion, was carried out in an absorber (diameter 55 mm, water column height 500 mm) equipped with magnetic stirrer for achieving a better interfacial contact and a sintered glass gas distributor (pore diameter 100 μm). The experiment was carried out until suspension pH reached 7.5. Then the suspension was filtered and solid residue dehumidified at 105°C. For the liquid phase, Ca^{2+} content (Vilbok [9]), TDS (total dissolved solids) and alkalinity [10] were determined. For the solid residue the content of free CaO (Reispere [11]) and CO_2 and pH of aqueous suspension of solid residue [12] were determined.

3 Results and discussion

Free CaO is the main binder of SO_2 and CO_2 in the ashes (Reactions 1 and 2). Although MgO does not take part in CO_2 binding under these conditions [13], it has a certain role in SO_2 binding (Kaljuvee et al. [13, 14]). Hence, the negative influence of coarser fractional composition of the samples and, correspondingly, the lower level of specific surface area (SSA) in the case of, for example, CFBC/BA (mean particle size 197 μm; SSA 2.1 m^2/g) and CFBC/INT (95 μm; 2.6 m^2/g) should be compensated by higher level of free CaO content – 11.9 and 18.9%, respectively (see Table 1).

PF/BA and PF/CA are both characterized by a quite high content of free CaO (23–25%), but they differ noticeably (5 times) in their SSA. PF/ESPA1 is characterized by fine fractional composition (d_{mean} = 24 μm), low content of free CaO, and despite a small particle size by a low value of SSA (0.61 m^2/g) and should show a modest binding activity. CFBC/ECO and CFBC/PHA are somewhere in the middle of the scale considering these parameters.

3.1 Transformations in the Gas-Solid System SO_2-Ash

After 30 minutes of the contact between solid and gaseous phases, SO_2-binding capacity of different CFBC ashes was about 26–30 mg SO_2 per 100 mg sample. 55–70% of this value was achieved during the first 2-min contact already (Fig. 1).

BC values for PF ashes differed from each other more considerably – from 10.4 to 23.1 mg SO_2 per 100 mg sample – being the highest for PF/ESPA1. During a 2-min contact, PF/BA bound 58%, while PF/ESPA1 and PF/CA only 33% of the total amount of bound sulphur dioxide. This data correspond to the data obtained in our earlier research (Kaljuvee et al. [13, 14]) and can be explained by the differences in chemical and fractional composition as well as in physical and chemical properties of the ashes. Thus, CFBC/BA and CFBC/INT have coarser fractional composition and lower SSA level, but higher free CaO content, while CFBC/ESPA has fine fractional composition (and high level of SSA), but low content of free CaO (see Table 1). Differences in *BC* values for PF/BA and PF/CA can be explained by their different SSA values (1.75 and 0.36 m^2/g, respectively).

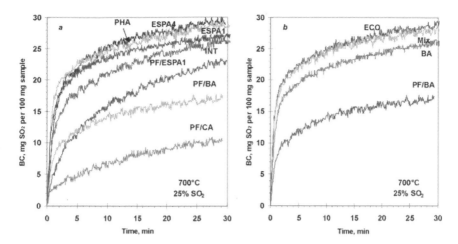

Figure 1:   SO$_2$-binding capacity (BC) of the ashes.

In the experiments of SO$_2$-binding MgO, free CaO and, partially, Ca-silicates can take part in the reactions with SO$_2$. So, in this case binding efficiency, BE was calculated on the basis of total content of CaO and MgO in the sample – $BE(CaO{\cdot}MgO)$. $BE(CaO{\cdot}MgO)$ values indicated the highest level of utilization of both oxides for ESPA4 (74%) and the lowest for BA and INT – about 41% (Fig. 2) BA was the most active among PF ashes, having an average SO$_2$-binding rate 0.10 mg SO$_2$ per mg sample $\cdot$ min^{-1}. Like other indicators of SO$_2$-binding ability, the values of $BE(CaO{\cdot}MgO)$ for PF ashes were much less than those for CFBC ashes: 46.9% (the highest) for ESPA1, 25.7% for BA and 15.4% for CA (Fig. 2).

For all the ashes studied preliminary grinding (up to grain size -45 µm) increased their binding ability. In general, it was noticed that for the ashes characterized by coarse fractional composition (CFBC/BA, CFBC/INT, PF/BA) and/or those having been allocated to high temperatures at PF (especially PF/CA) resulting in the formation of liquid phases hindering further diffusion of SO$_2$ into the particles.

Thus, grinding increased the BC and BE values for PF/CA 2.2 times, for CFBC/BA, CFBC/INT and PF/BA 55–70%, for CFBC/ECO and CFBC/PHA 20–30%, for CFBC/ESPA1 and PF/ESPA1 10–20%, and for CFBC/ESPA4 there was no change in these values. Among ground CFBC ashes, BC value was the highest for BA and INT (40–42 mg SO$_2$ per 100 mg sample), BE value for ECO –81.5% (Fig. 2). Among PF ashes, BC value was the highest for BA –30.2 mg SO$_2$ per 100 mg sample and BE value for ESPA1 –50.6 % (see Fig. 2).

These results are in good correlation with the increase in the SSA of the samples: for PF/CA 4.4 times (from 0.36 m^2/g to 1.45 m^2/g), for CFBC/BA, CFBC/INT, and PF/BA 1.5–2.3 times, for the other ashes on the level of 1.1 times, except for CFBC/ESPA4 and PF/ESPA1, for which even a small decrease in the SSA was observed. Comparing these results with those obtained with

Karinu limestone, the *BC* values of the more active ashes (CFBC/BA and INT) were only 1.4 times less, but the *BE* value of CFBC/ECO was on the same level than that of Karinu limestone (see Fig. 2).

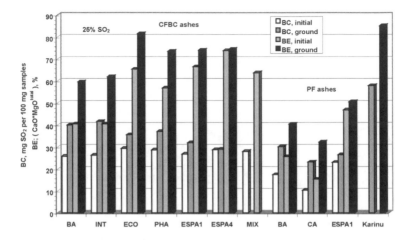

Figure 2: The influence of grinding on SO_2-binding capacity (*BC*) and efficiency (*BE/CaO·MgO*) at 700°C after 30-min contact.

3.2 Transformations in the gas–solid system CO_2–ash

CO_2-binding capacities of the ashes stay between 1 and 13 mg CO_2/100 mg (Fig. 3). Among CFBC samples, the highest binding capacity was calculated for bottom ash. Binding capacities of ECO, PHA and Mix were about 5 mg CO_2/100 mg, the lowest *BC* values were of INT and ESPA. Bottom ash of pulverized firing bound also about 5 mg of CO_2 per 100 mg sample; binding capacities of PF/CA and PF/ESPA were low.

Grinding of CFBC samples decreased their binding capacities noticeably, except for INT. Decrease in *BC* was from 12% (BA) to 60–70% (ECO, PHA and ESPA). This phenomenon is specific of CFBC samples, because *BC* of ground PF samples is about 75–85% higher as compared to the initial ones. This correlates with the increase in specific surface area, which is 4 times higher for ground PF/CA and 1.5 times higher for ground PF/BA. However, grinding increased also SSA of CFBC ashes. The decrease in *BC* was less for BA and INT, also the increase in SSA during grinding was higher for these samples – 2.3 and 1.6 times, respectively.

The effect of decarbonization temperature on *BC* was even more severe – the decrease in *BC* for the samples heated before binding to 700°C and 900°C was 67% – from 5.5 to 1.8 mg CO_2/100 mg.

Separate experiments were performed to estimate changes in SSA and CaO_f content during pre-treatment. It was found that a 30-min grinding increased SSA from 5.3 to 7.7 m^2/g, but heating of this ground sample to 900°C decreased its SSA to 3.1 m^2/g. Also the level of free CaO decreased remarkably during heating. Presumably, some secondary reactions take place during heating that reduce free CaO content in CFBC ashes being more intensive at higher temperatures.

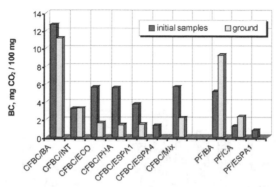

Figure 3: CO_2-binding capacities (BC) of the initial and ground to particle size less than 45 μm ash samples after 30-min binding.

Thus, as compared to PF ashes, CO_2 binding capacities of CFBC ashes are 2-3 times higher. The best results were obtained in both cases with bottom ashes. The method of grinding used improved noticeably binding parameters of PF ashes and reduced those of CFBC ashes. According to the results obtained, it can also be concluded that intensive contamination of heating surfaces in CFB boilers with secondary carbonaceous deposits should not be a serious problem.

3.3 CO₂-Ash-Water System

To characterize CO_2-binding in ash–water suspensions the following parameters were used. Carbonization extent was described by index N indicating the excess of the CO_2 amount entrained into suspensions over the stoichiometric ratio calculated according to Equation (1). Effectiveness of the carbonization process was described by the CO_2-binding degree (BD_{CO2}). BD_{CO2} shows which part of the theoretical ash-binding capacity is utilized. It was calculated basing on the changes in CO_2 content:

$$BD_{CO2} = \frac{CO_2}{CO_{2\,max}} \cdot 100 \ \%$$ (4)

CO_2 is analytically determined CO_2 content of the sample (%)

$CO_{2\,max}$ is the maximal possible CO_2 content of the sample (%), calculated on the basis of content of free or total CaO in the initial sample as follows:

$$CO_{2max} = \frac{CaO^i \cdot M_{CO_2} / M_{CaO} + CO_2^i}{100 + CaO^i \cdot M_{CO_2} / M_{CaO}} \cdot 100 \ \%$$ (5)

CaO^i and CO_2^i denote the content (%) of free or total CaO and CO_2 in the initial sample, respectively.

The results obtained are presented in Figure 4. Chemical analysis of the liquid phase indicated that after carbonization the suspensions of both kinds of ashes contain small amounts of alkaline components. In the suspensions of CFBC ashes, content of Ca^{2+} ion is near saturation point or even higher (in the case of intrex ash 1110 mg/l). In the suspensions of PF ashes, the concentrations of Ca^{2+} ions stay on a noticeably lower level (240–590 mg/l). The TDS values were proportional to the content of Ca^{2+} ion.

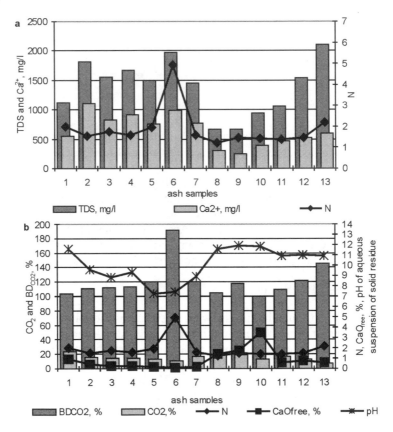

Figure 4: Ca^{2+} and TDS contents in suspensions of CFBC and PF ashes (a) and characteristics of carbonation process and its final product for CFBC and PF ashes (b): 1-CFBC/BA, 2- CFBC/INT, 3- CFBC/ECO, 4- CFBC/PHA, 5- CFBC/ESPA1, 6-CFBC/ESPA4, 7-CFBC/Mix, 8- PF/BA, 9- PF/SHA, 10- PF/ECO, 11- PF/CA, 12-PF/ESPA1, 13- PF/ESPA3.

Compared to PF ashes, CFBC ashes can be carbonized more deeply accompanied by lowering of pH of the solid residue in most cases below 9 (with the exception of CFBC/BA, full carbonization of which is inhibited due to coarse fractional composition. Higher pH value (10.9–11.9) is elicited by higher content of free CaO (0.57–3.47%) in solid residues of carbonized PF ashes. Apparently, in the case of nonporous PF ashes, some part of free CaO present is not accessible and therefore cannot take part in reaction under these conditions. The content of CO_2 is proportional to that of free CaO and CO_2 in the initial samples (see Table 1). Binding degrees that were calculated basing on content of free CaO in the initial samples were in CFBC and PF ashes predominantly over 100% (104.1–191.3% and 99.9–145.1%, respectively) because there can be other CO_2-binding compounds (MgO, Ca-silicates) present in oil shale ash of heterogeneous composition. In general, for total utilization of free CaO the CFBC ashes need less CO_2 per ton of ash as they contain less free CaO in the initial ash.

4 Conclusions

1) Oil shale ashes formed in boilers operating at different combustion technologies differ by their chemical reactivity towards acidic gases.
2) Sulphation of ashes in model conditions shows that SO_2-binding capacity of CFBC ashes at 700°C remained within a narrow range being the highest for PF/BA. SO_2-binding capacity of CFBC as well as of PF ashes is not completely utilized. Being characterized by a higher binding rate during the initial stage of gas–solid contact, CFBC ashes could be more promising as potential SO_2 sorbents for dry desulphurization of flue gases.
3) Transformations in the gas–solid system CO_2–ash, which are important in evaluating possibilities for formation of calcareous precipitations on heat transfer surfaces in boilers, were investigated. CO_2-binding ability of the ashes at 700°C was relatively low, being 2-3 times higher for CFBC ashes. Within 30 minutes 8–52% of free CaO present in the sample was utilized. According to these data, intensive formation of secondary carbonaceous precipitations on the heat transfer surfaces of the CFB boiler is not foreseen.
4) The results of the experiments on binding gaseous CO_2 by ash–water suspension indicated that, in the same conditions, CFBC ashes could be carbonized more intensively and deeply than PF ashes. Besides lime as a compound of the highest reactivity towards CO_2 also other components like MgO and Ca-silicates present in ash take part in CO_2-binding reactions. Owing to improved CO_2-binding rate, more intensive natural CO_2-mineralization of CFBC ashes as compared to PF ashes in open-air deposits as well as intensive binding of CO_2 from flue gases by aqueous suspension of ash are expected.

Acknowledgements

Authors express their gratitude to Estonian Science Foundation (Grant 6195), SC Narva Elektrijaamad and Nordic Energy Research Programme (Project "Nordic CO_2 sequestration") for partial funding of this work.

References

[1]   Ots, A. *Oil Shale Combustion Technology*: Tallinn. 768 pp. 2004. [in Estonian]

[2]   Ots, A., Arro, H., Jovanovic, L. et al. The Behaviour of Inorganic Matter of Solid Fuels during Combustion. *Fouling and Corrosion in Steam Boilers*: Beograd, 276 pp, 1980.

[3]   Kikas, V. Composition and binder properties of Estonian kukersite oil shale ash. *International Cement-Lime-Gypsum*, **50(2)**, pp. 112–126, 1997.

[4]   Kärblane, H. *Handbook of Plant Nutrition and Fertilization. Ministry of Agriculture of the Republic of Estonia*: Tallinn. 285 pp. 1996. [in Estonian]

[5]   Certificate of Authorship 280493 (USSR). *Method of Neutralization of Phosphoric Acid*.Tallinn Technical University. Veiderma, M. A., Vendelin, A. G., Kuusik, R. O., Kuusk, A. A.-M. Appl. 30.12.1968, No.1293097/23-26. Publ. in B. I., 1970 No.28. MKI C 05b 1/02. [in Russian]

[6]   Trikkel, A. *Estonian Calcareous Rocks and Oil Shale Ash as Sorbents for SO$_2$* / Academic Dissertation, Tallinn University of Technology: TTU Press, 70 pp, 2001.

[7]   Kuusik, R., Veskimäe, H., Uibu M. Carbon Dioxide Binding in the Heterogeneous Systems Formed by Combustion of Oil Shale. Transformations in the system suspension of ash – flue gases. *Oil Shale*, **19(3)**, pp. 277-288, 2002.

[8]   Kuusik, R., Uibu, M, Kirsimäe, K. Composition and physico-chemical characterization of oil shale ashes formed at industrial scale boilers with CFBC. *Oil Shale*, **22(4S)**, pp. 407-419, 2005.

[9]   Vilbok, H., Ott, R. Volumetric Analysis. *Instructions for Practical Works* / Tallinn Polytechnical Institute. – Tallinn, 1977. [in Estonian]

[10]   Water Quality. Determination of Alkalinity. International standard ISO 9963-1:1994(E).

[11]   Reispere, H. J. Determination of Free CaO Content in Oil Shale Ash / Tallinn Polytechnical Institute: Tallinn, No 245. pp. 73–76, 1996 [in Estonian]

[12]   Determination of the pH value. Regulation (EC) No 2003/2003 of the European Parliament and of the Council of October 13, 2003 relating to fertilisers / *Official Journal of the European Union*. pp 60-61. 2003.

[13]   Kaljuvee, T., Trikkel, A., Kuusik, R. Decarbonization of natural lime-containing materials and reactivity of calcined products towards SO$_2$ and CO$_2$. *J. Therm. Anal. Cal*, **64**, pp. 1229–1240, 2001.

[14]   Kaljuvee, T., Kuusik, R., Trikkel, A., Bender, V. The role of MgO in the binding of SO$_2$ by lime-containing sorbents. *J. Therm. Anal. Cal*, **80**, pp. 591–597, 2005.

The path of packaging waste in the secondary sector

B. A. Curran[1], D. W. Dixon-Hardy[1] & C. J. Barns[2]
[1]School of Process, Environmental and Materials Engineering,
University of Leeds, UK
[2]School of Mechanical Engineering, University of Leeds, UK

Abstract

The secondary packaging sector specified in various UK legislation states that a large sector of secondary packaging originates from supermarkets. This paper examines how supermarkets deal with secondary packaging waste that encourage recycling. Supermarkets generally have a policy to recycle cardboard, paper and plastic. Packaging is returned to a central depot and this is where the recycling or bailing occurs. Each supermarket chain has its own policy on recycling and waste disposal based on Government guidelines. Anecdotal evidence suggests that manufacturers of typical supermarket products take little interest in to what supermarkets do with their secondary packaging. This paper presents the current UK situation as to the path of supermarket packaging waste.

Keywords: packaging waste, secondary sector, supermarket, recycling, re-use.

1   Introduction

In the UK alone there is in the region of 400 million tonnes of waste (Chameleon Net [1]), produced each year and a quarter of which is from households, commercial and industry. The remainder is made up of construction and demolition wastes, mining and agricultural wastes, sewage sludge and dredge spoils. This is shown in Figure 1. The most favoured option for disposal of waste is to place it in landfill sites. Recently this option has been questioned as to being the best option as available landfill sites are becoming scarce. In 2003 there were around 2,300 landfills (Environment Agency [3]), and each one was in the region of 28,000 hectares, which in perspective is less than 0.2% of the land of England and Wales.

WIT Transactions on Ecology and the Environment, Vol 92, © 2006 WIT Press
www.witpress.com, ISSN 1743-3541 (on-line)
doi:10.2495/WM060141

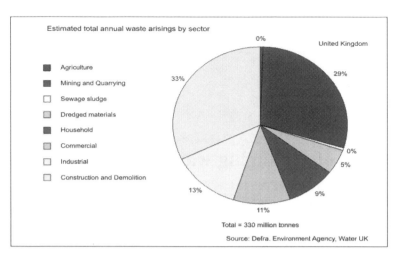

Source: Defra. Environment Agency, Water UK

Figure 1:   Waste production pie chart 2002/3 (Defra [2]).

The Waste Hierarchy (NWRA [4]), a pyramid diagram illustrating how to reduce waste, is used to analyse how to manage waste. The Waste Hierarchy is formed from the four principles set out in the Waste Strategy 2000 (Defra [5]), that are: -

- Reduction of waste produced
- Re-use any products where possible
- Recycle what cannot be re-used and recover energy from waste that cannot be recycled
- Disposal of the remaining waste i.e. landfill

By using the waste hierarchy (NWRA [4]), waste that is not re-used or recycled has to be disposed of and the obvious option is to send it to landfill. In 2002/03 nearly half of commercial and industrial waste was recycled (Figure 2), where 40% was sent to landfill (nearly 30 million tonnes of waste).

2   Definitions

Quote from (Oian et al. [7]) defining packaging:
" *'Packaging'* shall mean all products made of any materials of any nature to be used for the containment, protection, handling, delivery and presentation of goods, from raw materials to processed goods, from the producer to the user or the consumer. 'Non-returnable' items used for the same purposes shall also be considered to constitute packaging. "

Packaging defined above is then segregated in waste online (Environment Agency [3]), into three different categories:
- **Primary** packaging is handled by the consumer, as it is wrapping or a container for the product.

- **Secondary** packaging term is used for larger cases or boxes used to distribute and display the primary packaged product.
- **Transit** packaging is used to group the packages together to ease loading and unloading of the product in bulk form, i.e. wooden pallets, board and plastic wrapping and containers.

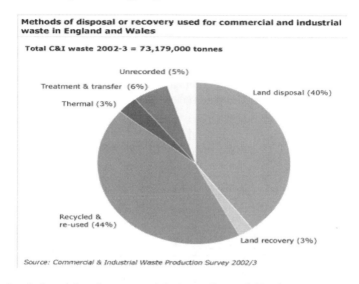

Figure 2: Industrial and commercial waste disposal (Environment Agency [6]).

Quote from (Oian et al. [7]) defining packaging waste:
" *'Packaging waste' shall mean and packaging or packaging material covered by the definition of waste in Directive 75/442/EEC, excluding production residues.* "

Quote from (NWRA [4]) defining waste:
" *'Waste' means any substance or object which the holder disposes of or is required to dispose of pursuant to the provisions of national law in force.* "

Quote from (Defra [5]) defining producer:
" *'Producer' means any person who, irrespective of the selling technique used, including by means of distance communication in accordance with Directive 97/7/EC of the European Parliament and of the Council of 20th May 1997 on the protection of consumers in respect of distance contracts.* "

3 Packaging and packaging waste directives

The European Packaging and Packaging waste directive '94 (Oian et al. [7]) outlines the aims to manage packaging and packaging waste in order to prevent production of packaging waste. Additionally it aims to re-use, recycle or recover packaging to reduce the amount sent to landfill. The directive '94 (Oian et al. [7]) was amended in 2004 (European Parliament and Council [8]) to update the national targets for recycling of packaging waste. The directive discusses national programs and projects to reduce the impact of packaging waste by putting responsibility on producers of the packaging. The European standards are to be amended so that packaging waste will have minimal environmental impact.

The 2004 Directive (European Parliament and Council [8]) states that by 30th June 2001 50% - 65% wt packaging waste should be recycled or have energy recovered from it, and by 31st December 2008 it should be at least 60%. The targets for recovery and recycling of the packaging waste are shown in Figure 3 as minimum and maximum targets for both 2001 and 2008. Where appropriate the Directive 2004 (European Parliament and Council [8]) recommends that energy be recovered from the material rather than being recycled due to cost benefit and environmental reasons.

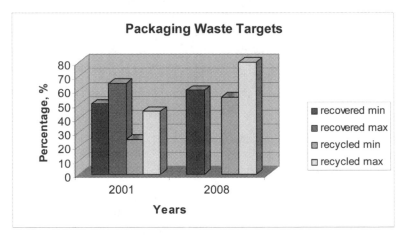

Figure 3:   Packaging waste targets for 2001 and 2008 (European Parliament and Council [8]).

The minimum targets increase over the seven years from 2001 to 2008 in accordance with the advancement in recycling across the board. The recovery minimum target has increased by 10% whereas the recycling minimum target has increased by 30%. Targets are set and reviewed every few years to assess if set targets are realistic, if not they can be amended accordingly. Supermarket chains are considered to be part of the producing chain of packaging waste so hence are responsible for its production. With supermarkets having to aim to achieve these targets incentives such as money for the tonnes of material that is recycled are given. This is 'accountable profit' (Morrisons [9]) for all the tonnes of packaging waste that is bailed and recycled they receive payment. They have to pay to send the remaining packaging waste to landfill.

From the pie charts in Figure 4 and Figure 5, data taken from Waste Online (Environment Agency [3]), the majority of the weight of packaging waste consists of paper and board material, 43%. Comparing this to the percentage of packaging waste, plastic takes up more than half and paper and board only 25%.

This shows that paper and board can account for the most weight of packaging waste produced and plastic only 20%, but can only account for a quarter of the packaging used for all products. Due to this the majority of supermarket chains only recycle plastic, and paper and cardboard. It can easily be separated and bailed and then recycled. With supermarkets paid by tonnage (Morrisons [9]) of material recycled, paper and cardboard are the obvious choice. With plastic, paper and cardboard removed from the waste, packaging waste sent to landfill should be reduced by 78% (see Figure 5).

To encourage the use of recycled material the 2004 directive (European Parliament and Council [8]) states that recycled packaging can be used to manufacture packaging and other products by improving the market conditions so these materials can be more economically viable and also review the regulations preventing the use of these materials. Raw materials are easy and readily available to use but making recycled goods cheaper will make them more attractive to packaging producers.

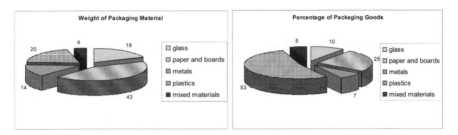

Figure 4:   Weight   of   packaging   Figure 5:   Percentage of packaging
             material.   goods.

4 Supermarket packaging recycling

Seven supermarkets were chosen to investigate, of which five are in the top seven supermarkets according to TNS (BBC [16]). The supermarkets looked at were Morrisons (Morrisons [9, 13]), Asda (Asda [17]), Safeway (Morrisons [9]), Tesco (Tesco [10, 11]), Iceland (BBC [14]), the Co-op (Co-op [12]) and J Sainsbury (J Sainsbury [15]). These are the largest of the supermarket outlets in the UK and hence have the largest amount of packaging waste to dispose of. Figure 6 below shows how the supermarkets selected have increased their recycling of packaging waste and Figure 7 shows the recyclables in 2005.

From many conversations with supermarket chains it was found out that each supermarket has a responsibility to dispose of their waste wisely as they have a large amount of packaging waste cardboard and plastic. The cardboard and plastic from each supermarket chain are to be bailed and recycled. Each supermarket chain has a number of depots around the country that deliver to the supermarket outlets and the bails are returned to the corresponding depot. The larger supermarket outlets can bail the cardboard and plastic on site and then send it on to the depot but the smaller outlets send it all flat packed to be bailed when it reaches the depot.

Figure 6:   Total recycling of packaging waste in supermarkets.

The graph in Figure 6 shows that both Morrisons and Tesco have greatly increased the amount of packaging waste that is recycled from 2004 to 2005. They have very similar trends on the graph but Tesco has a much larger amount of recycled packaging waste because it has the greater amount of outlets due to the fact that it dominates a much larger percentage of the market (BBC [16]), around 30.6% whereas it's closest rival Asda holds only 16.6% which is nearly half that of Tesco. Safeway only has the data for 2004 because its outlet stores were purchased by Morrisons and some were sold on again to other supermarket chains.

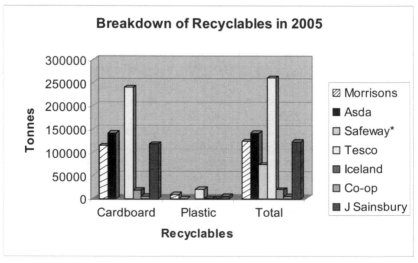

Figure 7:   Breakdown of supermarket recyclables in 2005.

Asda and J Sainsburys also have similar trends as well with a gentle increase in recycling each year. Figure 6 shows that all the supermarkets looked at are increasing the amount of packaging waste produced that is recycled. This shows that the supermarkets are aware of the problem with the amount of packaging waste that is sent to landfill. The breakdown of the recyclables from the supermarkets in 2005 are shown in Figure 7 except for Safeway that is for 2004 when it was bought out by Morrisons. The majority of tonnage of recyclables is cardboard (Figure 7) because plastic can have a large volume but a very low weight. Figure 4 and Figure 5 show that the majority of packaging material is plastic and the majority of packaging weight is from cardboard and paper. Therefore the two materials that the supermarkets will find most useful to recycle are plastic and cardboard because the most space in the packaging waste is taken up by the amount of plastic packaging waste and the added weight can be accounted for by the cardboard.

Figure 6 and Figure 7 both show that Tesco does the greatest amount of recycling but this is because it has the largest niche in the market. If the tonnes of packaging waste that is recycled is divided by the number of stores and the annual turnover then there will be a more realistic view as to how much relatively each supermarket is recycling. This is shown in Figure 8 and Figure 9 respectively, from this it is shown that per store Asda recycles the most packaging waste with Morrisons closely behind. Figure 8 shows that Tesco has a small amount of recycled packaging waste compared to the amount of stores that it owns. Iceland has a very small amount of recycled packaging waste but being a frozen food store it may have a different form of packaging waste to the other supermarkets, it is in the top seven supermarkets with a market share of 1.8% (BBC [16]). A more realistic view of which supermarkets are most affective at recycling is shown in Figure 9 because the tonnes of recycled packaging waste have been divided by the annual turnover of each individual supermarket.

Figure 8:   Tonnes of packaging recycled per store.

The data was obtained from their annual reports (Morrisons [9], Tesco [10], Co-op [12], J Sainsbury's [15] and Walmart [17]). This shows the amount of packaging waste that is recycled in comparison to the annual turnover of each supermarket. From Figure 9 you can see that Morrisons has increased the amount of packaging waste that is recycled by a large amount from 2004 to 2005 compared to its annual turnover. Each supermarket has increased the amount of packaging waste that is recycled in 2005 compared to 2004 except for Asda. Asda has steadily decreased the amount of packaging waste recycled compared to the annual turnover from 2003 to 2004 and 2004 to 2005.

Figure 9 clearly shows that Morrisons has the largest increase in recycling of its packaging waste. Whilst Tesco, Co-op and J Sainsbury have only increased the recycling of their packaging waste by a small amount.

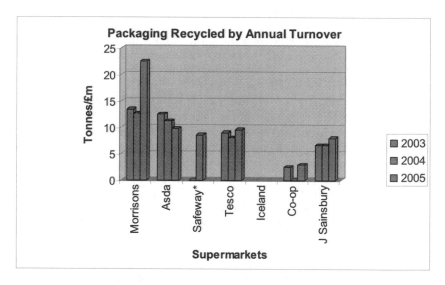

Figure 9: Tonnes of packaging recycled divided by the annual turnover.

5 Innovative ideas

Any packaging waste from the supermarket outlet that has been contaminated by produce has to be sent to landfill and can not be recycled. Tesco has created green trays (Tesco [10]), they are plastic trays that can be sterilised after use and then re-used. They replace the cardboard and other materials used for transporting and displaying the produce. In 2004 Tesco saved 4,000 tonnes of cardboard packaging in just replacing them with green trays. They have also developed the first degradable plastic bags (Tesco [10]). The plastic bags break down in 60 days to create carbon dioxide, water and mineral matter. There is no harmful residue unlike conventional plastic bags. They have offset 6,035 tonnes of conventional plastic carrier bags by the customers using in the region of 719 million of the degradable plastic carrier bags. They are now working with Waste and Resources Action Programme (WRAP) to aim to reduce product packaging by 10% whether this is in the primary, secondary or transit sector of packaging.

6 Conclusion

This paper shows that in the secondary sector of packaging waste the supermarkets are working towards reducing the amount of packaging waste that is produced and sent to landfill. The supermarkets looked at in this report have to pay to have the waste that they produce to be sent to landfill, so reducing the amount of waste produced would be beneficial to the supermarkets as it will reduce the cost of waste disposal. The waste is reduced by recycling the packaging waste in the form of cardboard and plastic. Comparing the supermarkets with the amount of packaging waste that is recycled difficult because each supermarket recycles in a slightly different way and all cover a different percentage of the market.

Comparing the supermarkets tonnes of packaging waste recycled per store showed which supermarkets were recycling the most in each store but it was not a fair comparison as the store sizes weren't taken into account. The larger store can have the facility to bail cardboard and plastic on site so that larger amounts of packaging waste can be recycled easily. The smaller stores may not have the room to store the packaging waste for recycling and have to throw it away. Comparing the supermarkets using their annual turnover was a fairer comparison as it showed per each £million annual turnover the supermarket made the amount of tonnes of packaging waste that was recycled was shown. This shows that the largest supermarket Tesco may have the greatest turnover and the largest sector in the market but it is not recycling as much packaging waste per £million as Morrisons or Asda. It shows that the most improved supermarket in 2005 is Morrisons. Asda, Tesco and J Sainsbury are all producing similar amounts of recycled packaging waste per £million. Further investigation is needed into why Morrisons and Tesco were able to increase the amount of packaging waste that was recycled much more than the other supermarkets.

There is a clear understanding that the supermarkets need to reduce the amount of packaging waste produced and sent to landfill. This report shows that they are working towards either reducing the amount of packaging waste produced and increasing the amount of packaging waste that is produced to be recycled. The next step is to reduce the amount of packaging waste produced. This can be done by designing packaging so that it can be re-used, or designed so that it can be disassembled and either the re-assembled to be used again or disposed of. Morrisons has offered to work on a further paper to look directly into this in their stores.

References

[1]   Chameleon Net, Packaging recycling information sheet from 2004 Waste Online, Design & development by Chameleon Net, archive 'Packaging Recycling'. Metadata Resource: 315

[2]   Defra, Environmental Agency e-Digest of Environmental Statistics. Published by Defra, Environment Agency, Water UK on 31/03/05, Data taken from the Municipal Waste Management Survey 2003/4, http://www.defra.gov.uk/environment/statistics/index.htm

[3]   Environment Agency on Landfills, What are Landfill Sites. Published by the Environment Agency – Environmental Facts and figures, http://www.environment-agency.gov.uk/yourenv/eff/1190084/resources_waste/213982/207743/?version=1&lang=_e

[4]   North West Regional Assembly (NWRA). Regional Planning, Transport and Sustainability. The four principals of The Waste Strategy 2000 Cm 4693-1, ISBN 0 10 146932 2

[5]   Defra, Waste Strategy 2000 for England and Wales Part 1 & 2, Department for Environment, Food & Rural Affairs. Cm 4693-1, ISBN 0 10 146932 2

[6]   Environment Agency – C&I Commercial and Industrial Waste Survey 2002/2003, www.environment-agency.gov.uk

[7]   Oian, X, Koerner, R. M, Gray, D.H, Geotechnical Aspects of Landfill Design and Construction. Published by Upper Saddle River, N.J. : Prentice Hall, c2002

[8]   European Parliament and Council Directive 2004/12/EC of 11 February 2004 on packaging and packaging waste

[9]   Morrisons 2006 Annual Report

[10]   Tesco Corporate Responsibility Review 2005

[11]   Tesco Corporate Responsibility Review 2004

[12]   United Co-operatives Limited, Our Stakeholder Report 2005

[13]   BBC, Safeway sales drop hits Morrisons, 04/06/2004 BBC News archive

[14]   BBC, Iceland sales back under pressure, 25/05/2004 BBC News archive

[15]   J Sainsbury Plc Annual Report 2005

[16]   BBC, Tesco's market share still rising, 08/02/2006 BBC News archive

[17]   ASDA WalMart Annual Report 2005

Section 4
Water and wastewater treatment

Synthesis, characterisation and application in coagulation experiments of polyferric sulphate

Z. I. Anastasios, V. Fotini & M. Panagiotis
Division of Chemical Technology, Department of Chemistry,
Aristotle University of Thessaloniki, Greece

Abstract

The process of coagulation is a core environmental protection technology, which is mainly used in the wastewater treatment facilities. Research is now focused on the development of new inorganic coagulants. A characteristic example is PFS (polyferric sulphate), a relatively new pre-polymerised inorganic coagulant with high cationic charge. It consists of medium and high molecular weight polymeric chains and it can be described with the chemical formula $[Fe_2(OH)_n(SO_4)_{(6-n)/2}]_m$. In this paper, the role of some parameters including temperature, types of chemical reagents etc in the preparation stages of PFS were investigated. Furthermore, the prepared PFS was characterized based on typical properties such as the percentage of the polymerised iron present in the compound etc. Furthermore, dynamics of coagulation process were examined by means of the Photometric Dispersion Analyzer (PDA). Finally, the coagulation performance of PFS in reducing the turbidity of synthetic kaolinite suspensions was studied.
Keywords: coagulation, polyferric sulphate, inorganic polymeric coagulant, pre-polymerised coagulant.

1   Introduction

The process of coagulation is widely used in the wastewater treatment facilities especially for the destabilisation of colloids suspensions and for the removal of suspended solids along with the removal of phosphate ions. It is known as a core environmental protection technology. Nearly all the colloids found in natural waters carry negative charge and therefore, they remain in suspension, due to the mutual electric repulsions. Therefore, the addition of a cation will result in colloidal destabilisation, as they specifically interact with the negatively charged colloids and neutralise their charge. Highly charged cation, such as Fe^{3+}, is

WIT Transactions on Ecology and the Environment, Vol 92, © 2006 WIT Press
www.witpress.com, ISSN 1743-3541 (on-line)
doi:10.2495/WM060151

regarded as one of the most effective cation for such a purpose. Therefore, there are a number of conventional coagulants based on iron such as the $Fe_2(SO_4)_3$ and the $FeCl_3$. However, current research focuses on the development of new type of coagulants, which will combine superior efficiency and lower operational cost as compared with the conventional coagulants.

A relatively new inorganic coagulant is the polyferric sulphate (PFS). PFS is a pre-hydrolysed coagulant, which can be described by the chemical formula $[Fe_2(OH)_n(SO_4)_{(6-n)/2}]_m$, where $n<2$ and $m>10$. It contains polynuclear complex ions, such as $Fe_2(OH)_2^{4+}$, $Fe_3(OH)_4^{5+}$, formed by OH bridges and a large quantity of inorganic macromolecular compounds. The molecular weight can be as high as 10^5 (Chang and Wang [1]). Due to the presence of polymeric species the PFS carries high cationic charge, which can improve the charge neutralising capacity and hence, it becomes more effective at a comparatively lower dose, than the conventional coagulants. (Jiang and Graham [2]). Furthermore, it has been reported by several researchers (Chang and Wang [1], Jiang and Graham [2], Jiang and Graham [3], Butler et al. [4], Li et al. [5]), that PFS exhibits a superior efficiency in the removal of chemical oxygen demand (COD), biochemical oxygen demand (BOD), turbidity and colour, than the conventional coagulants based on iron. Also, it can be used in a wide range of pH and temperature, due to its strong hydrolysis. Meanwhile, some preliminary toxicity studies suggested that drinking water treated with PFS is safe for consumption (Hendrich et al. [6]).

Several researchers have proposed different methods for preparing the PFS in laboratory scale (Jiang and Graham [7], Fan et al. [8], Cheng [9]). However, the role of certain parameters, such as the temperature, the type of chemical reagents in the various stages of preparation methods has not been thoroughly examined. Therefore, the aim of this paper is the investigation of the optimum experimental conditions for the synthesis of PFS. Furthermore, the coagulation performance of PFS is assessed by examining the efficiency in removing the turbidity of simulated kaolinite water suspensions. The specific conditions applied in the coagulation experiments were found by means of the Photometric Dispersion Analyser (PDA).

2 Experimental procedures

2.1 Mechanism of PFS preparation

The synthesis of PFS commences with the oxidation of ferrous sulphate (5.59×10^{-3} M as Fe) to ferric sulphate in highly acidic conditions (H_2SO_4, 96 wt%). The oxidising agent was nitric acid (HNO_3, 65 wt%).

$$FeSO_4 + 1/2SO_4^{2-} + \text{oxidising agent} \rightarrow 1/2Fe_2(SO_4)_3$$

When the amount of sulfuric acid is limited, the hydroxide ion will replace the sulphate ion in the hydrolysis stage and therefore, the polymerisation will occur:

$$Fe_2(SO_4)_3 + nOH^- \rightarrow Fe_2(OH)_n(SO_4)_{3-n} + n\,SO_4^{2-} \text{ (Hydrolysis)}$$
$$mFe_2(OH)_n(SO_4)_{3-n/2} \rightarrow [Fe_2(OH)_n(SO_4)_{3-n/2}]_m \text{ (Polymerisation)}$$

2.2   Synthesis of PFS

The synthesis of the PFS was performed following the proposed method of Jiang and Graham [7]. The investigation of optimum experimental conditions was conducted by altering certain experimental parameters, such as the temperature of each stage of the preparation method, as well as the type and the amount of chemical reagents used.

2.3   Characterisation of the PFS

2.3.1   Determination of total iron concentration
The total iron concentration was measured by means of Atomic Absorption Spectroscopy (Perkin-Elmer 2380) (Clesceri et al. [10]).

2.3.2   Determination of polymerised iron concentration
The measurement of the Fe species by the ferron-timed spectroscopy method has been reported previously (Jiang and Graham [7]). The ferron reagent (8-hydroxy-7-iodoquinoline-5-sulphonic acid) can form complexes with single ferric ion, monomeric and dimeric species within the reaction time of 1 min, while it can complex with medium and high molecular weight polymers with increasing reaction time. Precipitated ferric species will not react with ferron. Based on this principle, visible absorbance at 600 nm was measured as a function of time and each absorbance corresponded to the respective ferric concentration.

2.3.3   Determination of ferrous ion concentration
A potassium permanganate titration method was used to determine the ferrous ion concentration in the PFS. The method can be described by the following equation:

$$MnO_4^- + 5Fe^{2+} + 8H^+ \rightarrow Mn^{2+} + 5\ Fe^{3+} + 4H_2O$$

2.3.4   Determination of the ratio $r = [OH\text{-}]/[Fe^{3+}]$
The molar ratio of OH^- to the Fe^{3+} is called the r value and it can be directly related to the stability of the PFS. In particular, for high values of r, e.g. $r > 0,4$, the prepared PFS is unstable and precipitates may emerge. It is determined by adding excess hydrochloric acid (HCl 1 N) and potassium fluoride solution (KF 50% wt to a sample) and back titrating with sodium hydroxide (NaOH 1 N).

2.4   Photometric Dispersion Analyser (PDA)

The extent of aggregation, as well as the kinetics of coagulation was examined with the aid of Photometric Dispersion Analyser (PDA 2000, Rank Brothers Bottisham UK). The PDA instrument measures the ratio R, or the flocculation index FI, which are directly related with the mean concentration and size of dispersed particles. The test suspension (1.5L) was contained into a 2-L beaker and was constantly stirred by a JP SELECTA jar test device. The created microflocs passed through the measuring transparent plastic cuvette (3 mm diameter) with the aid of a peristaltic pump. The applied flow rate was 30 ml/min in order to secure laminar conditions throughout the experiment. The ratio R curves derived from the PDA instrument have the typical form shown in Figure 1 (Hopkins and Ducoste [11]).

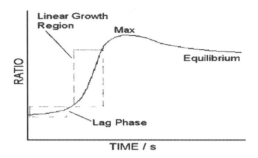

Figure 1:   Schematic representation of the typical ratio R curve.

The R curve can be divided into three parts: A) immediately after the coagulant addition there is a small change in the R values. During this part, which is called "lag phase", the destabilisation of the particles takes place. B) Then, the linear growth region part follows, where the collisions of previously destabilized particles result in aggregate formation. C) Finally, in the equilibrium phase the value of the R remains relatively constant as the rate of the formation and the breakage of flocs is rather equal. Generally, it can be concluded that the higher R values imply bigger particles size and therefore, a better separation by the application of subsequent sedimentation.

2.5 Jar test

The coagulation experiments were carried out using a jar test apparatus with six paddles (Aqualytic). Kaolinite water suspension was used as the test suspension, dosed with the appropriate amount of the coagulant. The conditions, under which the experiments were conducted, were determined from the study of coagulation dynamics in accordance with the relevant literature (Fan et al. [8], Jiang and Graham [2]). In particular, the fast mixing time was set to 3 min at a paddle speed of 300 rpm to allow the particles to be destabilized. The flocculation period was 30 min at a paddle speed of 35 rpm and the sedimentation period lasts for 45 min. After that period a supernatant sample (50 ml) was withdrawn for turbidity measurements. Turbidity measurements were carried out by using a HACH RATIO/XR Turbimeter.

3 Results and discussion

3.1 Characterisation of PFS

Table 1 shows the characteristics of the PFS produced, as well as the conditions under which their synthesis occurred. The use of weak base solution in contrast

Table 1:   Characteristics of the PFS produced according to different experimental conditions.

	Oxidation stage (i) oxidant (ii)duration (iii)temparature	Hydrolysis stage (i)base (ii)duration (iii)temperature	Aging (i)duration (ii)temperature	Fe_{total} (g/l)	Polymerization percentage (% of Fe_{total})	density (g/l)	pH	$[OH^-]/[Fe^{+3}]$
A	(i) HNO_3 (ii) 2 h (iii) 90° C	(i) $NaHCO_3$ 0.5N (ii) 1h (iii) 50° C	(i) 2 h (ii) 50° C	52.5	38	1182	1.5	0.25
B	(i) HNO_3 (ii) 2 h + 10 min (iii) 90°C	(i) NaOH 0.5N (ii) 1 h + 15 min (iii) 50° C	(i) 2 h (ii) 50° C	59.5	17	1191	1.1	0.12
C	(i) HNO_3 (ii) 2 h + 45 min (iii) 90°C	(i) $NaHCO_3$ 0.5N (ii) 1 h + 20 min (iii) 70° C	(i) 2 h (ii) 50° C	50.5	26	1138	1.2	0.16
D	(i) HNO_3 (ii) 2 h (iii) 90° C	(i) $NaHCO_3$ 1 N (ii) 1 h + 17 min (iii) 50° C	(i) 2 h (ii) 50° C	52.5	15	1198	1.4	0.21
E	(i) HNO_3 (ii) 1 h + 50 min (iii) 90° C	(i) NaOH 1 N (ii) 1 h + 2 min (iii) 50° C	(i) 2 h (ii) 50° C	50.5	15	1196	1.5	0.27
F	(i) HNO_3 (ii) 2 h + 45 min (iii) 90° C	(i) NaOH 1 N more base added (ii) 1 h + 2 min (iii) 50° C	(i) 2 h (ii) 50° C	25	11	1162	2	0.4

with the use of a strong base results in the increase of polymerisation percentage, of the ratio r = $[OH^-]/[Fe^{3+}]$ and of the product stability during the following 4-month period (precipitate appearance). Additionally, Table 2 shows a comparison between the PFS produced in the laboratory and those produced according to the relevant literature.

Table 2: Comparative table of the PFS produced in the laboratory and those produced according to the relevant literature.

	Fe_{total} (g/l)	Polymerized Fe (% of Fe_{total})	pH	$[OH^-]/[Fe^{+3}]$	density (g/l)	% turbidity removal
laboratory PFS	52.5	38	1.5	0.25	1182	95*
Jiang and Graham [7]	40	65.2	1	0.3	-	96**
Cheng [12]	150	-	0.56	0.4	1480	-

* kaolinite suspension, ** surface water with algae

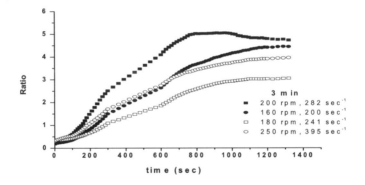

Figure 2: Impact of fast mixing speed (or velocity gradient G (s⁻¹)) on the coagulation process of kaolinite water suspension (5 mg/L); PFS(5 mg/l), pH =7.

3.2 Study of kinetics of the coagulation using the PDA

3.2.1 Effect of mixing speed (or velocity gradient G (s⁻¹)) in the kinetics of coagulation

The effect of speed during the fast mixing period on the coagulation process is shown in Figure 2. The mixing speed is expressed in rpm, as well as in velocity gradient units G(s⁻¹). As shown in Fig. 2, for the mixing speed of 200 rpm, the "lag" phase is limited and this results in a faster flocculation. Also, for the same speed a relatively higher speed of flocculation is achieved and the final flocs after the equilibrium of the system have a larger size. In the contrary, for the mixing speed of 160 rpm, the "lag" phase has a bigger duration and therefore the formation of flocs is delayed. The flocculation of suspended particles proceeds in lower speed and the final size of flocs is smaller. The same happens in the case of very high mixing speed (250 rpm). The small size of flocs is due to the fact that the smaller flocs, which are formed during the fast mixing period, are destroyed by the shear stress which develops during that stage and therefore, it is impossible to obtain a bigger size in the flocculation stage.

3.2.2 Effect of mixing time in the kinetics of coagulation process

Figure 3 shows the effect of mixing time in the kinetics of coagulation process. The speed in the fast mixing stage is 200 rpm, which found to be the optimum as shown in the previous section 3.3.1. The higher values of ratio, which correspond to the bigger size of formed flocs, were achieved, when the fast mixing time was 180 s. It is worth noting that the "lag" phase in both cases is similar and the fact which distinguishes the two curves, is the bigger size of the flocs in the slow mixing period.

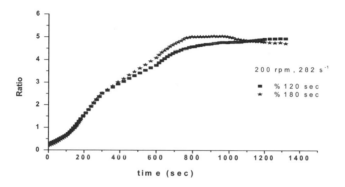

Figure 3:   Impact of time during the fast mixing period in kaolinite water suspension (5 mg/L); PFS (5mg/l), pII=7.

3.3   Jar test results

3.3.1 Comparative jar test results among the prepared PFS

Figure 4 summarises the comparative results among the different PFS. The test solution was kaolinite water suspension (5 mg/L), with initial turbidity 8 NTU and pH = 7. The polyelectrolyte used was the non-ionic Magnafloc LT20, for a dosage corresponding to the 1/10th of the coagulant dosage. The suspensions were stirred rapidly in 200 rpm for 3min, the flocculation period was 30 min (in 35 rpm) and finally the flocs were left to settle for 45 min. Figure 4 shows that higher turbidity removal was achieved when using PFS A, which is directly related to the higher degree of polymerization of that coagulant. PFS A and C, which are synthesized using $NaHCO_3$ (0.5 N) solution, showed better performance, than those produced with NaOH (0.5 N) (B, E, F).

In general, the polymeric species carry high cationic charge, because of their larger size, which results in increasing surface activity and thus, improving their ability to neutralize the charge of suspended particles. Therefore, the increase in the polymerization percentage results in the increase of coagulating performance (Jiang and Graham [7]).

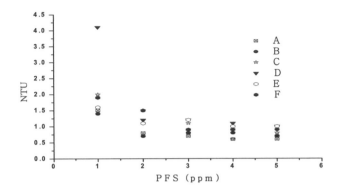

Figure 4:   Turbidity removal of kaolinite water suspension (5 mg/L) against the PFS dosage, for polyelectrolyte addition equal to 1/10th of the coagulant dosage.

3.3.2 Correlation between the degree of polymerisation and the removal of turbidity

From the conducted experiments it was shown that an increase in the polymerization percentage could enhance the ability of the PFS to remove turbidity, which is confirmed by the relevant literature as well (Jiang and Graham [7]). The above statement is shown in Figure 5.

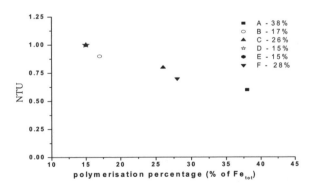

Figure 5:   Turbidity removal against the polymerisation percentage, for a coagulant dose of 5 mg/L and polyelectrolyte dose 0.5 mg/L.

3.3.3 Comparative results between PFS and FS

Furthermore, jar tests were conducted using the FS in order to compare its performance with the PFS produced in the laboratory. The ferric sulphate (FS) is a well-established conventional coagulant with a wide range of applications. The jar test conditions were similar with those used in previous experiments, with polyelectrolyte Magnafloc LT20 using a dosage of 1/10th of the coagulant dosage. The results are presented in Figure 6.

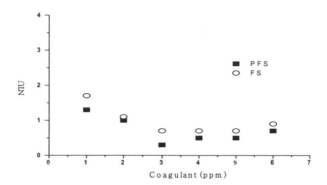

Figure 6:   Turbidity removal of a kaolinite water suspension (5 mg/l) against the two different coagulants: PFS and FS.

Figure 6 shows that the PFS exhibits a better performance in removing turbidity, than the FS. Although, a similar number of flocs were formed during the flocculation period of the jar test, the flocs of PFS were bigger in size and inevitably easier to settle. This is due to the fact that the polymeric species of PFS have the ability to form larger and therefore heavier flocs, which are easier to settle.

4   Conclusions

The use of a weak base (NaHCO$_3$) as a reducing agent results in the increase of polymerisation degree of PFS, of the ratio r, as well as in improving the stability of the product. Furthermore, the coagulating performance of prepared coagulants is enhanced.

High temperature (90°C) in the oxidation stage and intermediate temperature (50°C) in the hydrolysis and polymerisation stage results in increasing the degree of polymerisation along with its coagulating efficiency.

The presence of polymeric species is directly related to the coagulating efficiency of the PFS. In particular, the higher the polymerisation percentage the better the performance in removing the turbidity from kaolinite water suspensions.

PFS exhibits a higher coagulating efficiency compared with a conventional coagulant, e.g. Fe$_2$(SO$_4$)$_3$ under the specific experimental conditions.

Acknowledgments

Thanks are due to the Greek Ministry of Education for funding this research through the Pythagoras II program.

References

[1]   Chang Q., Wang H., "Preparation of PFS coagulant by sectionalised reactor, Journal of Environmental Sciences, 14(3),345-350, 2002
[2]   Jiang J-Q., Graham N.J.D., "Coagulation of upland coloured water with polyferric sulphate compared to conventional coagulants", *J. Water SRT – Aqua*, **45**(3), 143-154, 1996
[3]   Jiang J-Q. and Graham N.J.D.,"Observations of the comparative hydrolysis/precipitation behavior of polyferric sulphate and ferric sulphate", *Water Res.*, **32** (3), 930-935, 1998
[4]   Butler A.D., Fan M., Brown R.C., Cooper A.T., van Leeuwen J.H., Sung S., " Absorption of dilute SO2 gas stream with conversion to polymeric ferric sulphate for use in water treatment", Chemical Engineering Journal, **98**, 265-273, 2004
[5]   Li F., Ji G., Gi X., " The preparation of inorganic coagulant – Polyferric sulphate", *J. Chem. Tech. Biotechnol."*, **68**, 219-221, 1997.
[6]   Hendrich S., Fan M., Sung S., Brown R.C., Semakaleng R., Myers G., "Toxicity evaluation of polymeric ferric sulphate", *Int J. Environ. Technol. Managem.*, **1**, 464-471, 2001
[7]   Jiang J-Q., Graham N.J.D., "Preparation and Characterisation of an optimal polyferric sulphate (PFS) as a coagulant for water treatment.", *J.Chem.Technol.Biotechnol.*, **73**, 351-358, 1998
[8]   Fan M., Sung S., Brown R.C., Wheelock T.D. and Laabs F.C., "Synthesis, characterization and coagulation of polymeric ferric sulphate", *J. Env. Engineering*, **128** (6), 483-490, 2002
[9]   Cheng W.P., "Comparison of hydrolysis/coagulation behavior of polymeric and monomeric iron coagulants in humic acid solution", *Chemosphere*, **47**, 963-969, 2002
[10]   Clesceri L., Greenberg A. and Trussell R., "Standard methods for the examination of water and wastewater", 17th Ed., APHA-AWWA-WEF, Washington DC, 1989.
[11]   Hopkins D. and Ducoste J., "Characterising flocculation under heterogeneous turbulence", *J. Coll. Interf. Sci.*, **264**, 184-194, 2003
[12]   Cheng W.P., "Hydrolysis characteristics of polyferric sulfate coagulant and its optimal condition of preparation", *Colloids and Surfaces*, **182**, 57-63, 2001.

Performance prediction of a constructed wetland wastewater treatment plant

V. Tomenko[1], S. Ahmed[1, 2] & V. Popov[1]
[1]*Wessex Institute of Technology, Southampton, UK*
[2]*Department of Civil Engineering, Jamia Millia Islamia*
(Central University), New Delhi India

Abstract

Artificial neural network (ANN) models were developed to predict the performance of a constructed wetland wastewater treatment plant (CWWTP). The model assesses the Biochemical Oxygen Demand (BOD) concentration at outlet of a treatment plant. Training of ANN models was based on experimental results of a pilot plant study in India. The data used in this work were obtained under various hydraulic and BOD loading. Regular records of BOD were made at inlet, and outlet levels through various stages of the treatment process for over 18 months. The ANN-based models were found to provide an efficient and a robust tool in predicting CWWTP performance.
Keywords: neural networks, constructed wetland, model studies, prediction, optimization, biochemical oxygen demand.

1   Introduction

The proper operation and management of constructed wetland wastewater treatment plants (CWWTP) is receiving attention because of the rising concern about environmental issues and growing importance of sustainable and natural wastewater treatment techniques. Improper design and operation of a CWWTP may cause serious environmental and public health implications, as its effluent may contaminate receiving water body, causing severe aquatic pollution and spread various water born diseases. For proper design and assessment of quality of non-conventional wastewater treatment and thereafter to conserve the receiving water bodies, reliable prediction of effluent from Constructed Wetland (CW) is essential. A better control can be achieved by developing a mathematical tool for predicting the plant performance based on past observations of certain key parameters. However, modeling a CWWT is a difficult task due to the complexity of the treatment processes. The complex physical, biological and chemical processes involved in constructed wetland treatment process exhibit non-linear behaviors, which are difficult to describe by linear mathematical models. This paper presents predictive models based on the concept of neural networks (NN).

Artificial intelligence concepts have been successfully used in a range of engineering, environmental, and financial problems [1, 2]. The NN-based model that was applied to a pilot plant study of CW in India have performed consistently well in the face of varying accuracy and size of input data. Using these models, the planner and decision maker can easily make assessment of the expected plant effluent.

2 Constructed wetland pilot plant and experimental set up

Wetlands are considered as low-cost alternatives for treating municipal, industrial, and agricultural effluents. Constructed wetlands (CW) are preferred because of low maintenance, shock loading absorbance capacity and less energy consumption [3]. They may be classified as surface flow marshes, vegetated subsurface flow beds, submerged aquatic beds, and floating leaved aquatics [4] This new developing technology may offer a low cost and low maintenance alternative for treatment of domestic wastewater which is especially suitable for developing countries [5, 6].

The scheme of the units designed in New Delhi is presented in Fig. 1.

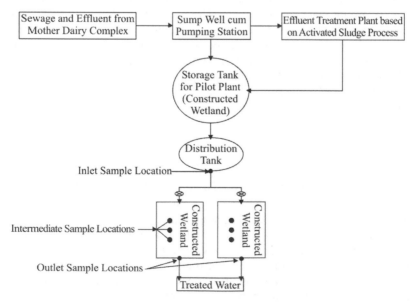

Figure 1:   Pilot plant scheme.

Construction was completed in August 2001. The initial sampling started after eight months i.e. April –2002 giving sufficient time for full growth of reeds since wetlands typically require a few months for vegetation and bio-film establishment according to [7]. Sizing of the wetland units was done by using basic relationship for plug flow reactor for the BOD loading as recommended by [8]:

$$A = \frac{Q_d \left(\ln C_0 - \ln Ct \right)}{K} \qquad (1)$$

where A is the plane area of the wetland unit, (m^2), K is the specific removal rate constant for given constituent at 200C, (d^{-1}), Q_d is the average flow rate of wastewater (m^3 d^{-1}), C_0 is the average BOD$_5$ of the influent (mg/l) and C_t is the average BOD$_5$ of the effluent (mg/l). Each unit has length = 6.6 m, width = 5.3 m. and depth = 0.6m.

Two parallel Subsurface Horizontal Flow Constructed wet land units A and B with same filler material were used in this study. The inlet zone (first 100 cm) of unit B has coarse sand filler matrix. Both were planted with Phragmites australis. Different hydraulic loading starting from 34 l/m^2/d to 200l/m^2/d were used in the experiments. The BOD loading of the pilot plant was varied from 45 mg/l to 1580 mg/l, so that it can be tested for wide range of conditions. The flow meters were provided before the inlet zone to record the average flow rate and total quantity of waste-water feeded in each bed. All precautions were taken for equal distribution of wastewater in each of the CW beds. Five different triplicate samples including three intermediate were collected for set of parameters. Intermediate samples were collected from different ports at downstream distance of 3.15 m, 4.15 m and 5.15 m respectively. Acrylic sheets of 45 cm height were embedded up to 30 cm in the soil filter at inlet zone to avoid any over flow condition in CWs.

Wastewater influent and effluent from the CW units were monitored and recorded after giving at least 15 days of acclimatization period to the CW. After fifteenth day, five daily samples were taken for given hydraulic and organic loading and mean values were obtained. In situ measurements for temperature and pH were also recorded at the influent, effluent and intermediate sampling points. The wastewater samples were not collected simultaneously but after giving due consideration of lack time for different hydraulic loading.

Samples were analyzed for Chemical Oxygen Demand (COD), Biochemical Oxygen Demand (BOD), Phosphate (PO$_4$) Faceal Coliform (FC) and Total Coliform (TC). All these analyses were conducted in accordance with [9]. In this study only BOD values were considered for development of ANN model.

3 Neural network modelling

During several last decades neural networks have been successfully applied for prediction and pattern recognition tasks. Their potential is especially high for the systems with complex non-linear behavior, when the laws governing the system are not known or poorly understood. Artificial neural networks are superior to other computational formalisms in the case of noisy and ill-defined data, massively parallel computation, collective effects and signal dynamics [10].

There is a wide variety of wastewater treatment problems, where neural networks have been applied successfully, including modelling of the physicochemical water treatment process [11], prediction of wastewater treatment plant performance [12], photocatalytic treatment of waste waters [8] etc.

The performance of the network strongly depends on the choice of process variables. The data available is also of great importance, especially in terms of representing the domain of experiment. Another important point is data preprocessing. Usually, the range of model variables varies and, therefore, some data preprocessing technique should be utilized.

The most widely applied NN for modeling of steady-state systems are of feed-forward type. In such networks, the signal propagates in one direction via weighted connections between neurons of different layers (Fig. 2). The network model includes three layers of neurons: input, hidden and output layer. The neurons of input layer do not perform any calculations; they just store the input variables and propagate them to neurons of hidden layer. Each neuron of hidden and output layers (Fig. 3) calculates the weighted sum of inputs plus bias as follows:

$$u_i = \sum_{j=1}^{N} w_{i,j} x_j + \theta_i \qquad (2)$$

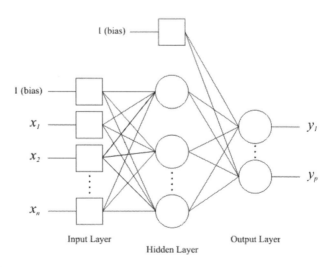

Figure 2:   Structure of feed-forward artificial neural network.

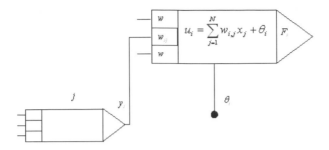

Figure 3: Scheme of hidden and output layer neurons.

The output of a neuron is generated by its activation function. The typical choice is sigmoidal function (equation (3)) or hyperbolic tangent (equation. (4)):

$$y_i = 1/\left(1 + e^{-\beta u_i}\right) \tag{3}$$

$$y_i = \left(e^{\beta u_i} - e^{-\beta u_i}\right)/\left(e^{\beta u_i} + e^{-\beta u_i}\right) \tag{4}$$

where β is the slope of a function, u_i is the weighted sum of inputs to the neuron. The outputs of hidden layer neurons are propagated as inputs to neurons of output layer. The network output, or more generally a series of outputs, represents the response calculated by the network given input.

Neural networks are trained by changing the weights in order to reach the convergence between response values y_i and experimental responses d_i. There are several algorithms, rigorously proven and heuristic, which can be adopted for training of feed-forward networks [13]. In the present study, back-propagation algorithm, which is the generalization of steepest descent method, was chosen.

For a given data set, back-propagation algorithm may proceed in two modes: sequential mode and batch mode. The sequential mode is also known as on-line or stochastic mode. In this mode weight adaptation is performed after the presentation of each input-output training pair. The cost function, which value should be minimized, is calculated for each training pair as follows:

$$E = \frac{1}{2}\sum_{i=1}^{p} e_i^{\;2} \tag{5}$$

where $e_i = \left(y_i - d_i\right)$ is the error between actual and desired outputs respectively. In the batch mode, which is utilized in this study, weights are updated after the presentation of all training pairs that constitute an epoch. For a particular epoch, the cost function is defined as the average squared error of equation (5), calculated for all training pairs:

$$E = \frac{1}{2N}\sum_{n=1}^{N}\sum_{i=1}^{p} e_i^{\;2} \tag{6}$$

The objective of the learning process is to adjust the free parameters (weights, biases) of the network to minimize E. In the process of learning, the adjustments of the weights are calculated for each training pair according to delta rule [14]:

$$\Delta w_{i,j}^z = -\eta \frac{\partial E}{\partial w_{i,j}} \tag{7}$$

where η is the learning-rate parameter, E is calculated by equation (5). The arithmetic mean of individual adjustments over the learning set gives estimate of the change that would minimize the cost function E and the weights are updated as follows:

$$w_{i,j}^z(t+1) = w_{i,j}^z(t) + \Delta w_{i,j}^z \tag{8}$$

For the neurons of output layer, the adjustments of weights are calculated in a straightforward manner:

$$\Delta w_{i,j}^z = -\eta \frac{\partial E}{\partial w_{i,j}^z} = -\eta \frac{\partial E}{\partial e_i} \frac{\partial e_i}{\partial y_i} \frac{\partial y_i}{\partial u_i} \frac{\partial u_i}{\partial w_{i,j}^z} = -\eta e_i y_i'(u_i) y_j^{z-1} = -\eta e_i y_i'(u_i) y_j^{z-1} \tag{9}$$

If neuron j is located in the hidden layer z of the network, there is no specified desired response for that neuron. Therefore, the error for such neuron is determined with the help of error signals of neurons of output layer. Taking into account:

$$\frac{\partial E}{\partial y_j^z} = \sum_{i=1}^{p} \frac{\partial E_i}{\partial x_i^{z+1}} = \sum_{i=1}^{p} \frac{\partial E_i}{\partial y_i^{z+1}} \frac{\partial y_i^{z+1}}{\partial u_i^{z+1}} \frac{\partial u_i^{z+1}}{\partial x_i^{z+1}} = \sum_{i=1}^{p} (y_i^{z+1} - d_i) y_i'^{z+1} (u_i^{z+1}) w_{i,j}^{z+1} \tag{10}$$

Adjustments of weights of neurons of hidden layer can be calculated in the following way:

$$\Delta w_{j,i}^z = -\eta \frac{\partial E}{\partial w_{j,i}^z} = -\eta \frac{\partial E}{\partial y_j^z} \frac{\partial y_j^z}{\partial u_j^z} \frac{\partial u_j^z}{\partial w_{i,j}^z} = y_j'^z(u_j^z) x_i^z \left[\sum_{i=1}^{p} (y_i^{z+1} - d_i) y_i'^{z+1} (u_i^{z+1}) w_{i,j}^{z+1} \right] \tag{11}$$

During the learning process, not only weights and biases are varied, but also the number of neurons of hidden layer, learning-rate parameter (η) and even the structure of the network itself (some connections can be eliminated). Generally, the number of neurons of hidden layer should be reduced to prevent network from over-fitting. On the other hand, elimination of too many neurons leads to reducing the performance of the network. Moreover, convergence of the solution is extremely sensitive to the adequate choice of η [13].

In general, back-propagation algorithm cannot be shown to converge. Therefore, the typical stopping criteria are the number of iterations or the sufficiently small gradient values [13].

4 Experimental results

NN model adopted contains three layers (input, hidden and output layers) as illustrated in Fig. 2. The aim of the learning procedure was to predict BOD_{out} as a function of input variables. Six input variables were chosen:

1) BOD_{in}, influent contamination (mg/l);
2) HL :, hydraulic Loading (l/m²/d);
3) OL :, organic loading (g/ m²/d);
4) t : temperature (⁰C);
5) v : velocity of the stream (m/s);
6) d : distance from the CWWTP influent (m).

The output variable (response) is the level of BOD_{out} after treatment. Therefore, the output layer of the neural network includes one neuron.

The distribution of the data pairs $\left(\left(BOD_{in}, HL, OL, t, v, d \right), \left(BOD_{out} \right) \right)$ into the learning set (LS) and test set (TS) was based on the experimental design and dataset properties. The LS and TS are comprised of 78 and 17 input-output pairs, respectively.

The convergence criterion utilized was the cost function, defined by equation (6). Since the determination of the adequate number of neurons (m) in hidden layer is important to prevent network from over-fitting, a number of learning procedures were performed in order to find the optimal combination of m and number of learning procedure epochs (NE). The number of neurons in the hidden layer varied between 6 and 14 and the number of epochs ranged from 500 to 1500. The values of cost function decreased as m increased from 6 to 12 for both data sets (LS and TS). For $m > 12$ E calculated for TS increased (the network became over-fitted). According to the results, m was chosen to be 12 and $NE = 800$. When the cost function reached its minimum value, the weights of the network were fixed.

Fig. 4 shows a comparison of calculated and experimental values of BOD_{out} for both sets. The agreement between actual and predicted values is adequate for both LS and TS ($R^2 = 0,997$ for LS and $R^2 = 0,995$ for TS).

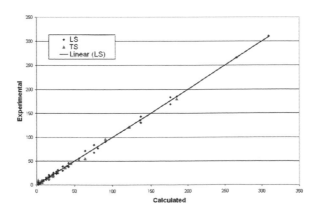

Figure 4:   Comparison of calculated and experimental values of BOD_{out}.

Given fixed values of HL, OL, t, v, d, Figs. 5, 6 and 7 respectively show BOD_{out} as a function of distance from the influent of CW for three initial values of BOD_{in} ($BOD_{in} = 45$, $BOD_{in} = 156$, $BOD_{in} = 784$). The approximation is accurate; therefore, the neural network model is able to describe adequately the process kinetics under different conditions (initial values of input variables).

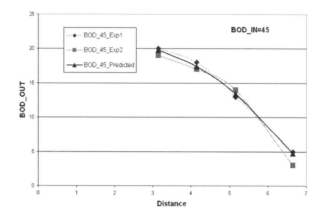

Figure 5: BOD_{out} for $BOD_{in} = 45$.

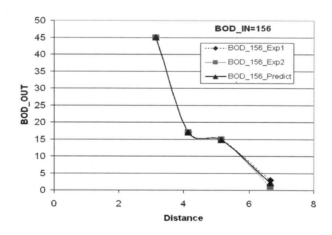

Figure 6: BOD_{out} for $BOD_{in} = 156$.

5 Conclusions

The neural network was adopted for prediction of constructed wetland contamination treatment performance for given hydraulic and BOD loading. The

NN model adequately describes the behavior of complex CWWTP within the range of different experimental conditions. Thus, if experimental data for a given system are available and cover the whole domain of interest, a simplified mathematical model of the wastewater treatment process may be obtained by utilizing feed-forward neural networks. Simulations based on the neural networks can then be performed to estimate the behavior of the system under different conditions.

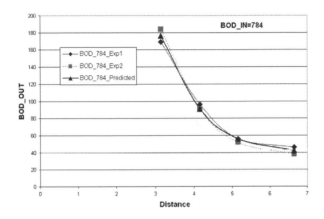

Figure 7: BOD_{out} for $BOD_{in} = 784$.

This new developing technology may offer a low cost and maintenance to domestic wastewater treatment, which is especially suitable for developing countries.

References

[1]   Jean P. Hybrid Fuzzy Neural Network for Diagnosis. Application of Anaerobic Treatment of Wine Distillery Waste Water Treatment in Fluidized Bed Reactor. *Water Science Technology*, Vol. 36 No 6-7 pp. 209-217, 1997.

[2]   Mohammed F. Integrated Waste Water treatment Plant Performance Evaluation Using Artificial Neural Network. *Water Science Technology*, Vol. 40 No 7 pp. 55-65, 1999

[3]   Vymazal J. The use of sub-surface constructed wetlands for wastewater treatment in the Czech Republic: 10 years experience. *Ecological Engineering* 18, pp. 633–646, 2002

[4]   Reed SC, Middle brooks EJ, Crites RW. *Natural systems for waste management and treatment.* New York, NY: McGraw-Hill, 1988.

[5]   SÇ Ayaz and I Akca. Treatment of wastewater by constructed wetland in small settlements. *Water Science and Technology* Vol. 41 No 1 pp 69–72, 2000.

[6]   Ahmed Sirajuddin, Application of Root Zone Treatment System for Dairy Wastewater. *Proceedings of the Mother Dairy International Conference on Constructed Wetlands for Waste Water Treatment in Tropical And Subtropical Regions.* Anna University, Channai, December 11-13 2002.

[7]   Billore SK, Singh N, Sharma JK, Dass P, Nelson RM. Horizontalsubsurface flow gravel bed constructed wetland with Phragmites karka in central India. *Water Sci Technol* 1999;40(3):163–71.

[8]   Wood A. Constructed wetlands for wastewater treatment— engineering and design considerations. *Proceedings of the International Conference on the use of Constructed Wetlands in Water Pollution Control.* Cambridge, UK, 24–28 September 1990.

[9]   APHA. *Standard methods for the examination of water and wastewater, 19th ed.* Washington, DC, USA: American Public Health Association/American Water Works Association Water Environment Federation; 1995

[10]   Kohonen T. *Self-Organizing Maps.* Berlin: Springer, 1995.

[11]   Gob S., Oliveros E., Bossmann S. H., Braun A. M., Guardani R., Nascimento C.A.O. Modeling the kinetics of a photochemical water treatment process by means of artificial neural networks. *Chemical Engineering and Processing* 38, 373–382, 1999.

[12]   Hamed M. M., Khalafallah M. G., Hassanien E. A. Prediction of wastewater treatment plant performance using artificial neural networks. *Environmental Modelling & Software* 19, 919–928, 2004

[13]   S. Haykin, *Neural Networks: a Comprehensive Foundation,* 2nd Ed. Upper Saddle River, NJ: Prentice Hall, 1999.

[14]   Widrow B., Hoff M.E. Adaptive switching circuits. *In 1960 IRE WESCON Convention Record*, pp. 96-104, 1960.

Economical and social aspects: evaluation of incorporation potential in municipal wastewater reuse procedure in Greece

S. Bakopoulou & A. Kungolos
Department of Planning and Regional Development,
University of Thessaly, Volos, Greece

Abstract

It is the objective of this study to evaluate the incorporation potential of economic and social factors in municipal wastewater reuse procedure, aiming at determining whether it is worthwhile for a municipality to construct the necessary advanced treatment systems or not. Advanced wastewater treatment is the additional treatment needed to remove any constituents remaining after conventional secondary treatment and it is considered necessary when there is plan for treated wastewater reuse. Economic and social factors, as well as the necessary environmental factors, could be used as the main parameters, which will serve in the development of an evaluation model. This model could be based in a cost-benefit analysis which, through a correlation procedure, can lead to an optimal solution adoption. This model could be useful for many municipalities in Greece, where wastewater reuse does not consist presently a common method of treated wastewater management.

Keywords: wastewater reuse, advanced wastewater treatment, irrigation.

1 Introduction

Water scarcity and deterioration in the quality of water resources in many countries have led to the recognition that water shortage and water pollution control should be solved by a careful water resource management that incorporates advanced technologies. Desalination of seawater as well as reclamation and reuse of municipal wastewater are the main strategies that have been proposed for investigation and application in many countries all over the world [1]. Treated wastewater could be reused for specific purposes, such as irrigation purposes, industrial uses as well as aquifer recharge purposes. However, the most common reuse application includes agricultural irrigation. On this occasion, wastewater can serve as a source of both water and nutrients, thus reducing fertilization costs. On the other hand, landscape irrigation and other non-potable urban uses are gaining interest in recent years because of modern technology developed specifically for such uses [2].

In Mediterranean area a lot of drainage periods have taken place in recent years, especially in countries like Greece and Portugal. In these countries, the main problem may not be scarcity of water in terms of average per capita, but the high cost of making water available at the right place, at the right time with the required quality [3]. This problem relates mainly to the fact that many municipalities do not show interest in developing advanced treatment systems, which will make secondary effluents suitable for irrigation, because of the economical cost. So, in these countries, more than anywhere else, an integrated approach for water resources management including wastewater reclamation and reuse is required.

Taking in mind the above information, this paper aims at presenting the main problems related to wastewater reuse practices in Greece. These data can assist in developing an evaluation model which could be used as a basis for determining whether it is worthwhile for a municipality to construct advanced treatment systems for making treated effluents suitable for irrigation or not. In Greece, there have been quite some studies regarding issues such as proposals for national guidelines development, but no study about economical evaluation of wastewater reuse procedure.

2 International current tendencies

In the Mediterranean basin, Israel has been a pioneer in the development of wastewater reuse practices. It is estimated that by the year 2040, treated sewage effluent will become the main source of water for irrigation in Israel and the Palestinian autonomous regions, supplying 1000 million m^3 (70%) out of the 1400 million m^3, that will be used for irrigation [4]. The Israel "example" was also followed by Cyprus, Jordan, Tunisia, Italy, and France. In these countries, full fledged national regulations set the basic conditions for a safe reuse of wastewater. In rest Mediterranean countries, lack of national regulations has led to the existence of relatively few efforts regarding developing of new projects in wastewater reuse sector. Spain is an exception, since the recent development of regional regulations has had a positive effect in wastewater reuse methods used in this country.

Other countries which have developed national or regional regulations regarding municipal wastewater reuse include USA, Japan and South Africa. USA and more specifically regions like California, Florida and Arizona are pioneers in using extremely modern technology of sewage wastewater recycling. Canada and China have supported in recent years a significant number of researches into treatment and reuse of domestic greywater for non-potable uses [5, 6].

In 1989 the World Health Organisation set up a research for determining guidelines and regulations for wastewater reuse. These guidelines and regulations were based mainly in findings of epidemiological researches, so the majority of them were referred only to limits like pathogenic microorganisms maximum number. Later, in 1992 the US Food and Agricultural Organism reviewed the previous limits and developed new limits-guidelines regarding chemical, physical and microbiological parameters [7, 8].

3   Present situation in Greece

In Greece, the distribution and availability of water sources deflect from statistically average values. The most significant problems are noticed in Aegean islands as well as in most of the eastern Greek regions. In these areas rainfalls are relatively few if we compare them with those noticed in western regions of Greece. Nevertheless, reuse of municipal wastewater has not been a common management practice in Greece in reverse with direct disposal in rivers, lakes or sea. Only a few hotel enterprises in coastal areas have used their secondary effluents for landscape irrigation purposes in recent years. As a conclusion, a lot of water quantities are not utilized while a great quantity of potable water is used in non-potable uses like landscape irrigation.

In Greece no guidelines or criteria for wastewater reclamation and reuse have yet been adopted. Secondary effluent quality criteria are used for discharging purposes (No E1b/221/65 Health Arrangement Action) and are independent of the disposal, reclamation and reuse effort. In 2000 the watering and sewage municipal enterprise of Larissa set up a study regarding determining of guidelines for wastewater reuse. The main intention of this study was the initiative estimation of qualitative criteria that should characterize the treated effluents [9]. Later on and more specifically in 2003 a research program regarding wastewater reuse took place under LIFE European Community Initiative in Thessaloniki area. The aims of this research project related to specific guidelines and regulations determination for wastewater reuse, as well as pilot wastewater use for agricultural irrigation and aquifer recharge [2, 10, 11]. The criteria and guidelines allocated in this project were rather specific and detailed compared to older studies.

The above studies set the basis for the legal solution of the problem by enabling the state to incorporate the proposed criteria in national regulations. Furthermore, a lot of studies regarding effects of wastewater irrigation in specific crops have taken place in recent years [12, 13, 14]. However, a crucial point, which should be paid attention to, represents the socio-financial problem regarding wastewater reuse. Many municipalities are not keen on constructing advanced wastewater treatment systems which will make the secondary effluent suitable for reclamation purposes. On the other hand, many Greek farmers are also negative in using such a water source. In an effort to solve these problems, we propose the development of an evaluation model which should incorporate the relative financial, social and environmental aspects regarding wastewater reuse and intend at determining if it is worthwhile for a municipality to construct advanced treatment systems or not.

4 Description of the evaluation model structure

The evaluation of advanced treatment systems viability could be done by running a model which will incorporate the relative economic, social and environmental aspects regarding wastewater reuse. The economic aspects can be expressed in financial terms directly, while the social and environmental aspects could be expressed in financial terms indirectly using specific techniques. The above model could depend on a cost-benefit analysis which leads to the optimal solution selection.

A cost-benefit analysis (CBA) estimates and totals up the equivalent money value of the benefits and costs of projects so as to determine whether they are worthwhile. In order to reach a conclusion, all aspects of the project, positive and negative, must be expressed in terms of a common unit. The most convenient common unit is money. This means that all benefits and costs of a project should be measured in terms of their equivalent money value. Furthermore, all benefits and costs should be expressed in terms of money value of a particular time. This is due to the differences in the money value at different times because of inflation.

Expression of environmental aspects in financial terms could be done by using specific techniques described in scientific literature like revealed preference techniques and stated preference techniques [15, 16]. In general, the financial rating of environmental aspects is based on public preference for a specific environmental quality level.

Social aspects could be evaluated by using appropriate questionnaires which will be allocated in farmers of the target area. This method aims at finding out whether the farmers of a specific region are eager to use treated wastewater for irrigation purposes or not. Another important parameter which will be of great importance is the money value that the farmers wish to pay. These data could be used as a benefit in our analysis. On the other hand, if the farmers do not present any intention to buy such a water resource, then we will have a cost in our analysis. All the essential information regarding costs or benefits will arise after the questionnaires have been statistically analysed. During this procedure, all the farmers will be well-informed for the real advantages and disadvantages of using wastewater as a water resource.

Another important benefit in our analysis relates to landscape irrigation and other non - potable urban uses. The use of potable water for irrigation of such landscapes (parks, streets, cemeteries) is a common practice in Greece. If these landscapes were irrigated by treated wastewater, then the potable water would be used in other beneficial uses. The example of Larissa municipality is representative of such an occasion. Larissa is a city located in the central-eastern part of Greece. Its population in 2001 reached close to 130000 inhabitants. Landscapes such as parks and squares cover a great surface of Larissa municipality area. Larissa wastewater treatment plant receives wastewater reaching up to 20000 m³ per day and 7200000 m³ per year. Larissa municipal enterprise of watering consumes every year about 500000 m³ for landscape irrigation. If this quantity were derived from recycled wastewater, then the 500000 m³ of potable water would be used in other uses such as drinking. The benefits arising from such a case are significant.

The main structure of our evaluation model is being described in Figure 1.

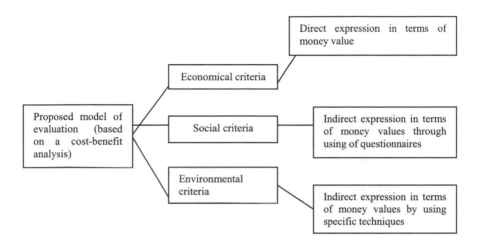

Figure 1:   Main structure of proposed evaluation model.

5   Conclusions

In this study, the current situation regarding reuse of municipal wastewater for irrigation purposes in Greece was examined. Our main conclusion relates to the fact that there has been an increasing interest in the above sector of waste management in recent years. Projects that are being carried out in this sector can contribute positively in the general strategy of sustainable management of water resources and protection of natural resources from essential pollution. By evaluating economical and social aspects in relation with the necessary environmental ones, one can solve significant problems that have been raised during last years and were well described above.

In conclusion we could say that informing and educating local authorities, workers in wastewater treatment plants and farmers for the real use of wastewater recycling is necessary. The media should also contribute in this effort by informing the local society for how important such water management could be for the protection of natural resources and the environment.

References

[1]   Brenner A., Shandalov S., Messalem R., Yakirevich A., Oron G., Rebhun M., (2000) "Wastewater reclamation for agricultural reuse in Israel: Trends and experimental results", *Water Air Soil Poll.*, 123: 167-182.

[2]   Andreadakis A., Gavalaki E., Mamais D., Noutsopoulos K., Tzimas A., (2003) "Proposal for national guidelines development regarding municipal wastewater reuse in Greece", in proceedings of the conference *Reclamation and Reuse of Wastewater*, Thessaloniki, 19-75. (in Greek)

[3]   Angelakis A.N., Marecos do Monte M.H.F., Bontoux L., Asano T., (1999) "The status of wastewater reuse practice in the Mediterranean basin: Need for guidelines", *Water Research*, 33 (10):2201-2217.

[4]   Haruvy N., Offer R., Hadas A., Ravina I., (1999) "Wastewater irrigation - Economic concerns regarding beneficiary and hazardous effects of nutrients", *Water Resources Management*, 13: 303–314.

[5]   Exall K., (2004) "A review of water reuse and recycling, with reference to Canadian practice and potential: Applications", *Water Qual. Res. J.*, 39(1): 13-28.

[6]   Chu J., Chen J., Wang C., Fu P., (2004) "Wastewater reuse potential analysis: implications for China's water resources management", *Water Research*, 38: 2746–2756.

[7]   WHO (1989) *Health guidelines for the use of wastewater in agriculture and aquaculture*, Report of a WHO Scientific Group, WHO technical report series 778, Geneva.

[8]   FAO (1992) *Wastewater treatment and use in agriculture*, M.B. Pescod, Irrigation and Drainage Paper 47, Rome.

[9]   Angelakis A., Tsagarakis K., Kotselidou O., Vardakou E. (2000) *A need for national guidelines development regarding reclamation and reuse of municipal wastewater in Greece: Preliminary estimation*, Union of Greek Municipal Watering and Sewage Enterprises, Larissa. (in Greek)

[10]   Tsiridis V., Petala M., Kungolos A., Samaras P., Sakellaropoulos G.P. (2003) "Reclamation of municipal wastewater using advanced treatment technology", in proceedings of the conference *Reclamation and Reuse of Wastewater*, Thessaloniki, 77-99. (in Greek)

[11]   Georgiadou M., Kakani M., Loubari E., Meladiotis I., Baliaka B., Moutsopoulos K., Naskos N., (2003) "Monitoring and recharge of aquifer with treated wastewater", in proceedings of the conference *Reclamation and Reuse of Wastewater*, Thessaloniki, 101-116. (in Greek)

[12]   Vakalis P. and Tsantilas Ch. (2002) "A study regarding effects of irrigation with treated wastewater in cotton and corn cultivations", *Agricultural Research*, 25(1): 13-20. (in Greek)

[13]   Tsantilas Ch. and Samaras V. (1996) "Use of treated wastewater coming from Larissa treatment plant for agricultural irrigation and fertilization", in proceedings of 2nd national conference *Land Reclamation Works, Water Resources Management, and Agricultural Engineering*, Larissa, 549-557. (in Greek)

[14] Paranychianakis N.V., Chartzoulakis K.S., Aggelides S. Amgelakis A.N., (2002) "Grapevine growth and nutrition as affected by irrigation with recycled water", in proceedings of regional symposium on *Water Recycling in Mediterranean Region*, Iraklio, Greece, 457-464.

[15] Bithas K. (2003) *An economic consideration of environmental protection procedure*, Tipothito Publications, Athens. (in Greek)

[16] Fletcher J., Adamowicz W., Graham T., (1990) "The travel cost model of recreation demand", *Leisure Sciences*, 12: 119-147.

Effect of ferrous iron on the settling properties of granular sludge in a UASB reactor

A. Vlyssides, E. M. Barampouti, S. Mai & A. Moutsatsou
National Technical University of Athens,
School of Chemical Engineering, Athens, Greece

Abstract

The effect of ferrous ion addition on the settling properties of the granules in an upflow anaerobic sludge blanket (UASB) reactor was investigated. A UASB reactor (35°C; pH=7) was operated for 3 months at a 20-h hydraulic retention time (HRT) at organic load from 2,0 to 20,0 gCOD·L^{-1}·d^{-1}. Ferrous iron as FeSO$_4$·7H$_2$O was fed to the reactor in a range of load from 0,03 to 0,28 g Fe^{+2}·L^{-1}·d^{-1}. After steady state conditions were achieved, the settling properties of granular sludge from the anaerobic sludge bed reactor were determined by using a novel upflow velocity test. From the results, it was concluded that iron promoted granulation. The U10% (10% of sludge washed out by applying upflow velocity U10%) was increased from 0,23 m.h^{-1} to 13,21 m.h^{-1}, while U30% from 0,87 m.h^{-1} to 29,11 m.h^{-1} and U60% from 6,63 m.h^{-1} to 49,82 m.h^{-1}.
Keywords: UASB, granulation, ferrous iron, settling properties.

1 Introduction

The concept of the upflow anaerobic sludge blanket (UASB) reactor was developed in 1970s. Today the UASB reactor has become the most popular high-rate reactor for anaerobic treatment of wastewater throughout the world [1, 2]. The UASB reactor has been used increasingly in recent years to treat a variety of industrial waste and municipal waste. The treatment capacity of UASB reactors depends on the amount of active biomass retained, as well as the contact between biomass and wastewater. The biosolids inside the UASB reactor can be either in granular or flocculent form [3]. Retention of an adequate level of methanogenic bacteria in UASB reactors gives good performance in terms of chemical oxygen demand (COD) removal and methane yield [1, 4].

WIT Transactions on Ecology and the Environment, Vol 92, © 2006 WIT Press
www.witpress.com, ISSN 1743-3541 (on-line)
doi:10.2495/WM060181

Granulation is the process in which suspended biomass agglutinates to form discrete well-defined granules. Microbial granulation is a complex process, involving different trophic bacterial groups, and their physico-chemical and microbiological interactions [5]. Many factors contribute, in one form or another, to the granulation process [6, 7]. Granulation may be initiated by bacterial adsorption and adhesion to inert matters, to inorganic precipitates [2, 8], and/or to each other through physico-chemical interactions and syntrophic associations [9]. These substances serve as initial precursors (carriers or nuclei) for further bacterial growth. These initial granules will grow continuously into compact mature granules, if favourable conditions pertaining to bacteria are maintained [10].

The granular form of the sludge offers various engineering advantages over the flocculent form, such as, the high solid retention time due to its excellent settling property, providing maximum microorganisms to space ratio, and application of higher loading rates as compared to UASB reactor with flocculent sludge. Hence, characteristics of the sludge developed are of vital importance for maximizing advantages of this reactor, and affecting the process economy [3].

Wastewater type rather than the loading rates and reactor design is reported to have strong influence on the bulk characteristics of sludge [11]. However, Liu and Tay [12] have stated that, the hydrodynamic conditions inside the reactor play crucial role in the formation of anaerobic granules. Applied loading rates are responsible for defining hydrodynamic conditions in the reactor. Based on the studies reported it could be said that, the characteristics of the sludge inside the reactor depend upon the operating conditions during primary start-up and granulation, apart from the characteristics of wastewater and inoculum used. For the same wastewater, the sludge characteristics differ with applied loading rates [3].

It has been shown that divalent metal ions, such as Ca^{2+} and Fe^{2+}, enhance the granulation [8]. Divalent ions were reported to play an important role in microbial aggregation. Elemental analysis of granules pointed to the importance of inorganic precipitates, such as calcium phosphate, while other pointed to the importance of iron in non-chelated form or in chelated form [13]. It was found that extracellular polymers prefer to bind to multi-valent metals due to the formation of stable complexes [14].

The aim of this work was to investigate the influence of ferrous iron on the settling properties of UASB granular sludge under increasing loading rates.

2 Materials and methods

2.1 Composition of feed

Synthetic milk wastewater used in this study was prepared by diluting fresh milk. The COD of the wastewater ranged from $2000 - 20000$ mg·L^{-1}. The total nitrogen and total phosphorous of the wastewater ranged from $100 - 1000$ mg·L^{-1} and $16 - 160$ mg·L^{-1} respectively. pH varied between 6,5-7,0.

2.2 Reactor operating conditions

A 20L UASB reactor was operated at 35°C. Hydraulic retention time was 20 h. The experimental schedule as far as COD load and ferrous ion load is concerned is shown in Figure 1. As it is obvious, the ratio of influent COD/ Fe^{2+} was maintained stable and equal to 0,014 $g \cdot g^{-1}$, which is found to be the optimum ratio (unpublished work).

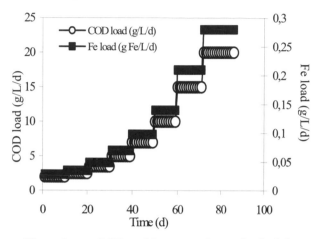

Figure 1:   COD and iron experimental schedule.

During the reactor operation, the COD conversion, gas production and pH of the effluent were measured daily. For each loading rate, the performance data mentioned above were collected until steady state conditions were obtained. The steady state condition in the biological system implied that the daily changes in the biogas production and effluent COD were not more than ±15% and ±16% respectively. After steady state conditions were achieved, granules from the sludge bed of the UASB reactor were sampled for the determination of their settling characteristics. All analysis for pollution parameters were conducted in accordance with the standard methods.

2.3 Settleability test

A simple upflow velocity test to characterize settling properties of granular sludges from anaerobic sludge bed reactors was used [15].

2.3.1 Apparatus
A sludge sample from the sludge bed reactor was placed in a 60 cm long and 1 cm I.D. glass column and subjected to gradually increasing upflow velocities. A variable speed peristaltic pump pumped water through the test column in an upflow mode. Ten different flow velocities in ascending order, were maintained for 5 minutes each. Granules exited the glass column via an overflow port. The fraction of sludge exiting at each particular velocity was collected on filter paper and the dry weight of each fraction was determined. Percentage of suspended

solids exiting the glass column plotted against upflow velocity gives the settling velocity profile of a particular sample. For the simultaneous filtration of granules that exited the glass column an 11,5 cm diameter Buchner funnel fitted with dried filter paper and connected to a vacuum flask was placed under the overflow port. A sketch of the apparatus is shown in Figure 2.

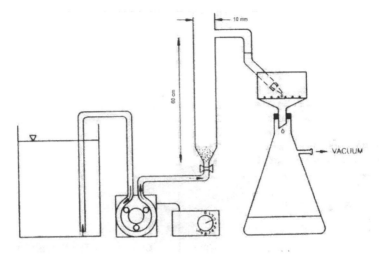

Figure 2:   Test apparatus.

2.3.2 Test procedure

The glass column was half filled with tap water. Approximately 5mL of concentrated sludge was transferred into the column while keeping the line connecting the bottom of the column with the pump clamped. The first filter paper was placed in the Buchner funnel, wetted and the vacuum started. The contents of the column were stirred to avoid the sludge moving in a slug, then the clump was removed and a pump was set to the desired velocity for 5 minutes. The column overflow was then switched over to a second filtration device while simultaneously increasing the flow velocity. This procedure (except for initial stirring) was repeated with subsequent higher flow velocities. Ten fractions were collected at upflow liquid velocities of 1,8, 2,2, 3,4, 5,1, 10,7, 16,8, 21,1, 30,9, 46,6, 99,6 m.h^{-1} respectively. Granules still remaining in the column after the highest flow velocity were collected as the last fraction (10^+). Next, total suspended solids values (TSS) of each fraction were determined.

3 Results

Plotting upflow velocity versus percent of TSS lost from the glass test device depicts a settling velocity profile of a particular sludge (Figures 3(a), 4(a)). In general this type of plot is rather descriptive, however it does give a basic understanding of the quality and nature of a particular sludge.

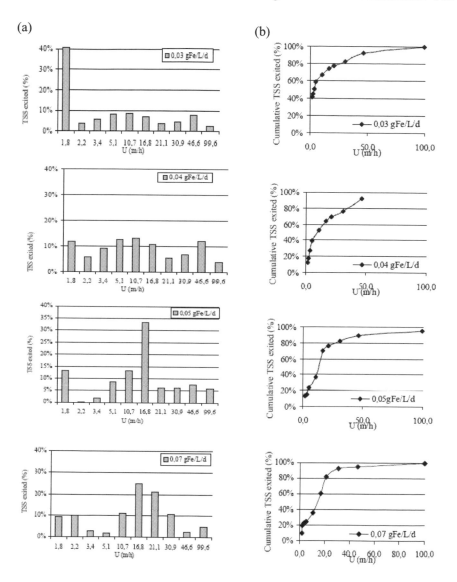

Figure 3:   Upflow velocities profile (a) and cumulative solids loss plots
             (b) for 0,03, 0,04, 0,05, 0,07 gFe·L^{-1}·d^{-1} iron loads.

As one might expect, a poor settling sludge would have a large proportion of
biomass exiting from the system at low upflow velocities, while a sludge with
good settling properties would remain within the test system at much higher
velocities. If the majority of sludge exited within a small range of upflow
velocities, we characterize it as a homogeneous sludge with poor, moderate or
good settling properties.

(a) (b)

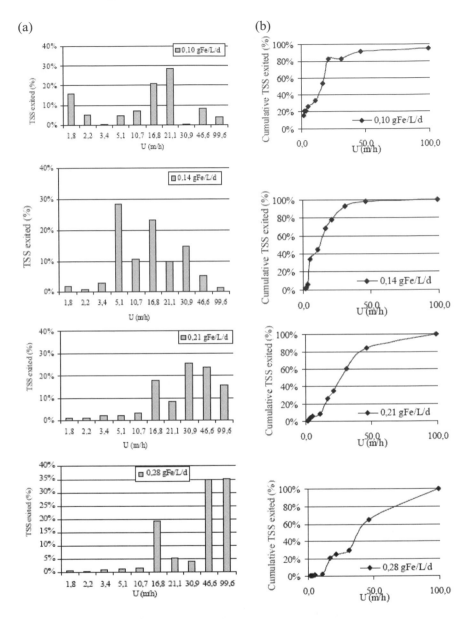

Figure 4:   Upflow velocities profile (a) and cumulative solids loss plots (b) for 0,10, 0,14, 0,21, 0,28 gFe·L^{-1}·d^{-1} iron loads.

A plot of cumulative percent of TSS lost from the glass test device versus upflow velocity proved to be more informative and an easier way to interpret the results. Figures 3(b), 4(b) are the mean cumulative profiles for the five runs for each iron load applied. From the cumulative plots, the upflow velocities corresponding to wash out of 10%, 30% and 60% of the sludge (U10%, U30%, U60%) were determined. The lower the upflow velocities U10%, U30% and U60%, the worse the settling characteristics of the sludge. Figure 5 represents the variations of U10%, U30% and U60% for each COD and iron load.

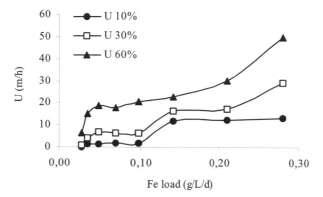

Figure 5:   Variations of U10%, U30% and U60% for each COD and iron load.

U10% increased from 0,23 m.h^{-1} to 13,21 m.h^{-1}, U30% from 0,87 m.h^{-1} to 29,11 m.h^{-1} and U60% from 6,63 m.h^{-1} to 49,82 m.h^{-1}.

4 Conclusions

According to Ghangrekar et al. [3] the settling characteristics of UASB anaerobic sludge deteriorate as the COD load increases. On the contrary, from the experiments described above, it is obvious that this is not the case. As the COD load was increased from 2,0 to 20,0 gCOD·L^{-1}·d^{-1}, the settling characteristics of sludge improved. This conclusion can be drawn from the values of the upflow velocities U10%, U30% and U60%. Specifically, U10% increased from 0,23 m.h^{-1} to 13,21 m.h^{-1}, U30% from 0,87 m.h^{-1} to 29,11 m.h^{-1} and U60% from 6,63 m.h^{-1} to 49,82 m.h^{-1}. This increase corresponds to a sludge bed with well-defined, larger granules which may resist to high hydraulic pressures. In other words, the addition of ferrous iron significantly contributes to the formation of granular sludge which a key function for the good operation of a UASB reactor.

It can also be noted that the characteristic upflow velocities increase more sharply as the ferrous iron addition gets greater. This increase may be attributed to the ferrous iron addition, which renders the granules larger.

Acknowledgements

This work was done under the operational Program for Educational and Vocational Training II (EPEAEK II) and particularly the Program

HRAKLEITOS. The project is co-funded by the European Social Fund (75%) and National Resources (25%).

References

[1] Hulshoff Pol, L. W. The Phenomenon of Granulation of Anaerobic Sludge. Ph.D. Thesis, Wageningen Agricultural University, The Netherlands, 1989

[2] Lettinga, G., Man, A.D., ver der Last, A.R.M., Wiegant, W., van Knippenberg, K., Frijns, J. and van Buuren, J.C.L., Anaerobic treatment of domestic sewage and wastewater. Water Sci. Technol. **27**, pp. 67–73, 1993.

[3] M.M. Ghangrekar, S.R. Asolekar and S.G. Joshi Characteristics of sludge developed under different loading conditions during UASB reactor start-up and granulation, Water Research, Vol. 39, Issue 6, pp. 1123-1133, 2005

[4] Fang, H.H.P., Chui, H.K., Li, Y.Y. and Chen, T., Performance and granule characteristics of UASB process treating wastewater with hydrolyzed proteins. Water Sci. Technol. 30, pp. 55–63. 1994.

[5] Schmidt, J.E. and Ahring, B.K., Granular sludge formation in upflow anaerobic sludge blanket (UASB) reactors. Biotechnol. Bioeng. 49, pp. 229–246, 1996.

[6] De Zeeuw W. J., Acclimatization of anaerobic sludge for UASB reactor start-up. Ph.D. Thesis, Wageningen Agricultural University, The Netherlands, 1984

[7] Fang, H.H.P., Chui, H.K. and Li, Y.Y., Effect of degradation kinetics on the microstructure of anaerobic biogranules. Water. Sci. Technol. 32, pp.165–172, 1995.

[8] Mahoney, E.M., Varangu, L.K., Cairns, W.L., Kosaric, N. and Murray, R.G.E., The effect of calcium on microbial aggregation during UASB reactor start-up. Water Sci. Technol. 19, pp. 249–260, 1987.

[9] Dolfing, J., Griffioen van, A.R.W. and Zevenhuizen, L.P.T.M., Chemical and bacteriological composition of granular methanogenic sludge. Can. J. Microbiol. 31, pp. 744–750, 1985.

[10] Thaveesri, J., Daffonchio, D., Liessens, B., Vandemeren, P. and Verstraete, W., Granulation and sludge bed stability in upflow anaerobic sludge bed reactors in relation to surface themodynamics. Appl. Environ. Microbiol. 61, pp. 3681–3686, 1995.

[11] Batstone and Keller, 2001 D.J. Batstone and J. Keller, Variation of bulk properties of anaerobic granules with wastewater type, Water Res. **35** (7), pp. 1723–1729, 2001

[12] Liu and Tay, 2002 Yu. Liu and J.-H. Tay, The essential role of hydrodynamic shear force in the formation of biofilm and granular sludge, Water Res. 36, pp. 1653–1665, 2002

[13]   Oleszkiewicz, J. A.; Romanek, A. Granulation in anaerobic sludge bed reactors treating food industry wastes.   Biological Wastes, 27(3), 217-35, 1989.

[14]   Rudd, T., Sterritt, R. and Lester, J., Complexation of heavy metals by extracellular polymers in the activated sludge process. J. Water. Poll. Contam. Fed. 56, pp. 1260–1268, 1984.

[15]   Andras E., Kennedy K.J. and Richardson D.A., Test for characterizing settleability of anaerobic sludge, Environmental Technology letters, Vol.10, pp.463-470, 1989.

Effect of the temperature on the performance of a sludge activated petrochemical wastewater treatment plant

S. A. Martínez[1], M. Morales[2], M. Rodríguez[1], R. Aguilar[1]
& D. Narváez[2]
[1]Universidad Autónoma Metropolitana –Azcapotzalco, México
[2]Petroquímica Morelos S.A. de C. V., México

Abstract

Petrochemical Morelos is located in an industrial zone among 50 refineries and petrochemical industries, in the coastal region of Mexico. All these industries produce 111 Mm^3/d of wastewaters which are discharged into the Coatzacoalcos River, previous treatment in their wastewater plants. However, the high temperatures in the region negatively affect the performance of the different wastewater plants increasing the discharge of pollutants into the river during the high temperatures in the year. Petrochemical Morelos has an activated sludge system to treat its wastewater flow which is about 7000 m^3/d. The aeration is supplied by a fine bubble diffusers system. Four compressors supply the air to the bioreactors. The high temperatures in the region and the compression effect on the air supply cause the temperature of the air exiting from the compressor to reach up to 82°C. As a direct consequence of the high air temperature, the temperature in the bioreactor reaches 32°C during the fall, whereas in the spring and summer, the bioreactor temperature reaches up to 41°C. The high temperatures have an adverse effect on the microbial activity and affect the performance of the biological process. In this study, the effect of temperature on the process is considered. A dynamic model, based on actual operation data, was validated at five scenarios presented during the year. The effect of temperature on μ_{max}, kla and kd, was incorporated in the mass balance equations of the model. Moreover, the model is applied to find the operating space of the process at different scenarios with the high temperatures, to reach the effluent quality standards required by Mexican environmental laws.
Keywords: activated sludge, modelling, petrochemical, wastewater, temperature.

WIT Transactions on Ecology and the Environment, Vol 92, © 2006 WIT Press
www.witpress.com, ISSN 1743-3541 (on-line)
doi:10.2495/WM060191

1 Introduction

Petrochemical Morelos is located in an industrial zone among 50 refineries and petrochemical industries, in the coastal region of Mexico. All these industries produce 111 Mm^3/d of wastewaters which are discharged into the Coatzacoalcos River, which in this part is an estuary, previous treatment in their wastewater plants. However, the high temperatures in the region affect negatively the performance of the different wastewater plants increasing the discharge of pollutants into the river during the high temperatures in the year. The Mexican petrochemical industry, Morelos S.A. de C.V., produces wastewater generated from the different chemical processes. The wastewater flow produced is about 7000 m^3/d and contains volatile organic carbon substances classified as toxics as 1,2 dichloroethane, chloroform and benzene, among others volatile compounds (VOC's). To comply with the effluent quality required by the Mexican environment legislation, SEMARNAP [4], Morelos petrochemical remove the different pollutants the wastewater in a treatment plant before being discharged into the Coatzacoalcos River. The treatment plant is located near the Mexican coast, where the mean weather temperature in the hottest months (April to August) is nearly 35°C and it can reach up to 38°C at extreme conditions. Such high temperatures affect the air temperature at the compressor exit producing a significant air temperature rise up to 82°C or more. When the air enters through the diffusers, it provokes an increase of the temperature within the bioreactor liquor. As a matter of fact, the actual temperature conditions within the bioreactor are 32°C in October-November reaching up to 41°C in August-September. Due to high temperature effect, the microorganism's activity is affected; hence temperature plays an important role in the performance of the biological system that must be considered in the dynamic modelling of the process. Some models have been developed to describe the effect of temperature on bacterial growth, Heitzer et al. [3], Raltowsky et al. [6], and Zwietering et al. [9]. The authors showed that at high temperatures the maximum specific growth rate (μ_{max}) is reduced.

The purpose of this study was to show to the other industries that discharge wastewaters, that it is possible to reduce the temperature effect on the wastewater treatment process to reduce de pollutants discharge to the estuary in this region of the Coatzacoalcos River, in order to preserve it, because is an important fishing zone for the inhabitants of the region. It was modelled the COD behaviour in the sludge activated treatment plant of the petrochemical at the different temperatures .in the region The temperature effect at the maximum specific growth rate, mass transfer coefficient for oxygen (kla) and death coefficient (k_d), were incorporated in the mass balance equations of the process.

2 Methods

Experimental data were obtained from the petrochemical biological wastewater plant. Different samples were taken out daily (after 8 hours), from the influent wastewater and the bioreactors during the period from October 2002 to

September 2003. The bioreactors capacity was 5000 m³ each. The flow was about 2300 m³/d to 2600 m³/d and the mean residence time in each bioreactor was about 2.0 days. The bioreactors operate with the bubble fine diffusers equipment (BFD). The power level in the chambers with the FBD system is about 0.04 HP/m³. It was considered that all the bioreactors were mixed flow reactors. The chemical oxygen demand (COD), biomass and dissolved oxygen were measured, APHA [8], in the plant's laboratory. Temperature and dissolved oxygen (DO) in the bioreactors were measured *in situ*. The mean of three replicates analyses was taken as the daily mean analysis per chamber. In order to obtain the results reported in the figures, the average of the daily mean analysis was obtained. The kinetic parameters were obtained in laboratory bioreactors following the method by Ramalho [7]. The eqns (7), (8) and (9) were used to evaluate the temperature effect on following parameters:

·   the maximum specific growth rate, was evaluated with eqn (7), Ratkowsky et al. [6], where b = 0.05 $K^{-1} h^{-0.5}$, Heitzer et al. [3], and c = 0.005 K^{-1} (which is a parameter to fit the experimental data to the model),

·   the death coefficient (k_d) with eqn (8), Ramalho [7].

·   the mass transfer coefficient for the oxygen   (*kla*) with eqn (9) Eckenfelder [2].

All of these were incorporated in the mass balance equations of the process.

The activated sludge petrochemical wastewater treatment process is shown in figure 1.

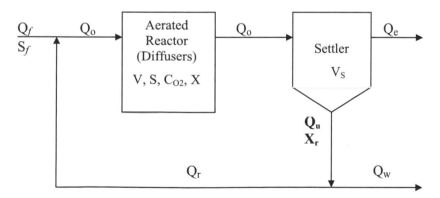

Figure 1:   The activated sludge petrochemical wastewater treatment plant.

The process is described by the following mass balance equations, Olsson and Newell [5], and the temperature effect on different parameters is considered. The bioreactor behaviour was assumed as completely mixed flow reactor.

In the reactor the balance eqns (1), (2), (3) are as follows:

$$\frac{dS}{dt} = \frac{Q_f}{V} S_f - \frac{Q_o}{V} S - \frac{\mu_{max}}{Y}\left(\frac{S}{K_s + S}\right)\left(\frac{C_{O_2}}{K_{OH} + C_{O_2}}\right) X + k_d (1 - f_n) X \quad . (1)$$

$$\frac{dX}{dt} = \frac{Q_r}{V}X_r - \frac{Q_o}{V}X + \mu_{max}\left(\frac{S}{K_s + S}\right)\left(\frac{C_{O_2}}{K_{OH} + C_{O_2}}\right)X - k_d X \quad (2)$$

$$\frac{dO_{O_2}}{dt} = \frac{Q_f}{V}C_{O_{2f}} - \frac{Q_o}{V}C_{O_2} - \frac{\mu_{max}}{Y_{O_2}}\left(\frac{S}{K_s + S}\right)\left(\frac{C_{O_2}}{K_{OH} + C_{O_2}}\right)X + kla(C_{O_{2sat}} - C_{O_2})(3)$$

It was assumed that there was no biomass at the overflow of the settler Dochain and Vanrolleghem [1].

In the settler eqns (4), (5), (6) are:

$$\frac{dX_r}{dt} = \frac{Q_U}{V_S}X_r - \frac{Q_O}{V_S}X \quad (4)$$

and

$$Q_O = Q_f + Q_r \quad (5)$$

$$Q_U = Q_w + Q_r \quad (6)$$

where:
t   = time (d)
Q_f = influent flow rate (m³/d)
Q_r = recycle flow rate (m³/d)
Q_W = waste flow rate (m³/d)
S_f = COD concentration in the influent (mg/L)
S   = COD concentration in the reactor (mg/L)
X   = biomass concentration in the reactor (mg/L)
X_r = biomass concentration in the settler or volatile solids suspended in the settler (SSVS) (mg/L)
C_{O2f} = dissolved oxygen concentration in the influent (mg/L)
C_{O2} = dissolved oxygen concentration in the reactor (mg/L)
C_{O2sat} = dissolved oxygen saturation concentration (mg/L)
μ_{max} = maximum specific growth rate (d⁻¹)

$$\mu_{max} = b^2(Tw - 285)^2(1 - e^{[c(Tw - 330.5)]^2}). \quad (7)$$

b= 0.05 (K⁻¹ h⁻⁰·⁵)
c= 0.005 (K⁻¹)
Tw = wastewater temperature in the reactor (°K)
K_s = 30 mg/L (substrate saturation coefficient)
K_{OH} = 0.2 mg/L (oxygen saturation coefficient)
f_n = fraction inerts on decay = 0.1

k_d = death coefficient (d⁻¹)

$$k_d = kd_{20} \, 1.05^{(T_w - 20)}$$ (8)

k_d = 0.03 (d⁻¹) = death coefficient at 20°C
Y = 0.67 = yield coefficient (mg biomass produced/mg COD consumed)
Y_{O2} = yield oxygen coefficient
        = 2.03 (mg biomass produced /mg O_2 consumed)
kla = mass transfer coefficient (d⁻¹)

$$kla = kla_{20} \, 1.02^{(T_w - 20)}$$ (9)

kla_{20} = mass transfer coefficient at 20°C (d⁻¹)
T_w = wastewater temperature in the reactor (°C)
V = 15000 m³ (reactor volume)
V_S = 750 m³ (settler volume)

3 Results and discussion

The model was validated with the actual COD data obtained from the wastewater treatment plant during a year, from October 2002 to September 2003. Due to the different processes in the Morelos petrochemical, the atmospheric condition, the wastewater composition and operation; flow rates always were subject to variation along the year, five average operation scenarios were used to validate the model. Table 1 shows the different scenarios used to validate the dynamical modelling

Table 1:   Scenarios presented during the wastewater treatment plant operation to validate the dynamical modelling

Parameter	Scenario 1	Scenario 2	Scenario 3	Scenario 4	Scenario 5
S_f (mg/L)	2500	2800	3100	2000	2600
T_w (°C)	32.5	38	36	39	41
Q_f (m³/d)	7300	7200	7600	7300	7400
Q_r (m³/d)	1500	2000	1500	2600	2600
Q_w (m³/d)	950	700	800	800	750

Due to the model is formed by four nonlinear differential eqns (1), (2), (3), (4), it was necessary to solved them with Runge-Kutta method.

In the Figure 2 it is shown that the model fits pretty well to the plant operation data trend during the different scenarios presented along one year operation of the wastewater petrochemical treatment plant. Based on this model, different estimations of COD in the bioreactor liquor were done at different operation conditions to find out the effect on the plant performance.

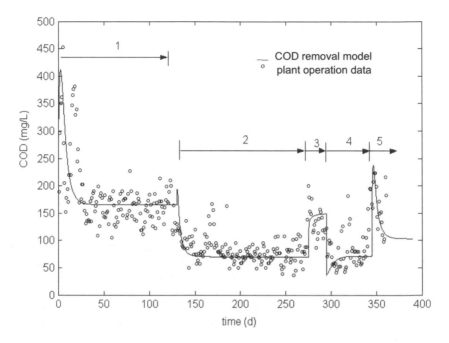

Figure 2:   COD removal model and actual plant operation COD data presented along the year at the five different scenarios.

The temperature in the reactor, the recycle flow rate (Q_r) and the waste flow rate (Q_W) were changed to evaluate their effect on the COD removal. It is important to point out that both flows rates can be easily changed in plant to control the biological process, by this reason these were chosen.

Based on the model, different simulations were run in order to find out the behaviour of the system under different operation conditions and to control the adverse temperature effect driving off the system (i.e. changing the flow rates Q_r and Q_W) to keep the COD concentration in the effluent at the permitted levels by the Mexican environmental regulations.

Figure 3 shows the COD in the bioreactor exit effluent when the wastewater treatment plant is operated at different Q_r, Q_W, and temperatures. As it can be seen, for the different Q_r and Q_W, the COD in the effluent has the lowest values at temperatures between 32°C to 37°C. This is because the microorganisms were acclimated to these conditions.

It is important to point out that at bioreactor temperatures lower than 30°C or higher than 37°C, the activity of the biomass decreases as predicted by eqn (8).

The high COD concentration in the effluent exit from the bioreactors caused by the adverse effect of the extremes temperatures on the activity can be reduced by controlling the waste flow rate. The COD in the effluent is reduced as the Q_W become lower. In the same figure it is possible to see that the effect of the lower or higher temperatures on the activity of the microorganisms is reduced by the reduction of the Q_W. This is because of the COD removal rate depends on the biomass concentration in the reactor (X), which increases in the bioreactor, due to less quantity of biomass is taken out of the system.

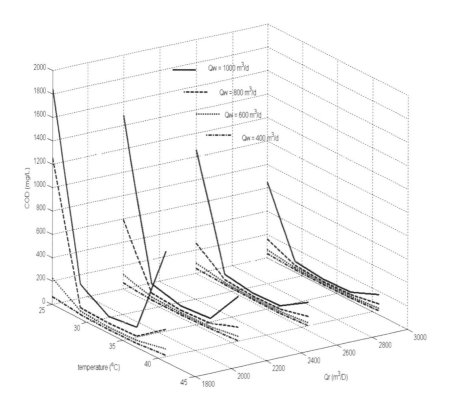

Figure 3:   COD bioreactor effluent at different operation temperatures, different Qr and Qw.

The figure 4 shows the COD curves at different Q_r (1800 to 3000 m³/d), when the plant is operated at a constant $Q_w = 1000$ m³/d. If it is made a comparison at different Q_r, it is observed that as the recycle flow rate increases, the temperature effect on the COD effluent is reduced. In case that the temperature increases at the highest levels (August to September), its effect on the process could be reduced by increasing the recycle flow rate.

Figure 5 shows the effect of different operation temperatures, Q_r and Q_w, on the COD of the bioreactor's effluent, when the system is operated at Q_w (1000

m^3/d and 400 m^3/d), different temperatures and recycle flow rates. It is important to observe that at Q_w (1000 m^3/d), more quantity of biomass is taken out from the bioreactor, hence the effect of the temperature on the COD removal is more important than if the system were operated at lower Q_w (400 m^3/d). On the one hand, the higher Q_w, the smaller the range temperature is at which the COD in the effluent is less than 150 mg/L, level permitted [4] by Mexican regulations, (e.g. from 29°C to 37°C at Q_r = 3000 m^3/d), but on the other when the system is operated at Q_w = 400 m^3/d, it is possible to operate the wastewater treatment plant in all the temperature range (from 25°C to 41°C) and if it were operated at low Q_r the pumping costs would be reduced.

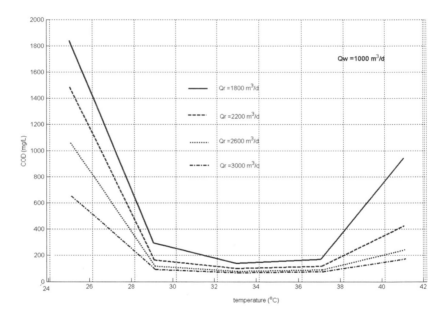

Figure 4:   COD bioreactor effluent at different operation temperatures, Q_r and Q_w.

It is important to point out that, as can be seen in figure 6, as the waste flow rate is reduced, the volatile solids suspended in the settler (SSVS) increase. In the same figure, it is shown that, at lower waste flow rates (Q_w = 400 m^3/d), the SSVS concentrations reach higher values (> 17000 mg/L), however at higher waste flow rates (Q_w = 800 m^3/d), the SSVS concentrations reach lower values (<13000 mg/L). The SSVS (X_r) is very important parameters that must be considered because at high SSVS concentrations the height of the blanket sludge could be increased and this would cause the microorganisms go out in the overflow rate. In addition, as can be seen in the same figure, it is possible to control the high X_r concentrations by changing the recycle flow rate.

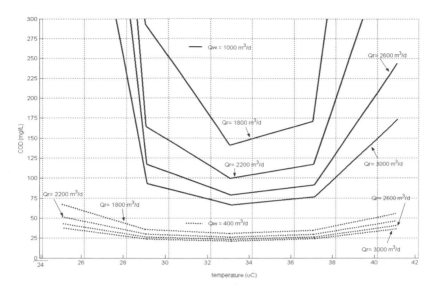

Figure 5: COD bioreactor effluent at different operation temperatures, Q_r and Q_w.

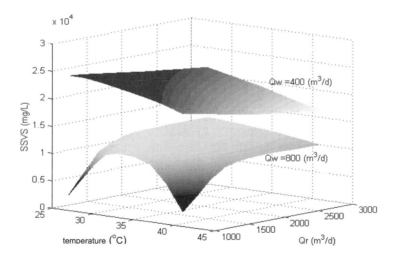

Figure 6: Volatile solids suspended in the settler (SSVS) at different temperatures and Q_r with $Q_w = 400$ m^3/d and $Q_w = 800$ m^3/d.

The dissolved oxygen in the reactor also is affected by the changes in temperature, Q_r and Q_w, however, after running the model, it was found that during all the scenarios presented along the year, the dissolved oxygen concentration in the bioreactor was > 2.5 mg/L, this means that the aeration system is able to keep the dissolve oxygen concentration at adequate level during the operation of the petrochemical wastewater treatment plant.

4   Conclusions

The performance of biological process depending on different temperatures and scenarios can be described by the dynamic model proposed in this work. The behaviour of the COD, biomass and oxygen in the reactor and also the biomass in the settler, can be known at different conditions. Moreover, based on the model, it is possible to find out the operating space of the biological wastewater treatment process to keep a good performance of the wastewater biological treatment process by means of the control of the recycle and waste flow rates.

This work could be used as an example to be done in other wastewater treatment plants of other industries in the region to reduce the discharge of pollutants to the estuary in this region of the Coatzacoalcos River, in order to preserve the aquatic life which is one of the most important sources of fishing in this region.

Acknowledgements

The authors wish to express their gratitude to Ing. Lorenzo Aldeco Ramírez, Subdirector de Operaciones de Pemex-Petroquímica, Ing. Guillermo García Reynaga, Auditor de Calidad, Seguridad y Protección Ambiental and Ing. Victor M. Herrero V., Subgerente de Calidad y Protección Ambiental, for the support given to conduct this study.

References

[1]   Dochain, D. & Vanrolleghem, P., Dynamic modeling and estimation in wastewater treatment process. IWA Publishing; U.K, 2001.
[2]   Eckenfelder, W., Industrial water pollution control. Environmental. Engineering Series. McGraw Hill, 200).
[3]   Heitzer, A., Hans-Peter E., Reichert, P. & Hamer, G., Utility of phenomenological models for describing temperature dependence of bacterial growth. Appl. Environ. Microbiol., 57, pp. 2656-2665, 1991.
[4]   NOM001-ECOL-1996, Ministry of the Environment, Natural Resources and Fisheries, SEMARNAP, México D.F., México, 1997.
[5]   Olsson, G & Newell, B., Wastewater treatment systems. Modelling, Diagnosis and Control, IWA Publishing: UK, 2001.
[6]   Raltowsky, D.A., Ross, T., McMeekin, T.A. & Olley, J., Comparison of Arrheius-type and Bélehrádek-type model for prediction of bacterial growth in foods. Journal of Appl. Bacteorology, 71, pp. 452-459, 1991.
[7]   Ramalho, R. Wastewater Treatment, Reverté: Spain, 1999.
[8]   Standard Methods for the Examination of Water and Wastewater 19th edn, APHA/AWWA/WEFederation, Washington DC, USA, 1995.
[9]   Zwietering, M.H., De Koos, J.T., Hasenack, B.E., De Wit, J.C. and Van 'Triet, K., Modeling of bacterial growth as a function of temperature. Appl. Environ. Microbiol., 57, pp. 1094-1101, 1991.

Enhancing the performance of a wastewater treatment plant processing wheat starch factory tailings

E. F. Pidgeon & J. N. Ness
Griffith School of Engineering, Griffith University, Australia

Abstract

The performance of a waste water treatment plant (WWTP) at a wheat starch factory (average flour throughput of 59 t/day) is reported in this paper. Most factory-generated wastestreams were processed through the onsite WWTP prior to being discharged to a municipal wastewater treatment facility for further treatment. One stream was trucked off-site as 'liquid fertiliser'. Over the two-year monitoring period, the onsite WWTP discharged an average of 430 kL of wastewater per day (range 323–3,264, standard deviation 198) with an average suspended solids (SS) concentration of 2,500 mg/L (range 610–11,700, standard deviation 1,568) and an average BOD_5 of 1,950 mg/L (range 140–13,000, standard deviation 1,855). Impacts to the WWTP as a result of water usage and wastewater generation practices within the manufacturing plant were investigated. The operation, effectiveness and maintenance of the system were assessed in terms of how water usage and practices within the starch recovery process affected effluent treatment performance. By-products of the treatment process included beneficial and non-beneficial outputs. For example, methane generated in the anaerobic digester was utilised in the boilers and offset the purchase of natural gas costs while excess anaerobic sludge posed a disposal problem, as there was no local market for this material. Performance inhibitors were identified during the study, as well as the consequences of failure of the anaerobic digester. Recommendations, based on theoretical background and practical experience are provided for enhancing this kind of treatment system and for reducing overall water usage in the starch factory.
Keywords: wheat starch, wastewater treatment, water usage, anaerobic treatment.

WIT Transactions on Ecology and the Environment, Vol 92, © 2006 WIT Press
www.witpress.com, ISSN 1743-3541 (on-line)
doi:10.2495/WM060201

1 Introduction

Traditionally, industrial processes were designed in the absence of consideration to wastes produced or the potential environmental impact that they may cause Unnikrishnan and Hegde [1]. As legislation was established and compliance issues arose, the wastewater treatment industry developed, with an end-of-pipe treatment focus that aimed at cleaning up the combined wastewater prior to discharge Savelski and Bagajewicz [2].

Problems of high COD wastewater and associated high disposal costs to the local municipal treatment works were being encountered in a starch manufacturing plant that had been designed and commissioned in the 1950s and to which, over subsequent years, various treatment plants had been added to meet compliance with the developing local environmental regulations. At the start of the project to investigate the performance of the wastewater treatment train, factory management had the attitude that the solution to the problem would be the addition of appropriate plant to "clean up" the final effluent before discharge to the municipal sewer. Opportunities to reduce water use and waste generation through process modification and changes to plant operational procedures had not been seriously investigated. This paper presents the observations and results of an investigation into the performance of the wastewater treatment plant (WWTP).

2 Plant description

2.1 Inputs and outputs for a wheat starch manufacturing processes

With the basic inputs of wheat flour and water, after the processing stages outputs included a range of saleable products as well as unwanted by-products and three tailing streams. During the study, the starch manufacturer had an average flour throughput of 59 t/d, an average freshwater usage of 8.6 kL/t flour processed, and approximately 75% of input water was discharged to sewer Pidgeon *et al.* [3]. Inputs and outputs for the site studied are shown in Figure 1.

Outputs, for which no commercial market existed, included by-products such as gums, pentosans and fibre as well as three tailing streams. These tailing streams were named according to the equipment from which they were generated and ended with an 'E' to represent 'effluent' (as opposed to the concentrate stream from the separator). By-products were combined and trucked off-site to a farm where the composite was used as liquid fertiliser – at a trucking cost for the starch manufacturer. The remaining tailing streams (QXE, SB80E and SD4E) were treated through the on-site WWTP prior to sewer discharge. Factory shutdowns or operations at a reduced flour throughput were sometimes necessary while a backlog of effluent was treated and discharged, as well as during periods of effluent treatment plant failure and subsequent recovery. Spillage of effluent onto the processing floor area was not uncommon. WWTP processing outputs included an effluent stream, anaerobic sludge and biogas. The biogas was used in the boiler and offset costs associated with natural gas purchases.

WIT Transactions on Ecology and the Environment, Vol 92, © 2006 WIT Press
www.witpress.com, ISSN 1743-3541 (on-line)

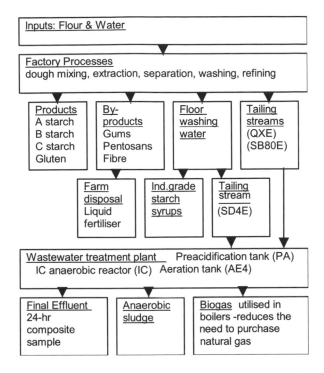

Figure 1:   Inputs and outputs for a wheat starch manufacturer.

2.2 Onsite wastewater treatment plant layout

When the investigation commenced, the onsite WWTP, as shown in Figure 2 consisted of a pre-acidification (PA) head tank, the PA tank, a mixing tank (where the combined preacidified starch tailings were mixed with a portion of recycled anaerobic effluent), an internally circulating (IC) anaerobic reactor and an aeration tank. Although controlled through a PLC/MMI, in times of crisis field adjustments were possible.

The pre-acidification (PA) tank doubled as a buffer tank and received unacidified tailing streams from the starch factory. With a 120 m³ total capacity, it was designed to operate at less than 50% volume to provide an additional 4.5 hour 'calamity' capacity at a typical inflow of 13m³/h. During times of extraordinary production conditions the maximum throughput of the on-site treatment plant was 26m³/h. The feed rate set from the PA tank to the Mix Tank was equal to the effluent discharge rate to sewer Aquatec-Maxcon Pty Ltd [4].

During the investigation the WWTP continuously evolved. The aeration tank was taken off-line and replaced with a temporary tank. An EIMCO Delta-Stak™ Clarifier was then installed and the aeration process was eventually removed altogether. The WWTP and its processing streams were continuously altered and experimented upon, with the aim of improving final effluent quality.

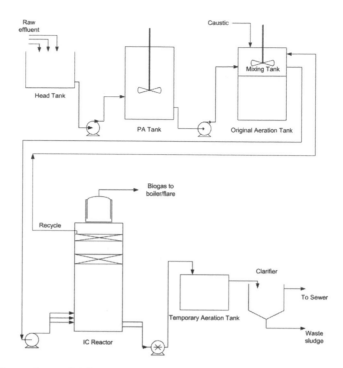

Figure 2:   Main components of the wastewater treatment plant.

2.3 Operation and maintenance of WWTP

The most important process parameters for operating and maintaining an effective WWTP at this site, according to Moosbrugger [5] are listed in Table 1.

For the operation of the IC reactor, decreases in gas production concurrent with VFA increases are signs of imminent process failure; 5 meq/L of VFA in the reactor was the operational guideline. Alkalinity provides a buffer against pH changes and therefore pH is more stable at higher alkalinities. Causes of excessive granule loss include COD loading increase; toxic spills in factory; and changes to substrate feed, pH, and temperature. The measurement of settleable solids was indicative of the situation in the IC reactor. Clarifier performance was dependent upon the solids settling rate EIMCO Process Equipment Company [6] and whilst reducing the solids in the final effluent, an unwanted sludge product was collected from the bottom of the cone and pumped to another sludge storage tank for later disposal.

3   Materials and methods

Background information was collected from the study site whilst also gaining practical experience in the day-to-day operations related to the starch factory and WWTP processes. In-house data, process flow sheets, performance reports and commissioning operational manuals were all reviewed. Research was undertaken that included the collection of WWTP process samples, sample analysis and data entry to continuously monitor performance trends. Personal communications with administrative and processing staff also assisted in gaining a deep understanding of the issue, including the realistic constraints for that particular site, wastewater generation practices, likely process inhibitors and wastewater by-products.

Table 1:   Important process parameters for WWTP operation.

Process Parameter	Design capacity, optimal range and other factors
COD-load to the IC (Measured on unfiltered sample from mix tank feed (PA effluent))	• 4.4 tonnes (metric) COD per day • COD load (t/d) = COD concentration [kg/m^3] * mix tank feed flow [m^3/d]
pH in IC reactor	6.8 to 7.2 optimal; Never <6.5 or >7.5
pH in mix tank	pH range 6.0 to 6.2 and adjusted to correct IC reactor pH if needed.
Mix tank temperature	36 ± 2 °C optimal 30 – 40 °C no serious impact on performance 25 – 30 °C anaerobic treatment efficiency decreases > 40 °C damaging to anaerobic biomass
VFA in IC reactor	Should not exceed 5 meq/L
Alkalinity in IC reactor	Should be > 50 meq/L.
Settleable solids in IC reactor effluent	1 min – should not exceed 1 mL (measured with Imhoff cone).
Suspended solids (SS) in PA and IC effluent	Used to assess changes in SS levels in the IC reactor
Gas production rate	Important indicator for 'health' of anaerobic reactor

All samples collected on site were analysed immediately according to appropriate standard methods. Samples were collected throughout the treatment train, including input streams (QXE, SD4E and SB80E), wastewater treatment effluent streams (PA, PA sediment, mix tank, IC and AE4) and final effluent stream (automatic sampler on magflow). Input streams and treatment processing streams were grab samples while the final effluent stream was a composite sample collected over a 24 hour period. Parameters tested included COD, pH, temperature, SS, total solids (%), volatile fatty acids (VFA), alkalinity, pH, temperature, SS, settleable solids (1 min and 60 min), and feed and discharge rates (m^3/h). All COD tests were COD total using a Merck COD testing kit and Solids (%) was measured using a Sartorius moisture analyser. Operating parameters measured and/or calculated for the WWTP during the course of the study are assessed against the design capacity and optimal operating conditions for effective operation of the WWTP.

4 Results and discussion

4.1 Factory generated waste streams

The measured characteristics of three tailing streams are shown in Table 2. QXE was a relatively dilute and oxygen depleted tailing stream containing a mixture of tailing and fractured B-starch particles, proteins, fats and water. With a flow rate of 12-14 m^3/h this represented around 60% of the WWTP influent volume. If the QXE were to be subjected to advanced treatment such as membrane filtration to obtain the water for reuse, this would effectively increase the hydraulic retention time (HRT) and the solids retention time (SRT) of the other two streams. This would, in effect, increase the organic loading rate (OLR) capacity of the WWTP. The SB80E stream contained a large proportion of gums and pentosans while the SD4E stream was what remained following final scavenging of solid particles from the floor washing wastewater.

Table 2: Characteristics of tailing streams generated from wheat starch manufacturing process (mean values followed by range in parenthesis).

Parameter	QXE	SB80E	SD4E
Flow rate range, kL/h	12-14	4-6	4-6
Temperature (°C)	Mean 28.5 (23.8-32.6)	Mean 30.0 (25-34)	Mean 25.6 (22.7-28.1)
pH	Mean 5.0 (3.73-6.11)	Mean 3.6 (3.48-3.87)	Mean 3.7 (3.38-4.16)
COD Total (mg/L)	Mean 12,020 (9720-15,300)	Mean 55,110 (22,120-99,800)	Mean 27,280 (23,040-33,800)
Suspended Solids (SS) (mg/L)	Mean 930 (610-2010)	Mean 44,840 (36,580-55,260)	Mean 11,600 (7770-15,480)
Solids (%)	Mean 0.6 (0-6.4)	Mean 1.7 (0-6.4)	Mean 0.8 (0-1.82)

4.2 Operational parameters for the WWTP during the study

Process parameters of importance to the effective operations of the WWTP were measured and data are presented in Table 3. By comparing the operational realities with the design or optimal operational values it becomes apparent that the IC reactor was regularly overloaded with organic matter, it was not always operating within the optimal pH range, temperature fluctuations occurred, alkalinity was often lower than recommended and settleable solids results indicate that biomass washout did occur. Any of the factors causing biomass loss could have been responsible as factors relating to organic overloading, substrate feed changes, pH changes, temperature changes and toxic spills to the factory floor were all known to occur during the study period.

The mean suspended solids levels in the PA and IC reactor were similar values demonstrating poor performance for the IC reactor in terms of SS reductions from the PA effluent. A high SS loading in the IC effluent was also an issue with respect to compliance with discharge requirements as outlined in Pidgeon *et al.* [3]. Given the excessive organic loading rate to the WWTP, the mean gas production rate of 45 m^3/h was low averaging 0.22 m^3 CH_4/kg COD removed. On days of moderate production (average flour throughput of 59 t/d) and good wastewater analytical results the biogas production rate easily achieved around 80 m^3/h.

Table 3:   Operating parameters measured at WWTP.

Process Parameter	Operational values
COD-load to the IC (tonnes COD/d)	Mean 6.9 tonnes/d, Std Dev 2.7, range 0.8-23.3
pH in IC reactor	Mean 6.9, Std Dev 0.4, range 5.3-10.6
pH in mix tank	Mean 6.5, Std Dev 0.5, range 3.5-10.6
Mix tank temperature (°C)	Mean 28.8, Std Dev 4.2, range 6.71-36.6
VFA in IC reactor (meq/L)	Mean 15, Std Dev 20, range 0-134
Alkalinity in IC reactor (meq/L)	Mean 38, Std Dev 15, range 0-112
Settleable solids in IC effluent (mL)	1 min: Mean 0.3, Std Dev 0.5, range 0-2 60 min: Mean 79, Std Dev 56, range 0.5-450
SS in PA tank and IC effluent (mg/L)	PA effluent: Mean 2825, Std Dev 2895, range 0-22,970 IC effluent: Mean 2700, Std Dev 1559, range 384-13,260
Biogas production rate (m^3/h)	Mean 45, Std Dev 21, range 3.2-112

The performance of the WWTP was usually measured as the amount of COD reduction achieved from the PA effluent to the IC effluent, expressed as a percentage. Over a monitoring period of 18 months, the average calculated reduction in COD concentration from the PA to the IC was 58% (std dev = 17). On average, the onsite WWTP discharged 430 kL of wastewater per day (range 323-3,264, standard deviation 198) with an average suspended solids (SS) concentration of 2,500 mg/L (range 610-11,700, standard deviation 1,568) and an average BOD_5 of 1,950 mg/L (range 140-13,000, standard deviation 1,855).

4.3 Wastewater generation practices and performance inhibitors

Factors identified as having the potential to adversely affect treatment plant performance included factory shutdown maintenance and cleaning issues, hydraulic and organic overloading, accidental inputs of sulphur and human operating errors.

Maintenance of equipment during shutdown periods often involved the stripping down of equipment, repairing, reassembly and degreasing and/or re-greasing. Sodium hypochlorite was also used as a cleaning agent throughout the factory following shutdown. Floor washing wastewater was collected in a sump and processed through the SD4 to scavenge solids. SD4E was then directed to the WWTP for treatment with the other processing streams. Sometimes this effluent stream had a strong odour of chlorine and/or solvents. Cleaner production principles and practices applied during maintenance events aimed at preventing these materials from entering into the floor washing wastewater would reduce inputs with reactor inhibitory consequences.

The IC anaerobic reactor was regularly overloaded in terms of organic and hydraulic inputs Pidgeon et al. [3]. It was designed to treat an average OLR of 4.4 tonne COD/d and an average flow rate of 13 m^3/h, however it typically operated with an average COD loading of 6.9 t/d at a flow rate of 18-20 m^3/h. By-products included beneficial and non beneficial products, these being methane and anaerobic sludge respectively. The negative impact of COD overloading is enhanced when raw inputs to the reactor are insufficiently pre-acidified. Experiments treating wheat starch wastewater found that acidogenic fermentation of the raw starch stream prevented the growth of nuisance anaerobic bacteria that caused sludge bulking Endo and Tohya [7] and positively influenced granular sludge bed formation and stability in UASB reactors Moosbrugger [8].

High total sulphide levels (up to 31 mg/L) were detected in samples of the final effluent discharged to sewer during the third and fourth months of monitoring while the PA tank sample had a zero value indicating that inputs of sulphur were occurring within the treatment train. Total sulphide (S^{2-}) had a maximum guideline of 5 mg/L in the Trade Waste Guide Industrial Liquid Waste Sewer Acceptance Criteria as it can cause corrosion and generate odours and gases in sewers, potentially compromising safety Brisbane Water [9]. Sulphide is also known to inhibit methane generation Metcalf and Eddy [10] due to the fact that sulphidogens and methanogens have physiological similarities and operate under similar optimum temperature and pH conditions Zhou and Fang [11]. According to Randall [12] sulphides are one of the contaminants typically present in spent caustic liquors that have been used to strip acidic components from gas streams. It was concluded that a spent caustic solution obtained from a local gas company, for stream pH adjustment purposes, was the most probable source of sulphur inputs. Following this discovery only new NaOH was used for pH adjustment purposes and the problem soon disappeared. However, corrosion had already rendered the aeration tank as unsuitable for use due to a combination of operating conditions within the WWTP, including the presence of sulphur, a low operating pH in the IC reactor and final aeration of the waste stream prior to discharge.

A human operating error which occurred late in the seventh month of testing caused the IC anaerobic reactor to be dosed with an excessive amount of NaOH and subsequently fail. A check on the PLC showed that the pH 'high' level setting had been altered to pH=10 during the evening. Although believed to be an accident, this uninformed change by some unknown person caused an immediate shutdown for production. The program did not include a register of persons logging onto or off the PLC system. It was a costly exercise to dispose of the highly alkaline waste in the reactor as well as the issue of re-establishing new biomass within the reactor. Upon start-up, production could only occur at a reduced rate to enable the biological activity within the reactor to increase progressively towards steady state conditions. As a consequence of this reactor failure, Mg (OH)$_2$ was used for pH adjustments following that time. A more secure and accountable computer program log-in was also set up to ensure that only those authorised could gain access to areas where process parameter set points were altered.

Good housekeeping practices using a cleaner production approach, the implementation of systems that minimise human errors and ensuring inputs do not contain inhibitory substances should enhance the performance of the WTTP studied.

5 Conclusions

Factors were identified as having the potential to adversely affect the WWTP operations and performance, these being overloading, housekeeping issues during maintenance and cleaning events, accidental inputs of inhibitory substances such as sulphur compounds in a spent caustic stream, and human errors. Preventing inhibitory substances from entering into the floor washing wastewater ensures that they are not in the SD4E stream that is input to the PA head tank. Human errors were addressed by increasing security measures where the settings could be changed on the PLC as well as changing to a less toxic compound being used for pH adjustment purposes.

The IC reactor was not always operated at recommended pH levels or with sufficient alkalinity to protect the sludge against shock associated with sudden changes. Operation at a pH less than the optimal range probably contributed to a reduction in methane generation and caused the formation of H$_2$S, which upon aeration is oxidised to sulphuric acid, and ultimately corroded the aeration tank. Alkalinity ties in with pH as it provides the stream with a buffering capacity against sudden pH changes. The unstable pH recorded during the study would, in part, be a symptom of operating at a lower than recommended alkalinity. Operation of the WWTP according to optimal pH and alkalinity values would increase biological activity causing an increase in methane generation, reduce the incidence of washout and ensure that H$_2$S does not dominate.

Washout of biomass did occur for the plant with factors such as organic overloading, sudden changes to pH and toxic inputs from floor washing water or other sources thought to be responsible. WWTP performance was generally poor with biogas generation averaging 45 m^3/h despite heavy organic loadings. The average calculated reduction in COD from the PA to the IC stages was 58%.

Enhancing the performance of this WWTP treating wheat starch factory tailings requires that the system be operated at optimal conditions for the biological component to optimally treat the waste inputs. Should the QXE stream that represents 60% of the WWTP influent be polished for reuse within the factory process this in turn would increase the HRT and SRT of the other streams treated through the WWTP. Water extracted for reuse prior to entry to the WWTP would result in less fresh water being required. As well as reducing overall water usage in the starch factory, this would simultaneously reduce the hydraulic flow rate to the WWTP. Effectively this would increase the OLR of the WWTP – perhaps even to the point where a portion of liquid fertilizer could be treated through the WWTP thereby saving trucking costs and converting more solids to methane for beneficial on-site use.

Acknowledgements

Funding and assistance in kind for this project were provided by Weston Bioproducts, a Division of George Weston Foods Limited, Australia. One of the authors (EFP) was supported by a scholarship from Weston Bioproducts during the course of this study.

References

[1]   Unnikrishnan, S. and D.S. Hegde, *An analysis of cleaner production and its impact on health hazards in the workplace.* Environment International, **32**(1), p. 87-94, 2006.

[2]   Savelski, M. and M. Bagajewicz, *On the optimality conditions of water utilization systems in process plants with single contaminants.* Chemical Engineering Science, **55**, p. 5035-5048, 2000.

[3]   Pidgeon, E., J.N. Ness, and J.A. Scott. *The application of membrane filtration technology for a wheat starch processing industry.* in *National Environment Conference.* Brisbane, 2003.

[4]   Aquatec-Maxcon Pty Ltd, *Love Starches Pty Ltd - Moorooka Site IC Reactor Trade Waste Plant Operating & Control Philosophy*, Aquatec-Maxcon Pty Ltd, Brisbane, p. 1-6, 1996.

[5]   Moosbrugger, R., *IC reactor COD overload protection & operating instructions*, unknown, Brisbane, p. 1-7, 1997.

[6]   EIMCO Process Equipment Company, *EIMCO Delta-Stak Clarifier Operating Instructions*, Salt Lake City, Utah, 1999.

[7]   Endo, G. and Y. Tohya, *Ecological study of anaerobic sludge bulking caused by filamentous bacterial growth in an anaerobic contact process.* Water Science and Technology, **20**(11/12), p. 205-211, 1988.

[8]   Moosbrugger, R.E. *Single and two stage UASB systems: preacidification and sludge granulation.* in *8th IAWQ International Conference on Anaerobic Digestion.* Sendai, Japan: IWA Publishing, 1997.

[9]   Brisbane Water, *The Trade Waste Guide: Protecting our waterways and Moreton Bay*, Brisbane City Council, Brisbane, p. 1-16, 1999.

[10] Metcalf & Eddy, *Wastewater Engineering Treatment and Reuse, Metcalf & Eddy, International Edition.* fourth ed, ed. G. Tchobanoglous, F.L. Burton, and H.D. Stensel. Sydney: McGraw-Hill, 2003.

[11] Zhou, G.-M. and H.H.P. Fang, *Competition between methanogenesis and sulfidogenesis in anaerobic wastewater treatment.* Water Science and Technology, **38**(8-9), p. 317-324, 1998.

[12] Randall, T.L. *Case studies on caustic sulfide wastewater wet oxidation treatment* in *Hazardous & Industrial Wastes, 26th Mid Atlantic Industrial Waste Conference.* Newark, DE: Technomic, 1994.

Data acquisition, validation and forecasting for a combined sewer network

É. Crobeddu & S. Bennis
École de Technologie Supérieure,
Département de Génie de la Construction, Québec, Canada

Abstract

The scope of this work is the development of a computerized tool, named PREVAL, for day-to-day sewer network management. This tool allows: 1) to simulate hydrographs and pollutographs at the outlet of an urban catchment basin and through control devices, 2) to make the validation and the filtering of measurement data by using univariate or multivariate methods based on material or analytical redundancy, 3) to provide hydraulic and hydrological models with reliable data for the calibration stage, 4) to evaluate combined sewer overflows in terms of quality and quantity and to check the conformity of the operation of overflow devices with respect to prescribed constraints. PREVAL is the result of a close cooperation between École de Technologie Supérieure, Hydro-Québec and various partners such as the cities of Montreal, Laval and Verdun.
Keywords: control, overflow structures, pollution, validation.

1 Introduction

The management of sullage discharge into a natural environment requires a qualitative and quantitative assessment of the outflow. The estimation of the quantity and quality of sullage is based on measurements and on calibrated hydraulic and hydrological models. The reliability of sullage discharge assessment can be warranted only by the validation of discharge flow rate data.

1.1 Quantitative assessment

The quantitative assessment consists in estimating the frequency and the volume of the sullage discharged into the natural environment by control devices. The

WIT Transactions on Ecology and the Environment, Vol 92, © 2006 WIT Press
www.witpress.com, ISSN 1743-3541 (on-line)
doi:10.2495/WM060211

estimation of the quantity and quality of sullage is based on measurements and on calibrated hydraulic and hydrological models. The direct measurement of discharge is obtained by flow rate or height measuring instruments (Bertrand-Krajewski et al. [6]) installed at the exit of control devices. Hydraulic and hydrological models provide theoretical estimations of sullage discharged into the natural environment. The theoretical computations require that the functioning of the control devices be known. The discharge laws are not always adapted to the control devices used in sewer networks (Zug et al. [10]). They usually necessitate specific calibration.

The validity of an assessment depends on the validity of the data obtained by measurements or by models. Measurement data may be affected by errors due to instrument disorders or an inadequate environment. Hydraulic and hydrological models require data which have been obtained and validated for their calibration. The experimental errors in pluviometer and hydrometer data are combined with the shortcomings of the model in producing the total errors of the forecasted values. The validation of the quantitative assessment of the discharge requires the validation of the data obtained by measurements or by models.

1.2 Review and validation of the data

Environmental regulation constraints and social pressure prompt managers to better know the working of their networks. The instrumentation of a network allows for an efficient control. It consists in providing the network with gauges and instruments for supervision. The obtained data are stored locally or centrally. This information is important but subject to errors. Among the numerous sources of errors, one may mention:

-   The lack of maintenance or improper conditions of utilization of the gauges causes biased and aberrant values in data series.
-   The problems of transmission and storage of information are causes of missing data.

The use of instruments produces a great quantity of information which cannot be dealt with by manual procedures.

The use of computerized methods is therefore a necessity. In practice however, the verification and validation of environmental data are either missing or are not subject to an automated procedure. The time needed for the validation and its cost become then very important. As a consequence, the validation of environmental data by the managers of urban sewer networks tends to be a summary and improvised procedure. Scientific and technological developments are recent, as far as the automated validation of urban environmental data is concerned. Most of them rely on the use of statistical tools (Berrada et al. [5]), (Bennis et al. [3]), (Bennis et al. [1]), (Bennis et al. [2]). Some of the validation procedures make a direct use of material and analytical redundancy, relying on the manager's know-how (Blanchet et al. [7]). The occurrence of missing data in chronological series is a common problem. Techniques for the rehabilitation of missing data have been developed (Bennis et al. [2]). Little research has been done as far as the automated review (pre-validation) of environmental data is concerned (Mourad and Bertrand-Krajewski [8]).

2 PREVAL program

2.1 Introduction

Managers need efficient and comprehensive tools in order to control discharge. These requirements are not simultaneously fulfilled by present management tools. PREVAL was developed to overcome this shortage by allowing forecasting and validating hydraulic data for the control of pollution discharged into the natural environment.

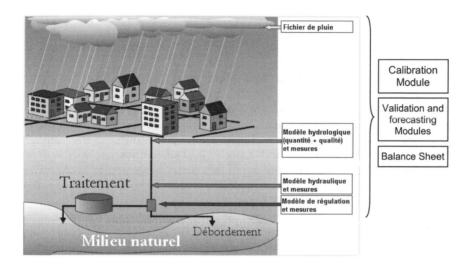

Figure 1:   Configuration of the urban sanitation system.

The tool consists of three modules : a hydrological module, a hydraulic module and a validation module. These modules may be used together or independently from one another.

2.2 Model design tools

2.2.1 Tools for hydrological models
The hydrological module represents globally the runoff phenomena on a drainage basin and the flow process in pipelines. The transportation of pollutants and in particular of suspended matter is obtained by a modified "Rating Curve" model (Temimi and Bennis [9]). This model takes into account the variable phase shifts often encountered between hydrographs and pollutographs. The generated hydrographs and pollutographs are useful for the evaluation of the hydraulic performance of an existing network from the viewpoint of overflow while they are required to generate the analytical redundancy needed for the validation of hydrological data.

2.2.2 Tools for hydraulic models

The purpose of the hydraulic simulation module is to investigate and follow up the flow process over weirs. Contrary to existing commercial programs, our purpose is not to develop a detailed model for a given system (sewer network + control devices). In our hydraulic module, a global approach is preferred to represent the various overflow devices which may be encountered in sewer networks. The Theoretical study has been completed by laboratory tests in order to validate the selected models.

2.3 Analysis and computation tools

2.3.1 Validation and forecasting modules

Validation is an essential step for the control of the pollution discharged by overflow weirs. The validation module includes innovating mathematical procedures allowing automated data processing. Two types of validation methods are integrated for the corrections needed by the noisy and aberrant values of data series. The univariate validation methods consist of statistical procedures adapted to single data series (Berrada et al. [5]). Multivariate validation methods, on the other hand, allow processing several strongly correlated data series (Bennis and Kang [2]). The multivariate methods are especially well adapted to the handling of data series generated by analytical or material redundancy. The validation module includes methods for the estimation of missing data (Bennis et al. [3]). It possesses several powerful filters (standard regression, Kalman filter, principal component analyses, ARIMA). The validation module includes forecasting methods for hydraulic data. Forecasting is obtained by the ARMAX and ARMA models.

2.3.2 Calibration and balance-sheet modules

The calibration of the parameters pertaining to the hydraulic and hydrological models is obtained with the help of the calibration module. This module uses standard regression tools or the Kalman filter. The balance-sheet module comprises statistical procedures for the analysis of overflow frequency as well as of the overflowing masses and volumes. It is thus possible to insure the follow-up of the performance of a sewer network system over long periods.

3 Testing of the PREVAL program

PREVAL has been tested on Sector 1 of the combined sewer network of the borough of Verdun. This portion the Verdun basin extends over an area of 177 ha and has 40% permeability. The test of PREVAL proceeded in three phases. The first phase consisted in producing a hydrological simulation and in using the simulation results for the validation of the measured data. In the second phase, the validated measured data were used to determine the volumes of water and the quantities of pollutants overflowing into the natural environment.

3.1.1 Hydrological simulation and validation

PREVAL was used for the calibration of the hydrological model. The quality of the simulation has been evaluated on the basis of four performance indicators:

- The NASH coefficient   : $NASH = 1 - \dfrac{\sum\limits_{i=1}^{n}\left(Q_i^{Measured} - Q_i^{Simulated}\right)^2}{\sum\limits_{i=1}^{n}\left(Q_i^{Measured} - Q_i^{Mean}\right)^2}$

- The volume ratio   : $RV = \dfrac{total\ volume\ of\ simulated\ runoff}{total\ volume\ of\ meaured\ runoff}$

- The peak ratio   : $RP = \dfrac{Q_{Max}^{Simulated}}{Q_{Max}^{Measured}}$

- The peak flow synchronism : ΔT = time delay between measured and simulated peak flow rates.

The results of the simulations are shown in table 1.

Table 1:   Results of hydrological simulations obtained by PREVAL.

Events	RV	RP	NASH	ΔT (min)
30-09-1999	1,11	0.90	0,89	5
13-10-1999	1,1	1.231	0.84	5
16-08-2000	1,0	0.99	0.81	5
23-08-2000	0,83	0.98	0.70	10

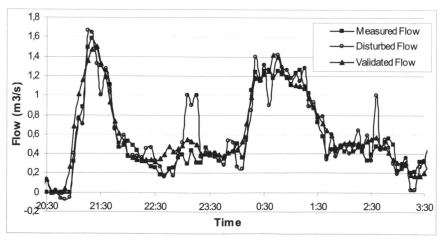

Figure 2:   Measured, noisy and validated hydrographs of October 13 1999.

The hydrological model used in PREVAL appears to be a satisfactory tool for the simulation of hydrographs at the exit of an urban basin. It is therefore possible to rely on simulated hydrographs for the multivariate validation of measured flow rates. The events of October 13 1999 and of August 16 2000 have been selected for the sake of demonstration. White noise having a variance respectively of 10% and 20% has been added to the measured flow values. Moreover, aberrant values having W, U and A shapes have been inserted to simulate temporary gauge failures. A multivariate validation was thereafter applied using PREVAL. The validation results are shown in figure 2 and figure 3.

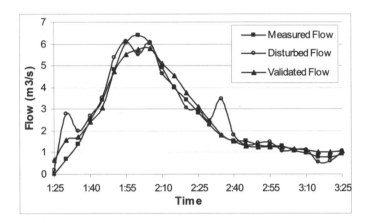

Figure 3:   Measured, noisy and validated hydrographs of August 16 1999.

The PREVAL program has attenuated the noise and suppressed aberrant values. PREVAL corrected the noisy data sequence on the basis of the information provided by the simulated flow rates. Validation which would normally require a manual procedure based on empirical assumptions can now be automated and is governed by physical and statistical criteria

3.1.2 Hydraulic simulation

A control system similar to the one found at the exit of the Verdun basin has been configured in PREVAL. It consists of two rectangular flood-gates (0,914m wide, 0,914m high) controlling the flow toward the station. It is located at the end of a conduit having a 2m wide rectangular section and a 0,1% slope. The excess overflow is discharged into the natural environment by a pumping station equipped with 6 pumps in parallel, each discharging $0,6m^3/s$. The pollutograph of the suspended matter concentrations for the event of August 16 2000 has been measured at the upstream end of the control system. PREVAL was able to compute the hydrograph and the pollutograph for the overflow discharged into the natural surroundings for a given gate closure scenario during the rainfall event. The results of the hydraulic simulation by PREVAL are shown in figure 4, taking into account the measured flow rates and the previously obtained noisy and validated data.

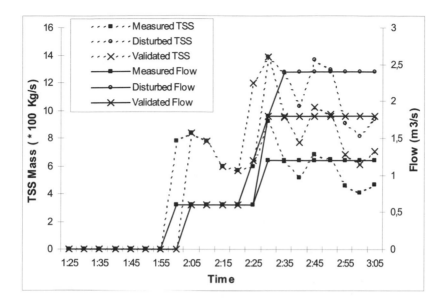

Figure 4:   Hydrographs   and   pollutographs   of   discharge   into   natural
environment

Figure 4 shows that the noisy measured flow data produced an overestimation of the flow rates and pollutant loads of the flow discharged into the natural surroundings. Validated flow rates produced discharge loads and flow rates closer to the true values.

4   Conclusion

PREVAL is a complete and modular tool satisfying present needs of sewer network managers. In differs from other programs found on the market by the fact that the routines for the simulation of hydraulic phenomena, for validation, for forecasting and for balance-sheet calculations are included in a single program. It offers great flexibility due to its modular structure. Each module can be used independently from the others during an analysis, according to the user's specific needs. PREVAL is a particularly efficient tool for the preparation of reliable balance sheets dealing with pollutant discharge by overflow weirs. It is also an efficient tool for the follow-up of the performance and functioning of a sewer network. The application of PREVAL to the Verdun site was a conclusive and promising exercise. The hydraulic phenomena have been correctly simulated and the validation of the measured data has allowed one to realistically assess the amount of pollution discharged to the natural environment.

References

[1]   Bennis, S., Berrada, F. and Bernard, F. (2000a). Méthodologie de validation des données hydrométriques en temps réel dans un réseau d'assainnissement urbain. *Revue des Sciences de l'Eau*, 13(4), 483-498.

[2]   Bennis, S., Berrada, F. and Kang, N. (1995). Improving single variable and multivariable techniques for estimating missing hydrological data. *Journal of Hydrology*, 191(1-4), 87-105.

[3]   Bennis, S., Cote, S. and Kang, N. (1996). Validation des données hydrométriques par des techniques multivariées de filtrage. *Canadian Journal of Civil Engineering*, 23(1), 218-230.

[4]   Bennis, S. and Kang, N. (2000b). Multivariate technique for validating historical hydrometric data with readundant measurements. *Journal of Nordic Hydrology*, 31(2), 107-122.

[5]   Berrada, F., Gagnon, L. and Bennis, S. (1996). Validation des données hydrométriques par des techniques univariées de filtrage. *Revue Canadienne de Génie Civil*, 23, 872-892.

[6]   Bertrand-Krajewski, J. L., Laplace, D., Joannis, C. and Chebbo, G. (2000). *Mesures en hydrologie urbaine et assainissement*. Tec & Toc, Lavoisier, Paris.

[7]   Blanchet, F., Breuil, B. and Viola, A. (1998). Aquaval: Un système d'acquisition et de validation automatique des mesures en réseau d'assainissement. Exemple d'application sur le département de la Seine Saint-Denis. *Novatech 1998. 3e conférence internationale sur les nouvelles technologies en assainissement pluvial*, Lyon 4-6 mai 1998, 155-162.

[8]   Mourad, M. and Bertrand-Krajewski, J. L. (2002). A method for automatic validation of long time series of data in urban hydrology. *Water Science and Technology*, 45(4-5), 263-270.

[9]   Temimi, M. and Bennis, S. (2002). Prévision en temps réel des charges de polluants dans un réseau d'assainissement urbain. *Revue des Sciences de l'Eau*, 15(3), 661-675.

[10]   Zug, M., Vazquez, J., Bellefleur, D. and Issanchou, E. (2001). Les déversoirs d'orage: Connaît-on les ouvrages de nos réseaux et comment ils fonctionnement ? *4ième conférence internationale NOVATECH*, Lyon, 205-212.

Section 5
Resources recovery

Resource recovery from wastewater containing hazardous oxoanions by hydrothermal mineralization

T. Itakura[1], R. Sasai[2] & H. Itoh[2]
[1]Department of Applied Chemistry, Graduate School of Engineering, Nagoya University, Japan
[2]EcoTopia Science Institute, Nagoya University, Japan

Abstract

We developed a new treatment method for wastewater containing various harmful ions such as arsenite, arsenate, boric, fluoride and fluoroboric ions by hydrothermal mineralization using $Ca(OH)_2$ as a mineralizer. As a result, complete recovery of these ions was attained regardless of the initial concentration and oxidation number of these harmful ion species in wastewater. Therefore, the present hydrothermal treatment using $Ca(OH)_2$ mineralizer is recommended as one of the most effective techniques to remove these ions from wastewater and recover them as recyclable resource.
Keywords: hydrothermal mineralization, arsenite, arsenate, recovery.

1 Introduction

Various oxoanions such as arsenite and boric acid have high toxicity against human health and the environment. They are important resource, however, for plating or advanced material manufacturing industries, from which the wastewater containing these oxoanions is generated everyday. Several methods to remove them have been already reported by using adsorption, electro-coagulation, membrane and biological techniques [1–5]. But, these methods have the following problems. (1) Removal yield is low. (2) The applicable concentration range is narrow. In addition, used adsorbent or collected residues are still hazardous wastes, so that they must be treated by proper method, though it is very difficult to convert them to recyclable resource in various industries. These problems are caused by the difficulty to recover these oxoanions as stable

WIT Transactions on Ecology and the Environment, Vol 92, © 2006 WIT Press
www.witpress.com, ISSN 1743-3541 (on-line)
doi:10.2495/WM060221

solid precipitates with low solubility in water. The establishment of recycling system of these hazardous compounds will be one of the world-important issues, especially, in Japan, which is poor in natural mineral.

In the present study, the recovery method of boron, fluorine and arsenic from wastewater containing fluoride, boric, fluoroboric, arsenite and arsenate ions were developed to produce reusable minerals by the hydrothermal treatment, which was analogous to the formation process of minerals in nature [6].

2 Experimental

Model synthetic wastewaters containing 1–3000 ppm of boric, fluoroboric, arsenite and arsenite ions were prepared by dissolving B_2O_3 (Wako Pure Chemical Industries, Ltd.), hydrofluoric acid (48 wt percent, chemical supplier: *ditto*), fluoroboric acid (48 wt percent, *ditto*), As_2O_3 (*ditto*), Na_2HAsO_4 (*ditto*) in distilled and deionized water. These model wastewaters (30 ml) were sealed in a pressure vessel lined with fluorocarbon resin together with reagents. Mineralizer $Ca(OH)_2$ was added into the vessel and in some cases, H_3PO_4 or H_2O_2 was also added in order to increase recovery yield. Hydrothermal treatments were carried out by leaving the vessel in a dry oven for 2 – 36 h at 100 – 200°C. After the hydrothermal treatment, the vessels were cooled down in atmospheric air for 1 h. Precipitates obtained by the hydrothermal treatment were filtered and collected.

The precipitates were identified by X-ray diffraction (XRD: RIGAKU Rint-2500) using CuKα radiation. The microstructural observation and qualitative element analysis of the precipitates were performed by scanning electron microscopy (SEM: JEOL JSM-T20) equipped with energy dispersive X-ray spectrometer (EDS: JED-2140). Thermometric analysis of the precipitates was carried out by thermogravimetry and deferential thermal analysis (TG-DTA: RIGAKU Thermo Plus2 TG8120). Quantitative element analysis of the solvent obtained after hydrothermal treatment was carried out by the inductively couple plasma-atomic emission spectrometry (ICP-AES: Perkin-Elmer Optima3300DV). Concentration of fluoride ion in the treated-water was measured by ion chromatograph (Shimadzu, Shim-pack IC-A3) with conductivity detector (Shimadzu, CDD-10A). Quantitative analysis of the arsenate in water was carried out by molybdenum blue method. Oxidation to determine the total arsenic content in the solvent was carried out by hydrothermal oxidation in concentrated HNO_3 aqueous solution (0.2 dm^3 / 10 dm^3 of treated-water) at 200°C for 12 h.

3 Results and discussion

3.1 Recovery of F from wastewater containing fluoride ion

Figure 1 shows the ion chromatograms of the model wastewater before and after the hydrothermal treatment using 0.5 g of $Ca(OH)_2$ as mineralizer at 200°C for 4h. The peak of fluoride ion was observed in the chromatogram of wastewater before the treatment and its concentration was 18 ppm. On the other hand, the peak disappeared after the treatment. This suggests that the fluoride ion in the wastewater is completely consumed to form precipitates under the hydrothermal conditions using Ca(OH)$_2$ mineralizer. XRD patterns of the precipitates before and after the treatment exhibited that they consisted of CaF$_2$. However, the concentration of fluoride ion in the treated-water was below enough compared with the concentration that is calculated from the solubility of CaF$_2$. It is considered that all fluoride ions in the model wastewater would be precipitated as CaF$_2$ under the hydrothermal condition. Additionally, it is expected that CaF$_2$ produced under hydrothermal condition may prevent the redissolution in the cooling process.

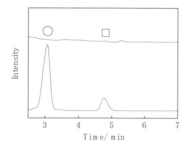

Figure 1:   Ion chromatograms of the wastewater before (a) and after (b) the hydrothermal treatment. ○; F$^-$, □; CO$_3^{2-}$ (F-; 7000 ppm, Ca(OH)$_2$; 0.5g, 200°C, 4 h).

Result of SEM observation of the precipitates obtained before and after the hydrothermal treatment showed that the crystallinity and crystal size of CaF$_2$ increased dramatically by the hydrothermal treatment. Thus, the recovery of fluorine from wastewater was achieved by decreasing dissolution rate of CaF$_2$ at room temperature because of decrease in specific surface area. Therefore, the present hydrothermal mineralizing treatment can recover fluorine completely from wastewater using the minimum amount of Ca(OH)$_2$ required to form CaF$_2$.

3.2 Recovery of B from wastewater containing boric acid

Figure 2 shows the result of hydrothermal mineralization treatment for the model wastewater containing 500 ppm of boron in case of adding Ca(OH)$_2$ and H$_3$PO$_4$. It is found that boric acid in the model wastewater decreases considerably in these treatments. However, the concentration was still higher than 100 ppm in case of using only Ca(OH)$_2$. The reason may be caused by redissolution of the precipitate during the cooling process after hydrothermal treatment. On the other hand, the concentration of boron in the treated-water decreased to ca. 5 ppm in case of using both Ca(OH)$_2$ and H$_3$PO$_4$. XRD pattern showed that the mineral formed in this process is Ca$_2$B$_2$O$_5$·H$_2$O (parasibirskite) and Ca$_5$(PO$_4$)$_3$(OH) (hydroxyl apatite). In order to clarify the crystallization mechanism of calcium phosphate, the variation of diffraction patterns during hydrothermal treatment was examined in detail. Diffraction peaks of $CaHPO_4 \cdot H_2O$ observed before the treatment disappeared gradually with an increase in treatment time. New diffraction peaks of both $CaHPO_4$ and $Ca_{10}(PO_4)_6 \cdot 5H_2O$ appeared after the hydrothermal treatment for 6 h. Then, the diffraction peaks originated from only $Ca_{10}(PO_4)_6 \cdot 5H_2O$ was observed, when treatment time became longer than 12 h. These results indicates that the $CaHPO_4 \cdot H_2O$ contained in the precipitate before the treatment converts into $Ca_{10}(PO_4)_6 \cdot 5H_2O$ via $CaHPO_4$ during longer hydrothermal treatment time. In contrast, the required treatment time to crystallize the $Ca_2B_2O_5 \cdot H_2O$ from the model wastewater in case of using only $Ca(OH)_2$ was 6 h, which was shorter than that for formation of $Ca_{10}(PO_4)_6 \cdot 5H_2O$. Figure 3 shows the SEM photograph obtained by hydrothermal mineralization treatment in using both $Ca(OH)_2$ and H_3PO_4. The fine particles of hydroxyl apatite, which would be separated out and grown on the precipitate of $Ca_2B_2O_5 \cdot H_2O$ and residual $Ca(OH)_2$, can be seen. Therefore, the capsulation with dense intercepting layer of $Ca_{10}(PO_4)_6 \cdot 5H_2O$ is considered to prevent the redissolution of $Ca_2B_2O_5 \cdot H_2O$ into water.

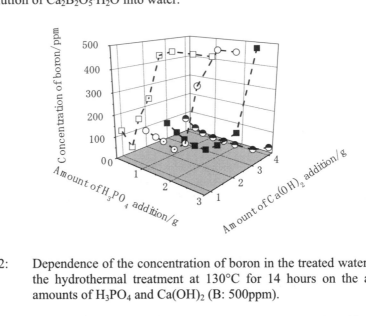

Figure 2: Dependence of the concentration of boron in the treated water after the hydrothermal treatment at 130°C for 14 hours on the added amounts of H_3PO_4 and $Ca(OH)_2$ (B: 500ppm).

3.3 Recovery of B and F from wastewater containing fluoroboric acid

Figure 4 shows the treatment time dependence of the B and F concentrations in the wastewater treated at 150°C. The significant enhancement of recovery yield of fluorine was observed at 2 h and it was completed by 4 h. However, the recovery yield of boron was only 30% at 2 h, and then gradually increased. XRD patterns of precipitates before the hydrothermal treatment showed only the diffraction peaks of $Ca(OH)_2$. On the other hand, the diffraction peaks of CaF_2 and $Ca_2B_2O_5 \cdot H_2O$ were observed after the treatment and the intensities of diffraction peaks of $Ca_2B_2O_5 \cdot H_2O$ increased up to 24 h. Thus, it is expected that the decomposition of fluoroboric acid takes place during the treatment, and the recovery of F and B is achieved by forming CaF_2 and $Ca_2B_2O_5 \cdot H_2O$, respectively, in the same manner as the case of wastewater containing fluoride or boric ion only. Therefore, the thermal decomposition of BF_4^- would occur at the initial stage of treatment (within 4 h), and then the mineralization reaction between Ca^{2+}, and F^- / $B(OH)_4^-$ would be followed.

Figure 3:   SEM photograph of precipitate obtained by the hydrothermal treatment at 130°C for 14 hours ($Ca(OH)_2$: 3.0g, H_3PO_4:1.5g, B: 500ppm).

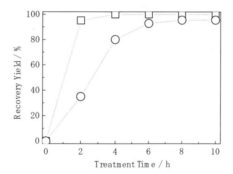

Figure 4:   Dependence of recovery yield of B and F in the treated-water on treatment time at 150°C (BF_4^-: 8000 ppm, $Ca(OH)_2$: 1.0g). ○: B, □: F.

The optimal conditions to recover both F and B from model wastewater containing 8000 ppm fluoroboric ion were at 200°C for 36 h, when the concentrations of F and B were 0.3 ppm and 20 ppm, respectively. As described in the previous section, we observed that $Ca_2B_2O_5 \cdot H_2O$ redissolved in aqueous solution during the cooling process in the case of the treatment for boric acid and the boron concentration in the treated-water was ca. 100 ppm on account of its solubility. However, the boron concentration in the case of fluoroboric acid solution was reduced down to ca. 20 ppm even in the absence of the inhibition reagent against redissolution, phosphoric acid. Therefore, it is considered that the coexistence of CaF_2 would affect the increase in recovery yield of boron.

SEM photographs of a bulky precipitate obtained by the hydrothermal treatment were shown in Figure 5. Three layers were observed in the overview photograph (Figure 5-a). From the results of EDS and XRD analyses, the first surface layer (Figure 5-b) was CaF_2, the second layer (Figure 5-c) was the mixture of CaF_2 and $Ca_2B_2O_5 \cdot H_2O$, and the third layer (Figure 5-d) was the mixture of $Ca_2B_2O_5 \cdot H_2O$ and residual $Ca(OH)_2$. As a result of detailed analysis, the formation of $Ca_2B_2O_5 \cdot H_2O$ layer in this study would have started in an earlier time range of 2 – 4 h and completed for 8 – 10 h by heterogeneous nucleation on the surface of $Ca(OH)_2$, after which the suspended CaF_2 wrapped over $Ca_2B_2O_5 \cdot H_2O$ because of the slow sedimentation rate of CaF_2 fine particles. Possibly this dense sediment layer would play a role to inhibit the redissolution of $Ca_2B_2O_5 \cdot H_2O$ into aqueous media.

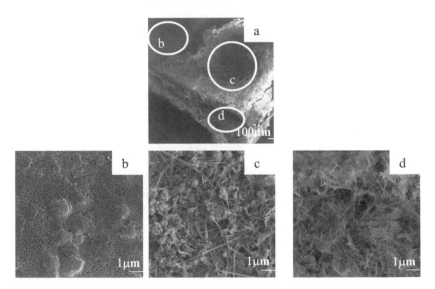

Figure 5: SEM photographs of the precipitates obtained by the hydrothermal treatment. a; over view of precipitate, b; first layer, c; second layer, d; third layer (BF_4^-: 8000 ppm, $Ca(OH)_2$: 1.0g, 150°C, 24 hours).

3.4 Recovery of As from wastewater containing arsenite and arsenate ions

Treatment time dependence of As concentration in the model wastewater containing 2000 ppm of arsenite (AsO_3^{3-}) treated at 100 - 150°C with or without 5% of H_2O_2 is shown in Figure 6. Precipitate containing As was observed by adding $Ca(OH)_2$ to the model wastewater, when the concentration of As was reduced to 4 ppm. However, this does not meet the standard of discharged water in Japan (0.1 ppm). When the hydrothermal mineralization was performed in using this model wastewater added by $Ca(OH)_2$ only, As concentration showed concave curve against treatment time. The optimum treatment condition showed the minimum As concentration of ca. 0.4 ppm, which was one tenth of that before the hydrothermal treatment. However, this concentration is still higher than the standard of discharged water. On the other hand, the addition of H_2O_2 was found effective to reduce the As concentration down to 0.02 ppm under optimal hydrothermal condition, which is lower than the standard of discharged water. This result verifies that the hydrothermal mineralization is an effective method to recover arsenite (AsO_3^{3-}) from aqueous media as precipitate, even when the model wastewater contains large amount of arsenite. It is suggested, therefore, that the As recovery mechanism in this treatment is considerably different from that of the conventional lime precipitation method.

Various analyses on the obtained precipitates were carried out in order to clarify the mechanism of the As recovery by the hydrothermal mineralization. From XRD, SEM-EDS and TG-DTA analyses, the precipitate obtained only by addition of $Ca(OH)_2$ was identified as $Ca_3(AsO_3)_3(OH)·4H_2O$. The same analysis of the precipitate after the hydrothermal treatment using $Ca(OH)_2$ showed that the crystal water was eliminated. The solubilities of $Ca_3(AsO_3)_3(OH)·4H_2O$ and $Ca_3(AsO_3)_3(OH)$ to water were estimated 13.79 mg As / 100 dm^3 and 29.10 mg As / 100 dm^3, respectively by a simple solubility test. Therefore, the concave tendency in Figure 1 would be caused by the elimination of crystal water from $Ca_3(AsO_3)_3(OH)·4H_2O$ with an increase in the treatment time. On the other hand, XRD patterns of the precipitates obtained by hydrothermal treatment with $Ca(OH)_2$ and H_2O_2 exhibited $Ca_5(AsO_4)_3(OH)$ as final product. Thus, the coexistence of H_2O_2 with $Ca(OH)_2$ immediately would give rise to the oxidation of arsenite ion and produce simultaneously arsenate apatite which is known as one of the insoluble natural mineral. Therefore, it is concluded that the hydrothermal mineralization with $Ca(OH)_2$ and H_2O_2 is effective to reduce the concentration of As in the wastewater with arsenite ion (AsO_3^{3-}), which is usually difficult to remove from wastewater.

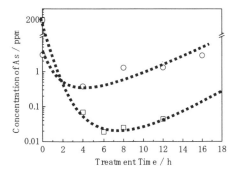

Figure 6: Dependence of the concentration of As in the treated water on treatment time. $Ca(OH)_2$; 0.36g. ○; 150°C, without H_2O_2, □; 150°C, with 5% H_2O_2.

Figure 7 shows the initial arsenite (AsO$_3$$^{3-}$) concentration dependence of arsenite in the treated-water using Ca(OH)$_2$ and 3% of H$_2$O$_2$ at 100°C for 12h. The residual As concentration was reduced at less than 0.1 ppm except for the case of the wastewater containing 2000 ppm of As. Moreover, it was mostly independent on the initial concentration. This result suggests that the As concentration after the treatment may be determined by only the solubility of the precipitate produced during the hydrothermal mineralization, if the amount of H$_2$O$_2$ is enough to convert the arsenite ions (AsO$_3$$^{3-}$) into the arsenate ions (AsO$_4$$^{3-}$). Whereas, this hydrothermal conditions for model wastewater containing 2000 ppm of arsenite (AsO$_3$$^{3-}$) could not sufficiently decrease the residual As concentration. However, the addition of 5% of H$_2$O$_2$ achieved the As concentration reduction less than 0.1 ppm, even when As concentration was 2000 ppm. This result shows that the lowest limit of added H$_2$O$_2$ amount may be fixed by the initial concentration of arsenite dissolved in water. Figure 8 shows the As recovery from model wastewater containing arsenate (AsO$_4$$^{3-}$) or mixture of arsenate (AsO$_4$$^{3-}$) and arsenite (AsO$_3$$^{3-}$) by the treatment with Ca(OH)$_2$ and 3% of H$_2$O$_2$. The As in the model wastewater was completely reduced for 12 h by the treatment. In addition, the treatment for model wastewater containing the mixture of arsenate and arsenite decrease As concentration significantly with 3% of H$_2$O$_2$ addition. These results support the speculation for addition of H$_2$O$_2$ because the maximum concentration of arsenite in this mixed solution was 1000 ppm. The concentration of As in the treated water depends only on the solubility of arsenate apatite when enough amounts of H$_2$O$_2$ and Ca(OH)$_2$ were added. Therefore, it was elucidated that the hydrothermal mineralization treatment could recover As regardless of the initial concentration and oxidation number of As.

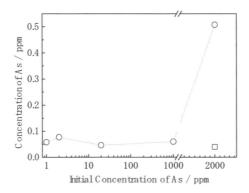

Figure 7:   Dependence of the concentration of As in the treated-water on initial concentration of arsenite . Ca(OH)$_2$; 0.36g, 100°C, ○;3% H$_2$O$_2$, □; 5% H$_2$O$_2$

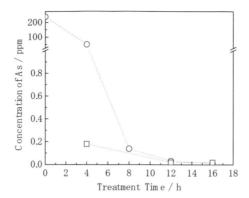

Figure 8:   Dependence of the concentration of As in the treated-water on treatment time.   Ca(OH)$_2$; 0.36g, 3% H$_2$O$_2$, 100°C, ○; AsO$_3^{3-}$ 1000ppm, AsO$_4^{3-}$ 1000ppm, □; AsO$_4^{3-}$ 2000ppm,

4   Conclusions

The hydrothermal mineralization treatment can recover boron, fluorine and As from model wastewater containing fluoride, fluoroboric, arsenite and arsenate ions.   All concentrations of these harmful elements in the synthetic model wastewater were reduced down below the standard of discharged water in Japan. The minerals formed in this treatment had the same composition as natural ones. Thus, they can be reused easily in the production processes of pure raw materials from natural minerals.   Furthermore, this treatment is independent on the initial concentration and oxidation number of ions.   Therefore, the present hydrothermal mineralization treatment can be used for various kinds of wastewaters.

References

[1]   Kartinen, E. O.; Martin, C. J., An overview of arsenic removal processes. *Desalination* 1995, 103, (1-2), 79-88.
[2]   Al Rmalli, S. W.; Harrington, C. F.; Ayub, M.; Haris, P. I., A biomaterial based approach for arsenic removal from water. *Journal of Environmental Monitoring* 2005, 7, (4), 279-282.
[3]   Ho, L. N.; Ishihara, T.; Ueshima, S.; Nishiguchi, H.; Takita, Y., Removal of fluoride from water through ion exchange by mesoporous Ti oxohydroxide. *Journal of Colloid and Interface Science* 2004, 272, (2), 399-403.
[4]   Celik, M. S.; Hancer, M.; Miller, J. D., Flotation chemistry of boron minerals. *Journal of Colloid and Interface Science* 2002, 256, (1), 121-131.

[5]   Redondo, J.; Busch, M.; De Witte, J. P., Boron removal from seawater using FILMTEC (TM) high rejection SWRO membranes. *Desalination* 2003, 156, (1-3), 229-238.

[6]   Itakura T., Sasai R., Itoh H., Precipitation recovery of boron from wastewater by hydrothermal mineralization. Water Research 2005, 39 (12), 2543-2548.

Metal content and recovery of MSWI bottom ash in Amsterdam

L. Muchová & P. C. Rem
Delft University of Technology,
Faculty of Civil Engineering and Geosciences, The Netherlands

Abstract

Incineration reduces the mass of municipal solid waste (MSW) by 70% to 80%, and it reduces the volume by 90%. The resulting fractions of the incineration are bottom ash, fly ash and flue gas residue. Bottom ash is by far the largest residue fraction. About 1.1 million tons of bottom ash are produced in the Netherlands and about 20 million tons in Europe, every year. The production of bottom ash is rising because MSW is increasingly incinerated. Bottom ash is land filled in many European countries. However, the material is suitable as a building material from a civil engineering viewpoint, e.g. for embankments and foundations of roads.

Bottom ash contains a considerable amount of non-ferrous and ferrous metals that should be removed for such an application. The recovery of these metals improves the engineering and environmental properties of the ash, and creates a financial benefit. Conventional dry physical methods recover only a small part of the metal value from the ash. This study gives a mass balance for the metal recovery plant of the Amsterdam incinerator (AEB). The results are based on research experiments performed in a new pilot plant for the wet physical separation of bottom ash. The recovery of ferrous and non-ferrous metals is above 70%. The metals that are found in the ash pay for a substantial part of the separation process.

Keywords: bottom ash, metal content, physical separation.

1 Introduction

Bottom ash from the incineration of household waste is usually treated by magnetic and eddy current separation (for the coarse particle fraction) or it is not treated at all. Therefore, this type of bottom ash represents a large potential for

WIT Transactions on Ecology and the Environment, Vol 92, © 2006 WIT Press
www.witpress.com, ISSN 1743-3541 (on-line)
doi:10.2495/WM060231

the recovery of valuable secondary ferrous and non-ferrous metals. The new pilot plant in Amsterdam can recover both the coarse and the fine ferrous and non-ferrous metals by wet physical separation.

The concentration of ferrous and non-ferrous metals in MSWI bottom ash varies for each country and for each incinerator. The raw Amsterdam bottom ash contains 13% ferrous metals and 2,2% of non-ferrous metals. The bottom ash from the incinerator of Rotterdam contains 1,2% of aluminium, 0,2% of copper, 0,2% of heavy alloys and 9,9% of ferrous metals [1]. Chimenos et al. [2] investigated two kinds of bottom ashes from different locations in Spain. The concentration of non-ferrous metals found in the Spanish bottom ash varies between 2% and 5% and the concentration of ferrous metals varies between 5% and 12%.

Physical separation is the most attractive option for the metals recovery from bottom ash. The dry mechanical processing of incineration slag from household waste was tested at the RWTH Aachen for a particle size below 40 mm [3]. The Aachen process applied eddy current separation, jigging and dense-medium separations. The jig processed the 10-40 mm fraction and the 0,5-4 mm fraction. The heavy medium separation was used for the 10-20 mm fraction and for the 20-40 mm fraction. Each fraction was smelted and the results are shown in Table 1. The process from RWTH shows the concept of multi-step separation of bottom ash. The separation plant in Amsterdam uses dry separation at the first step. The second part of the plant uses wet separation to separate the fraction below 40 mm [4]. The advantage of wet separation compared to dry separation is a higher recovery.

Table 1:   Composition and melting recovery of heavy and light product from RWTH Aachen, (in % by wt.).

	<4mm (light)	4-40mm (light)	<10mm (heavy)	10-40mm (heavy)
Al	95	94	2,4	0,3
Heavy non-ferrous metals	2,5	4	70	89
Ag			1,5-2	
Melting recovery	74,4	88,4	75	88

2   Separation plant in Amsterdam (the Netherlands)

The former dry physical separation plant in Amsterdam processed approximately 865 000 tons of bottom ash in 2005. In the same year, the new wet separation pilot plant was built and optimised to recover more ferrous and non-ferrous metals from the pre-separated bottom ash. Table 2 shows the separated amounts for relevant fractions in 2005, excluding the metals separated by the new wet separation plant.

2.1 Metal content in Amsterdam bottom ash

The total bottom ash treatment in Amsterdam consists of 3 steps. The first step is the former bottom ash treatment where the bottom ash is screened and shredded to below 40 mm. The main purpose of this step is to remove the ferrous scrap and the very coarse non-ferrous metals. At the second step the fraction between 0-40 mm is separated by an eddy current separator. The eddy current separator removes the +15 mm non-ferrous metals. The residue from the eddy current separator continues to the new wet physical separation pilot plant where the 0-20 mm ferrous and non-ferrous metals are removed. The system of three separation steps is shown in Fig.1.

Table 2:   Fractions produced in the existing bottom ash dry separation plant.

	Ton /Year
Total household waste	865.000
Bottom ash	205.000
Ferrous scrap from bottom ash	15.200
Non-ferrous scrap from bottom ash	2.100

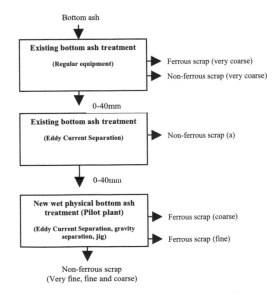

Figure 1:   Scheme of the separation system in the AEB Amsterdam.

2.2 The new wet separation plant for the 0-40 mm fraction

The wet pilot plant separates the pre-treated bottom ash based on the size distribution and the density of the particles. The non-ferrous fractions are removed by eddy current separation, density separation [5] and jigging. The resulting fractions are the coarse non-ferrous (6-20 mm), fine non-ferrous (2-6 mm) and the very fine non-ferrous (<2 mm) product. The 2-6 mm aluminium scrap is separated from the heavy non-ferrous by density separation.

The grade and the recovery of magnetic, non-ferrous and aluminium fraction is shown in Table 3.

Table 3:   The grade and recovery of the magnetic, non-ferrous and aluminium fractions from the wet separation pilot plant.

	Magnetic fraction*		Non-ferrous fraction		Aluminium fraction	
	Grade (%)	Recovery (%)	Grade (%)	Recovery (%)	Grade (%)	Recovery (%)
6-20 mm	62	90	87	77	n.d.	n.d.
2-6 mm	n.d.	n.d.	80	81	82	83
<2mm	98	29	29	40	n.r.	n.r.

* the magnetic fraction means all magnetic particles removed by magnetic separator.
n.d. – not determined; n.r. – not recovered.

The wet separation pilot plant has a capacity of 50 t/h. The grade and the recovery of the fine and the coarse fraction were measured with the full capacity of the plant. The recovery of the very fine non-ferrous fraction was performed in the laboratory with a magnetic separator that was not optimised. The result from this experiment was added into Table 3. The very fine fraction has to be improved in the future.

2.3 Experimental result: the grade and recovery of Amsterdam's bottom ash

The grade and recovery of the combined separation (wet and dry physical separation) is shown in Table 4. The ferrous concentration of the 6-20 magnetic fractions was estimated as 20%.

Table 4:   The grade and recovery of dry separation and wet separation for ferrous scrap, non-ferrous and aluminium fraction.

	Ferrous scrap		Non-ferrous fraction		Aluminium fraction	
	Grade (%)	Recovery (%)	Grade (%)	Recovery (%)	Grade (%)	Recovery (%)
Dry physical separation -existing plant	n.d.	75	75	35	n.d.	n.d.
Wet physical separation (average) - new plant	83	8	80-87	38	82	83
Total		83		73		

n.d. – not determined.

2.4 Precious metals in the Amsterdam's bottom ash

The content of precious metals in the fine and very fine non-ferrous concentrates removed by the wet physical separation was measured. The non-ferrous metals from the fine and very fine fraction were smelted. The smelts were analysed by drilling holes of 6 mm diameter through the centre. The drillings were milled and then analysed by XRF. Table 5 shows the average result from the XRF. The results are based on four smelts for each fraction.

Table 5:   Precious metal content of fine and very fine fraction analysed by XRF (in ppm).

Element	Fine fraction	Very fine fraction
Ag	2700	n.d.
Au	82	109
Pt	14	n.d.

n.d. – not determined.

The experimental work with the smelting and analysing by XRF was done by our own procedure with smelting samples of 2 kg each. An external smelter laboratory was contacted to correlate our result with a bigger sample of 300 kg. The result from the smelter measured a silver concentration of 3700 ppm and a gold concentration of 100 ppm. The differences between the results are believed to be due to the heterogeneity of the material.

3 Conclusion

The physical separation of bottom ash is successful in recovering metals from the ash at high grade. The second advantage of the separation is minimising the heavy metals in the residue which then is more suitable for the building industry. The physical separation plant in Amsterdam can remove non-ferrous metals with 73% recovery and 83% recovery of ferrous scrap. The grades of the fractions satisfy the limits of secondary smelters.

The coarse non-ferrous mixture from the wet physical separation pilot plant has a similar quality to the products of dry processing lines for bottom ash. The prices of the ferrous and non-ferrous metals rapidly increased the last year. The prices offered by sink–floaters for this kind of mixture are about 640 euro/t of non-ferrous at grades better than 75% metal. Table 6 shows the price of the ferrous, non-ferrous and aluminium scrap removed by the physical separation from the Amsterdam's bottom ash. In Table 6 a value of 640 euro/ton for the non-ferrous fraction was used for the ferrous scrap 140 euro/ton and for aluminium 700 euro/ton.

Table 6:   Prices of the metals removed from the separation plant in Amsterdam in 2005.

	Ferrous scrap (euro/year)	Non-ferrous scrap (euro/year)	Aluminium scrap (euro/year)
Dry physical separation-existing plant	1.596.000	1.102.500	n.d.
Wet physical separation (average) -new pilot plant*	566.836	1.000.992	155.400
Total	2.162.836	2.103.492	155.400

n.d. – not determined.
* The prices for the metals (separated by pilot plant) are established based of the grade from the experimental result not based on the yearly production. The precious metals were excluded from the calculation.

References

[1]   Waterman, M., Van Houwelingen, J. Economische terugwinning van aluminium uit bodemassen van afvalverbrandingsinstallaties (AVI's). Aluminium 5, 1997.

[2]   Chimenos, J.M., Segarra, M., Fernandez, M.A., Espiell, F., Characterization of the bottom ash in municipal solid waste incinerator. Journal of hazardous materials 64, 211-222, 1999.

[3]   Schmelzer, G., Wolf, S., Hoberg, H. New wet treatment for components of incineration slag. AT Aufbereitungs Technik, 37, 1996.

[4]   Rem, P. C., De Vries, C., Van Kooy, L., Bevilacqua, P., Reuter, M. The Amsterdam pilot on bottom ash, Minerals engineering, 17, 2004.

[5]   Van Kooy, L., Mooij, M., Rem, P., Kinetic gravity separation. Physical Separation in Science and Engineering, Vol. 13, No. 1, pp.25-32, 2004.

Effective utilization of waste: development of CH_4 dry reforming catalysts from spent nickel metal hydride battery for resource recovery

T. Kanamori, R. Hayashi, M. Matsuda & M. Miyake
Department of Material and Energy Science,
Graduate School of Environmental Science, Okayama University,
Tsushima-Naka, Okayama, Japan

Abstract

The resource recovery of nickel metal from the spent nickel metal hydride (Ni-MH) battery was investigated by using the CH_4 dry reforming, aiming at reusing it as a raw material of the Ni-MH battery. From the results, a compound identified as a single phase of NiO by XRD was successfully prepared from the spent Ni-MH battery by a series of chemical processes using HCl and NH_3 aq. solutions followed by calcinations. The resulting NiO exhibited excellent CH_4 conversion in the CH_4 dry reforming, i.e., the CH_4 conversion higher than 96% continued for 50 h when the flow rates of injection gases were controlled at $CH_4/CO_2/Ar = 10:10:80$ ml·min^{-1}. Although the XRD pattern revealed the reduction of NiO to Ni^0 and the deposition of carbon after the CH_4 dry reforming, the deposition of carbon could be suppressed by changing the injection gas ratio. The resource recovery of nickel metal was, therefore, concluded to be possible through the CH_4 dry reforming over NiO prepared from the spent Ni-MH battery.
Keywords: resource recovery, nickel metal hydride secondary battery, nickel metal, CH_4 dry reforming, nickel-based catalyst.

1 Introduction

Resource recovery of waste is one of important subjects for sustainable development. Technological development of resource recovery has, however, fallen behind, compared with that of creation of high functional materials using pure raw materials. In the battery field, nickel metal hydride (Ni-MH) batteries

WIT Transactions on Ecology and the Environment, Vol 92, © 2006 WIT Press
www.witpress.com, ISSN 1743-3541 (on-line)
doi:10.2495/WM060241

with high power and rechargeable features have been developed and widely used as power sources of cell phones, digital cameras, hybrid cars, etc. On the other hand, about four hundreds tons of spent Ni-MH batteries have been disposed in Japan even though they contain valuable transition and rare-earth elements. It is, therefore, desired to develop environmentally-friendly resource recovery technologies of spent Ni-MH batteries.

CH_4 dry reforming, Eq. (1), which produces synthesis gas with lower H_2/CO ratio than steam reforming, Eq. (2), attracted a great deal of attention, because the syntheses of liquid fuels such as methanol and dimethyl ether require the synthesis gases with lower H_2/CO ratio.

$$CH_4 + CO_2 \rightarrow 2H_2 + 2CO \qquad H_2/CO = 1 \qquad (1)$$
$$CH_4 + H_2O \rightarrow 3H_2 + CO \qquad H_2/CO = 3 \qquad (2)$$

Furthermore, CH_4 in natural gas and CO_2 in flue gas can be effectively used in the CH_4 dry reforming. The natural gas is cleaner energy source than oil and coal, since it contains CH_4 as a main component, little NO_x and no SO_x.

It has been known that noble metal (Pt, Rh, Ru, Pd and Ir) supported catalysts had good catalytic performance for the CH_4 dry reforming [1–9]. However, the development of cheap metal catalyst has been desired, because the noble metal catalyst is not suitable for utilizing on a large scale. Supported Ni-based oxides [10, 11] and lanthanoid perovskite type oxides [12–15] are promising as the alternative catalysts for this reaction. We took note of the CH_4 dry reforming, referring to previous reports, because spent Ni-MH batteries contained Ni as a main element. In the present study, we report the resource recovery process of Ni from Ni-MH batteries by using the CH_4 dry reforming, aiming at reusing it as a raw material of Ni-MH batteries.

2   Experimental

Anode material of spent Ni-MH battery, which was provided by Mitsui Mining and Smelting Co. Ltd., Japan, was employed in the experiment. Nickel component was separated from the anode material according to a procedure as shown in fig. 1.

The anode material was characterized by powder X-ray diffraction (XRD; Rigaku RINT2100/PC) with monochromated CuKα radiation and an X-ray fluorescence technique (XRF; Rigaku RIX3000). The separated specimens were identified by XRD, and their chemical compositions were analyzed by an inductively coupled plasma atomic emission spectrometer (ICP-AES; Seiko Instruments SPS7700). Thermal analysis (TG-DTA; Rigaku TG 8120) was carried out to estimate amounts of deposited carbon over the samples.

The CH_4 dry reforming was performed at 780°C under atmospheric pressure, using a vertical fixed-bed reactor of quartz tube with 12 mmφ in internal diameter. The resulting powder sample of 1 g was held in place by glass wool. The flow rates of injection gases were controlled in the range from $CH_4/CO_2/Ar$ = 10:10:80 to 6:14:80 ml·min⁻¹. The vent gases were cooled with an ice-cooler to remove H_2O produced by side reactions. The gaseous products were analyzed by an on-line gas chromatograph (Shimadzu GC-8A) equipped with a thermal conductivity detector.

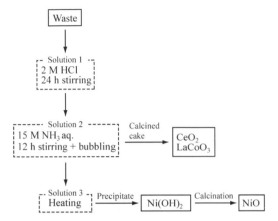

Figure 1:   Flow chart of separation process from anode material of spent Ni-MH battery.

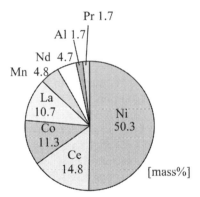

Figure 2:   Chemical compositions of anode material of spent Ni-MN battery.

3   Results and discussion

3.1 Sample preparation process

The chemical compositions, as shown in fig. 2, reveal that the anode material of spent Ni-MH battery contains Ni as a main element, Ce as a second element and much Co, La, Mn and Nd next to them. Main crystalline phases were identified as LaNi$_5$ by XRD. In the beginning of the recovery process, pre-experiments were carried out for the anode material of spent Ni-MH battery to search optimum conditions. As a result, the proper recovery process was determined as shown in fig. 1.

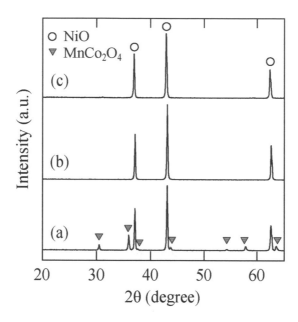

Figure 3:   XRD patterns of products obtained with (a) stirring for 12 h, (b) stirring for 24 h and (c) stirring and bubbling air for 12 h.

The anode material of 10 g was treated with 2 M (mol· dm^{-3}) HCl solution of 200 ml for 24 h with stirring (solution 1). The anode material was completely dissolved into HCl solution, when HCl solutions with higher concentration than 2 M were used. Next, 15 M NH$_3$ aq. of 300 ml was added to "solution 1" to form Ni-ammonium complexes as well as precipitate dissolved metal ions as their hydroxides except nickel ions, and the solution (solution 2) was stirred for fixed time. After filtration, the obtained reddish purple filtrate was slowly heated around 90°C to decompose the Ni-ammonium complexes and to precipitate Ni component as Ni(OH)$_2$. NH$_3$ gas, which came out from the filtrate by heating, was recovered by passing in water. The obtained precipitates were calcined at 1000°C for 1 h to dehydrate into oxides. The XRD patterns of the calcined precipitates are presented in figs. 3a and b. The resulting oxides were identified as NiO, accompanied with a small amount of MnCo$_2$O$_4$ when the stirring time was 12 h, whereas they were identified as a single phase of NiO when the stirring time was 24 h. This suggests that a long time stirring helps manganese and cobalt ions   precipitate as hydroxides and/or oxides through contacting with air. Thus, solution 2 was stirred with bubbling air to precipitate manganese and cobalt ions for 12 h, and the filtrate was heated around 90°C. Consequently, the resulting oxides were identified as a single phase of NiO, as shown in fig. 3c. Namely, the treatment with bubbling air successfully prevented MnCo$_2$O$_4$ from mixing with NiO. The chemical compositions of the products by

ICP-AES are listed in table 1. Although manganese component was detected in the products obtained from solution 2 with stirring for 12 and 24 h, it was hardly detected in that with stirring and bubbling air for 12 h. Therefore, the stirring time with bubbling air was decided on 12 h. A slight amount of lanthanum was contained in every final product. It was considered that the resulting NiO formed a solid solution with CoO such as (Ni, Co)O containing a small amount of lanthanum.

Table 1:   Chemical compositions of products obtained from solution 2.

Stirring time	Metal content (mol%)			
	Ni	Co	Mn	La
12 h	82.6	13.9	3.5	0.02
24 h	84.8	15.1	0.07	0.03
12 h with bubbling air	85.7	14.3	0	0.03

The filter cake of "solution 2" was calcined at 1000°C and identified by XRD. As a result, it was found the calcined filter cake was a mixture of CeO_2 and La-based perovskite. Considering the chemical compositions of anode material as shown in fig. 2, both of them were supposed to form solid solutions.

3.2 CH₄ dry reforming

The CH_4 dry reforming reaction temperature was examined, considering the temperature ranges based on the free energy changes of the main and side reactions. As the estimation indicates that the temperature range from 700 to 835 °C is appropriate for the CH_4 dry reforming, the reaction temperature was decided on 780°C in the present study. This temperature is consistent with those reported by a lot of researches [4, 5, 7, 8, 14].

Figure 4 shows the CH_4 conversion over NiO prepared from the anode material as a function of reaction time. Although NiO showed excellent performance, the CH_4 conversion slightly decreased from ca. 98 to ca. 96% for 50 h reaction, when the flow rates of injection gases were controlled at $CH_4/CO_2/Ar = 10:10:80$ ml·min⁻¹. The XRD pattern of NiO after the CH_4 dry reforming is displayed in fig. 5a. The XRD pattern exhibited that the NiO was reduced to Ni^0, and furthermore carbon coexisted with Ni^0 after the CH_4 dry reforming. No XRD peaks due to other than Ni^0 and carbon are observable, although Co component was contained in the prepared NiO. The appearance of carbon suggests that the CH_4 dry reforming reaction (1) shifts to reaction (3), which is a production reaction of carbon and water.

$$CH_4 + CO_2 \rightarrow 2C + 2H_2O \tag{3}$$

Next, the flow rates of injection gases were changed to $CH_4/CO_2/Ar = 8:12:80$ and 6:14:80 ml·min⁻¹ in order to suppress reaction (3). Decreasing CH_4 flow rate and increasing CO_2 flow rate were expected to restrain the carbon deposition and accelerate reaction (4), which is a consumption reaction of carbon.

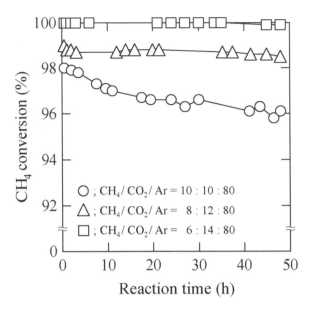

Figure 4: CH$_4$ conversion over prepared NiO at various gas flow rates as a function of reaction time.

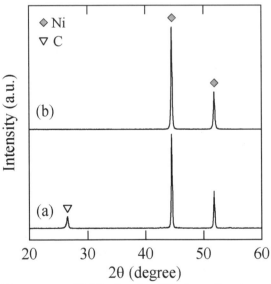

Figure 5: XRD patterns of NiO after the CH$_4$ dry reforming at CH$_4$/CO$_2$/Ar = (a) 10:10:80 and (b) 6:14:80 ml·min^{-1}.

$$C + CO_2 \rightarrow 2CO \qquad (4)$$

As a result, the CH$_4$ conversions were improved to higher than 98%, as seen in fig. 4. The XRD peaks due to Ni^0 were only detectable on the XRD pattern of the product after the CH_4 dry reforming, as seen in fig. 5b. Thermal analyses exhibited that amounts of deposited carbon in the products decreased from ca. 60 to 1.5 mass%. Namely, these results revealed that changing the flow rates of the injection gases from $CH_4/CO_2/Ar = 10:10:80$ to 8:12:80 and 6:14:80 ml·min^{-1} was effective in the stimulation of the CH_4 dry reforming reaction (1) as well as the inhibition of carbon and water production reaction (3).

In conclusion, the resource recovery of nickel metal was possible through the CH_4 dry reforming over NiO prepared from spent Ni-MH battery.

Acknowledgements

This work was supported by Grant-in-Aids for Scientific Research on Solid Waste Management (K1615 and K1731) from Ministry of the Environment, by a Grant-in-Aid for Scientific Research (B) (no. 17350101) from the Japan Society for the Promotion of Science, and in part by the Okayama University 21st Century COE Program "Strategic Solid Waste Management for Sustainable Society".

References

[1]   Rostrup-Nielsen, J. R. & Hansen, J. H. B., CO_2-reforming of methane over transition metals. *J. Catal.*, **144**, pp. 38–49, 1993.

[2]   Erdohelyi, A., Cserényi, J. & Solymosi, F., Activation of CH_4 and its reaction with CO_2 over supported Rh catalysts. *J. Catal.*, **141**, pp. 287–299, 1993.

[3]   Bitter, J. H., Seshan, K. & Lercher, J. A., Deactivation and coke accumulation during CO_2/CH_4 reforming over Pt catalysts. *J. Catal.*, **183**, pp. 336–343, 1998.

[4]   Mattos, L. V., Rodino, E., Resasco, D. E., Passos, F. B. & Noronha, F. B., Partial oxidation and CO_2 reforming of methane on Pt/Al_2O_3, Pt/ZrO_2, and $Pt/Ce–ZrO_2$ catalysts. *Fuel Process Technol.*, **83**, pp. 147–161, 2003.

[5]   Tsyganok, A. I., Inaba, M., Tsunoda, T., Hamakawa, S., Suzuki, K. & Hayakawa, T., Dry reforming of methane over supported noble metals: a novel approach to preparing catalysts. *Catal. Commun.*, **4**, pp. 493–498, 2003.

[6]   Paturzo, L., Gallucci, F., Basile, A., Vitulli, G. & Pertici, P., An Ru-based catalytic membrane reactor for dry reforming of methane—its catalytic performance compared with tubular packed bed reactors. *Catal. Today*, **82**, pp. 57–65, 2003.

[7]   Nagaoka, K., Okamura, M. & Aika, K., Titania supported ruthenium as a coking-resistant catalyst for high pressure dry reforming of methane. *Catal. Commun.*, **2**, pp. 255–260, 2001.

[8]   Stagg-Williams, S. M., Noronha, F. B., Fendley, G. & Resasco, D. E., CO_2 Reforming of CH_4 over Pt/ZrO_2 catalysts promoted with La and Ce Oxides. *J. Catal.*, **194**, pp. 240–249, 2000.

[9]   Ferreira-Aparicio, P., Rodríguez-Ramos, I., Anderson, J. A. & Guerrero-Ruiz, A., Mechanistic aspects of the dry reforming of methane over ruthenium catalysts. *Appl. Catal. A-Gen.*, **202**, pp. 183–196, 2000.

[10]   Irusta, S., Cornaglia, L. M. & Lombardo, E. A., Hydrogen production using Ni–Rh on La$_2$O$_3$ as potential low-temperature catalysts for membrane reactors. *J. Catal.*, **210**, pp. 7–16, 2002.

[11]   Inui, T., Spillover effect as the key concept for realizing rapid catalytic reactions. *Stud. Surf. Sci. Catal.*, **77**, pp. 17–26, 1993.

[12]   Guo, J., Lou, H., Zhu, Y. & Zheng, X., La-based perovskite precursors preparation and its catalytic activity for CO$_2$ reforming of CH$_4$. *Mater. Lett.*, **57**, pp. 4450–4455, 2003.

[13]   Parvary, M., Jazayeri, S. H., Taeb, A., Petit, C. & Kiennemann, A., Promotion of active nickel catalysts in methane dry reforming reaction by aluminum addition. *Catal. Commun.*, **2**, pp. 357–362, 2001.

[14]   Wu, Y., Kawaguchi, O. & Matsuda, T., Catalytic reforming of methane with carbon dioxide on LaBO$_3$ (B = Co, Ni, Fe, Cr) catalysts. *Bull. Chem. Soc. Jpn.*, **71**, pp. 563–572, 1998.

[15]   Batiot-Dupeyrat, C., Valderrama, G., Meneses, A., Martinez, F., Barrault, J. & Tatibouët, J. M., Pulse study of CO$_2$ reforming of methane over LaNiO$_3$. *Appl. Catal. A-Gen.*, **248**, pp. 143–151, 2003.

Recycling of poly lactic acid into lactic acid with high temperature and high pressure water

M. Faisal[1,2], T. Saeki[1], H. Tsuji[1], H. Daimon[1] & K. Fujie[1]
[1]Department of Ecological Engineering,
Toyohashi University of Technology, Japan
[2]Department of Chemical Engineering, Syiah Kuala University,
NAD, Indonesia

Abstract

The converting of poly L-lactic acid into its monomer (lactic acid) was performed using high temperature and high pressure water at the temperature range of 250-350°C and for a period of 30 minutes. The results show that temperature and reaction time affected the recycling efficiency and optical selectivity. Under the tested conditions, the highest amount of lactic acid (yield of 90.7-92.5%) was obtained at temperature of 250°C and reaction time of 10-20 min. The remains were an un-harmful byproduct. A study on the effect of temperature on the optical purity of lactic acid product demonstrated L-lactic acid was predominant at a low range temperature of 250-260°C. At higher temperature, racemic mixture was observed. Based on our experimental results, the method was found to be simple and promising for recycling of poly lactic acid into its monomer, lactic acid.

1 Introduction

The replacement of highly biodegradation-resistant and petroleum-based plastics with biodegradable ones, such as poly lactic acid (PLA), is of interest to researchers. PLA can be synthesized from renewable resources (e.g. corn, wheat, potato or sugar beat) by fermentation. Due to its biocompatibility and biodegradability, PLA has widespread potential use in medicine, agriculture, and packaging applications [1,2]. Recently, Cargill Dow LLC has started to operate the PLA plant producing of 140,000 t/year in November 2001. They are planning to increase annual output up to 450,000t in 2010 [3].

WIT Transactions on Ecology and the Environment, Vol 92, © 2006 WIT Press
www.witpress.com, ISSN 1743-3541 (on-line)
doi:10.2495/WM060251

However, biodegradable plastics also pose some adverse environmental risk such as pollution in waterways due to high BOD concentration resulting from the breakdown of starch-based biodegradable plastics. PLA releases carbon dioxide and methane during its biological breakdown phase, migration of plastic degradation by products (i.e., contaminant additives and modifiers) and slow degradation rate [4] where its microbial degradation is limited to a few species of microorganism, such as *Amycolatopsis* and *Streptomyces* strain [5,6]. In fact, the annual production of PLA increases markedly [3]. Any large-scale consumption of PLA products will bring the associated problem of an excess of PLA waste, which will difficult to treat by biodegradation either in composting plants or in the natural environment [7].

Consequently, PLA should be recycled into its monomer, lactic acid, instead of disposing it, and afterward can reproduce the same polymer. The recycling is supposed to lead the simultaneous solution of energy, resources, CO_2 and waste problem. Moreover, minimizing resource consumption as well as efficient utilization of renewable resources should be established for the sustainable system. Recycling of PLA oligomers to produce cyclic dimer (L-lactide) in the presence of catalyst has been reported [8]. However, this method is requiring the catalyst. Moreover, the yield of L-lactide was low at a short reaction time. A high temperature and high pressure (HTHP) water treatment, an environmentally friendly recycling process which only uses water is effective for recovery of lactic acid from PLA without addition of any catalysts.

In recent years, the use of HTHP water treatment process for thermal decomposition [9-11], organic synthesis [12,13] and recovery of useful materials from various organic wastes [14-18] has attracted the interest of many researchers. This technique was also applied to chemical recycling processes such as hydrolysis of polyethylene terephthalate (PET) into ethylene glycol and terephthalic acid [19]. Park *et al.* [20] showed the ability of supercritical and near-critical water to break down the styrene-butadiene rubber (SBR) into a range of lower molecular weight organic compounds for potential recovery. Thus, water under HTHP conditions provides excellent properties (i.e. high ion product, low solvent polarity, high solubility for oil, etc.) for hydrolysis and offer opportunities to adjust the reaction environment to optimal conditions for the chemical transformation of interest [21].

In this work, our effort is to present a way to prepare lactic acid from poly L-lactic acid (PLLA) under a promising technique. Optimum condition of recovery, effect of high temperature on DL type of lactic acid will be discussed. The result is significant for development of recycling technology and biomass efficiency.

2 Experimental

All experiments were performed using SUS 316 batch reactor (8.2 ml volume, 12.7 mm OD, 10.2 mm ID, and 100 mm length). The reaction conditions were over a temperature range of 250 to 350°C at the corresponding saturated vapor pressures and time ranging from 1 to 30 minutes. During a typical run, about

0.24 g PLLA (LACTY 5000, Shimadzu Corp., weight average molecular weight (Mw) = 4.4x 10^5, Mw/ number average molecular weight (Mn) = 2.1, initial fraction of L-lactic acid = 95%) and 4.8 g of distilled water (weight ratio of 1:20) were charged into the reactor. The reactor was then sealed, and then the air inside was replaced by Argon gas. Subsequently, the reactor was immersed into the preheated molten salt bath (TSC-B600, Taiatsu Techno) containing a mixture of potassium nitrate and sodium nitrate set at the desired temperature. After the desired reaction time had elapsed, the reactor was plunged into a water bath to bring them quickly to room temperature, thus, effectively ceasing any occurring reactions. The reaction products were analyzed using an organic acid analyzer (LC-10A, Shimadzu Corp.) with an ion-exclusion column (Shim-Pack SCR-102H, Shimadzu Corp.) and electroconductivity detector (CDD-6A, Shimadzu Corp.). The optical isomer of lactic acid produced was also analyzed using these conditions: analytical column; Sumichiral OA-5000 (5μm particle size, 4.6 mm i.d. x 150 mm long, Sumika Chemical Analysis Service Ltd. Japan), column temperature; 30°C, mobile phase; 1mM copper (II) sulfate in water, flow rate; 1mL/minutes, injection volume 20μL. The total organic carbon (TOC) was analyzed using TOC analyzer (TOC-VE, Shimadzu Corp.).

3 Results and discussion

3.1 Temperature and time dependence on lactic acid production

The effect of temperatures on thermal recycling of PLLA into lactic acid and other organic acids (i.e. pyrubic, succinic, propionic, acetic and formic, acids) at 10 min of reaction time is demonstrated in Figure 1. The products were mostly lactic acid with small amount of by products. As can be seen from the Figure, the amount of lactic acid was found to increased with the increased of reaction temperature up to 250°C then decreased as temperature rises 350°C. It was observed that in the low temperature range of 200-230°C and short reaction time (5-10 min), the thermal depolymerization of PLLA results only the reduction of molecular weight, suggesting thermal energy was insufficient to produce lactic acid. At these conditions, acetic and formic acid which are the major intermediate products prior to complete degradation to volatile carbon and water [22] were not observed, suggesting that the depolymerization occurs without the formation of volatile product. Our observation showed that at a low temperature ranging from 240-250°C succinic acid was observed, while at higher temperatures, propionic and acetic acid were observed.

Figure 2 shows the effect of reaction times on thermal recycling of PLLA at 250°C. Results demonstrated that from 8-30 min of reaction time, formation of lactic acid do not change markedly, indicating that the depolymerization reaction occurs relatively slow at these conditions. It was observed also that the formation of other low molecular weight organic compounds decreased as time progresses. Based on Figures 1 and 2, the results suggest that both temperature and time affect the amount of lactic acid obtained.

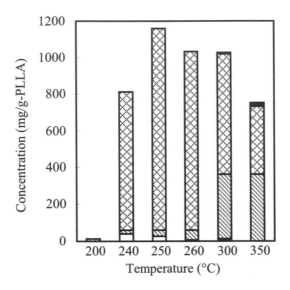

Figure 1:   Effect of temperature on thermal recycling of PLLA at 10 min of reaction time.

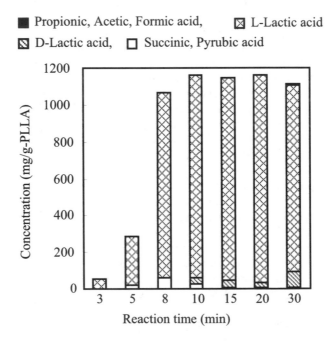

Figure 2:   Effect of reaction time on thermal recycling of PLLA at 250°C of poly lactic acid at 250°C.

The recycling efficiency, strongly depend upon the reaction conditions. Table 1 summarizes the results of the recycling efficiency of PLLA at various conditions. As can be seen, at 250°C and 20 min of reaction time, 92.5 % of PLLA was converted into lactic acid. Presumably, the ion product of water is good enough for lactic acid recovery under this temperature. The ion product of water goes through a maximum (about 10^{-11} (mol/L)^2) at about 250°C under a pressure of 4 MPa. Large amount of dissociated ions and high dielectric constant might also promote depolymerization process. Then, the yield of lactic acid decreased at 30 min. At a higher temperature of 350°C, the degradation rate of PLLA was fast and small amount of propionic and acetic acid were observed. This result suggests that, further decomposition of acetic acid might have occurred producing carbon monoxide and eventually carbon dioxide.

To obtain additional information concerning the recycling of PLLA, further experiment on the effect of ratio water to sample (PLLA) were also done with the ratio of 10:1 (water:PLLA). Results show that at 10 min of reaction time the yield of lactic acid was found lower than that of the ratio of 20:1. Small amount of un-reacted material was observed. Probably, at this condition the amount of water produce insufficient of OH radical concentration for degradation of PLLA.

The conversion yield of PLLA into lactic acid could be improved by manipulating some conditions. Reaction time, temperature and pressure have been found as key parameters on the recycling of PLLA. The addition of selected catalyst might also improve the yield of lactic acid. The using of agitated reactor will affect the mass transfer inside the reactor which is might influence the yield of lactic acid. Kao et al. [23] reported that the rate of PET hydrolysis increased with the agitation speed and approach a constant value when the agitation speed was larger than 400 rpm. In addition, our preliminary experiment on the effect of reactor material on the reaction also showed that reaction rate in SUS 316 material was slightly higher than that of Hastelloy C-22 and Inconel 625.

Table 1:   Recycling efficiency of PLLA into lactic acid at various conditions.

Reaction time (min)	Recycling efficiency (%)				
	240°C	250°C	260°C	300°C	350°C
5	11.9	21.6	53.9	67.4	55.5
8	35.2	80.1	70.8	71.9	46.7
10	61.7	90.7	84.8	83.8	42.0
15	85.5	91.1	89.2	83.0	30.1
20	-	92.5	91.5	82.5	22.0
30	-	88.4	80.1	79.9	12.7

3.2 Effect of temperature and reaction time on DL-types of lactic acid

Since L-type of lactic acid is a significant important compare to D-type or racemic mixture, the measurement of the effect of HTHP water process on DL-types of lactic acid is highly useful. Figure 3 demonstrates the course of the selectivity to L-lactic acid at the conditions employed in this work. The L fraction of lactic acid was calculated based upon the initial L-fraction in the sample, ca.95%. As can be seen, the selectivity value for L-lactic acid at short reaction time and temperature range of 250-260°C were found at around 1 indicating the insignificance change in optical purity occurred during recycling process. Moreover, the selectivity of L-lactic acid remained unchanged for a period of 30 minutes at the temperature of 250°C. These findings suggest that the degradation of PLLA at these conditions results in only L-lactic acid and water-soluble oligomers. As expected, at a higher temperature, the selectivity decreased gradually with reaction time. For instance, at 300°C and 20 minutes of reaction time, the selectivity was reduced to 0.55 indicating the racemization occurred resulting in the partial loss of optical activity. In other words 45% of L-enantiomer was converted to D-lactic acid. This result reflects that L-lactic acid changed into D-lactic acid at high temperature through the contribution of radical-homolysis and enolization pathways as suggested by Tsuji *et al.* [24]. These findings reveal that the temperatures ranging from 250-260°C are most favorable for high yield of L-lactic acid.

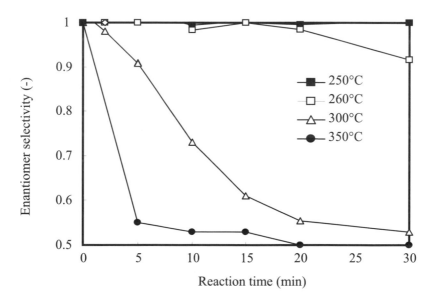

Figure 3:   Course of enantiomer selectivity to L-lactic acid at various temperatures.

3.3 Carbon balance

The PLLA degradation process was also investigated by TOC measurements. Figure 4 demonstrates the effects of time and temperature on the course of TOC reduction. The value 1 means the theoretical TOC value when all of the PLLA chains are degraded completely to produce lactic acid, water-soluble oligomers and other organic acids. A significant increase of the TOC with increasing reaction temperature and time was found as expected. As can be seen in Figure 4, at 8-15 minutes of reaction times and temperatures of 250-300°C, the amount of TOC achieved a maximum and reached a theoretical value suggested that all of the PLLA chains were degraded into its monomer, water-soluble oligomers and other organic acids. The TOC value then remained unchanged even when the reaction was continued for 30 minutes, which was in contrast with the lactic acid formation (excluding the result at 250°C). These occurred presumably due to the formation of low molecular carboxylic acids (from lactic acid) such as acetic acid. As expected, at higher temperature of 350°C, TOC decreased significantly with reaction time exceeding 5 minutes. This suggests that volatile carbon such as CO_2, CO and CH_4 were formed by thermal decomposition of lactic acids through decarboxylation and the dehydration route of decomposed compounds [11].

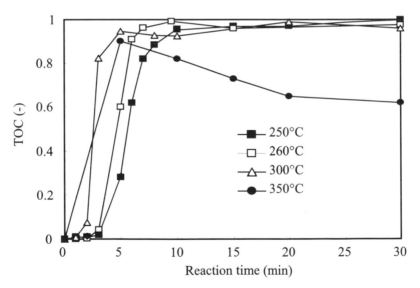

Figure 4:   Course of TOC reduction at various conditions.

4 Conclusion

Recycling of PLA was investigated in a pressurized batch reactor at subcritical conditions. The recycling efficiency was strongly dependence upon the operating conditions of temperature, reaction time and ratio water to sample. The monomers recovered mainly consist of lactic acid. Our observation shows that the recycling to lactic acid appears predominant at lower temperatures, while at the higher temperatures was favorable to thermal degradation into other low molecular weight carboxylic acids. Our observation suggests that HTHP water process has been found to affect the optical isomer of lactic acid. At low temperatures, it was observed that monomer recovered mainly consist of L-lactic acid.

Acknowledgement

The authors are grateful to The 21st Century COE Program at the Toyohashi University of Technology (Ecological Engineering for Homeostatic Human Activities) for financial support of this research.

References

[1]   Datta, R. and S.P Tsai; "Lactic acid production and potential uses: a technology and economic assessment", Am. Chem. Soc., Chp 12, pp. 225-235, 1997

[2]   Chiellini, E. and R. Solaro; "Biodegradable polymer materials", Adv. Mater., 8, pp. 303-313, 1996

[3]   Verespej, MA.; "Winning technologies: polylactic polymer", Industry Week, Dec 11, 2000

[4]   Brown, D.; "Plastics packaging of food products: the environmental dimension", Trends Food Sci. Technol., 4, pp. 294-300,1993

[5]   Pranamuda, H. and Y. Tokiwa; "Degradation of poly(L-lactide) by strains belonging to genus Amycolatopsis", Biotechnol. Lett., 21, pp. 901-905, 1999

[6]   Pranamuda, H., Y. Tokiwa, and H. Tanaka; "Polylactide degradation by an Amycolatopsis sp.", Appl. Environ. Microbiol., 63, pp. 1637-1640, 1997

[7]   Fan, Y., H. Nishida, T. Mori, Y. Shirai and T. Endo; "Thermal degradation of poly (L-lactide): Effect of alkali earth metal oxides for selective L,L-lactide formation", Polymer, 45, pp.1197-1205, 2004

[8]   Noda, M. and H. Okuyama; "Thermal catalytic depolymerization of poly (L-lactic acid) oligomer into LL-lactide: effects of Al, Ti, Zn and Zr compounds as catalyst", Chem. Pharm. Bull., 47, pp. 467-471, 1999

[9]   Goto, M., T. Nada,   S. Kawajari, A. Kodama and T. Hirose; "Decomposition of municipal sludge by supercritical water oxidation", J. Chem. Eng. Japan, 30, pp. 813-818, 1997

[10]   Goto, M., T. Nada, A. Ogata, A, Kodama and T. Hirose; "Supercritical water oxidation for the destruction of municipal excess sludge and alcohol distillery wastewater of molasses", J. Supercrit. Fluids, 13, pp. 277-182, 1998

[11]   Li, L., J.R. Portela, D. Vallejo and E.F. Gloyna; "Oxidation and hydrolysis of lactic acid in near – critical water", Ind. Eng. Chem. Res., 38, pp. 2599-2606, 1999

[12]   Savage, P.E.; "Organic chemical reactions in supercritical water", Chem. Rev., 99, pp. 603-621, 1999

[13] Holliday, R.L., B.Y.M. Jong and J.W. Kolis; "Organic synthesis in subcritical water oxidation of alkyl aromatic", J. Supercrit. Fluids, 12, pp. 255-260, 1998

[14] Kang, K., A.T. Quitain, H. Daimon, R. Noda, N. Goto, H. Hu and K. Fujie; "Optimization of amino acids production from fish entrails by hydrolysis in sub-and supercritical water", Can. J. Chem. Eng., 79, pp. 65-70, 2001

[15] Shanableh, A.; "Production of organic matter from sludge using hydrothermal treatment", Wat. Res., 34, pp. 945-951, 2000

[16] Daimon, H. K. Kang, N. Sato, and K. Fujie; "Development of marine waste recycling technologies using sub-and supercritical water", J. Chem. Eng. Japan, 34, pp.1091-1096, 2001

[17] Quitain, A.T., M. Faisal, K. Kang, H. Daimon, and K. Fujie; "Low-molecular-weight carboxylic acids produced from hydrothermal treatment of organic wastes", J. Hazard. Mater., 93, pp. 209-220, 2002

[18] Quitain, A.T., N. Sato, H. Daimon and K. Fujie; "Production of valuable materials by hydrothermal treatment of shrimp shells", Ind. Eng. Chem. Res., 40, pp. 5885-5888, 2001

[19] Arai, K. and T. Adschiri; "Importance of phase equilibria for understanding supercritical fluid environments", Fluid Phase Equilibria., 158, pp. 673-684, 1999

[20] Park, Y., J.N. Hool, C.W. Curtis and C.B. Roberts; "Depolymerization of styrene-butadiene copolymer in near critical and supercritical water", Ind. Eng. Chem. Res., 37, pp. 1228-1234, 2001

[21] Akiya, N. and P.E. Savage; "Kinetics and mechanism of cyclohexanol dehydration in high – temperature water", Ind. Eng. Chem. Res., 40, pp. 1822-1831, 2001

[22] Mishra, V.S., V.V. Mahajani and J.B. Joshi; "Wet Air Oxidation", Ind. Eng. Chem. Res., 34, pp. 2-48, 1995

[23] Kao, C. B. Wang and W. Cheng; "Kinetic of hydrolytic depolymerization of melt poly (ethylene terepthalate)", Ind. Eng.Chem.Res., 37, pp.1228-1234, 1998

[24] Tsuji, H., I. Fukui, H. Daimon and K. Fujie; "Poly (L-Lactide) XI. Lactide formation by thermal depolymerisation of poly (L-lactide) in a closed system", Polym. Degrad. Stab., 81, pp. 501-509, 2003

Section 6
Waste incineration
and gasification

Waste incineration in
Swedish municipal energy systems

K. Holmgren
Linköping Institute of Technology, Linköping, Sweden

Abstract

Waste is widely used as a fuel in the Swedish district heating (DH) systems, thereby linking waste management and the energy system. This paper summarizes earlier studies by the author on the role of waste as a fuel in DH systems. The method used is case studies of three Swedish municipalities that utilise waste in their DH systems. Economic optimisations of the DH systems are made using the linear programming model MODEST, and environmental effects in terms of carbon dioxide emissions are assessed. It is economically advantageous to use waste as a fuel due to regulations in the waste management sector and high taxes on fossil fuels. There can be a conflict between combined heat and power (CHP) production in DH systems and waste incineration, since the latter can remove the heat sink for other CHP plants in combination with low electrical efficiency in waste incineration plants. CHP is the main measure to decrease carbon dioxide emissions in DH systems on the assumption that locally produced electricity replaces electricity in coal condensing plants. It can be difficult to design policy instruments for waste incineration due to conflicting goals for waste management and energy systems. To put costs on environmental effects, so called external costs, is one way to include them but the method has drawbacks, for example the limited range of environmental effects included. Comparing the energy efficiency of material recovery and energy recovery from waste incineration is one way to assess the resource efficiency of the waste treatment methods.
Keywords: district heating, energy recovery, combined heat and power, material recovery, waste incineration, modelling, policy instruments, waste management.

 WIT Transactions on Ecology and the Environment, Vol 92, © 2006 WIT Press
www.witpress.com, ISSN 1743-3541 (on-line)
doi:10.2495/WM060261

1 Introduction

Waste management and energy systems are closely linked in Sweden, since heat from waste incineration is an important contributor to the overall supply in the district heating (DH) networks. Regulations in one sector have impacts in the other. This paper will summarize findings from earlier studies by the author of the role of waste incineration in the Swedish district heating systems, including the following: the profitability of using waste as a fuel for energy utilities; investigating the role of DH system as a user of various kinds of waste heat, i.e. from industries, waste incineration, and combined heat and power (CHP); analysis of some policy instruments affecting the waste management and DH sector; comparing energy and material recovery from the point of view of energy efficiency; and internalising the external costs in a DH system with the emphasis on waste incineration in order to obtain a socio-economic perspective on using waste as a fuel.

2 Methodology

The method used is case studies, where three Swedish municipalities of various sizes were analysed, all with the main common feature, viz. that they use waste as a fuel for DH production. A model of the features of the DH system, such as conversion units, heat demand, and fuels used, was built in an optimisation tool called MODEST (Model for Optimisation of Dynamic Energy Systems with Time-dependent components and boundary conditions) (Henning [1]). It is a linear programming model that minimises the cost of supplying heat demand during the analysed period. Electricity sales are treated as income. Scenarios are chosen to reflect issues to be investigated. After modelling, the results were analysed, and have been presented and discussed with the utilities operating the DH systems.

3 Waste management with emphasis on waste incineration

The EU's waste policy is founded on the waste hierarchy, described in the Sixth Environmental Action Programme from the European Commission [2] and states that first comes waste prevention, then recovery (reuse, material and energy recovery where material recovery, including biological treatment is preferred to energy recovery) and finally disposal, where landfill and waste incineration without energy recovery are included. Swedish waste policy is based upon this hierarchy. The strategy for Swedish waste management can be found in (Swedish Environmental Protection Agency [3]).

Energy recovery by waste incineration is the treatment method for almost half of all municipal waste today, as can be seen in Figure 1 (Swedish Association of Waste Management [4]).

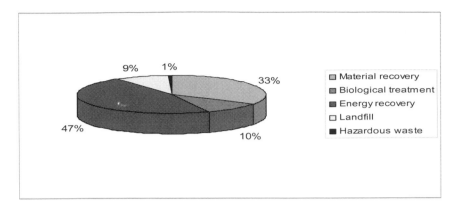

Figure 1:   Treatment methods of municipal waste in 2004, total amount 4.2 million tons.

3.1 Waste as a fuel in district heating systems

In Sweden, with extensive district heating (DH) systems that supply 40% of the total heating demand of buildings and premises, heat supply from waste incineration has a substantial share of the total DH supply of about 15% (Swedish Energy Agency [5]). Today, there are 29 waste incineration facilities in Sweden, including 14 hot water boilers and 15 combined heat and power (CHP) plants producing about 8.6 TWh heat and 0.74 TWh electricity (Swedish Association of Waste Management [4]). In 2004, these facilities treated about 1.95 million tons of municipal waste and 1.2 million tons of other waste, mainly from the manufacturing industry.

Capacity for waste incineration is currently increasing and is forecast to expand from 2.8 Mton in 2002 to 4.9 Mton in 2008, if all planned projects are carried out (Swedish Association of Waste Management [6]) resulting in a total of 40 waste incineration plants. Despite these investments there will still be a lack of treatment capacity. Quantities of waste are also increasing, between 1985 and the present by approximately 2-3% per year. If this trend is not broken, additional waste treatment capacity will also be needed after 2008.

CHP is recognised as one measure to decrease carbon dioxide emissions in the European Union. A directive is in place that aims to create a framework for promoting cogeneration [7]. However, electricity production in the Swedish DH networks is low. The total delivered heat in 2003 was 47.5 TWh, and the amount of electricity produced in the DH system was 6.0 TWh [8]. There may be a conflict between CHP and waste incineration, since waste-fired CHP plants have low electrical efficiency. This is due to the many impurities in the fuel; the temperature of the steam in the boiler cannot exceed 400°C without entailing high maintenance costs due to corrosion. However, electricity production at waste incineration plants is forecast to increase from 0.7 to 1.7 TWh between 2002 and 2010 (Swedish District Heating Association [9]). The study does not clarify the reason for the increase in electricity production at waste fired CHP plants, but it is reasonable to believe that it is a result of the higher electricity prices that are anticipated when Swedish electricity prices are harmonized with those in continental Europe (Trygg and Karlsson [10]). A proposed tax on incinerated waste, which is designed to promote CHP production, is probably also a factor (Ministry of Finance [11]).

Awareness of increased global warming makes it vital to analyse emissions of carbon dioxide for the DH systems, as has been done in studies by the author. Most municipal waste is of biological origin, but part is of fossil origin, such as plastic waste. The figure used for carbon dioxide emissions from municipal waste is around 90 kg/MWh$_{th}$ (Swedish Association of Waste Management [12]), as compared to 280 kg/MWh$_{th}$ for oil. An important assumption when analyzing emissions of carbon dioxide is that locally produced electricity replaces electricity produced in coal condensing power plants, with emissions of carbon dioxide of around 950 kg/MWh$_{el}$ (electrical efficiency 35%, emissions from coal 335 kg/MWh$_{th}$), which is the marginal producer of power in the European power system (Swedish Energy Agency [13]).

Other environmental issues associated with waste incineration include flue gas emissions and ashes from the incineration. The flue gases consist of hazardous substances such as heavy metals (e.g. lead, cadmium and mercury), dioxins, dust, and also substances that cause acidification. Today, waste incineration facilities have advanced flue gas cleaning systems and the emission of hazardous substances has decreased dramatically since the 1980s. However, dioxins and heavy metals end up in the flue gas ashes. These ashes constitute about 4% of the weight of the municipal waste and are classified as hazardous waste. These ashes have to be landfilled safely in order to prevent leakage. The bottom ash is about 19% of the weight of the municipal waste and is mostly landfilled, even if it might be used for road construction and covering landfills (Swedish Association of Waste Management [14]). In Holmgren and Amiri [15] some environmental effects other than carbon dioxide emissions are taken up. The debate around waste incineration has shifted from being a problem with emissions to whether it is suitable when aiming at a sustainable society with high resource efficiency. This question will be analysed to a certain extent in the study, where energy recovery and material recovery are compared in an energy efficiency perspective (Holmgren and Henning [16]), described in Section 4.4.

3.2 Main policy instruments influencing waste incineration

The main policy instruments affecting waste incineration include the introduction of a tax on landfill in 2000, at present 46.3 €/ton (Ministry of Finance [17]) and a ban on landfill of combustible waste from 2002, and from 2005 also of organic waste (Ministry of the Environment [18]). Carbon dioxide taxes for fossil fuels for heating purposes are around 0.1 €/ton, where heat from CHP and for industrial consumers have deducted levels (Ministry of Finance [19]). There is no carbon dioxide tax on waste. Electricity production is not taxed, but consumption is. The government has proposed a tax on incinerated waste (Ministry of Finance [11], further explained in Section 4.3. Policy instruments are examined in more detail in for example Holmgren [20].

4 Results from case studies

4.1 Profitability of waste

Earlier studies by the author on the role of waste as a fuel in DH systems concern the existing waste incineration plant in the city of Linköping (Holmgren and Bartlett [21], investment in a waste incineration plant in the city of Skövde (Holmgren and Gebremedhin [22]) and also additional investment in Linköping (Holmgren and Henning [16]). The studies show that in Skövde, the introduction of a waste incineration plant is profitable and in Linköping, investment in extended capacity is also profitable. What fuels the waste replaces depend on the configuration of the existing system. In Skövde it is mostly wood chip, since the new waste incineration plant replaces the wood chip boiler as base supplier of heat. In Linköping, there are large savings in biomass fuel, but the largest savings come from oil. The main measure to lower carbon dioxide emissions is to utilize CHP in the DH systems.

4.2 The role of a DH system as user of various forms of waste heat

DH systems have a big advantage since they can utilize heat that would otherwise be of limited use. This is considered in a study of the DH system in Göteborg, which uses various kinds of waste heat; from industries, waste incineration, and combined heat and power (Holmgren [20]). The base load of heat supply is from oil refineries in the vicinity of the city and heat from a waste-fired CHP plant. Other heat sources are a natural gas fired CHP plant, heat pumps and hot water boilers utilising pellet, natural gas, and oil. The utility is currently investing in an additional natural gas-fired CHP plant, where the plant's profitability is dependent not only on electricity prices and policy instruments, but also on the utilisation of the heat in the DH system. The issue analysed is whether the various types of waste heat will "compete" with each other. An important assumption in this study is the realisation of an integrated European electricity market, which will mean higher electricity prices than are traditional in Sweden (Trygg and Karlsson [10]). The new natural gas fired CHP plant seems to be a beneficial investment, since the operating cost is reduced by €150 million over a ten-year period (no investment costs included), in light of these electricity prices. The conclusion is that there is space in the system for all these different waste heat sources, since the new CHP plant mainly affects production from hot water boilers and heat pumps and to a lesser degree waste heat from industries and the existing natural gas fired CHP plant.

Economic findings are that heat from waste incineration is advantageous and that a decrease in waste heat from industries would raise the operating cost considerably, even if more electricity were produced in that scenario and income generated from electricity sales is thus higher, due to increased use of more expensive heat supply from hot water boilers and heat pumps. The study also shows the importance of using the DH system for electricity production to control carbon dioxide emissions, since electricity produced locally can replace marginal power producers in continental Europe and therefore decrease carbon dioxide emissions. The sensitivity analysis shows the substantial impact of the assumed electricity price on the new CHP plant's operating time and hence its profitability.

The findings in this study contradict an earlier study that showed that increased waste incineration reduces electricity production since it removes the heat sink for CHP plants (Holmgren and Bartlett, [21]). An overall study of the DH systems in Sweden (Sahlin et al. [23]) also showed a decrease in electricity production due to greater waste incineration capacity. This shows that conditions vary in different DH systems.

4.3 Analysis of two policy instruments

Policy instruments have a substantial impact on the waste management and energy sectors. In Holmgren and Gebremedhin [22], the impact on the investment in a waste incineration plant of the introduction of a tax on incinerated waste is analysed. Tax levels of 11 and €42.5 per ton were analysed; these were the levels proposed in an earlier government investigation (Ministry of Finance, [24]). The conclusion was that at the tax level of €11 per ton, the investment was still profitable for the utility, but at the €42.5 per ton level, the investment was not profitable. The prerequisite for the results is naturally that the utility cannot raise the gate fee for receiving the waste. The results indicate, however, that at the higher tax level, other treatment options begin to be of interest. The recently proposed tax on incinerated waste (Ministry of Finance [11]) puts different tax levels on waste for CHP plants, around €7 per ton, and for hot water boilers around €47 per ton, in contrast to the above mentioned investigation.

The tax on incinerated waste is discussed in Holmgren [25], together with the green electricity certificate system. This system gives producers of electricity from renewable sources a certificate (Ministry of the Environment [26]). Consumers need certificates in relation to their consumption, thus creating a demand. Municipal waste is not included in the approved sources. The discussion concerns the difficulty in designing policy instruments for waste incineration due to its double function; as a waste treatment method and as a supplier of electricity and/or heat. The goals for waste management and energy system conflict, which makes it a complex affair to design policy instruments that affect waste incineration. In Holmgren [25] it is shown that when designing the proposed waste incineration tax, the energy system perspective was given prominence; the tax on incinerated waste had to harmonise with the taxation on other fuels used in the DH systems and incentives for CHP production were deemed important. No incentives were given to encourage more biological treatment and material recovery, except for plastic waste, even if that is a waste management goal. On the contrary, when designing the electricity certificate system and excluding municipal waste from approved sources, the perspective of waste management was put before the goals of the energy system. Including municipal waste would provide a greater incentive to produce more electricity in waste incineration plants but it would also steer waste of biological origin towards incineration, which would be inconsistent with Sweden's waste management goals, which state that at least 35% of biodegradable waste should be biologically treated by 2010 (Swedish Environmental Protection Agency [3].

4.4 Comparison between energy recovery and material recovery from an energy efficiency viewpoint

A study of the energy efficiency of material recovery and energy recovery is presented in Holmgren and Henning [16]. Material recovery saves virgin material, and also energy, since production processes that use recovered material are less energy intensive than processes that use virgin material, whereas energy recovery saves other fuels that differ from energy system to energy system. This study analyses two Swedish municipalities. The operation of the DH systems is optimized in two scenarios; with or without waste incineration. The study also shows the fuels used and the amount of electricity produced in the DH systems.

The fractions of glass and metal do not give any heat contribution when incinerated, but save varying amounts of energy when material recycled. The combustible fractions are more complicated to compare since they can be recovered in both ways. The study shows that even if there is a DH system able to utilise the heat, paper and plastics should be material recycled, whereas cardboard and biodegradable waste is more suited for energy recovery through waste incineration. These calculations were made on the assumption that biomass is a limited resource, and when saved eventually saves oil somewhere else in the system. Furthermore, in the calculations, electricity was multiplied by 2.5 based on the assumption that electricity is produced in a condensing plant with an electrical efficiency of 0.4. That makes it extra important to produce CHP from waste incineration and also to consider how electricity intensive the various material production processes are.

4.5 Internalising external costs of a DH system

An external cost can be defined as "when the social or economic activities of one group of persons have an impact on another group and when that impact is not fully accounted, or compensated for, by the first group" (European Commission [27]). Electricity and heat production give raise to several negative external effects, such as climate change, acidification and health impacts (also positive external effects can occur, such as local employment). The costs for these effects should be internalized in the price for energy supply; otherwise a suboptimal consumption of energy occurs from a socio-economic perspective. The internalization can be made by for example taxes, subsides and fees. However, it is difficult to estimate the cost of external effects. Several attempts have been made to estimate the external costs of energy supply, where the ExternE project within the European Union is one of the most comprehensive. For a more thorough summary of external costs, see Carlsson [28].

A study is carried out to investigate whether external cost is a suitable method to assess the environmental impacts of waste incineration in a DH system (Holmgren and Amiri [15]). The aim is also to include more environmental effects, in addition to the emissions of carbon dioxide, than in earlier studies. The drawback is that only emissions to air is included in the external cost data, and the issues of residual products, such as ashes, and the efficient use of resources are not addressed; these are essential issues since the control of emissions to air has improved in recent years and has partly shifted the problems to the ashes. One major advantage is that it is a way to incorporate environmental effects in the existing systems and models and comparisons with for example environmental tax levels can easily be made. The main result from the study is increased CHP production in the DH system when including external costs, due to the assumption that this electricity replaces electricity produced in coal condensing power plants.

5   Conclusion

It is important to sort municipal waste in order to treat the different fractions of the waste according to the most preferable method for that fraction, and a variety of treatment methods are needed to avoid landfill. When choosing a waste treatment method, the connection to the technical energy system is important; whether it is possible to utilise the heat from the waste incineration, what other energy carriers are used, and so on. The structure of the energy system affects the consequences of choices made in the waste management system.

Waste management legislation banning landfill of combustible and organic waste and the taxes on landfill make waste competitive as a fuel in the analysed DH systems. Heat can be sold to DH customers, making incineration of waste competitive compared to other waste treatment methods. An additional effect is the favourable taxation of waste as a fuel and high taxes on fossil fuels. However, a tax on incinerated waste has recently been proposed which will alter the economic conditions for waste incineration.

Waste incineration can make it less viable to produce CHP in DH networks and this can be seen as a conflict between the need to treat waste in an acceptable way and the goal of more CHP production in the energy system. CHP is the main measure to decrease carbon dioxide emissions from DH systems on the assumption that locally produced electricity replaces electricity produced in coal condensing power plants.

Policy instruments have a significant impact on the systems. It is complicated to design policy instrument for waste incineration since the goals for waste management and the energy systems are conflicting.

The concept of external costs is one way to include environmental impact in calculations of profitability. One of its weaknesses, however, is that it is difficult to assess the cost of environmental impacts and the limited number of impacts included.

Acknowledgments

The work was carried out under the auspices of The Energy Systems Programme, which is financed by the Swedish Foundation for Strategic

Research, the Swedish Energy Agency and Swedish Industry. The author is grateful to Maria Danestig for valuable comments on the paper.

References

[1]   Henning, D. Cost minimisation for a local utility through CHP, heat storage and load management. International Journal of Energy Research 22:691-713; 1998.

[2]   European Commission. Environment 2010: Our Future, Our Choices – The Sixth Environmental Action Programme. (COM (2001) 31 final. Brussels, Belgium; 2001.

[3]   Swedish Environmental Protection Agency. Strategi för hållbar avfallshantering. (Strategy for sustainable waste management, in Swedish), Stockholm, Sweden; 2005.

[4]   Swedish Association of Waste Management. Svensk avfallshantering 2005 (Waste Management in Sweden 2005, in Swedish); 2005.

[5]   Swedish Energy Agency. Energy in Sweden 2004. Eskilstuna, Sweden; 2004.

[6]   Swedish Association of Waste Management. Avfallsförbränning. Utbyggnadsplaner, behov och brist. (Waste incineration. Expansion plans, capacity need and lack thereof, in Swedish) RVF-report 04:02; 2004.

[7]   European Union. Council Directive 2004/8/EC on 11 February 2004 on the promotion of cogeneration based on a useful heat demand in the internal energy market and amending Directive 94/43/EEC. Brussels, Belgium; 2004.

[8]   Swedish District Heating Association. Statistik 2003. (Statistics 2003, in Swedish) Sweden; 2005.

[9]   Swedish District Heating Association. Kraftvärme och dess kopplingar till elcertifikatsystemet. (Combined heat and power and the connection to the electricity certificate system, in Swedish); 2005.

[10]   Trygg, L. & Karlsson, B.G. Industrial DSM in a deregulated European electricity market – a case study of 11 plants in Sweden. Energy Policy 33:1445-1459; 2005.

[11]   Ministry of Finance. BRASkatt – beskattning av avfall som förbränns (GOODtax? – taxation on incinerated waste, in Swedish) SoU 2005:23. Stockholm, Sweden; 2005.

[12]   Swedish Association of Waste Management. Förbränning av avfall. Utsläpp av växthusgaser jämfört med annan avfallsbehandling och annan energiproduktion (Waste incineration. Greenhouse gas emissions compared to other waste treatment and other energy production, in Swedish). RVF report 2003:12, Malmö, Sweden; 2003.

[13]   Swedish Energy Agency. Marginal elproduktion och CO2-utsläpp i Sverige (Marginal electricity production and CO2-emissions in Sweden, in Swedish). ER 14:2002. Eskilstuna, Sweden; 2002.

[14]   Swedish Association of Waste Management. Förbränning av avfall: En kunskapssammanställning om dioxiner, (Waste-to-energy: An inventory

and review about dioxins, in Swedish) *RVF-report 01:13*, Swedish Association of Waste Management, Malmö, Sweden; 2001.

[15] Holmgren, K. & Amiri, S. Internalising external costs of electricity and heat production in a municipal energy system – the case of waste incineration. Manuscript intended for journal publication.

[16] Holmgren, K. & Henning, D. Comparison between material and energy recovery of municipal waste from an energy perspective. A study of two Swedish municipalities. Resources, Conservation and Recycling, 43:51-73; 2004.

[17] Ministry of Finance. Lag (2005:962) om ändring i lagen (1999:673) om skatt på avfall (Law (2005:962) on changes in the law (199:673) governing waste tax, in Swedish); 2005.

[18] Ministry of the Environment. Förordning (2001:512) om deponering av avfall (Ordinance (2001:512) on landfill of waste, in Swedish); 2001

[19] Ministry of Finance. Lag om skatt på energi (1994:1776). (Law (1994:1776) on tax on energy, in Swedish), Stockholm, Sweden: 1994.

[20] Holmgren K. The role of a district heating network as a user of waste heat supply from various sources – the case of Göteborg. Accepted for publication in Applied Energy.

[21] Holmgren, K. & Bartlett, M. Waste incineration in Swedish municipal energy systems – modelling the effects of various waste quantities in the city of Linköping. In: Afghan NH, Bogdan Z, Duic N. Editors. Sustainable development of energy, water and environment systems. Proceedings of the Conference, 2-7 June 2002, Dubrovnik, Croatia; 2004.

[22] Holmgren, K. & Gebremedhin, A. Modelling a district heating system: introduction of waste incineration, policy instruments and co-operation with an industry. Energy Policy, 32:1807-1817; 2004.

[23] Sahlin, J., Knutsson, D. & Ekvall, T. Effects of planned expansion of waste incineration in the Swedish district heating systems. Resources, Conservation and Recycling 41:279-292; 2004.

[24] Ministry of Finance. Skatt på avfall idag - och i framtiden (Tax on waste today – and in the future, in Swedish) SoU 2002:9, Stockholm, Sweden; 2002.

[25] Holmgren, K. Energy recovery from waste incineration: linking the energy systems and waste management. Submitted for publication in Conservation and Recycling of Resources, Nova Publisher, Editor: Frank Columbus.

[26] Ministry of the Environment. Lag (2001:113) om elcertifikat. (Ordinance (2003:120) on electricity certificates, in Swedish), Stockholm, Sweden; 2001.

[27] European Commission. ExternE-Externalities of Energy. Vol.1-6, EUR 16520EN, Directorate-General XII, Luxembourg; 1995.

[28] Carlsson, A. Considering External Costs – Their Influence on Technical Measures in Energy Systems. Dissertation No 766, Linkoping Institute of Technology, Linkoping, Sweden; 2002.

"Flameless" oxyfuel combustion development for process improvement, emission reduction in furnaces and incinerators

W. Blasiak[1] & J. von Schéele[2]
*[1]School of Industrial Engineering and Management,
Royal Institute of Technology (KTH), Sweden
[2]Linde AG, Gas Division, Sweden*

Abstract

In recent years, the focus for the development of combustion technology focus has been set on the following main aims: fuel consumption reduction, nitrogen oxides emission reduction, increased productivity and product quality. Fuel consumption reduction has been reduced by as much as 30–40%, and also CO_2 emission reduction was achieved by replacing combustion air with oxygen. To achieve very low emission of nitrogen oxides (NO_x) the new combustion technology is characterised by: lower temperature of flame, more uniform temperature distribution and reduced concentration of oxygen as well as nitrogen inside the combustion chamber. As in this combustion technique a flame is replaced by a large chemical reaction zone and thus is often not visible the process was named as "flameless" combustion. "Flameless" combustion process that use oxygen, so called oxyfuel combustion, as well as its technical application is the subject of this work. The work presents a description and main features of the "flameless" oxyfuel combustion, results of laboratory tests of a new type of burner, REBOX®, as well as examples of industrial applications including waste incineration are included.

Keywords: oxyfuel combustion, flameless combustion, NO_x reduction, CO_2 reduction, incineration improvement

1   Advantages of combustion with use of oxygen enriched air

Progress in combustion technologies made the use of oxygen for combustion possible in industrial processes drastically reducing CO_2 and NO_x emissions, as

WIT Transactions on Ecology and the Environment, Vol 92, © 2006 WIT Press
www.witpress.com, ISSN 1743-3541 (on-line)
doi:10.2495/WM060271

well as allowing for higher productivity at highest possible level of product quality. Industrial applications of the most advanced product of the oxyfuel technology development, so called "flameless" oxyfuel combustion, are already well documented and provide enormous future potential for this technology in various thermal processes.

Oxygen enrichment of air decreases the amount of nitrogen which is a ballast component in air combustion. In this work for simplicity oxygen enriched air, atmospheric air with higher oxygen content than the standard 21 vol.%, is called the oxidiser. Since the economy of oxyfuel combustion is dependent on the cost of oxygen, the availability of low cost oxygen is critical to the application economy as well. However, development of lower cost oxygen separation techniques allows at present use of oxidiser with almost 100% content of oxygen [1, 13].

Rising oxygen concentration in the oxidiser leads to:

- higher process efficiency, thus fuel consumption reduction and therefore CO_2 emission reduction, Practical experience demonstrates that the thermal efficiency for airfuel installations is found in the interval 25-60%, but for oxyfuel it is 75–90%.
- Higher flame temperature.
- Heat transfer enhancement, thus higher productivity.
- Reduction of NO_x emission.

2   Influence of flame properties on heat transfer in heating furnaces

Change of oxygen content in the oxidiser also influences heat transfer conditions because of changes in flame temperature, in-furnace gas composition, and in-furnace gas flow rate generated during combustion and flame volume. An increase of oxygen fraction in the oxidiser changes the flue gas composition rising concentration of components governing the flue gas emissivity but decreasing total amount of flue gases because of lack of nitrogen. Therefore, participation of convective heat transfer in total heat transfer mechanism is slightly reduced in comparison to air combustion. In case of air combustion the main components determining gas emissvity are H_2O, CO_2, CO and some fuel molecules. With temperature increase, with the effect of dissociation, another components such as OH, H and O should be also taken into account. It is known that emissivities of gas components vary with wavelength and temperature. Thus, absorption and emission of flue gases takes place in specific ranges of the spectrum [2]. An increase of fraction of radiation by CO_2 and H_2O results in more uniform radiation of the flame. An increase in temperature causes mainly an increase of heat transfer by short wavelengths. Increase of the short wavelength radiation is caused by the presence of OH, H, O and CH.

Heating rate in industrial heating furnaces is a function of the following parameters:

- Flame temperature and its emissivity.
- Flame shape.

- Firing rate.
- Heat sink temperature, and its emissivity.
- Walls temperature and its emissivity.

A simple overall heat transfer model in the industrial furnace was set up to analyze influence of these parameters on heat flux between flame and heat sink [3].

   In this work the flame volume and its emissivity are the main parameters discussed. Flame 5's irradiative characteristics at different firing rates and emissivities are considered as well as the effect on heat sink- and furnace wall temperatures. In this heat transfer model, the furnace volume has been divided into the combustion region (major chemical reaction zone) and the rest of the furnace that is called a 'combustion product region. Therefore, the combustion chamber can be separated as the combination of a well-stirred reactor (reaction zone) and a plug reactor (the reactor zone). Emissivities of heat sink and furnace walls are assumed: $\varepsilon_s = 0.8$ and $\varepsilon_w = 0.85$ respectively. Heat sink temperature, T_s is assumed equal to 1000 K and the furnace wall temperature, T_{wall} is set equal to 1273 K. In the analysis for simplicity the flame diameter is kept constant.

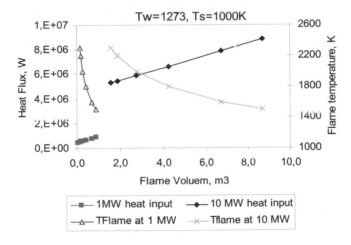

Figure 1:   Effects of the flame volume on the net heat flux from the flame to the heated sink for two different heat inputs; (T_w = 1273 K, T_s = 1000 K, Q_o = 1 MW) [3].

2.1 Flame volume versus heat flux

In this work, influence of the flame volume on the total heat flux between the flame and heat sink is presented. Effects of the flame volume on the net heat flux at different emissivities of the flame are shown in Figure 1. Heat transfer from the flame to the load is studied as a function of the flame volume for different heat inputs. In this work, the flame emissivity is assumed constant and equal to 0.25. Figure 1 shows the effect of the flame volume on the heat transfer rate from the flame to the load (heat sink). The enhancement of heat flux is increased with the flame volume at different heat inputs. It can be understood as: on the one hand, a larger flame volume leads to a larger flame area, which is favourable for the heat transfer. On the other hand, the flame temperature, T_{flame} decreases when the flame volume increases. This leads to a decrease of the heat radiation from the flame. Since the heat transfer from the wall to the heat sink occupies a larger proportion, the effect of flame area increase on the heat transfer plays a larger role than that of the flame temperature.

3 General features of "flameless" oxyfuel combustion

Oxidation reactions during "flameless" combustion as well as flame properties depend on the method of mixing between the oxidizer and fuel. From the oxidizer-fuel mixing point of view combustion process can be classified into three classes:

I. Combustion with high fraction of oxygen (up to 100%) in the oxidizer directly mixed with fuel directly at neighbourhood of the burner outlet. Fuel mixes directly with the oxidiser and some small amount of flue gases (recirculation). The degree of the flue gas recirculation depends on the burner design. Combustion is very intensive and takes place inside a small volume. Because volumetric heat load is very high the flame is very well seen as it radiates intensively. The flame is stabilised aerodynamically and "attached" to the burner outlet.

II. Combustion with use of atmospheric air can be assumed as less intensive and taking place in a large volume. Also, in this case the fuel mixes directly with the oxidiser and some small amount of flue gases (recirculation). The degree of the flue gas recirculation depends also on the burner design. Volumetric heat load and intensity of radiation are lower because the flame is diluted by nitrogen present in air. Flame is less luminous with less visible borders also however stabilised aerodynamically and "attached" to the burner outlet. In such combustion air preheating up to 400–500°C is used and air or fuel staging is used for low NO_x combustion effect.

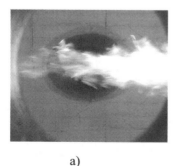
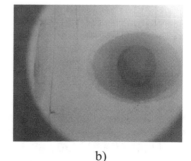

a)   b)

Figure 2:   Flame appearance during LPG combustion: a) conventional oxyfuel combustion b) "flameless" oxyfuel combustion [9].

III. Combustion in highly preheated air (above 1000°C), in pure oxygen or in slightly preheated air enriched with oxygen characterised by even less intensive combustion taking place in very large volume. To obtain such combustion phenomena, the fuel and oxidiser should be injected separately in order to delay mixing. The degree of the flue gas recirculation is very high and consequently fuel is mixed with the oxidiser at very low level of oxygen concentration. Oxygen is diluted in the mixture of flue gases and air of high temperature. Volumetric heat load of the flame and its luminous radiation is very small and therefore the flame is often not visible ("flameless" combustion, Figure 2) but present, occupying a much larger volume of the furnace chamber compared to combustion phenomena described in points I and II. The flame is stabilised by means of temperature field and is not longer "attached" to the burner outlet. Often the flame is characterised as a "lifted flame".

4   Development and study of the oxyfuel "flameless" combustion burner

In order to realize a new mode of oxyfuel combustion that is "flameless", a new type of burner was built. The burner was subject of this study and results in a comparison with other burners are presented below.

Flameless Oxyfuel Combustion burner, REBOX®-W, is shown in Figure 3. This burner uses commercial oxygen as oxidizer and gives a flameless combustion in the furnace. The high velocity of oxygen at the exit of the nozzles causes excellent internal mixing and accounts for the flameless combustion. The features of this burner are: simple construction, operation and very small in size. The single-flame HiTAC (High Temperature Air Combustion) burner used for the study was a REGEMAT® burner. It is a REGEMAT® 350 FLOX burner regenerative type that heats combustion air to 950°C. For efficient extraction of the heat from the regenerators, there are two sets that get alternatively heated and cooled by passing of flue gases and combustion air respectively. This is done with a cycle time of 10 seconds. During operation, 80% of flue gas is extracted through the regenerator. This burner is characterized by a single flame created by one fuel nozzle, surrounded by air inlets and flue gas outlets. This single flame develops along the axis of the fuel-jet nozzle during cooling and heat periods of the regenerators. Fuel is supplied continuously through the same nozzle and in this way a single flame can be formed with a permanent position. This position remains almost unchanged between heating and cooling periods, as the regenerators are located around the nozzle. The effect of oxygen enhancement on a HiTAC burner was studied by adding oxygen to air in the REGEMAT® burner. The mole fraction of oxygen (Ω) in the oxidizer was $\Omega = 29.2$. This is a retrofit arrangement. The conventional Oxyfuel combustion was carried out by modifications to the REBOX®-W burner.

A cold air burner was used in the study. The data obtained was used as a reference and all sets of measurement were not performed as it exhibited far inferior performance in Comparison to the above burners.

Figure 3:   Burner (REBOX® type) working according to oxyfuel "flameless" mode [16].

4.1   Results of measurements: flame shape and flame volume

To describe the flame boundary as per [10], the oxidation mixture ratio was defined according to equation (1). The oxidation mixture ratio allows estimating the combustion progress and can be calculated as mass fraction of oxygen to mass fraction of oxygen and the sum of oxygen needed to complete combustion at any point in combustion chamber, as follows:

$$R_o = \frac{m_O}{m_O + \sum_c s_o m_{F,c}} \tag{1}$$

where $s_O = n_O M_O / n_F M_F$. This ratio will have the value $R_O = 1$ when it is at air inlet or combustion is completed. Thus the flame volume defined by means of the oxidation mixture ratio was assumed to be inside the space limited by the following relationship:

$$0 < R_O \leq 0.99 \tag{2}$$

For fuel inlet the $R_O = 0$. $R_O = 0.99$ is assumed to indicate a flame border. Flame length was calculated as the distance between the burner's face and the axial location of the oxidation mixture ratio equal to 0.99. The calculated length was REBOX®-W 1.833 m, REGEMAT®1.534 m, REGEMAT®-Enhanced 2.044 m and Conventional oxyfuel 1.379 m.

5   NO$_x$ reduction mechanism

Most of NO generated during gas fuels combustion is from the high temperature of oxidation of atmospheric nitrogen. The NO formed is called "thermal NO" and its formation rate is an exponential function of flame temperature and a square root function of oxygen concentration. Thus, the formation of thermal NO can be controlled by controlling the flame temperature and the oxygen concentration. The main primary control strategies for reducing the formation of NO can be specified as follows:

- Stoichiometry based combustion control methods designed to control the mixing of fuel and oxidiser to reduce the concentration of oxygen in the flame zone,
- Dilution based combustion control methods designed to reduce the flame temperature in the flame zone by introducing inert gases.

For flameless combustion technology, the basic reason for thermal NO reduction is due to lower flame temperature and due to reduced oxygen concentration. The lower flame temperature is achieved by dilution of combustion zone with use of very intensive internal flue gas recirculation. It results also in bigger flame volume and thus also reduction of oxygen partial pressure. The NO_x emissions are as low as 2 mg/MJ in case of oxyfuel flameless combustion realized by means of the REBOX® type burners.

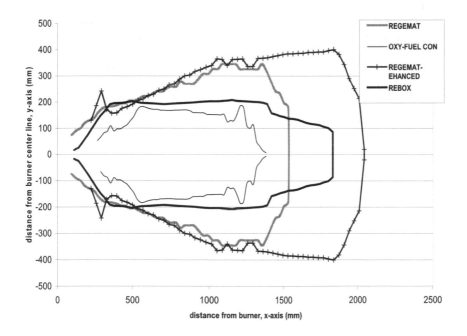

Figure 4:   Flame boundary in the central plane across the furnace [11].

6   Oxyfuel "flameless" combustion applied to incinerators

Grate fired furnaces are still a widely used technology for solid fuel combustion. Because the mixing process is usually poor in the fuel bed on the grate and in the gas phase region above the grate, complete combustion in a conventional grate fired boiler or incinerator is difficult to achieve. Since clean combustion of wastes has become an essential task due to the more strict requirements for environmental protection, it is desirable to use oxyfuel flameless combustion systems to improve wastes combustion in grate-fired furnaces.

The control of flow pattern and mixing processes inside combustion chambers of waste incinerators is necessary to achieve efficient combustion with low emissions of pollutants. There are three main types of waste incinerators used to burn municipal solid wastes (MSW) and biomass wastes. These can be subdivided into grate firing incinerators, fluidized bed incinerators and fixed bed incinerators. Normally, the combustion chamber of the incinerator consists of a lower and an upper furnace. The lower furnace is used to mix wastes with primary combustion air and therefore is also called the primary combustion zone. Mixing processes between primary combustion air and wastes occur on the grate or in the fluidized bed.

The thermo chemical processes occurring inside the primary combustion zone produce bed-off gas, which then reacts in the upper furnace of the waste incinerator [2]. Chemical composition of the bed-off gas depends on the amount and temperature of primary air. The bed-off gas leaving the lower furnace must react efficiently in order to complete combustion, reduce NO and SO_3 along the height of the upper furnace.

There are several ways to control mixing in the upper furnace e.g. changing the upper furnace geometry or controlling the mixing by means of number of nozzles placed on the walls of the upper furnace at various levels. The most common approach to complete combustion in the upper furnace and reduce pollutants (NO_x, CO dioxins, furans and particulates) is using number of air nozzles placed on the walls at various levels called the multi-nozzle system. Low temperature air is commonly used although secondary high temperature air (above $1000\,^{\circ}C$) has been already applied as well.

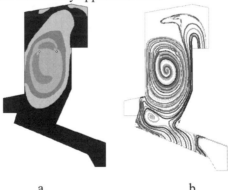

a.   b.

Figure 5:   Schematic idea of creation of oxyfuel flameless secondary combustion zone in municipal solid waste incinerator. A. Contours of mass fraction of oxygen B. Path lines after introduction of oxygen with use of perforated tubes [15].

Implementation of oxyfuel flameless second stage combustion (Figure 5) is an alternative that can dramatically improve performance of waste incinerators reducing all pollutants by controlling: temperature, residence time and stoichiometry. Use of well-stabilised large volume flameless combustion zone in the upper furnace of the waste incinerator will also stabilise thermal processes occurring in the bed of wastes on the grate. Oxygen could be for example injected with use of Ecotubes placed above the grate as schematically is shown in Figure 5 [14, 15].

7 Summary

Flameless oxyfuel combustion technology is characterised by: high efficiency, low formation of pollutants as NO_x, CO, soot, particulates. Flameless oxyfuel combustion is applied, with great success, in industrial installations including reheat and annealing furnaces at steel mills belonging to the following companies: Ascométal, Outokumpu Stainless, Scana Steel and Uddeholm Tooling. Additionally, this technology has been successfully installed for vessel pre-heating, for ladles and converters [16]. If applied as a secondary stage in waste incinerators it can be of great help to reduce not only NO_x, CO and particulates but also dioxins and furans. A large flame characterising the flameless combustion and high heat flux density is of great importance as well stabilising waste thermal decomposition in the primary combustion zone on the grate or in fluidised bed.

References

[1] Baukal C. E., Oxygen-Enhanced Combustion, CRC Press LLC, 1998, ISBN 0-8493-1695-2
[2] Rudnicki Z., Radiacyjny przepływ ciepła w piecach przemysłowych, Wydawnictwo Politechniki Śląskiej, Gliwice 1998.
[3] Blasiak W., Yang W., Wikström P., Jiang S., Heat transfer enhancements for energy savings in industrial furnaces, ISRN KTH/MSE-05/12-SE+ENERGY/RAPP, Stockholm, May 2005.
[4] Blasiak W., Yang W., Rafidi N., Physical properties of a LPG flame with high-temperature air on a regenerative burner, Combustion and Flame 136 pp: 567–569, Published by Elsevier Inc. (2004).
[5] Yang W., Blasiak W., Chemical flame length and volume in LPG combustion using high temperature and low oxygen concentration oxidizer, Energy and Fuels, Vol.18, 2004, pp. 1329-1335.
[6] Tsuji, H., Gupta, A.K., Hasegawa, T., Katsuki, M., Kishimoto, K., Morita, M., *High Temperature Air Combustion, From Energy Conservation to Pollution Reduction*, CRC Press LLC, New York, 2003
[7] Blasiak, W., Szewczyk, D., Mörtberg, M., Rafidi, N., Yang W., *Combustion Tests in a Test Furnace Equipped With High Temperature Air Combustion Mode*, Industrial Heating Journal, ISSN 0454-1499, Vol. 41 (2), pp 66-73, Japan 2004.
[8] Blasiak W., Narayanan K., Yang W., Evaluation New Combustion Technologies for CO2 and NO_x Reduction in Steel Industries, Air

Pollution 2004, Twelfth International Conference on Modelling, Monitoring and Management of Air Pollution, 30 Jun -2 July 2004, Rhodes, Greece.

[9] Yang W., Blasiak W., Ekman T., Numerical and thermodynamic study on oxyfuel flameless combustion, 8th International Conference on Energy and Clean Environment, Clean Air 2005, Lisbon, 27-30 June, 2005

[10] von Schéele J., Blasiak W., Oxyfuel solutions for lowered environmental impact of steel production, International Conference on Clean Technologies in the Steel Industry, 6-8 June, 2005, Balatonfüred, Hungary

[11] Narayanan K., Yang, W., Ekman T., Blasiak W., Combustion characteristics of flameless oxyfuel and oxygen-enriched-High Temperature Air Combustion, 6th HiTACG 2005, Sixth International Symposium on High Temperature Air Combustion and Gasification, 17-19 October 2005, Essen, Germany.

[12] von Schéele J., Vesterberg P, Ritzen O, Invisible Flames for Clearly Visible Results, Nordic Steel & Mining Review, pp 16 -18, 2005

[13] Kobayashi H, Boyle J.G., Keller J.G., Patton J.B., Jain R.C., Technical and economic evaluation of oxygen enriched combustion systems for industrial furnace applications, pp. 153 -163

[14] Blasiak W., W Yang., Combustion improvement system in boilers and incinerators, International Joint Power Generation Conference, IJPGC'03, June 16-19, 2003 (paper IJPGC2003-40141), Atlanta, Georgia, USA

[15] Blasiak W., Fakhrai R., Residence time and mixing control in the upper furnace of boilers and incinerators, International Conference on Incineration and Thermal Treatment Technologies, IT3, May 13-17, 2002, New Orleans, Louisiana, USA

[16] REBOX, http://www.rebox.info

Heavy metal behaviour during RDF gasification

G. Compagnone, P. De Filippis, M. Scarsella, N. Verdone & M. Zeppieri
Università degli Studi di Roma, "La Sapienza",
Dipartimento di Ingegneria Chimica, dei Materiali,
delle Materie Prime e Metallurgia, Roma, Italy

Abstract

Solid wastes, and especially RDF in dealing with the growing presence of organic compounds, mainly plastics, can be considered an important source of energy owing to their inexpensiveness and large availability. However, the presence of heavy metals in the waste can result in a recycling problem when thermal treatment is involved. In fact, small amounts of these metal species can be found in the gaseous stream, and this is particularly true for gasification processes. This work investigates the possibility of using theoretical calculations to evaluate the amount of volatile metal species in the gas stream, obtained by a gasification process, in order to select the adequate gas cleaning facilities. To this aim, the chemistry and volatility of the heavy metals As, Cd, Cr, Cu, Hg, Mn, Ni, Pb, Sb, Sn and Zn, commonly present in RDF, were investigated both theoretically and experimentally at different gasification conditions. A theoretical approach at equilibrium conditions based on thermodynamic data was performed by means of the total Gibbs free energy minimization method. The equilibrium distribution of the trace element species formed under reducing conditions in the 700-1300 K temperature range was calculated. The theoretical results are in substantial accordance with experimental data obtained using a bench scale gasification reactor. This study demonstrates that is possible to predict with reasonable conditions the operating conditions of a RDF gasifyier in order to obtain a syngas with a controlled content of polluting compounds.
Keywords: gasification, RDF, trace metals, thermodynamic simulation.

WIT Transactions on Ecology and the Environment, Vol 92, © 2006 WIT Press
www.witpress.com, ISSN 1743-3541 (on-line)
doi:10.2495/WM060281

1 Introduction

Solid wastes, in dealing with the growing presence of organic compounds, especially plastics, can be considered an important source of energy since they are inexpensive and easily obtainable materials [1, 2]. However, many municipal waste streams contain heavy metal species that during the thermal treatment may change their physical and chemical form and in some amounts volatilize and be released to the atmosphere together with small particulate matter. The fate of heavy metals and the consequent environmental impact of their compounds is affected by several parameters such as: the operating conditions, the composition of the inlet feed, the physical characteristics of the metal and its compounds. In this contest, the study of the behaviour of the trace elements (e.g. Pb, Cd, Sb) contained in waste as RDF has considerable importance [3]. In fact, even if these compounds are present in the wastes in very low concentration, they may cause various environmental or technological problems when released into the atmosphere [4]. Knowledge of the partition of the volatile and condensed heavy metals species during a thermal process is important in order to assess the related emission potential and to develop suitable systems for the reduction of volatile species [5-9]. The objective of this study is therefore to investigate the equilibrium distributions of the more relevant trace elements present in a typical urban waste during gasification processes and to compare simulation results with those obtained experimentally from a lab scale fixed bed gasifyer. The focus has been fixed on the gasification of refused derived fuel (RDF) that owing to its relatively constant composition and good transportation and storage characteristics constitutes a suitable feed in gasification processes for the production of fuels and/or energy [10].

2 Thermodynamic model

In order to investigate the fate of several trace elements in thermal conversion, the equilibrium composition of the resulting heterogeneous chemical system was calculated at different operating conditions according to a method based on the total Gibbs free energy minimisation [11-14]. The total Gibbs function was calculated assuming as reference state of the molecular species the gas state at the pressure of 1 bar and temperature of 298 K, considering the syngas as an ideal gas mixture and assuming that each solid species forms a separate phase. Under these conditions, the expression of the Gibbs free energy of the system can be written as:

$$nG = \sum_{i=1}^{N} n_i \Delta G_{fi}^o + \sum_{i=1}^{N} n_i RT \ln P + RT \sum_{i=1}^{N} n_i \ln \frac{n_i}{\sum_{i=1}^{N} n_i} + \sum_{j=1}^{M} n_j \Delta G_{fj}^o \qquad (1)$$

where nG is the total Gibbs energy (kJ), n_i is the number of moles of the i-th component in gaseous phase, n_j is the number of moles of the j-th component in separate solid phase, N is the total number of the gaseous species, M the total number of solid species and solid separate phases, $\Delta G^\circ{}_{fi}$ and $\Delta G^\circ{}_{fj}$ are the formation Gibbs free energies of the species i and j (kJmol^{-1}), respectively, R is the universal gas constant (kJK^{-1}mol^{-1}) and T is the absolute temperature of the system (K). The constrains introduced by the material balance of each element k are expressed by the relations:

$$\sum_{l=1}^{N+M}\sum_{k=1}^{K}\left(n_l a_{lk} - A_k\right) = 0 \qquad (2)$$

where a_{lk} is the number of atom masses of the k-th element in the l-th chemical species at the equilibrium condition, A_k is the total number of atom masses of the k-th element and K is the number of atomic species.

To minimize equation (1) under the constrain (2), the method of Lagrange's undetermined multipliers was adopted [15]. The software developed in this work is based on the SOLGASMIX-PV code [16].

As a thermodynamic approach, the method used in this work does not take into account the kinetic aspects of the reactions, physical adsorption, chemisorption and capillary condensation phenomena. The gas was assumed ideal and all condensed phases were considered as pure. The presence of ash forming elements, i.e. Al, Ca, Fe, K, Mg, Na and Si, was not considered although these compounds could influence the distribution and the composition of the metal species. No mixing models, in order to describe non-ideal behaviour of the system, were used. In spite of the above mentioned limitations, the method of determining stable phases, assuming global equilibrium, is at present one reliable computational possibility for generating knowledge about the chemistry of heavy metals in combustion systems.

Table 1:   Main physico-chemical characteristics of the RDF.

Bulk density (kg m^{-3})	380 ± 10
Humidity (mass %)	8.1 ± 0.3
Ashes (mass % dry basis)	16.3 ± 0.5
Lower heating value (MJ kg^{-1})	17.8 ± 0.3
Element (mass %)	
Carbon	55.1 ± 0.3
Hydrogen	11.5 ± 0.2
Oxygen	13.7 ± 0.7
Nitrogen	3.1 ± 0.2
Sulfur	0.04 ± 0.01

3   Experimental

3.1 Materials

The RDF used in this work was supplied by a local waste treatment company. Because its heterogeneity, the RDF was grounded to a grain size less than 1 mm before its characterization and use. The main physico-chemical characteristics and the chemical analysis of the processed RDF are reported in Tables 1 and 2, respectively.

Table 2:   RDF trace elements content.

Element	$(mg\ kg^{-1})$
Cl	1007
F	9.59
Br	94.97
As	0.7
Cd	0.8
Cr	76
Cu	320
Mn	66
Hg	0.5
Ni	166
Pb	880
Sn	1.1
Sb	81
Zn	445

3.2 Experimental equipment and procedures

The experimental apparatus used in this work is composed of two main sections: reaction and product collection. The reaction section is a classical updraft gasifyer reactor. The reactor consists of a 45 mm i.d., 550 mm length stainless steel tube equipped with a grid on the bottom side. The collection system consists of a high temperature filter stage (necessary to remove the particulates) followed by a water-cooled exchanger (for separating water and condensable fractions) and by a final gas filter.

The RDF feed was introduced in the reactor through a piston system at regular intervals of time in order to assure the continuity of the process. Because its low bulk density and cohesive characteristics, and in order to reduce the dragging phenomena in the reactor, the RDF powder was pelletized before its use as feed. The pellets of cylindrical shape (about 10 mm diameter and 6 mm height, with a mass of approximately 0.4 g) were produced by compression. An oxygen-steam mixture was introduced through the bottom of the reactor.

The selected operating conditions were: 700 ± 20°C for the gasification reaction; oxygen/carbon ratio equal to 0.2 mol/mol and a steam/charge ratio of about 0.5 g/g in order to control the reaction temperature. The runs were performed at a constant feed-rate and at atmospheric pressure. The duration of each test was two hours; the amount of sample feed in each run was about 800 g.

The off-gas, mainly constituted by hydrogen, carbon monoxide, carbon dioxide, gaseous hydrocarbons, volatile and non-volatile substances, steam and char, was sent to the collection system. The condensed fractions together with slag and fly ashes recovered on the final filter were analyzed for metal content.

Before starting each test, the equipment was preheated to the desired temperature. To assure safe working conditions, the feed was introduced before starting the oxygen-steam injection. This way, the charge undergoes immediate pyrolysis, producing char that in turn falls on the bottom of the reactor and reacts with the oxygen-steam introduced through the grid. This procedure avoids the formation of explosive mixtures in the gas phase.

4   Results and discussion

To compare the experimental and simulation results, the waste composition considered as the basis for the equilibrium calculations was that reported in table 1, while the oxygen/carbon and the oxygen/steam ratios were those fixed for the experimental test. The parameter of the simulation was the temperature, ranging from 700 to 1200 K. The equilibrium calculations were performed considering 11 trace elements: As, Cd, Cr, Cu, Hg, Mn, Ni, Pb, Sb, Sn, Zn, whose content in the RDF is reported in Table 2.

The behaviour of the different trace metals can be classified on the basis of the calculated equilibrium composition profiles in the temperature range considered in this study: those that can exist only in condensed phases, those that can exist only in gas phase and those that are distributed between the phases.

Cr, Cu and Ni are the trace metals in the first group, because in the considered temperature range they can exist only as condensed species, respectively $Cr_2O_3(cr)$, $Cu(cr,l)$ and $Ni(cr)$.

The second group includes As, Cd, Hg, Sb and Sn, that in the considered temperature range exist only as gaseous species. For arsenic the dominant compound between 700 and 840 K is $As_4(g)$ while above 840 K the stable form is $AsO(g)$. Several other gaseous species exist in all the temperature range, but in negligible amounts. For tin, in the temperature interval between 700 and 965 K $SnCl_2(g)$ is the most stable compound, while $SnS(g)$ is the major species formed above 965 K. Other gaseous species exist, also for tin, in all the temperature range, but their amount is negligible if compared with the two main species. For mercury the only stable species in the examined temperature interval is $Hg(g)$, while for antimony the dominant compounds is SbS(g) up to 1100 K and SbO(g) at higher temperatures.

Mn, Pb and Zn belong to the last group where both gas and condensed phases coexist. For manganese, MnS(cr,l) for temperatures below 1100 K and also MnO(cr) for temperatures above 800 K coexist with minor amounts of $MnCl_2$(g). For Pb and Zn the equilibrium composition profiles for the related species are reported in Figure 1 and 2 respectively, where the details relative to minor species are also reported in an enlarged scale.

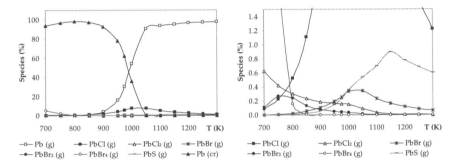

Figure 1:   Equilibrium composition profiles for Pb.

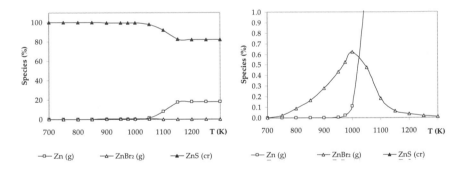

Figure 2:   Equilibrium composition profiles for Zn.

To validate the obtained equilibrium data, the attention has been focused on the behaviour of Cr, Cu, Mn, Ni, Pb, Sb and Zn, because only for these metals significant amounts in the RDF used for experimental gasification tests were found. At the end of each test a mass balance for each metal was carried out. Only those results showing a discrepancy in the metal balance less than 10% were retained.

A comparison between the experimental and calculated amounts of the species formed by the selected elements in both the condensed and gaseous phases is shown in Table 3, where the calculated data are reported for the two extremes of the experimental temperature range (680 and 720°C).

Table 3:   Comparison between experimental and calculated data.

	Condensed phase (%)		Gas phase (%)	
	Experimental	Calculated	Experimental	Calculated
Cr	99 ± 1	100.0	1 ± 1	0.0
Cu	71 ± 5	100.0	28 ± 3	0.0
Mn	99 ± 1	99.6 ÷ 98.3	1 ± 1	0.4 ÷ 1.7
Ni	76 ± 6	100.0	23 ± 3	0.0
Pb	63 ± 2	78.2 ÷ 36.1	36 ± 1	21.8 ÷ 63.9
Sb	11 ± 2	0.0	88 ± 3	100.0
Zn	97 ± 2	99.6 ÷ 99.4	2 ± 1	0.4 ÷ 0.6

A good agreement of the experimental results with the predicted ones can be observed for Cr, Mn, Pb and Zn, while for Cu, Ni and Sb the measured partition does not agree with that calculated at the equilibrium condition. Antimony is the only trace metal showing measured content in the condensed phase greater than the calculated one. This result could indicate that for Sb species the thermodynamic equilibrium is not reached.

The thermodynamic analysis performed in a limited temperature range could be responsible of the bigger amount of nickel and copper experimentally found in the gas phase. In fact, the thermodynamic model was applied just in the reaction zone, i.e. in the temperature range 680 – 720°C. Then the observed behaviour could be explained considering that volatile carbonyl complexes should be formed at the lower temperatures of the upper reactor section, where drying and pyrolysis of the feed in reduced atmosphere happen. Due to the reactor countercurrent geometry the volatile species of these metals, if formed, could be directly fluxed by the off-gas, so they are prevented from reaching the high temperature reaction zone where decomposition happens. Finally, also the possibility that a fraction of these elements can be entrained by the off-gas as a small particulate matter should be considered.

5   Conclusions

The developed thermodynamic model based on the free Gibbs energy minimization has proved to be a useful tool to qualitatively and quantitatively predict the amount of metallic compounds in the gas stream from a waste gasifier. This approach has shown a limitation with respect to some of the considered trace metals, namely Sb, Ni and Cu. However, such a limitation can be attributed to the reactor geometry, rather that to the validity of the used approach. An explanation of the deviation of the predicted partitions from the measured ones is presented. This study demonstrates that the thermodynamic approach can confidently estimate the influence of different gasification operating conditions on the fate of volatile metallic compounds. The good general agreement between simulations results and experimental measures shows that the predictive procedure based on the characterization of the thermodynamic favourite species is a reliable method for an environmentally sound conduction of waste gasification processes.

References

[1]   Borgianni C., De Filippis P., Pochetti F., Paolucci M., Gasification process of wastes containing PVC. Fuel, 81, pp. 1827-1833, 2002.
[2]   Belgiorno V., De Feo G., Della Rocca C., R. M. A. Napoli, Energy from gasification of solid wastes. Waste Management, 23, pp. 1-15, 2003.
[3]   Chen J. C., Wey M. Y., Su J.L., Two stage simulation of the major heavy-metal species under various incineration conditions. Environment International, 24(4), pp. 451-466, 1998.
[4]   Linak W. P., Wendt J. O. L., Toxic metal emission from incineration: mechanism and control. Prog. Energy Combust. Sci., 19, pp. 145-185, 1993.
[5]   Frandsen F., Johansen K. D., Rasmussen P., Trace elements from combustion and gasification of coal. An equilibrium approach. Prog. Energy Combust. Sci., 20, pp. 115-138, 1994.
[6]   Somoano M. D., Martinez Tarazona M. R., Trace element evaporation during coal gasification based on a thermodynamic equilibrium calculation approach. Fuel, 82, pp. 137-145, 2003.
[7]   Sorum L., Frandsen F. J., Hustad J. E., On the fate of heavy metals in municipal solid waste combustion. Part I: devolatilisation of the heavy metals on the grate. Fuel, 82, pp. 2273-2283, 2003.
[8]   Sorum L., Frandsen F. J., Hustad J. E., On the fate of heavy metals in municipal solid waste combustion. Part II: from furnace to filter. Fuel, 83, pp. 1703-1710, 2004.
[9]   Borgianni C., De Filippis P., Pochetti F., Paolucci M., Prediction of syngas quality for two-stage gasification of selected waste feedstocks. Waste Management, 24, pp. 633-639, 2004.
[10]   Cozzani V., Nicolella C., Petarca L., Rovatti M., Tognotti L., A fundamental study on conventional pyrolysis of a refuse-derived fuel. Ind. Eng. Chem. Res., 34, pp. 2006-2020, 1995.
[11]   Trouvé G., Kauffmann A., Delfosse L., Comparative thermodynamic and experimental study of some heavy metal behaviors during automotive shedder residues incineration. Waste Management, 18, pp. 301-307, 1998.
[12]   Verdone N., De Filippis P., Thermodynamic behaviour of sodium and calcium based sorbents in the emission control of waste incinerators. Chemosphere, 54, pp. 975-985, 2004.
[13]   Barin I., Knacke O., Thermochemical Data of Pure Substances, VCH Verlagsgesellschaft: Weinheim, 1989.

[14]   Chase M. W. (ed), Nist-JANAF Thermochemical Tables, 4th Ed. American Institute of Physics: Woodbury, New York, 1998.

[15]   Perry, R.H., Green, D. W. (eds), Perry's Chemical Engineers' Handbook, 6th Ed. McGraw-Hill: Singapore, 1984.

[16]   Besmann, T.M., SOLGASMIX-PV, a computer program to calculate equilibrium relationships in complex chemical systems. ORNL/TM-5775, Union carbide Corp., Nucl. Div., Oak Ridge Natl. Lab., 1977

Biomass gasification for farm-based power generation applications

C. C. P. Pian[1], T. A. Volk[2], L. P. Abrahamson[2], E. H. White[2]
& J. Jarnefeld[3]
[1]*Alfred University, Alfred, NY, USA*
[2]*SUNY College of Environmental Science & Forestry, Syracuse, NY, USA*
[3]*NY State Energy Research and Development Authority, Albany, NY, USA*

Abstract

Willow biomass crops have been shown to be a good fuel for farm-based power
production using advanced gasification technology. The fuel gas can be used for
generating electricity, using microturbines modified to operate on low-BTU gas,
or for other farm energy needs. Willow biomass was found to make an excellent
fuel for ash-rejection gasifiers with a predicted net gasification efficiency of
about 85 percent. The main drivers to the cost-of-electricity were found to be
associated with the harvesting, handling, and transporting of the willow biomass,
accounting for 40 to 50 percent of the annual operating cost. In the present
study, analysis showed that developing a method to co-gasify willow with
various amounts of low-cost wastes, such as dairy farm animal waste, can be an
excellent way to reduce the fuel cost, to increase the overall fuel availability and
help work around problems resulting from seasonal availability of bioenergy
crops. Co-gasification of dairy farm wastes along with willow offers an
economical way to dispose of the wastes and manage nutrient flows on a dairy
farm. The power generated from the animal waste can be used on the farm or
sold to offset the cost of waste treatment.
*Keywords: willow biomass, Salix, dairy-farm animal wastes, co-gasification, air-
blow gasification, farm-based power system.*

1   Introduction

A study was recently carried out to investigate the feasibility of using an
advanced gasifier to convert willow biomass crops into a fuel gas that could be

used for farm-based power production [1]. In the proposed power system, the gasifier product gas can be either utilized in microturbines for generating electricity or used for other farm energy needs. This paper reports on the extension of the original willow biomass study to include the co-gasification of willow with dairy-farm animal wastes as a way to lower the cost of energy production, increase the overall fuel availability, and to manage nutrient flows on a farm.

Willow is one of the more promising short-rotation woody crops for energy use in the near future in the northeastern and north-central United States. In New York, the federal and state governments have made a concerted effort to develop willow crops as an alternative renewable energy source. The cultiva-tion and use of willow for the production of bioenergy or bioproducts provide numerous potential environmental and rural development benefits. These benefits include reduction and replacement of fossil fuel usage, reduced imports of fuel and export of dollars, new uses for agricultural lands currently out of production, and creation of jobs in rural New York [2]. The near term use for willow biomass crops is co-firing with coal in existing utility power plants and direct-fired biomass facilities. Co-firing results in consistently reduced SO_x emissions, and reduced NO_x emissions under certain operating conditions. Gasification offers an alternate method of converting willow biomass to energy that should produce substantially less net CO_2 emission, more fuel flexibility and provide opportunities for distributed power generation.

Study results indicate that willow biomass makes an excellent gasifier fuel because of its low ash content and high ash-fusion temperature. The predicted gasification efficiency of the willow-fueled gasifier is about 85 percent [1]. Although willow would make an excellent gasifier fuel, it is still a relatively expensive fuel because it is a cultivated crop. In order for willow to be a commercially viable bioenergy fuel, it must be able to compete in a market where there is already an abundance of low-cost and negative-cost waste fuels. In the willow gasification system, the fuel cost accounts for 40 to 50 percent of the annual operating costs. For this reason, continuous efforts have been made to reduce the cost of willow biomass crops by improving yields and reducing production costs; and adding value to the multiple environmental and rural development benefits associated with the crop through programs such as the Conservation Reserve Program, federal biomass tax credits, green pricing premiums, renewable portfolio standards, and agricultural tax assessments on land used for growing willow crops.

Another issue with the willow fuel is its seasonal availability. Willow bio-mass crops are typically harvested in late fall and during the winter. Storage and scheduling of fuel will be needed in order to use willow for year-round power generation. Co-gasifying willow along with agricultural wastes can be a way to increase the overall fuel availability, ease the effects of seasonal variability of willow supply, and lower the cost-of-electricity (COE) of the power system.

Dairy farm animal waste is a possible fuel for co-gasification with willow biomass. In a previous study, we demonstrated that dairy farm wastes can be an excellent biomass fuel for our advanced gasifier [3]. A major problem facing the dairy industry is the proper disposal/utilization of livestock waste. Dairy farms have historically imported three times more nutrients than they export. Field spreading of manure, to recycle the nutrients to the crop fields, can cause soil to become saturated with some nutrients resulting in non-point source pollution. As stricter environmental regulations are applied, farms need a way to reduce the amount of manure they spread in a cost effective manner. Gasification offers an economical way to dispose of the wastes and to remove the nutrients from the farm. The power generated from the animal waste can be used on the farm or sold to offset the cost of waste treatment.

In the proposed co-gasification scheme, shown in Fig. 1, willow biomass will be made available to the dairy farm operator for use as animal bedding. In exchange, the liquid portion of the animal manure will be transported to the willow farm for use as fertilizer. The solid portion of the manure and the soiled bedding from the dairy farm are used to augment the willow biomass fuel for the gasifier. The nutrients that are concentrated in the combustion wastes of the gasifier are in forms that can easily be transported off the farms and sold as fertilizers. The export of nutrients from the dairy farm, in the form of liquid manure to the willow fields and solid manure to the gasifier/power system, will help reduce or eliminate field spreading of manure, which currently cost New York state dairy farms an averaged of about $50/cow/year in labor and equipment. The proposed arrangement will eliminate the need for dairy farms to purchase wood chips and sawdust for beddings, resulting in additional saving of approximately $50-100/animal/year [4]. In order for this scheme to be feasible, the willow production fields will need to be in close proximity to the dairy farms to minimize transportation costs, or strategically located on the dairy farms and integrated directly into the dairy farm operations to further reduce non-point source pollution.

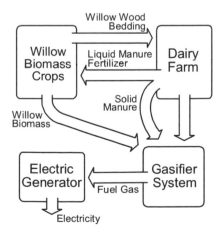

Figure 1:   Proposed willow/dairy farm-based power system.

The particular gasification system we have selected for our application was originally developed using high-temperature preheated air to convert coal and waste-derived fuels into synthetic fuel gas and value-added byproducts. This advanced gasifier system, known as MEET (Multi-staged Enthalpy Extraction Technology), has several unique attributes that are advantageous for biomass power generation. The gasification technology and its attributes are described in Section 2. The gasifier performance and cost analysis results, as well as implementation issues associated with the willow-biomass/manure-waste co-gasification scheme, are reported in Section 3.

2 Description of gasification system

The basic scheme of the gasification system is shown in Fig. 2. Solid fuel is gasified using high-temperature, preheated air in a reactor vessel to produce a flammable raw synthetic gas. The inorganic ash residue from the gasification reactions is extracted from the gasifier either as a molten slag (slagging mode of operation) or as ash (non-slagging or dried-bottom operation), depending on the selected operating temperature of the gasifier. Upon exiting the gasifier, the fuel gas is cooled in a heat recovery boiler and then cleaned, using conventional gas cleanup technology. A small fraction of the product gas is diverted to the preheater where it is used for heating the gasification air. Depending on the intended application of the plant, the rest of the cleaned fuel gas is utilized for a variety of possible downstream processes.

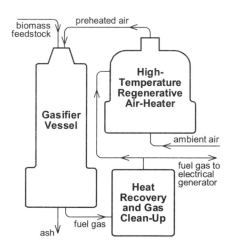

Figure 2:   Basic scheme of the MEET gasification system.

The gasifier vessel consists of an entrained-flow section, followed by a particulate-removal section. In the slagging MEET gasifier, the gasification temperature is maintained above the fluid temperature (FT) of the ash and a high-temperature ceramic pebble-bed filter is used for particulate removal.

When the gasifier is configured in the dried-bottom mode, the operating temperature is maintained at a value below the ash's initial deformation temperature (IDT) and an aerodynamic particulate removal device is used for ash rejection. Gasifier operations in the temperature range between the IDT and FT are avoided to prevent ash agglomeration and fouling problems. A compact regenerative heater supplies the high-temperature air for the gasifier. As mentioned previously, the source of heat for this preheater comes from the combustion of a small amount of the fuel gas produced by the gasifier.

The use of high-temperature air in the MEET process increases the yield of the gasification process over those of conventional air-blown gasifiers [5]. This is because preheated air increases the heating rate of the incoming fuel and the volatile yield from the fuel particles is greatly enhanced under these high-temperature, rapid-heating conditions. Another reason for the increased yield is that higher caloric-valued syngas can be obtained using preheated air as the gasifying agent. The gasifier can be operated at substantially more fuel-rich conditions to achieve the given gasification temperature, leading to higher caloric-valued product gas.

In addition to higher gasification yields, the MEET gasification system has several other features that are advantageous for biomass power generation applications. These attributes include fuel flexibility, compact size, ability to operate on low-rank solid feedstock and transportability. The MEET gasifier technology and the current status of its development are described in Pian and Yoshikawa [5].

3 Performance and COE estimates of the willow/manure gasification system

The ultimate and ash analyses of the biomass feedstock used in this study are shown in Table I. The willow wood data are based on the averages of a large group of SUNY College of Environmental Science & Forestry (ESF) crops harvested in the winter of 1999 [1]. The variations in composition between the different wood samples were small, on a dried basis. However, large variations in moisture content were typical due to differences in harvesting and storage conditions of the test samples.

The manure samples were collected from Sun Rich Farms, located in Albian, New York [3]. The ultimate and ash analyses of the manure waste are similar to those for grass and hay, except for the higher ash and SiO_2 contents, which presumably are caused by the soil that was picked up during manure collection or ingested by the cow while feeding.

The solid fuel's ash IDT and FT temperatures are important parameters needed for configuring our gasifier and for determining its operating condition. The values of these temperatures, under both oxidizing and reducing conditions, were determined using a hot-stage microscopic (HSM) technique [6]. Measured results of these temperatures are shown in Fig. 3 for varying mass fractions of willow biomass in the overall gasifier feedstock. The ash fusion temperature of the fuel decreases with increasing fractions of manure. This trend is due to the higher concentrations of phosphorus, potassium and sodium in the manure ash, which act as fluxing agents, as well as higher concentration of SiO_2, which can form a eutectic with CaO that lowers the fusion temper-ature. Comparison of ash fusion temperature measurements made in reducing and oxidizing conditions also showed minimum differences between the values. This behavior for the willow/manure ash, which has very low ferric concentrations, is similar to that measured for low-iron coal ashes [7]. Based on the HSM results one can conclude that the ash-rejection MEET gasifier should not encounter any ash fouling problems when operating at 900°C for any mixture ratios of willow and manure.

Table 1:   Analyses of SUNY-ESF willow biomass crops and Sun Rich Farms manure wastes.

Ultimate Analysis of Willow*		Ash Analysis	
Carbon (mass %)	49.71	Al_2O_3 (mass %)	0.20
Hydrogen	5.97	CaO	41.92
Oxygen	42.16	Fe_2O_3	0.73
Nitrogen	0.50	MgO	3.52
Sulfur	0.03	ZnO	0.35
Ash	1.62	P_2O_5	9.05
		K_2O	13.22
		SiO_2	3.65
HHV (cal/g)	4,798	Na_2O	0.28
		SO_3	2.40
Ultimate Analysis of Manure**		Ash Analysis	
Carbon (mass %)	44.65	Al_2O_3 (mass %)	1.51
Hydrogen	5.85	CaO	32.13
Oxygen	38.18	Fe_2O_3	0.93
Nitrogen	2.05	MgO	9.84
Sulfur	0.31	MnO	0.17
Ash	8.96	P_2O_5	13.04
		K_2O	9.41
		SiO_2	19.83
HHV (cal/g)	4,352	Na_2O	5.95
		SO_3	5.57
		TiO_2	0.11

* Dried basis.
** Dried basis. The as-received manure, after separation of liquid fertilizer by mechanical press, contained 69.57 mass % moisture.

The gasifier performances were estimated using a previously developed gasifier model [5]. The model assumes chemical equilibrium conditions and uses the minimization of Gibb's free energy method to determine the thermodynamic properties at the state points. Validation of model results was carried out previously by comparisons with coal-fired gasifier test data.

Figure 4 shows the estimated net gasifier conversion efficiencies and costs-of-electricity for the co-gasification of willow with various amounts of manure wastes. An air-preheat temperature of 1230°C and a fuel moisture content of 10% are assumed. The gasifier is configured in the non-slagging mode with an operating temperature of 900°C to minimize ash fouling problems. The net gasifier conversion efficiency is defined as the higher heat value of the product fuel gas, after subtracting that portion of the fuel gas used to fire the air preheater, divided by the thermal input of the solid fuel. The net conversion efficiency for the gasifier operating on willow is 85 percent and on manure is 80 percent [1]. For co-gasification operations, the gasifier performance decreases with increasing mass fraction of manure waste in the total fuel. This behavior can be attributed to several factors. The heating value of the manure waste is approximately 10% lower than that of the willow biomass. Consequently, as the mass fraction of manure waste increases, the gasifier must operate at a less fuel-rich condition to maintain the 900°C gasification temperature; larger fractions of the incoming fuel must be used to raise the gas temperature, resulting in a lower caloric-valued product gas. With increasing manure mass fraction, greater amount of the product fuel gas is used in the air preheater, also contributing to lowering the net gasifier efficiency.

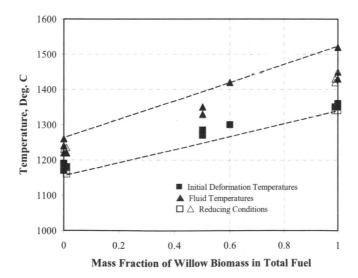

Figure 3:   Fusion temperatures of willow/manure ash determined from hot-stage microscope data.

For our cost analysis, we selected a gasifier system with a capacity for processing approximately 160 kg of dried willow biomass and/or manure wastes per hour. For a gasifier of this scale, one reactor vessel can process the yearly crop yield of a 106-ha willow farm or the animal wastes produced yearly by a

400-animal dairy farm. For larger applications, multiple units of this reactor will be used. A MEET gasifier of this size has been previously built and tested at the Mississippi State University by Pian and Yoshikawa [5]. This refractory-lined vessel, with exterior dimensions of approximately 1-m diameter and 3-m length, can be mounted on a flatbed truck for mobile applications.

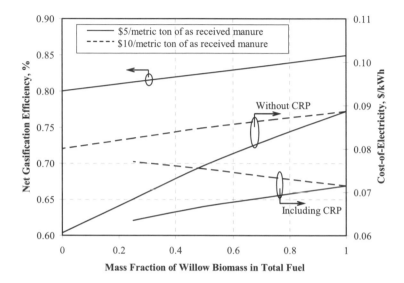

Figure 4:   Net efficiency and cost-of-electricity vs. mass fraction of willow biomass.

Microturbines, modified to operate on low-calorific fuel gas, will be used to generate electrical power for our small on-farm applications. FlexEnergy developed these 30-kW microturbine units in partnership with Capstone Turbine Company. Conversion efficiencies (fuel gas-to-electricity) of the order of 22% have been achieved during initial demonstration tests using low-calorific fuel gas [1]. If one assumes all the fuel gas available from the biomass gasification are used to generate electricity, then five microturbine units are required to support each gasifier unit.

The COE for the willow-fueled power system was estimated previously to be $0.09/kW-hr and a capital cost of about $2,833/kW [1]. The system consisted of a MEET gasifier operating in the dried-bottom mode and five microturbines. The COE estimate assumed a 25-year lifespan for the capital equipment, an annual 6% interest rate, an averaged 85% availability, and a willow delivered cost of $49.6/metric ton of dry wood. The COE also included an operation and maintenance cost of $0.012/kWh. The willow delivery cost was determined from actual field experiences accumulated by SUNY-ESF [8]. When the contributions from the Conservation Reserve Program (CRP) are included as part of producing the willow biomass, then the delivered price was reduced to $31.7

per metric ton of dry wood and the COE would decrease to $0.073/kWh. The CRP program contributes funding through the US Department of Agriculture, which partially pays for the site preparation and planting costs, as well as an annual rent payment to the farm landowner. Possible economic benefits from the sale of process heat, recovered solids/nutrients, or other "green energy" price premiums or tax incentives were not factored into the cost analysis.

Figure 4 shows the potential to lower COE by co-gasifying willow with various amounts of dairy-farm manure. Also shown are the sensitivities of estimated COE's to the assumed manure costs and the impacts of CRP contributions. Variations in the assumed manure cost can be used to account for project specific factors such as manure preparation, handling or transportation costs. The contribution of co-gasification to lower COE increases with increasing willow-fuel cost and decreases with increasing manure cost. Also, co-gasification can be a viable method to keep the willow-gasification power system commercially competitive after the various tax incentives have expired.

Usage of willow as bedding for dairy cows shows no evidence of being very different than any other wood bedding, assuming similar particle size and moisture content. However, dairy operators are reluctant to switch to willow for bedding in place of their current practice unless they can be convinced the practice is safe and beneficial to them. The willow must be proved acceptable by demonstration of use and the cost must be shown to be substantially lower than the $50/cow/year price of current wood bedding. Several New York-based dairy farms have agreed to be involved with trial demonstration tests, using willow bedding in heifer pens. Initial test results indicated the willow would need to be chipped finer, to the consistency of sawdust, and dried to a moisture content of 20 to 45%. Both requirements can be easily met. The particle size can be changed by new forage harvester systems that are being developed for willow biomass crops. Lower moisture content is obtainable by using the waste heat of the gasifier system to dry the willow or letting the willow sit on the field or in storage for a longer period of time. The benefits to the dairy farms are numerous. In addition to lower cost for bedding, and environmental and financial benefits from better nutrient management, the energy produced from the manure waste can contribute to lower farm operating cost; and in many cases become another source of revenue for the farms. In an earlier manure-gasification case study, we have found that the production of fuel gas from gasification of farm animal waste from a particular upstate NY dairy farm can generate almost four times the amount of energy necessary for energy self-sufficiency [3].

4 Summary

Willow biomass makes an excellent fuel for ash-rejection gasifiers because of its low ash content and high ash-softening temperature. The predicted gasification efficiency of the willow-fueled gasifier is about 85 percent. Higher fuel costs and seasonal variation in availabilities are some of the drawbacks associated with using willow wood for energy production. Co-gasifying willow wood with biomass wastes in a flex-fuel gasifier can be an effective way to lower the net cost of the fuel and ease the effects of seasonal variability in the willow supply.

The performance and operating characteristics of the MEET gasifier running on various ratios of willow biomass and dairy-farm livestock waste were investigated. The gasification efficiency decreased slightly, from 85 to 80 percent, with increasing amount of manure in the fuel mix. The contribution of co-gasification to lower COE increases with increasing willow-fuel cost and decreases with increasing manure cost.

Co-gasification of willow with manure waste also benefits the dairy farm operators, by offering an economical way to manage nutrient flows on the farm and reduce non-point source pollution. The power generated from the animal waste can be used on the farm or sold to offset the cost of waste treatment. In order for this scheme to be feasible, the willow production fields will need to be in close proximity to the dairy farms to minimize transportation costs, or strategically located on the dairy farms and integrated directly into the dairy farm operations.

References

[1]   Martin, J.R., Pian, C.C.P., Volk, T.A., Abrahamson, L.P., White, E.H., & Jarnefeld, J., Recent results of willow gasification feasibility study, Second International Energy Conversion Engineering Conference, Providence, RI, American Institute of Aeronautics and Astronautics, paper no. AIAA-2004-5650, 2004.

[2]   Volk, T.A., Verwijst, T., Tharakan, P.J., & Abrahamson, L.P., Growing Energy: Assessing the sustainability of willow short-rotation woody crops, *Frontiers in Ecology and the Environment*, **2(8)**, pp. 411-418, 2004.

[3]   Young, L. & Pian, C.C.P., High-temperature, air-blown gasification of dairy-farm wastes for energy production, *Energy*, **28(7)**, pp. 655-672, 2003.

[4]   Wright, P., Personal communication, 2004, Department of Biological and Environmental Engineering, Cornell University, Ithaca, NY.

[5]   Pian, C.C.P. & Yoshikawa, K., Development of high-temperature air-blown gasification systems, *Bioresource Technology*, **79(3)**, pp. 231-241, 2001.

[6]   Boccaccini, A.R. & Hamann, B., Review of in situ high-temperature optical microscopy, *Journal of Material Science*, **34(22)**, pp. 5419-5436, 1999.

[7]   Ergun, S., Coal Classification and Characterization, *Coal Conversion Technology*, Eds. C.Y. Wen & E.S. Lee, Addison-Wesley, Reading, MA, pp. 46-47, 1979.

[8]   Tharakan, P.J., Volk, T.A., Lindsey, C.A., Abrahamson, L.P. & White, E.H., Evaluating the impact of three incentive programs on the economics of co-firing willow biomass with coal in New York State, *Energy Policy*, **33(3)**, pp. 337-347, 2004.

Section 7
Waste pre-treatment,
separation and transformation

Decomposition and detoxification of DXNs adsorbed on various solid wastes by microwave plasma treatment

R. Sasai[1], M. Jin-no[2], A. Satoh[2], Y. Imai[2] & H. Itoh[1, 2]
[1]Division of Environmental Research, EcoTopia Science Institute, Nagoya University, Japan
[2]Department of Applied Chemistry, Graduate School of Engineering, Nagoya University, Japan

Abstract

Microwave plasma treatment of the various solid wastes adsorbed with DXNs was investigated to develop an effective and novel method for in situ detoxification of DXNs in solid wastes such as incinerator fly ashes of the municipal wastes (IFA) or used activated carbons (AC). It was found that the microwave plasma treatment could completely decompose the adsorbed DXNs into inorganic gases such as Cl_2, CO_2 and CH_4 with low plasma power (250 W) and short treatment time (< 15 min). Moreover, this decomposition behaviour did not depend on the kinds of solid wastes adsorbed with DXNs. In the case of IFA, the treatment time needed was slightly affected by the amount and/or chemical species of inorganic salt (CuO, $CuCl_2$, $Ca(OH)_2$, $CaCl_2$, etc.) contained in the IFA at low plasma power region (< 100 W), but this change was negligible at higher plasma power (> 250 W). The AC obtained after the microwave plasma treatment exhibited almost the same characteristics of specific surface area, pore size distribution, and pore volume. Furthermore, the weight loss of AC due to burning and/or ablation could be scarcely observed. This shows that the microwave plasma can selectively decompose the organic species adsorbed on solid matter. Thus, the present microwave plasma treatment was found very useful not only as a decomposition method of the toxic organic compounds on solid, but also as a recycling method of the used inorganic solid adsorbents.
Keywords: microwave plasma treatment, DXNs, incinerator fly ash, used activated carbon, detoxification, recycling.

WIT Transactions on Ecology and the Environment, Vol 92, © 2006 WIT Press
www.witpress.com, ISSN 1743-3541 (on-line)
doi:10.2495/WM060301

1 Introduction

Dioxin chemicals (DXNs) are generic name of the polychlorinated dibenzo-*p*-dioxin (PCDD), polychlorinated dibenxofurane (PCDF) and coplanar polychlorinated biphenyl (co-PCB). They are typically toxic and hazardous compounds, the discharged amounts of which are strictly controlled all over the world. Today, it is well known that most of the DXNs are emitted from the incinerator of the municipal wastes in Japan and they exist in the exhaust gas produced during incineration. The exhaust gas is detoxified by the electric ash precipitator or adsorbent such as activated carbon (AC), and then released to the atmosphere. These fly ash or adsorbent is discharged as the incineration wastes containing DXNs. Therefore, a proper treatment should be carried out to detoxify these condensed wastes. Usually, these hazardous wastes have been landfilled in the final disposal site after detoxification and/or stabilization treatment such as incineration at very high temperature or solidification into cement. However, these methods give rise to other serious problems like high-energy consumption, high maintenance fee, high CO_2 emission, etc. Then, several detoxification technologies have been developed and practically applied in a few cases. The thermal dechlorination and vitrification methods are well known as the detoxification process of DXNs containing IFAs [1-3]. In the former process, the detoxification of DXNs occurs by the substitution of chlorine with hydrogen at 350-550°C under poor oxygen atmosphere, but DXNs are resynthesized by *de novo* reaction during the cooling step [4]. Furthermore, DXNs cannot be detoxified completely even by heating, because the dechlorination occurs only on the surface of fly ash. In the latter process, DXNs are decomposed almost completely during the melting process of IFA at higher temperatures of 1200-1600°C in the vitrification process. Resynthesis of DXNs does not occur during this treatment, although a large-scale furnace with a considerably high-energy consumption and cost for the maintenance of the facility must be required.

Activated carbon is frequently used as adsorbent for removing the DXNs from the polluted exhaust gas, which is emitted after removal of the suspended particulate matter by the electric ash precipitator. Today, the used activated carbon with DXNs is not recycled, because there is no effective recycling process of the activated carbon adsorbed with such toxic stable organic molecules. Most of the current recycling methods proposed for the used activated carbon possess some serious problems such as the performance degradation, high energy consumption and low recovery rate.

To solve the above mentioned problems, it is essential to develop a novel process for complete and selective decomposition of DXNs adsorbed on IFA and AC at lower temperature and with higher efficiency. Here, we adopted a non-equilibrium microwave plasma process with a high reactivity at relatively low ambient temperature. Recently, the microwave plasma process has been applied to decomposition and detoxification of various hazardous gases, such as chlorofluorocarbon, NO_x, CO_2, volatile organic compounds and DXNs. However, there is no application of the microwave plasma treatment to the decomposition and detoxification of hazardous organic compounds adsorbed on solid matter. In this study, we attempted both detoxification of IFA adsorbed with DXNs and recovery of AC adsorbed with DXNs by the microwave plasma irradiation.

Table 1:   Element composition of IFAs.

	K	Al	Mg	Cu	Pb	Fe	Ca	Cd
A	65.6	57.6	12.2	0.31	0.60	8.28	445	0
B	44.8	26.0	5.6	0.74	3.99	7.75	325	0
C	34.8	48.7	13.6	0.34	0.60	4.13	253	0

* Unit is mg/g.

2 Experimental

2.1 Materials

Three IFAs received from different emission sources, were used as IFA samples (an average element composition of these IFAs is shown in Table 1). The amount of DXNs contained in these IFAs was (A) 0.17, (B) 1.56, and (C) 3.21 TEQ-ng/g, respectively. Granular activated carbon for adsorbing DXNs (DXN-4, Ajinomoto Fine-Techno. Co. Inc.) was used. In the microwave plasma treatment for activated carbon, two kinds of non-toxic polychrolinated dibenzo-*p*-dioxin (PCDD: Scheme 1) were used as a model DXN molecule.

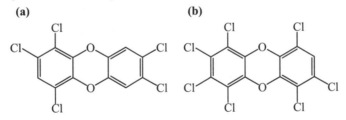

Scheme 1:   Chemical structure of PCDD ((a) 5CDD and (b) 7CDD).

2.2 Preparation of activated carbon with DD

Granular activated carbon with ca. 200 ng-PCDD/g-AC (AC/PCDD) was prepared by adding AC to 20 cm³ of PCDD diethylether solution, and then evaporating diethylether at room temperature. This as-prepared AC/PCDD was dried at 110°C for 3 h in vacuo.

2.3 Microwave plasma decomposition treatment

Figure 1 shows the experimental apparatus for the microwave plasma treatment of solid matter adsorbing DXNs. This apparatus consists of the following four parts: (1) decomposition and detoxification quartz reactor, (2) microwave plasma generator and (3) trap for the exhaust gas. A quartz crucible with a cap, in which the sample was charged, was held on a pedestal in a quartz reactor. A constant flow rate (70 sccm) of nitrogen/oxygen mixed gas was streamed at a controlled pressure (15 Torr) during the treatment. The generation of microwave plasma was carried out by the activation of the mixed gas at the microwave power with the frequency of 2.45 GHz from the magnetron of microwave generator. The contaminants in an exhaust gas generated during the treatment were collected in a series of taps for gas, water and organic solvent.

The IFAs with DXNs (1.5 g) were treated at 100 and 150 W of the microwave power for less than 30 min under N_2 atmosphere. The AC/PCDD (1.5 g) were treated at 250 W of the microwave power for less than 5 and 15 min under the nitrogen/oxygen mixed gas (oxygen content: $O_2/(N_2+O_2) = 0$ or 0.2).

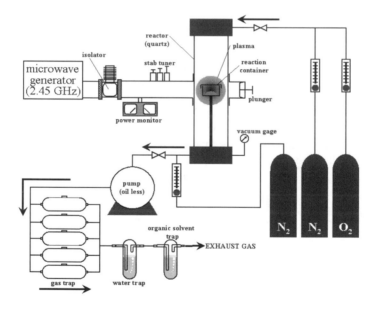

Figure 1:   Schematic illustration of the apparatus for the microwave plasma treatment of the solid matter adsorbing DXNs.

2.4 Characterization

2.4.1 IFAs with DXNs

Crystalline phase variation of the IFAs before and after the microwave plasma treatment at various conditions was examined by the X-ray diffraction (XRD) analysis (Rigaku: RINT-2000). Residual concentration and homologue distribution of DXNs in the IFAs before and after the microwave plasma treatment were evaluated by the gas chromatography-mass spectroscopy (GC-

MS, JEOL: JMS-700D) analysis of the DXN solution, which was concentrated and extracted by the prescribed method.

2.4.2 AC with DD

Residual concentration of PCDD in the AC before and after microwave plasma treatment was estimated from the high resolution GC-MS (JEOL: JMS-700D) spectra of the extracted fractions prepared by the legally assigned technique in Japan. Organic species in the exhaust gas or those dissolved in the liquid trap (in water and toluene) were identified by the GC-MS. Inorganic species in the exhaust gas were analyzed by the GC-TCD (Shimadzu: GC-8A) using the column (Shimadzu GLC: SHINCARBON ST). Inorganic ionic species were measured by the ion-chromatography (Shimadzu: LC-10ADsp) with the electric conductivity detector (Shimadzu: CDD-10Avp) and the anionic exchangeable column (Shimadzu: Simpack IC-A3/IC-GA3).

Characterization of AC before and after the microwave plasma treatment was carried out by the measurement of the specific surface area, pore size distribution and pore volume using nitrogen adsorption technique.

3   Results and discussion

3.1 Microwave plasma treatment of IFAs with DXNs

Figure 2 shows the treatment time dependence of the removal rate of DXNs (a) and TEQ decrease of IFAs (b), which were obtained after the microwave plasma treatment at 100 W. Total amount and TEQ value of DXNs contained in IFAs decreased exponentially with an increase in the treatment time, and then, the microwave plasma treatment for 30 min could remove most of DXNs from IFAs regardless of the kind of IFA. However, the difference of the removal rate of DXNs among IFAs was observed, when the microwave plasma treatment was carried out for 5 min. Especially, in IFA-(B), the removal rate did not change, and the TEQ value increased by the microwave plasma treatment for 5 min. These results indicate that the amount of DXN species with high toxicity will increase due to the microwave plasma treatment. To clarify the reason of this behaviour, the distribution of DXNs in IFA-(B) treated for 5 min was measured. Table 2 shows that the concentrations of hepta- and octa-chloro-DD (7CDD and 8CDD) decreased by the microwave plasma treatment, but those of the hexa-, penta-, and tetra-chloro-DD (6CDD, 5CDD and 4CDD) increased. This means that the dechlorination of DXNs occurs by the microwave plasma treatment under these conditions, but no decomposition of the skeleton structure of DXNs. However, the increasing amount of 4CDD, 5CDD and 6CDD was not the same as that of 7CDD and 8CDD. The *de novo* synthesis and the chloride addition reaction will be considered as for the production mechanism of 4CDD, 5CDD and 6CDD. Generally, it is known that the highly chlorinated DXN species forms by *de novo* synthesis. As highly chlorinated DXN species did not increase in the present case, the increase of the lower chlorinated DXN species will be caused by the chlorine removal reaction. This means that the reaction occurs preferentially towards lower chlorinated DXN species. Generally, the reason for the dechlorination and chlorination is a catalytic reaction of the inorganic compounds containing Cu, Ca, etc. during the incineration of municipal wastes. Thus, it is considered that such inorganic species contained in IFA will affect the dechlorination and chlorination reaction of DXNs in IFAs. Moreover, the inorganic compound composition on IFA surface will be important, because the dechlorination and chlorination reaction will occur on the surface of IFA particles. The element composition on the surface of IFAs is shown in Table 3. The amount of Cu and Ca on surface of IFA-(B) is found smaller than those of IFA-(A) and -(C). Moreover, the analysis of the main compounds of Cu and Ca in IFAs exhibited the presence of hydroxide and chloride. Although the percentage of Cu compounds did not vary by IFA species, the percentage of Ca compounds in IFA-(B) was higher than those in IFA-(A) and -(C). This means that the IFA-(B) contains large amount of chloride that would be a chlorine source for DXNs. Therefore, the excess increase of 4CDD, 5CDD and 6CDD in IFA-(B) observed after the microwave plasma treatment for 5 min may be caused by the chloride species of Ca existing on the surface of IFA-(B) particle. On the other hand, it is well known that Cu species play a role of catalyst for decomposition or dechlorination of DXNs during incineration. The amount of Cu on surface of IFA-(B) was smaller than that of other IFAs. This fact verifies that the efficiency of the decomposition or dechlorination reaction is the lowest in these IFAs. Thus, it is considered that the remarkable difference observed after the microwave plasma treatment of IFAs for 5 min can be related to the amount of Ca and Cu species on the surface of IFA particle.

Table 2:   Distribution of DXNs in IFA-(B) after the microwave plasma at the treatment for 5 min.

	[DXNs] (pmol/g)	
	untreated	treated
4CDD	41	49
5CDD	41	45
6CDD	45	51
7CDD	57	53
8CDD	70	61

Table 3:   Element composition on the surface of IFAs.

	K	Al	Mg	Cu	Pb	Fe	Ca	Cd
A	7.8	3.2	1.8	3.0	1.1	0.6	32.0	0
B	5.4	1.6	0.8	1.8	1.7	0.4	25.4	0
C	5.7	5.0	2.3	3.7	1.1	0.6	37.7	0

* Unit is atomic percentage.

When the microwave plasma power increased up to 150 W, the remarkable difference among three IFAs after the microwave plasma treatment was not observed (cf. Figure 3). This would be caused by higher decomposition rate due to the plasma active species with higher energy. These results indicate that the treatment of IFA with DXNs at higher microwave power efficiently decomposes DXNs contained in IFAs to non-toxic inorganic gases such as CO_2 and CH_4 regardless of the chemical composition of IFAs. It is concluded, therefore, that the microwave plasma treatment is very effective method for complete decomposition of DXNs contained in IFAs, besides this treatment does not select the kinds of IFA species.

3.2 Microwave plasma treatment of activated carbon with DXNs

In Table 4, the concentration of DXNs in AC added 1, 2, 4, 7, 8-pentachlorodibenzo-p-dioxin (5CDD) or 1, 2, 3, 4, 6, 7, 9-heptachlorodibenzo-p-dioxin (7CDD) before and after the microwave plasma treatment at 250 W is shown. Here, the 4CDD and 6CDD observed before the treatment will be impurity of the 5CDD and 7CDD, respectively. The DXNs concentration in both AC rapidly decreased with an increase in the treatment time. In the case of AC with 5CDD, no 6CDD, 7CDD and 8CDD species were observed after the microwave plasma treatment for 5 or 15 min. This means that the addition reaction of chlorine to skeleton structure of 5CDD does not occur, and thus, only the dechlorination and/or decomposition of 5CDD skeleton structure occur. On the other hand, the production of 4CDD and 5CDD species was observed after the microwave plasma treatment of the AC with 7CDD. This result indicates that the dechlorination reaction of 7CDD occurs dominantly during the present treatment. Besides, the decomposition reaction of skeleton structure of DXNs occurs because the total amount of DXNs adsorbed on AC is also decreasing. These results lead to the conclusion that the microwave plasma treatment is very effective to remove and decompose the DXN species adsorbed on AC, just as the fly ash with DXNs.

Table 4: Concentration of DXNs in AC before and after the microwave plasma treatment at 250 W.

| | added DXN species | | | | | |
| | 5CDD | | | 7CDD | | |
	0 min	5 min	15 min	0 min	5 min	15 min
4CDD	7050	134	30	0	50	0
5CDD	306914	238	73	0	70	10
6CDD	0	0	0	18752	109	58
7CDD	0	0	0	253984	205	106
8CDD	0	0	0	0	0	0

unit is pg/g

In the recovery case of polluted AC, not only the removal and decomposition of DXNs, but also the maintenance of AC characteristics such as adsorption ability and porosity is very important. In Table 5, the specific surface area of AC before and after the microwave plasma treatment at 250 W is shown. The surface area of AC with DXNs after the microwave plasma treatment was almost the same as that of the virgin AC. Moreover, the pore size distribution and volume of AC after the treatment were also the same as those of the virgin AC. This shows that the microwave plasma treatment does not seriously affect the properties of AC. Therefore, the present microwave plasma treatment is very effective to recover the AC polluted by DXNs.

Table 5:   Specific surface area of AC before and after treatment at 250 W.

virgin (before)	after			
	5CDD		7CDD	
	5 min	15 min	5 min	15 min
1125	1126	1208	1043	1171

4   Conclusions

The present microwave plasma treatment is one of the powerful techniques for the detoxification of the incineration fly ash and activated carbon polluted with DXNs. In our data, the DXNs including or adsorbing IFAs and AC particles were perfectly decomposed by the microwave plasma treatment with very low energy consumption (*i.e.*, < 250 W within 15 min). In addition, the present microwave plasma treatment brought about no degradation of the adsorption ability and pore structure of AC. Therefore, the microwave plasma treatment is effective to recover the waste AC, which is used for cleaning up the exhaust gas after the incineration. Further development of a large-scale treatment apparatus must be required as a future work for the practical application of our work, because our apparatus can treat only small amount of the solid matter with DXNs (< ca. 10 g).

Acknowledgement

We thank Mr. K. Ooba, Nagoya City Environmental Science Research Institute, for analyses of the concentrations and the homologue distribution of DXNs, and useful suggestions and discussion on the decomposition of DXNs.

References

[1] Miyata, H., *Dioxin*, Iwanami-shinsho: Tokyo, 1999.
[2] Kaneko, H., *Hibai-taisaku*, NTS: Tokyo, 1998.
[3] Kojima, K., Okajima, S., Ozaki, H., *Proc. 13th Annual Conference of The Japan Society of Waste Management Experts*, p. 702, 2002.
[4] Wang, Y-F., Lee, W-J., Chen, C-Y., Hsieh, L-T., *Environ. Sci. Technol.*, **33**, p. 2234, 1999.

Development of a way to recycle waste glasses: preparation of porous materials from cathode-ray-tubes and/or packaging glasses at the end of their life time

M. Cambon & B. Liautard
Technological Team of Research ERT 3,
Laboratoire de Physico-chimie de la Matière Condensée,
Montpellier II University, France

Abstract

French and European legislation concerning waste requires studies to be performed on the recycling of products issued from commercial electronic devices. Cathode ray tubes (CRT), and in particular glasses which constitute them, are of major interest. The chemical composition of these glasses is very important. Screen glass is composed of 8-12 wt% BaO and 6-10 wt% SrO and the cone glass which has a lower wall thickness than that of the screen contains 19-23% PbO. Previous characterizations of these special glasses led us to propose a way of recycling in which used CRT glass could be treated in order to respond to the legislation. Foam glass could be a possible solution to convert CRT glass into a material with very promising properties. A systematic study of the process parameters showed the possibility of modifying the properties of the porous material. The density can vary from 0.4 to 0.8 $g.cm^{-3}$ and the mechanical stress with a uniaxial compressive loading from 20 to 60 MPa. The materials have good insulation properties: thermal conductivity K inferior to 0.25 $Wm^{-1}K^{-1}$ and ε_r between 2.1 and 3.1 at 25°C. They are non-combustible (like the bulk glass), they resist corrosion in any environment and they present a low thermal expansion coefficient.
Keywords: cathode ray tube glass, recycling, cellular materials, microstructure, porosity, mechanical properties, electrical and thermal conductivity.

WIT Transactions on Ecology and the Environment, Vol 92, © 2006 WIT Press
www.witpress.com, ISSN 1743-3541 (on-line)
doi:10.2495/WM060311

1 Introduction

A cathode ray tube (CRT) typically represents 42% of the TV or monitor weight. CRTs may contain high levels of lead oxide and other undesirable metal oxides [1, 2]. At the end of their life time, when they are broken, CRTs can release lead into the environment thus making them a harmful material [3]. While CRT glass may be disposed of in hazardous waste landfills, recycling is the preferred management option for end-of-life CRTs [4, 5].

2 Composition of CRTs

A CRT is composed of two different types of glass. One, used for the cone section, is characterized by high levels of lead oxide and the other, used for the screen, is typically a lead-free glass that contains high levels of barium oxide [6]. Previous characterizations of waste CRT glasses have been reported. Chemical compositions and physical and chemical properties (the density, ρ, the thermal expansion coefficient, α and glass transition temperature, Tg) have been determined for each glass. There is a variation in the composition between glass made by different manufacturers, but, we showed surprisingly, that the values of parameters like ρ, α and Tg do not vary significantly for various glass samples, as shown in table 1. The results indicated that the recycling process can be carried out without taking the origins of each CRT into account [7, 8].

Table 1:   Properties of bulk glasses.

Characteristics	Color screen CRT glass	Color cone CRT glass	Black & white CRT glass
ρ [g.cm^{-3}]	2.80	3.0	2.70
$\alpha_{150-350°C}$ [10^{-6} K^{-1}]	10.5	10.5	10.5
Tg [°C^{-1}]	520	480	470

3 Experimental procedure

The CRT market overview summarizes several current recycling options as well as future market opportunities, including closed [9] and open-loop recycling [10].

   In our study, an open loop recycling process is chosen: recycling waste glass into an expanded glass, a useful product with excellent mechanical and insulation properties [11].

3.1 Processing

Various mixtures of glass powders, consisting of cone, screen or a 2:1 ratio of screen to cone glass by weight, were prepared with reducing agents (SiC or TiN). After the heat treatment of disc shapes obtained by the uniaxial dry pressing of these mixtures, expanded products were obtained in pebble form.

Samples prepared using different processing conditions, as shown in table 2, were characterized. The influence of the processing parameters on foam glass microstructure and consequently on its physical and chemical properties is studied.

Table 2:   Sample compositions and amount of reducing agent.

Sample	Glass	Reducing agent
S1	Cone	5 wt% SiC
S2	1/3 cone – 2/3 sreen	5 wt% SiC
T1	Cone	4 wt% TiN
T2	1/3 cone – 2/3 screen	4 wt% TiN

3.2 Results and discussion

The samples were characterized by measuring density with helium pycnometry. The microstructures were examined with scanning electron microscopy. Moreover, in order to valorize foamed glass technology, insulation and mechanical properties of our samples were determined by laser flash experiments, by impedance spectrometry and by compression testing.

3.2.1 Density-porosity

Helium pycnometry is a technique to measure the true density of solids. This technique is suitable for measuring the density of porous solids [12].

Since helium, which can enter even the smallest voids of pores, is used to measure the volume per unit weight, the final result gives information about the total porosity:

$$\% \ Porosity = \left(1 - \frac{bulk \ density}{powder \ density}\right) \times 100$$

Samples were characterized with helium pycnometry using a Micromeritics AccuPyc 1330. Open and closed porosities were obtained with mercury porosimetry using a Micromeritics Autopore II 9220 instrument.

Table 3:   Densities and porosities.

Sample	Composition	T (°C)	t (min)	Density (kg m^{-3})	Porosity (%)
SA	S1	750	120	460	84.5
SB	S2	850	60	375	79.5
TA	T1	750	120	378	86.1
TB	T2	850	60	499	78.0
TC	T2	750	120	878	67.9

Results presented in table 3 show that porosity depends on the reaction processes (temperature and time effects) that occur during expansion. Higher porosities are obtained with samples obtained from cone only. This phenomenon is due to the composition of CRT glasses: the lead oxide seems to further the expansion process. Reducing agents act preferentially on the lead oxide components of glasses and consequently a gaseous phase (CO_2 or N_2) is obtained yielding foam glasses [13]. Low density samples (ρ = 0.375 g cm^{-3}) can be prepared with controlled process parameters.

3.2.2 Microstructural characterization

Microstructure plays an important role on a wide variety of behavior (mechanical, thermal, electrical, etc.). Microstructure is largely developed during processing [14, 15].

Sample morphology was determined with a scanning electron microscope (SEM HITACHI S4500). The SEM microstrucure of samples containing 4 or 5 wt.% of pore-forming agents are shown in fig. 1 and fig. 2.

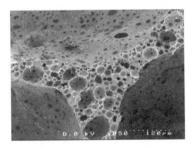

Figure 1:   Electron micrograph of a foam glass sample prepared with 4 wt.% TiN with a magnification of × 250.

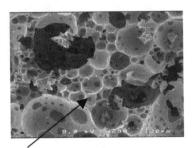

Connecting channels

Figure 2:   Electron micrograph of a foam glass sample prepared with 5 wt.% SiC with a magnification of × 250.

It was shown that large pores can be formed. A heterogeneity of the pore size distribution was observed particularly in the case where the reducing agent TiN is used (double distribution of heterogeneity). Generally, the emitted gas (CO_2 or N_2) created pores in expanded samples: an increase in the amount of reducing agent produced an increase in the diameters of the pores (up to 300 µm). These pores are interconnected in both samples: smaller dark zones on the micrographs were identified as channels connecting the cells [16]. In a previous paper [13], this observation was able to provide an explanation for the difference between the observed pore sizes and the pore size distribution obtained by mercury porosimetry.

3.2.3 Thermal conductivity

The laser method was used to determine the thermal conductivities of the samples at room temperature.

In the laser flash method, one surface (at x = 0) of a small disc shaped sample (10 mm diameter) of L = 2 mm thickness is irradiated by a laser pulse (0.5 ms) and resulting temperature rise at opposite surface (x = L) is used to calculate the thermal diffusivity α of the sample material. Thermal conductivity may be calculated from measurements of thermal diffusivity, specific heat and bulk density. This method is relatively fast and requires a small amount of material. The relationship between lambda (thermal conductivity) and alpha is given by:

$$\lambda = \alpha.C_p.d$$

where C_P is the specific heat measured using a differential scanning calorimeter (Netzsch DSC 200) and d the density.

The results for the thermal conductivity are shown in table 4 and the variation of the thermal conductivity with porosity is shown in fig. 3. The thermal conductivity increases with decreasing porosity. This linear decrease is very interesting; whatever mixtures of glass powders you take, a controlled porosity, i.e. a controlled processing, directly yields the thermal conductivity of the expanded materials. Samples with conductivity values lower than 0.25 $Wm^{-1}K^{-1}$ are classified as insulating materials. All our foam glasses can be considered as insulating materials.

Table 4:   Thermal conductivities obtained from the laser flash method.

Sample	C_P (J kg^{-1} K^{-1})	α (m^2 s^{-1})	K (W m^{-1} K^{-1})
SA	800	4.73 x 10^{-7}	0.10
SB	800	4.74 x 10^{-7}	0.14
TA	800	4.73 x 10^{-7}	0.08
TB	800	4.74 x 10^{-7}	0.19
TC	800	4.75 x 10^{-7}	0.24

3.2.4 Electrical properties

In order to fully characterize the electrical properties of a sample, the impedance response must be measured over a wide range of frequencies so that the entire distribution of relaxation times can be captured. Impedance properties were measured using a dielectric spectrometer (Novocontrol BDS 4000) in the frequency range from 0.01 Hz to 1 MHz at room temperature. The values of bulk resistance, and electrode dimensions, were used to calculate the dc conductivity, σ_{dc}, for all foam glasses. The permittivity of samples appears to tend towards a limiting value (from 2.1 to 3.1F m^{-1} with increasing thickness of material) determined by the porosity (air/vitreous matrix ratio) [17]. In the case of electrical properties, the pore diameter has a negligible effect on the permittivity of expanded samples.

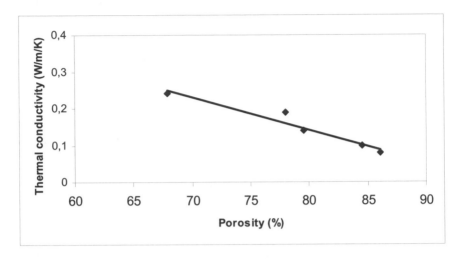

Figure 3:   Thermal conductivity as a function of porosity for various samples.

3.2.5 Compression testing

Square foam glass samples of 12.5mm length and 5mm x 5mm section area were subjected to uniaxial compressive loading. All porous specimens showed similar characteristic compressive behavior during load application [18].

Compression test data results are summarized in table 5. In order to show that there is a compromise between mechanical properties and the porosity level, samples with lower porosities were prepared (only from screen glasses for example). Thus, compressive strengths were plotted as a function of porosity (fig. 4). According to Weibull statistics [8], the maximum bending stress level is 60 MPa for samples with 40% porosity. Some foam glasses have microstructural heterogeneities, which lead to a reduction in their mechanical performance. The curve indicates that the ultimate strength of the material is a power law function of porosity.

Table 5:   Results of the failure stress obtained from the compression tests.

Sample	Porosity (%)	σ (Mpa)	K (Gpa)
SA	84.5 (1.7)	4	5.4 (0.1)
TA	86.1 (1.7)	4	0.4 (0.1)
TC	67.9 (1.4)	24	1.9 (0.2)
Screen + SiC : SC	3.7 (0.1)	267 (17)	5.4 (0.1)
Mixed + SiC : SD	46.5 (0.9)	60	4.4 (1.0)
Screen + TiN : TD	50.1 (1.0)	99 (11)	4.7 (0.8)

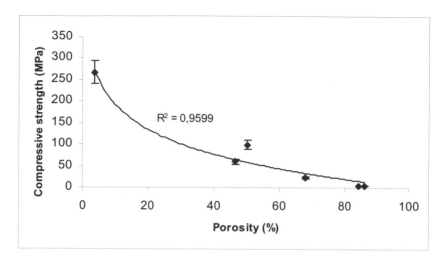

Figure 4:   Compressive strength as a function of porosity for various samples.

4   Conclusion

The objective of this work was to implement a way to recycle glasses issued from CRT glasses. We showed that the process used to obtain expanded glasses from this raw material source is controlled. We demonstrated that the process parameters (temperature, time, amount of reducing agent and its nature) could modify the properties of the porous material. These parameters modify the microstructure of the expanded product and consequently its physical properties like thermal, electrical and mechanical properties.

We must find a compromise between elaboration process and expected properties. In fact, we show that this material is totally in keeping with the general pattern of waste management. It can find applications in civil engineering and in insulation (sound proofing, thermal, electrical, etc.). The failure mechanisms of strength tested foams can also be studied with a view to manufacture of ultra-lightweight structures such as sandwich panels for example.

References

[1]   Palm, V., Environmental hazards connected to the compositions of cathode-ray tubes and cabinets, *In: Report Swedish Environmental Research Institute,* Stockholm, Sweden, 1995.

[2]   Andreola, F., Barbieri, L., Corradi, A., Lancellotti, I., Falcone, R., Hreglich, S., Glass-ceramics obtained by the recycling of end of life cathode ray tubes glass. *Waste Manage,* **25**, pp. 183-189, 2005.

[3]   Musson, E., Yong-Chul, J., Tonwsend, T.G., Il-Hyun, C., Characterization of lead leachability from cathode-ray-tubes using the toxicity

characteristic leaching procedure. *Environ. Sci. Technol.*, **34**, pp. 4376-4381, 2000.

[4] Smith, A.S., Recycled CRT panel glass as an energy reducing fluxing body additive in heavy clay construction products, *In: Creating Markets for Recycled Resources, Waste and Resource Action Programme (WRAP)*, Oxon, United Kingdom, 2004.

[5] Turmel, J.M., Rocherulle, J., Grange, P., Razafindrakoto, J., Verdier, P., Laurent, Y., European Patent 0871520, 1998.

[6] Méar, F., Yot, P., Cambon, M., Ribes, M., The characterization of waste cathode-ray tube glass, *Waste Management*, (in press), 2006.

[7] Méar, F., Yot, P., Cambon, M., Liautard, B., Cathode ray tube (CRT) glasses valorization. Part 1: characterization of the deposit, *Verre Review*, **9 (1)**, pp. 33-41, 2003.

[8] Méar, F., Yot, P., Cambon, M., Ribes, M., Properties and structural characterization of foam glass elaborated from cathode ray tube. *Advances in Applied Ceramics*, **104 (3)**, 2005.

[9] Piers, J., Peelen, J., US Patent 5725627, 1998.

[10] Garnier, C., Verdier, P., Razafindrakoto, J., Laurent, Y., WO Patent 9831639, 1998.

[11] Cambon, M., Yot, P., Méar, F., Liautard, B., Foam glass: a very promising way of valorization for cathode ray tubes. *Proc. of the 19th International Conference on Solid Waste Technology and Management*, Philadelphia PA, USA, pp. 1097-1106, 2004.

[12] Méar, F., Yot, P., Cambon, M., Caplain, R., Ribes, M., Characterisation of porous glasses prepared from Cathode Ray Tube (CRT), *Powder Technology*, **162,** pp. 59-63, 2006.

[13] Méar, F., Yot, P., Cambon, M., Ribes, M., The changes in lead silicate glasses induced by the addition of a reducing agent (TiN or SiC), *Journal of non-crystalline solids*, **351**, pp. 3314-3319, 2005.

[14] Cot, L., Ayral, A., Durand, J., Guizard, C., Hovnanian, N., Julbe, J., Larbot, A., Inorganic membranes and solid state sciences, Solid State Sciences, **2**, pp. 313-334, 2000.

[15] SanMarchi, C., Mortensen, A., Deformation of open-cell aluminium foam, Acta Materialia, **49,** pp. 3959-3969, 2001.

[16] Desforges, A., Arpontet, M., Deleuze, H., Mondain-Monval, O., Synthesis and functionalisation of polyHIPE beads, Reactive and Functional Polymers, **53 (1),** p. 241, 2004.

[17] Bansal, N.P., Zhu, D., Thermal conductivity of zirconia-alumina composites, Ceramics International, **31 (7),** pp. 911-916, 2005.

[18] Tasserie, M., Ph. D. thesis, Rennes University, France, 1991.

Experimental characterization of municipal solid waste bio-drying

E. C. Rada[1,2], M. Ragazzi[1], V. Panaitescu[2] & T. Apostol[2]
[1]Department of Civil and Environmental, Trento University, Italy
[2]Polytechnic University of Bucharest, Romania

Abstract

The bio-mechanical treatment of Municipal Solid Waste (MSW) has been adopted in Europe either as a pre-treatment before landfilling or as a pre-treatment before combustion. In this frame, the bio-drying process concerns the aerobic bioconversion applied mainly to MSW residual of selective collection. The aim of this process is the exploitation of the biochemical exothermic reactions for the evaporation of the highest amount of the humidity in the waste, with the lowest consumption of organic carbon. The obtained material can be easily refined to produce Refuse Derived Fuel. The present paper reports original assessments of process parameters characterizing the MSW bio-drying. In particular, outputs of a few pilot scale experimental runs have been elaborated in order to assess the following overall process parameters: m^3_{AIR} kg^{-1} of waste, m^3_{AIR} kg^{-1} of consumed volatile solids, m^3_{AIR} kg^{-1} of initial volatile solids, m^3_{AIR} kg^{-1} of organic fraction in the waste. Additionally, the assessed volatile solid dynamics during the bio-drying process are presented. These data are not generally available in the literature. Concerning the organic fraction contents in the waste suitable for bio-drying, usually its application is in the range of about 30-50%. The reasons are: a) lower values can give limited results in term of Lower Heating Value (LHV) increase of the bio-dried material; b) the application to waste with higher organic fraction content has the limitation of starting with very low LHV affecting the final characteristics of the bio-dried material. In this organic fraction range the air flow-rate is significant: it can vary between 6 and 10 m^3/kg_{MSW} that is similar to the off-gas generable from the incineration of the same waste.
Keywords: bio-drying, design, energy, MSW, pre-treatment.

WIT Transactions on Ecology and the Environment, Vol 92, © 2006 WIT Press
www.witpress.com, ISSN 1743-3541 (on-line)
doi:10.2495/WM060321

1 Introduction

The bio-mechanical treatment of Municipal Solid Waste (MSW) is an increasing option in Europe either as a pre-treatment before landfilling or as a pre-treatment before combustion. A process suitable for the last case is bio-drying. This process concerns the aerobic bioconversion applied to MSW residual of selective collection. Anyway it can be adopted also for treating MSW as is and contaminated organic fractions (under-sieve from mechanical selection, etc.). The aim of this process is to exploit the biochemical exothermic reactions for the evaporation of most of the initial humidity of the waste, with the lowest consumption of volatile solids. The obtained material can be easily converted in Refuse Derived Fuel (RDF), by a post-refinement with inert separation (European Commission [1]).

In November 2002 a study on the MSW bio-drying process began as a PhD activity in the Power Faculty of the Technical University of Bucharest (Romania). In September 2003 an international scientific collaboration between the Technical University of Bucharest and the University of Trento, Italy, was signed in order to go on with the development of the topic in the frame of a co-supervised doctorate (Rada [2]), as in Trento a bio-drying pilot plant was available at the Environmental and Civil Department of Trento University. This biological reactor was optimized during the PhD research. The reactor was used for several bio-drying runs from 2003 to 2005. The present paper reports original assessments of process parameters characterizing the bio-drying process. These data are not generally available in the literature as they are part of the know-how of the few companies proposing bio-drying (Ragazzi and Rada [3]).

2 Materials and methods

The biological reactor (Figure 1) used for the runs in the University of Trento, Environmental and Civil Department, is an adiabatic box of about 1 m³ with a leachate collection system.

Figure 1:   Bio-reactor used for the experimental runs.

The process air is filtered before entering into a blower. After a further filtration, the process air enters into an electro-valve installed to regulate the flow and, finally, in a flow-meter. After this path, the air is introduced in the biological reactor through a steel diffuser placed at the bottom. The air crosses upwards the waste from the lower part, activating the biological reactions and goes out of the biological reactor from the upper part, to be discharged into the atmosphere (in real scale, an air treatment line must be implemented to guarantee an acceptable environmental impact). When performing a run, the adopted biological reactor is placed on an electronic balance for monitoring the waste mass loss during the bio-drying process. For monitoring the temperature during the bio-drying process, it was decided to place a few temperature probes in different positions: one on the diffuser of the biological reactor (to measure the air temperature at the inlet), one on the piping of discharge (to measure the temperature of the process air at the outlet) and other probes on the vertical (to measure the average temperature of the waste). All these equipments are connected to a data acquisition system.

Starting from the waste characterization of each performed run, the overall composition necessary for the mass and energy balances was assessed. Basing on the performed experimental runs and using the parameters measured during those ones, a bio-drying model (Rada et al. [4]) was used. Thanks to energy and mass balances, it can describe the dynamics of the calorific value during the bio-drying process: both for the bio-dried material and for the RDF obtainable after a post-treatment (by separation of glass, metals and inert). By this approach, direct design and management parameters can be taken. The input data of the model are: the initial mass and the material and ultimate composition of waste sent to bio-drying, the amount of air and air temperature at the inlet and outlet of the biological reactor and the weight loss during the bio-drying process.

Several runs have been performed during the overall research in order to be sure about the replicability of the process (Rada et al. [5]). In the present paper four representative runs are discussed referring to different organic fraction concentrations. The outputs of the experimental runs have been elaborated in order to assess the following overall process parameters: m^3_{AIR} kg^{-1} of waste, m^3_{AIR} kg^{-1} of consumed volatile solids, m^3_{AIR} kg^{-1} of initial volatile solids, m^3_{AIR} kg^{-1} of organic fraction in the waste. The initial organic fraction concentrations were: 8%, 29%, 50% and 100%. In the first case, the MSW is typical of a region where a very high organic fraction selective collection is performed. The last case was developed in order to study in details the behavior of the organic fraction alone (with no bulky agent). The main management criteria were to keep the process air temperature below 60°C and to guarantee an adequate oxygen supply. No water addition was made.

3   Results and discussion

In Figure 2 the dynamics of the measured air flow for the four runs is presented. A typical lasting of the bio-drying process is 12-14 days (Rada et al. [4]), thus the comparison will be made choosing 312 hours as a representative retention time.

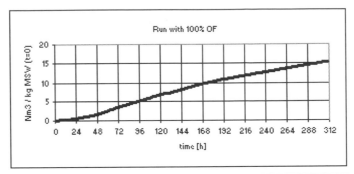

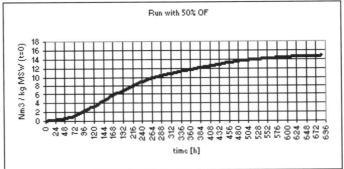

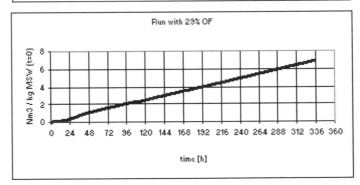

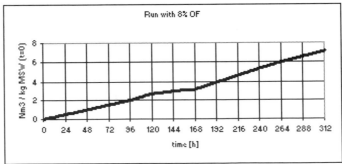

Figure 2:   Dynamics of cumulative air flow-rate during four runs.

As shown in Figure 2, the cumulative curves of air flow after 312 hours point out that the process is "air consuming": the values are similar to the one necessary for MSW incineration.

The main consequence of the process is an increase of the Lower Heating Value (LHV), as shown in Figure 3.

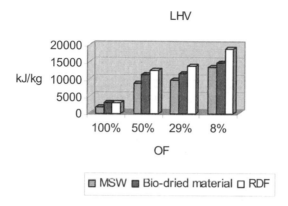

Figure 3:   LHV variations assessed from a bio-drying model (Rada [2]).

In the cases of 50% and 29% of organic fraction contents (OF), the LHV changed respectively from 8600 to 12500 kJ/kg and from 9600 to 13700 kJ/kg (there is no energy generation as the biodried mass is lower than the initial one). In order to have higher values an additional post-refinement is necessary (screening) that has the disadvantage of generating residues to be landfilled. On the contrary, the preliminary post-refinement generates only streams of materials to be recycled (glass, metals, inert). The case of 100% of organic fraction content is useful to demonstrate that if the target of a treatment is the production of a good RDF, it is important to apply the strategy only to waste having already a good energy content. The case of 8% of organic fraction content shows as a limited content of putrescible material cannot give significant results in term of LHV increase after bio-drying: only 9%. The post-refinement allows interesting results, but could be directly applied in this care, to the residual MSW avoiding the cost of bio-drying.

The exothermy of the process is a consequence of the volatile solids oxidation. The cited model (Rada *et al.* [4]) allows one to assess the dynamics of the volatile solid consumption as sum of C, H, O, N consumed, as reported in Figure 4. When bio-drying lasts longer than two weeks (see run with 50% org. fraction) the volatile solid consumption slows down. The reason is related to the decrease of the water content in the waste that causes a limiting effect in the process. It must be pointed out that, differently from composting and bio-stabilisation, bio-drying is performed without water addition in order to optimise the energy balance. The consumption of volatile solids in the case of 8% of organic fraction content is very low, but this depends on the unsuitability of bio-drying to the treated waste.

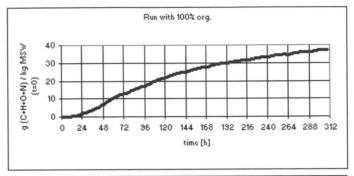

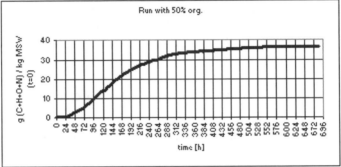

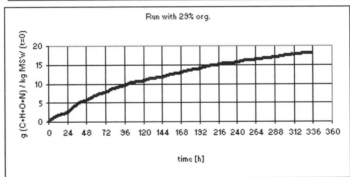

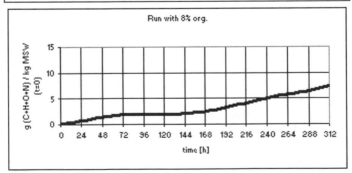

Figure 4:   Assessed volatile solids consumption dynamics.

In Tables 1 and 2, some parameters useful for a deeper understanding of bio-drying are reported. The very high values of the ratio $m^3_{AIR}/kg_{\Delta VS}$ demonstrate that the process is not oxygen limited. Indeed the management of the process is based on the regulation of air flow-rate for keeping the temperature lower than 60°C.

The ratio m^3_{AIR}/kg_{VSinit} allows one to make a few considerations:
a) the case of 100% OF shows values double than a typical composting process; the reason is related to the absence, in the studied case, of bulky agent usually added to guarantee an adequate porosity, the bulky agents contributes to the VS amount with slowly biodegradable materials;
b) the remaining cases show values lower than the one of a composting plant; this is a consequence of the lower amount of putrescible volatile solids available in case of MSW bio-drying (see the column $kg_{VSp\ init}/kg_{MSW}$ in Table 2).

The ratio $m^3_{AIR}/kg_{VSp\ init}$ gives an idea of the specific effects of the exothermy of the process: the cases 100% OF, 50% OF and 132% OF show values the same order of magnitude. On the contrary the case of 8% OF was characterised by the highest value. This does not depend on the heat generated from the biochemical oxidation, but to the need of increasing the air flow to see some effects of water removal (Rada [2]): the risk of bio-drying MSW with very low OF content is to operate the plant similarly to a thermal drying with a high air-flow rate (dewatering the waste by physical phenomena and not biological ones).

Table 1:   Parameters characterising bio-drying (part 1 of 2).

Run	OF (%)	m^3_{AIR}/kg_{MSW}	$kg_{\Delta VS}/kg_{MSW}$	$m^3_{AIR}/kg_{\Delta VS}$	kg_{VSinit}/kg_{MSW}
1	100%	15.4	0.037	412.7	0.168
2	50%	10.9	0.033	330.8	0.384
3	29%	6.4	0.018	361.7	0.452
4	8%	7.1	0.008	921.0	0.537

Table 2:   Parameters characterising bio-drying (part 2 of 2).

Run	OF (%)	m^3_{AIR}/kg_{OF}	kg_{VSinit}/kg_{MSW}	m^3_{AIR}/kg_{VSinit}	$kg_{VSp\ init}/kg_{MSW}$	$m^3_{AIR}/kg_{VSp\ init}$
1	100%	15.42	0.168	91.94	0.168	91.94
2	50%	21.80	0.384	28.42	0.084	129.98
3	29%	22.24	0.452	14.26	0.049	132.48
4	8%	88.87	0.537	13.24	0.013	541.45

4   Conclusions

The results of the international collaboration which supported the presented research have allowed one to have a deeper knowledge of the process both in a country where bio-drying is already performed in real scale (Italy) and in a country where the process was unknown (Romania). Indeed the results of the runs with 29% and 50% of OF can be representative of the behavior of an Italian and Romanian waste respectively. Presently bio-drying is recognized as a process available for waste management (European Commission [1]), but few free research has been developed on it. That makes it important the development of researches, like the one presented, able to generate design and management parameters.

An additional important aspect of the collaboration is the possibility of avoiding the monopolization of the technology in the country where bio-drying has to be introduced. That will allow one to keep low the cost of the process in a country at low income as Romania presently is.

References

[1] European Commission, Directorate General Environment. Refuse Derived Fuel, Current Practice and Perspectives (B4-3040/2000/306517/Mar/ E3) - Final Report, 2003.
[2] Rada E. C. Municipal Solid Waste bio-drying before energy generation, PhD Thesis, University of Trento and Politehnica University of Bucharest, 2005.
[3] Ragazzi M, Rada E. C. Dewatering and high temperature drying. Proceedings of the Workshop Research in the Waste Area, Towards the FP7, Brussels, 31st January 2006.
[4] Rada E. C., Taiss M., Ragazzi M., Panaitescu V., Apostol T. Un metodo sperimentale per il dimensionamento della bioessiccazione dei rifiuti urbani Journal RS - Rifiuti Solidi, year XIX, N.6, pp. 346-353, ISSN: 0394-5391, 2005.
[5] Rada E. C., Ragazzi M., Panaitescu V., Apostol T. An example of collaboration for a technology transfer: municipal solid waste bio-drying. Proceedings of the Tenth International Waste Management and Landfill Symposium, Sardinia 2005.

Composting of biodegradable municipal waste in Ireland

N. M. Power & J. D. Murphy
*Department of Civil, Structural and Environmental Engineering,
Cork Institute of Technology, Cork, Ireland*

Abstract

The implementation of the Landfill Directive (96/31/EC) will lead to a massive reduction in quantities of Biodegradable Municipal Waste (BMW), which may be landfilled. In 2016, an estimated 2.6 million tonnes of BMW will require diversion from landfill in Ireland. Recycling of dry paper, reuse of textiles and home composting will divert significant quantities of BMW. However, to divert the quantities required to comply with the Directive a significant proportion of the Organic Fraction of Municipal Solid Waste (OFMSW) will require treatment in centralised biological facilities.

A survey was undertaken of composting facilities in Ireland to ascertain present capacity, present technology employed and types of feedstock composted. A generic model (consisting of in-vessel composting followed by aerated static pile composting) was generated from technical, economic and environmental data obtained from the survey. The generic model is utilized to ascertain the effect of economies of scale, the effect of gate fees on potential profit per tonne feedstock and the greenhouse gas contribution of the composting process. The model was investigated for four scenarios ranging from small (11ktpa) to very large (220ktpa). The economic analysis indicated that a potential for profit of €52.5/t-€65.4/t was achievable without sale of compost. The direct greenhouse gas production equated to $566kgCO_2/t$. However considering the "do-nothing" scenario of landfill, $1,175kgCO_{2\ equiv}/t$ is avoided. Thus 1 tonne of BMW saves $609kgCO_{2\ equiv}$.

Keywords: composting, BMW, OFMSW, gate fee, greenhouse-gas analysis.

WIT Transactions on Ecology and the Environment, Vol 92, © 2006 WIT Press
www.witpress.com, ISSN 1743-3541 (on-line)
doi:10.2495/WM060331

1 Introduction

BMW is waste, which can undergo aerobic or anaerobic decomposition such as OFMSW, paper, card and textiles. However not all waste types are suitable for composting, textiles are slow to biodegrade and as such are difficult to compost. Paper is made up of 39% wet paper and 61% dry paper [1]. Dry paper is suitable for recycling, but wet paper is more suitable for composting.

The implementation of the Landfill Directive [2] will lead to a significant reduction in the quantities of BMW consigned to landfill. In 2003 Ireland produced 3 million tonnes of municipal solid waste (MSW), of which 2 million tonnes may be classed as BMW [3]. Between 2002 and 2003 household waste grew at a rate of 4.5%pa. If this growth rate continues to 2016, 3 million tonnes of biodegradable waste will be generated [4]. In 1995, the base year for the Landfill Directive [2], 1.3 million tpa of BMW were generated in Ireland. Therefore in 2016, only 450,000tpa of BMW may be consigned to landfill. Thus an estimated 2.6 million tonnes of BMW will require diversion from landfill.

Figure 1 shows the breakdown of BMW; approximately 220kg/pe of BMW may be collected for biological treatment.

2 Composting industry in Ireland

A survey of existing facilities was carried out [5]. In total 17 facilities were surveyed. The questionnaire was aimed at assessing the technical, economic and environmental parameters of composting.

It was found that the compost produced at the facilities is not sold, it is either given away for free or the operators of the facilities have a demand for the compost. Only two of the facilities surveyed located a market for the compost. Thus the gate fee for accepting waste is often the only source of revenue for many of the composting facilities. The fee charged for accepting waste ranged from €60-155/t.

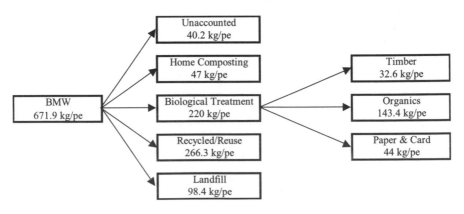

Figure 1:   Breakdown of biodegradable municipal waste.

WIT Transactions on Ecology and the Environment, Vol 92, © 2006 WIT Press
www.witpress.com, ISSN 1743-3541 (on-line)

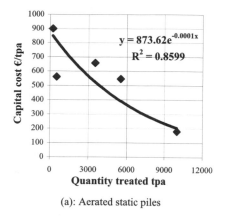

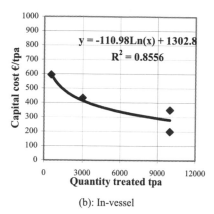

(a): Aerated static piles   (b): In-vessel

Figure 2:   Capital cost/tpa of (a) aerated static piles and (b) in-vessel composting.

The two most common methods of composting in Ireland were aerated static piles and in-vessel composting. Surprisingly the capital cost/tpa of in-vessel is cheaper than that of aerated static piles if less than 6ktpa is treated (figure 2).

Green waste does not require in-vessel composting. However waste that contains meat, such as OFMSW, must be composted in a closed reactor, which cannot be by-passed (in-vessel composting) in order to comply with the Animal By-Product Regulations (1774/2002) [6].

3   Scenarios investigated

The technology modelled employs two weeks retention time in in-vessel composting followed by eight weeks retention time in aerated static piles. Land prices, planning or waste permit costs are not considered. The scenarios to be modelled are:

1a.   Small region: 50,000pe, 11ktpa, Waterford City;
1b.   Medium region: 120,000pe, 26.4ktpa, Cork City;
1c.   Large region: 400,000pe, 88ktpa, Cork City and County;
1d.   Very large region: 1,000,000pe, 220ktpa, Dublin.

The scenarios are applicable to specific geographical locations in Ireland. Scenario 1d. is based on the population of greater Dublin. However it is assumed that the region will be served by two facilities treating 110ktpa.

4   Technical analysis

The scenarios investigated range greatly in size, thus the effects of economies of scale may be explored. Table 1 shows the technical analysis of the four scenarios under investigation. From the survey it was estimated that each tonne of biowaste treated via in-vessel composting followed by aerated static piles, required 35kW$_e$h and 2 litres of diesel. The analysis shows each tonne of BMW treated generates 488kg of compost.

Table 1:   Technical analysis of composting scenarios.

Technical parameters	Scenarios			
	1a	1b	1c	1d
Population equivalent	50,000	120,000	400,000	1,000,000
BMW collected kg/pe	220	220	220	220
BMW treated tpa	11,000	26,400	88,000	220,000
Dry solids content of BMW (47.9%) tpa	5,269	12,646	42,152	105,380
Volatile solids content (37.3% wet) tpa	4,103	9,847	32,824	82,060
Destrution of volatiles at facility (50%) tpa	2,052	4,924	16,412	41,030
Dry soids out tpa	3,218	7,722	25,740	64,350
Compost produced (60% dry solids) tpa	5,363	12,870	42,900	107,250
Electricity required (35kWeh/t)	385,000	924,000	3,080,000	7,700,000
Diesel required (2l/t)	22,000	52,800	176,000	440,000

5 Economic analysis

In figure 1b, eqn (1) is generated which allows calculation of the capital cost of an in-vessel composting facility with a capacity less than 10ktpa. The facilities that participated in the survey had a capacity of less than 10ktpa. This highlights the relative "youth" of the industry in Ireland. A number of proposals are mooted for larger developments. Detailed analysis of facilities and discussion with developers allowed generation of capital cost data for larger developments as outlined in eqn. (2).

$$y = -110.98\text{Ln}(x) + 1302.8 \quad (<10\text{ktpa}) \tag{1}$$
$$y = 5495.6x^{-0.3058} \quad\quad (>10\text{ktpa}) \tag{2}$$

where,

y is the capital cost of the composting facility in units €/tpa

x is the quantity of waste to be treated in units of tpa

The operating costs generated from the survey for in-vessel composting followed by aerated static piles ranged from €20-60/t, with a mean of €35/t. This excluded the cost of electricity and fuel. In table 2 the price of diesel is €1/l and electricity costs are based on a standing charge of €7.88/month, 48,000kW$_e$h at €0.1419/kW$_e$h and any remaining units at €0.1209/kW$_e$h. The gate fee for accepting biological waste ranged from €60-155/t. For this analysis it was estimated that a gate fee of €100/t was attainable. The effect of sale of compost on the economic viability of the project is investigated. Either no market is available for the compost produced or a sale price of €40/t for the compost produced is obtained (sale price of compost at one facility). From table 2 the potential profit ranges from €52.5/t - €65.4/t with a market for compost and €33/t - €45.9/t without a market for compost. The economies of scale are evident.

6   Sensitivity analysis

6.1   Critical variables

Table 2 assumed that a gate fee of €100/t was attainable and a market for compost of €40/t may be obtained. It also estimated operating costs of €35/t. A sensitivity analysis was carried out to determine the effects of these variables on the potential profit.

Table 2:   Economic analysis of composting scenarios.

Economic parameters	Scenarios			
	1a	1b	1c	1d
Liabilites				
Capital cost	3,520,000	6,468,000	14,960,000	35,200,000
Annual liabilities				
Cost of capital (r=5%, N=20 years)	282,454	519,009	1,200,429	2,824,539
Operating costs (€35/t)	385,000	924,000	3,080,000	7,700,000
Fuel (2l/t *€1/l)	22,000	52,800	176,000	440,000
Electricity	47,649	112,814	373,475	932,033
Total annual liabilities	737,103	1,608,623	4,829,904	11,896,572
Annual assest				
Gate fee (€100/t)	1,100,000	2,640,000	8,800,000	22,000,000
Sale of compost (€40/t)	214,520	514,800	1,716,000	4,290,000
Total annual assest	1,314,520	3,154,800	10,516,000	26,290,000
Potential profit				
With sale of compost (€)	577,417	1,546,177	5,686,096	14,393,428
(€/t)	52.5	58.6	64.6	65.4
Without sale of compost (€)	362,897	1,031,377	3,970,096	10,103,428
(€/t)	33.0	39.1	45.1	45.9

6.2   Sale price of compost

For the sensitivity analysis the price of compost ranged from €0/t (no market for heat) to €50/t. Due to the significant quantity of biodegradable waste that will require biological treatment (1 million tpa in 2016), a significant quantity of compost or similar material such as digestate will be produced. Thus it is unlikely that a market of €50/t will be attainable at all facilities. Figure 3 shows how the potential profit changes with a change in the compost sale price.

It was found that a market for compost significantly increases the potential for profit. In scenario 1a the potential profit rises from €33/t with no market for compost to €57.4/t if compost is sold for €50/t.

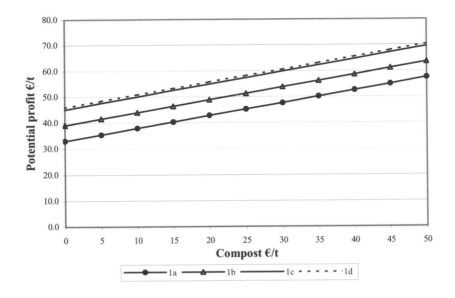

Figure 3:   The change in potential profit based on a change in compost sale price.

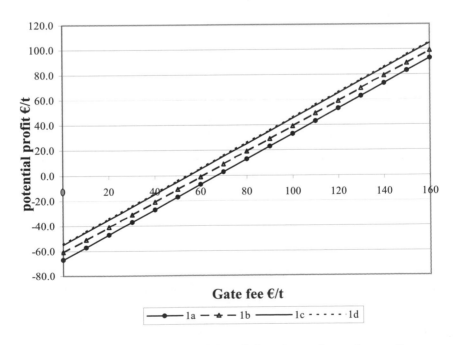

Figure 4:   Change in potential profit based on a change in gate fee.

6.1 Gate fee

The gate fee for accepting biodegradable waste in Ireland ranges from €60/t-€155/t; the analysis in table 2 is based on a gate fee of €100/t. Changes in gate fee result in changes in the potential profit as highlighted in figure 4. If no market for compost exists the gate fee required to break even would range from €54.1/t – €67/t.

6.2 Operating costs

The operating costs of in-vessel composting ranged from €20-60/t: the analysis in table 2 is based on an operating cost of €35/t. Figure 5 represents how the potential for profit decreases with increasing operating costs. The operating costs analysed ranged from €10-70/t. The corresponding potential profit ranged from €12.2/t-€72.2/t in scenario 1d with no market for compost.

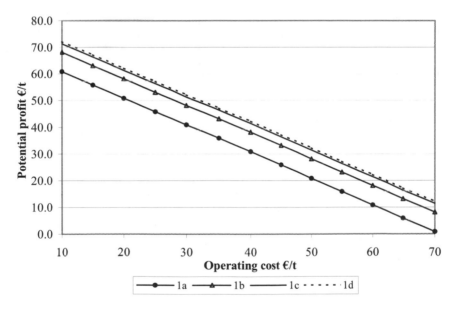

Figure 5:   Change in potential profit based on a change in operating cost.

7 Environmental analysis

It is estimated that on a dry basis BMW contains 47.7% carbon, 6.4% hydrogen, 38.7% oxygen, 2.1% nitrogen, 0.3% sulphur and 5.1% ash [adapted from 7]. Analysis of components of BMW that are modelled in the compost analysis leads to a moisture content of 52.1% [5]. Then the BMW may be described as $C_{19}H_{88.5}O_{40.5}N_{0.72}$. Complete composting of BMW is described by eqn. (3) [7].

$$C_aH_bO_cN_d + [(4a + b - 2c - 3d)/4]O_2 = aCO_2 + [(b-3d)/2]H_2O + dNH_3 \qquad (3)$$

From eqn (3) it may be calculated that 1 tonne of BMW may generate 836kgCO$_2$ if 100% destruction is experienced. However an OECD report on sewage sludge suggests that only 65% of volatiles are destroyed over time. If this figure were taken for BMW then 1t of BMW would emit 543kgCO$_2$. In Ireland electricity come from various source such as coal, natural gas, oil peat and renewable sources. However on average 651gCO$_2$/kW$_e$h is released. Work by Murphy et al. [9] estimates that diesel emits 2.69kgCO$_2$/l. The analysis of greenhouse house gas may be viewed in table 3.

Table 3:   Environmental analysis of composting scenarios.

Environmental analysis	Scenarios			
	1a	1b	1c	1d
Gross greenhouse gas production				
Degradation of BMW (543kg CO$_2$/t)	5,973	14,335	47,784	119,460
Electricity imported (651g CO$_2$ /kW$_e$) [8]	251	602	2,005	5,013
Diesel utilised (2.69kg CO$_2$/l) [9]	6	13	44	110
Total gross greenhouse gas (t CO$_2$ equivalent)	6,229	14,950	49,833	124,583
(kg CO$_2$ equivalent/ t BMW)	566	566	566	566
Net greenhous gas production				
Gross geenhouse gas	6,229	14,950	49,833	124,583
Landfill gas (1,175kg CO$_2$/t BMW)	12,925	31,020	103,400	258,500
Total net greenhouse gas (t CO$_2$ equivalent)	-6,696	-16,070	-53,567	-133,917
(kg CO$_2$ equivalent/ t BMW)	-609	-609	-609	-609

From table 3 it may be noted that every tonne of BMW treated via composting will release 566kgCO$_2$. However as this waste is composted it is diverted from landfill and landfill gas is not produced. From Box 1 it can be seen that 1m^3 of landfill gas emits on average 4.857kgCO$_2$ if the landfill gas is collected and combusted. If 65% of the volatiles in the BMW were destroyed in the landfill then 1t of BMW would produce 242m^3 of landfill or 1,175kgCO$_2$. The net greenhouse gas production indicates that every tonne of BMW treated via composting saves 609kgCO$_2$.

Box 1:   Greenhouse-gas emissions per m3 of landfill gas.

CH$_4$ + 2O$_2$ = CO$_2$ + 2H$_2$O	
1 mole CH$_4$	1 mole CO$_2$
1kg CH$_4$	2.75kg CO$_2$
Density of CH$_4$ = (16/22.412) = 0.714kg/m^3	
1m^3 CH$_4$	1.963kg CO$_2$

Landfill gas contains
55.5% CH$_4$ (60% is combusted) = 0.555*0.6*1.963kg CO$_2$ = 0.654kg CO$_2$
          (40% is emitted)   = 0.555*0.4*0.714kg/m^3*21(GWP) = 3.329kg CO$_2$
44.5% CO$_2$ (100% is emitted) = 0.445*(44/22.412)   = 0.874kg CO$_2$
1m^3 of landfill gas emits   = 4.857kg CO$_2$ equivalent

8 Conclusions

- Composting reduces the mass of BMW to 48.8% of the starting material and produces compost that may be used in agriculture in accordance with the Animal By-Products Regulation.

- The composting process has a net energy input of $35kW_eh/t$ treated and $2l/t$ of diesel.

- Economies of scale are important. The potential for profit varies from €33/t to €45.9/t for facilities treating 11ktpa to 220ktpa.

- A market for compost can aid the financial viability of a composting facility. If no market for compost is located the gate fee is the only revenue available to the facility. The potential profit varies from €33/t without sale of compost to €52.5/t with sale of compost for a plant treating 11ktpa.

- A sensitivity analysis shows that a change in the sale price of compost, gate fee attainable and operating cost has a significant effect on the potential profit of a composting facility.

- The greenhouse gas analysis shows that 1 tonne of BMW treated via composting produces $566kgCO_2$. However as this waste is no longer consigned to landfill $1,175kgCO_2$ equivalent is avoided. Thus 1 tonne of BMW saves $609kgCO_2$ equivalent.

Acknowledgements

Niamh Power is funded by the Irish Research Council for Science, Engineering and Technology under the Embark Initiative.

References

[1] Murphy J.D., McKeogh E., Technical, economic and environmental analysis of energy production from municipal solid waste, *Renewable Energy*, 29 pp. 1043-1057, 2004.

[2] Directive 1999/31/EC of the Council of the European Union of the 26th of April 1999, on the landfill of waste. Official Journal of the European Parliament L 182/01 June 1999.

[3] Environmental Protection Agency, *National Waste Database 2003 Interim Report*. P.O. Box 3000, Johnstown Castle Estate, Co. Wexford, Ireland, 2004.

[4] Central Statistics Office, *Population and Labour Force Projections 2006-2036* available at Central Statistics Office, Information Section, Skehard Road, Cork.

[5] Power N., *The potential for CH₄ enriched biogas as a transport fuel in Ireland*, Masters thesis, Cork Institute of Technology, unpublished, 2006.

[6]   Regulation no 1774/2002of the European Parliament and of the Council of 3 October 2002 on the laying down health rules concerning animal by-products not intended for human consumption. Official Journal of the European Parliament L273/1. November 2002.

[7]   Tchobanoglous G., Theisen H. & Vigil S., *Integrated Solid Waste Management*, International Edition, McGraw-Hill, 1993.

[8]   Howley M. & Ó Gallachóir B., *Energy in Ireland 1990-2003 trends issues and indicators*, Sustainable Energy Ireland, January 2005.

[9]   Murphy J.D., McKeogh E. & Kiely G., Technical, economic and environmental analysis of biogas utilisation, *Renewable Energy* 77 pp. 407-427, 2004.

Separation of divalent metal ions using *Pandanus Amaryllifolius Roxb* (Pandanus) leaves: desorption study

M. Z. Abdullah & K. P. Loo
Chemical Engineering Programme, Universiti Teknologi Petronas, Perak, Malaysia

Abstract

Desorption by dead biomass has been studied on *Pandanus Amaryllifolius Roxb* (Pandanus leaves) by conducting batch experiments. The recovery of heavy metals such as lead and copper ions from biomass was examined using a variety of desorbing chemicals. This study aims to discover the best chemical which is able to leach the metal effectively with highest desorbing capacity. The results showed that HCl at pH 2 and 3.0mM EDTA at pH 4.58 were effective in desorbing the copper and lead ions from the biomass. The recovery of copper is very feasible since over 90% of copper was removed from the biomass. The percentage of lead recovery is about 70%. In contrast, Na_2CO_3 and NaOH are not effective in desorbing both of the metals. The results indicated that low pH is preferable for desorbing the metal ions. The binding ability of HCl is explained using ion-exchanging principle. More concentrated protons are able to replace those ions thus regenerating the biomass. EDTA is functioning as polydentate ligands, which appear to grasp the metal between the six donor atoms. It was suggested that recovery of metal ions is mainly due to the strength of bonding between the fraction of functional group of biomass and metal ions. Recovery of the deposited metals can be accomplished because they can be released from the saturated biomass in a concentrated wash solution, which also regenerates the biomass for reuse. Desorbing chemicals such as HCl and EDTA have proved successful for desorbing the metal ions. Thus, biosorption of heavy metals by biomass will be emerged as one of the alternative technology in removing the heavy metals.

Keywords: Pandanus Amaryllifolius Roxb, copper, lead, biosorption, adsorption, desorption.

 WIT Transactions on Ecology and the Environment, Vol 92, © 2006 WIT Press
www.witpress.com, ISSN 1743-3541 (on-line)
doi:10.2495/WM060341

1 Introduction

The current technologies for heavy metals removal from industrial effluents appear to be inadequate and expensive. Thus, research on biomass including bacteria, plant, algae, fungi and yeast are found to be capable in adsorbing heavy metals. Biosorption mechanisms involved in the process might include ion exchange, complexation, chelation, adsorption and precipitation [1]. The active sites present on cell wall can be very different according to the nature of the biosorbent: carboxylic, phosphate, sulphate, amino, amide, and hydroxyl groups are most commonly found [2].

Interest in using plants for environmental remediation is increasing due to their nature ability in capturing the heavy metals. Moreover, plant species that have been found growing on heavy metal-contaminated soils have a tolerance to the toxic effect of heavy metals. It might due to the evolution of chemical functional groups that reduce the toxic effect of heavy metals.

According to Guangyu and Viraraghaven, both living and dead biomass possess biosorption capacity. The performance of living biomass in binding metal ions depends not only on the nutrient and environmental status, but also cell age. Living cells are subjected to toxic effect of certain level of heavy metals, thus resulting in cell death. Non-viable and dead biomass is preferred in removing metal ions as they can also be regenerated and reused easily [3].

Recovery of the deposited metals can be accomplished because they can be released from the saturated biomass in a concentrated wash solution, which also regenerates the biomass for reuse. Several desorbing chemical such as acid and chelating agent have been proved successfully in desorbing the metal ions [4]. Thus, biosorption of heavy metal by biomass has emerged as one of the alternative technology in removing the heavy metals.

Different affinities of metal ions for biosorbent result in certain degree of metal selectivity on the uptake. Likewise, selectivity may be achieved upon the elution-desorption operation, which may serve as another means of eventually separating metals from one another if possible.

Batch experimentation was used to determine the desorption capacity using 0.1M HCl on the metal bounded to African alfalfa shoot [5]. The saturated biomass was exposed to 2 ml of 0.1M HCl, equilibrated by rocking for 5 minutes, and then centrifuged. Supernatants were collected for analysis using Atomic Absorption Spectrometer (AAS).

From the pH profile studies of heavy metal binding by African alfalfa shoot; heavy metal uptake was low at low pHs. It is then suggested that the protons would displace the adsorbed heavy metal ions by lowering the pH. It was reported that lead and cadmium achieved the most percentage of recovery (more than 99%) followed by zinc. By using acid at this low strength, the biomaterial is not destroyed and it can be reused. Alfalfa biomass was also performed in immobilized column experiment. The results showed that almost 90% of lead and zinc was removed from the column. For cadmium, recovery was almost 70%. No chromium was recovered.

The batch and continuous experiment above has indicated that the metal ion with positive two charge have good recoveries. It might due to the metal ion being easier to be displaced by proton ion. The chromium, which has +3 charge, might have a stronger bond with the binding sites of the biomass and it may require a stronger acid concentration or other stripping agent to remove the metal ions.

Adsorption and desorption characteristic of biosorption process consists of biomass of marine algae *Sargassum baccularia*, cadmium ions and desorbing agents hydrochloric acid (HCl) and ethylenediaminetetraacetic (EDTA) were investigated using a batch reactor system [6]. It was found out that both desorbing agents are capable of stripping the cadmium. HCl at pH 2 could adsorb more than 80% of cadmium. Almost complete recovery of cadmium was achieved by using 3.2 mM EDTA solution. However, HCl was found to reduce cadmium uptake capacity by 56% while the reduction of EDTA was nearly 40% over the five adsorption-desorption cycles. The loss of biomass using HCl is 30% as the EDTA caused a biomass loss of 16%. The results showed that EDTA is a better desorbing agent than HCl.

A study on the packed bed sorption of copper using spent animal bones stated the use of various concentration of H_2SO_4 as desorbing chemical. A solution of 50 mM was found out to be suitable for this process and the efficiency of bones-packed column did not change significantly after four sorption and desorption cycles [7].

Studies were conducted to investigate the removal and recovery of copper (II) ions from aqueous solutions by bacteria *Micrococcus.sp*. Sodium polyphosphate (0.1M) and nitric acid (0.05M) were the most efficient desorption media, recovering more than 90% of initial copper adsorbed, while the distilled, deionized water control demonstrated no copper desorption [8].

Rcutilization of non-viable biomass of the cyanobacterium *Phormidium laminosm* for metal biosorption was studied [9]. They reported that the biosorbed heavy metal could be desorbed by washing the biomass with diluted acids but not with NaOH, NaCl, $CaCl_2$ or ultrapure water. Desorption of biosorbed metal with 0.1 M HCl allow the reuse of biosorbent for at least five biosorption and desorption cycles without decreasing its biosorption capability.

After the biosorption of toxic heavy metal from a large volume of low concentration aqueous waste, the heavy metal could be eluted from the saturated biomass by desorption or an elution process. Aldor et al investigated equilibrium cadmium desorption from protonated *Sargassum* by various elutants. It was established that that the mineral acids, particularly 0.1 M HCl or H_2SO_4 are efficient in metal elution. The biomass damage was limited during the acid wash.

This biosorption study will focus on desorption of divalent metal ions i.e. Pb^{2+} and Cu^{2+} from *Pandanus Amaryllifolius Roxb* (Pandanus) leaves using various desorbing chemicals. Recovery of the deposited metals is accomplished by using the suitable desorbing chemical for reuse of biomass. The possibility of regeneration of loaded biosorbent is crucial in keeping the process cost down and to opening the possibility of recovering the metal extracted from the liquid phase.

 WIT Transactions on Ecology and the Environment, Vol 92, © 2006 WIT Press
www.witpress.com, ISSN 1743-3541 (on-line)

2 Methodology

2.1 Biomass preparation

The biomass of *Pandanus Amaryllifoliusm Roxb* (pandanus leaves) was collected at Bota Kanan area in the state of Perak, Malaysia. The biomass was washed thoroughly with distilled water, dried in the oven at 60°C for 48 hours and ground using the laboratory grinder for 3 minutes until fine particles were achieved. It was then sieved into particle size ranges of 600 μm and 1.18mm which were used throughout the experiments.

2.2 Biosorption experiment

2.0 g biomass was added into 50 ml, 40-ppm metal ion solution, containing either Pb (II) or Cu (II) salt in different Erlenmeyer flasks. The solution was left to mix on a rotary shaker at 125 rpm for 24 hours to achieve equilibrium and then sieved to separate the filtrate and the biomass. The metal ion concentrations were analysed using the atomic absorption spectrophotometer (AAS) while the biomass were subjected for the desorption experiments.

2.3 Desorption experiment

The biomass that were previously separated were added into four different Erlenmeyer flasks containing 25.0 ml of various desorbing solutions i.e. 0.1 M HCl, 0.1 M NaOH, 0.1 M Na_2CO_3 and 3.0 mM EDTA, respectively. The mixture was left to mix on a rotary shaker at 125 rpm for 24 hours to achieve equilibrium. The mixtures were then filtered and the filtrates were analysed using the AAS.

3 Results and discussion

3.1 Biosorption experiment

The results from adsorption experiment showed that the adsorption of the copper and lead ions are feasible using pandanus leaves.

Figure 1 shows the heavy metal ion uptake by *Pandanus Amaryllifolius Roxb* while Figure 2 indicates the percentage removal of heavy metal ion by the biomass. The results denote that higher lead adsorption is corresponding to the stronger affinity of lead towards pandanus leaves in a metal ion solution which follow the trend of $Pb^{2+} > Cu^{2+}$.

The higher adsorption affinity of lead agrees with the selectivity of the biomass. Sadowski also observes this phenomenon in the uptake of lead and copper by microorganism *Nocardia Sp.* [10]. The author stated the adsorption of Pb^{2+} and Cu^{2+} are mainly due to its electrostatic attraction. The electronegativity of an atom in a molecule is related to its ionization energy and electron affinity. An atom, which has a very negative electrons affinity and high ionization energy will both attract electrons from other atoms and resists having its electron attracted away. Therefore, higher electronegativity of metal ion is preferred by biomass.

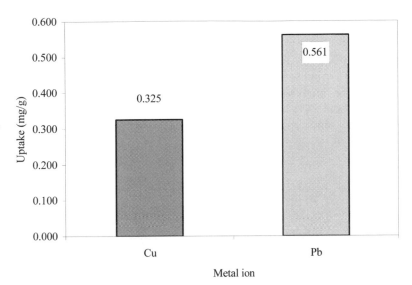

Figure 1:   Heavy metal ion uptake by *Pandanus amaryllifolius Roxb.*

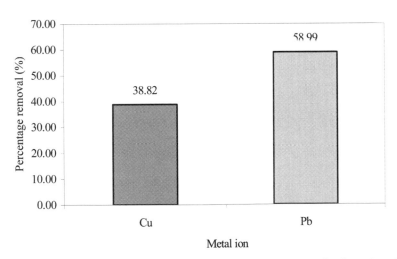

Figure 2:   Percentage of heavy metal ion removal by *Pandanus Amaryllifolius Roxb.*

By ionic radius comparison, lead ion, which has a smaller hydrated diameter possess the higher electronegativity. It has greater tendency to accept electron and share the electron with the weekly acidic cationic biomass, which is an electron donor.

3.2 Desorption experiment

Screening for the effective solution to regenerate saturated pandanus leaves is the main objective of the study. Figure 3 shows adsorbed metal ions could be removed by lowering the pH. HCl at pH 2 and 3.0mM EDTA at pH 4.58 were effective in adsorbing the copper ions from the biomass. It was found that recovery of copper ion is very feasible using HCl since complete removal is observed. EDTA could also desorbs more than 90% of copper ions initially loaded onto the biomass. The percentage of lead ion recovery using both chemicals is about 74% and 64% using HCl and EDTA respectively, which is expected lower compared to copper removal due to the greater bond experienced between the biomass and metal ion.

Table 1:   pH of desorbing solution.

Desorbing solution	pH
0.1 M HCl	2.00
3.0 mM EDTA	4.58
0.1 M NaOH	11.20
0.1 M Na$_2$CO$_3$	12.07

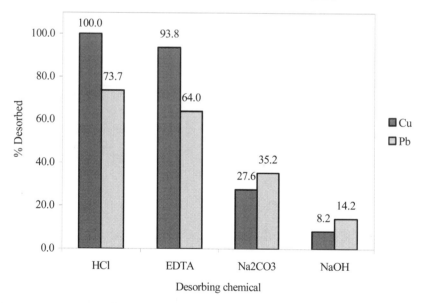

Figure 3:   Percentage metal ions being desorbed after contact with desorbing chemicals.

On the contrary, Na$_2$CO$_3$ and NaOH are not effective in desorbing copper and lead ions as indicated in Figure 3. The recovery of copper ion is as low as 8% and 27% using Na$_2$CO$_3$ and NaOH, respectively. The low percentage of lead ion recovery also showed that high pH of chemical is not suitable in desorbing the metal ions. Basic solution like NaOH and Na$_2$CO$_3$ is widely used to pre-treatment biomass to increase the adsorption capability, as they are able to remove surface impurities and expose the available binding sites. However, the pre-treatment using alkali resulted a great loss of biomass [3]. From the observation, the filtrate of copper and lead after desorption were green in colour. It was suggested biomass might be washed out during the desorption process.

The binding of metal ions is an ion exchange reaction where the metal ions are reversibly removed from solution and transfer to the biomass. Ion exchange reversible reaction is given as eqn (1).

$$(B-H) + Cu^{2+} \rightleftharpoons (Cu-B) + H^+ \tag{1}$$

where B is the functional group of the biomass.

For a weakly acidic cationic exchanger, the adsorption of divalent metals occurred according to eqn (2).

$$2RCOOH + Cu^{2+} \rightleftharpoons RCOO_2Cu + 2H^+ \tag{2}$$

where RCOOH is a resin.

The biomass can be considered as weakly acidic cationic exchanger. The resin is compared with the organic components of pandanus leaves which is consists of carbon and hydrogen compound. Studies found that organic constituents of 2-acetyl-1-pyrroline and ethyl formiate are found in Pandanus leaves [11].

Regeneration of biomass is typically conducted by adding a regenerating solution. Since the reaction is reversible, the more concentrated H$^+$ are able to replace those ions thus regenerating the biomass. The reversible reactions associated with ion exchanging can be reverted to left with consequent release of Cu^{2+} ions if strong acid is mixed with biomass.

HCl is among the strong acids that act as proton donor, which would displace the adsorbed metal ions. In an aqueous solution of HCl, this acid consists entirely of H$_3$O$^+$ and Cl$^-$ ions based on the dissociation in eqn (3).

$$HCl \text{ (aq)} + H_2O \text{ (l)} \rightarrow H_3O^+\text{(aq)} + Cl^- \text{ (aq)} \tag{3}$$

Therefore, 0.1M HCl, which is complete, ionized, was proven to be effective in desorbing chemical for copper and lead.

Other concentrated solutions other than acids can be used as regenerating solution. From Figure 3, EDTA is also capable in desorbing the metals. EDTA is a chelating agent that has two or more donor atoms that can simultaneously coordinate to a metal ion. EDTA is functioning as polydentate ligands, which appear to grasp the metal between the six donor atoms. The EDTA has two nitrogen atoms that have unshared pairs of electrons. These six donor atoms are located far apart that the ligand can wrap around a metal ion with the two nitrogen atoms simultaneously complexing to the metal in adjacent positions.

The [EDTA]$^{4-}$ ion contains one EDTA ligand in the octahedral coordination sphere of Cu^{2+}. Due to metal ions (particularly transition-metal ions) have empty valence orbitals, the ions act as Lewis acids (electron pair acceptors). The ligand, which has unshared pairs of electrons, is functioning as Lewis base (electron-pair donors). Thus, metal ion and ligand is bonded as the result of their sharing a pair of electrons that was initially on the ligand.

In general, chelating agents form more stable complexes than do monodentate ligands such as NH_3 and Cl^-. The formation for $[CuEDTA]^-$ can be represented in eqn (4).

$$[Cu(H_2O)_6]^{2+} \text{ (aq)} + [EDTA]^{4-} \text{ (aq)} \leftrightarrow [CuEDTA]^- \text{ (aq)} + 6H_2O(l) \qquad (4)$$

Although lead being adsorbed the most, the metal recovery using HCl and EDTA is lower than copper. This result showed that higher uptake of ions is not necessary followed by higher desorption. Biomass in general consists of a large number of functional groups or ligands that may interact with metal ions through variety of mechanism. Fraction of these ligands present on a particular biomass might have stronger bond with particular metal ion.

Hashim et al. stated the strength of bonding between metal ion and biomass can be determined using Langmuir model. The dissociation constant in the denominator of the model is able to indicate the strength of bonding [12]. In this study, lead ion proved to possess greater bond with the functional group of the biomass which is reflected in the higher amount of adsorption. Thus, desorbing chemical might experience intricacy to overcome the bonding and desorbs the metal ions from the biomass. However, lead ions demonstrate higher recovery in acidic medium, as HCl and EDTA are able to overcome the bonding, compared to in basic medium.

In ion exchange point of view, this phenomenon showed that for the same concentration of different ions, the relative preference of the biomass depends primarily on ionic size. Among ions having the same charge, the ions having the smallest hydrated diameter (largest unhydrated diameter tends to have smaller hydrated diameter) are preferred by the biomass. Typical order of ion exchange preference by the biomass is as follow:

$$Pb^{2+} > Hg2^+ > Ca^{2+} > Ni^{2+} > Cd^{2+} > Cu^{2+} > Zn2^+ > Fe^{2+} > Mg^{2+} > Mn^{2+}$$

4 Conclusion

The experimental results has shown that the recovery of heavy metal divalent ions is very feasible using HCl and EDTA as more than 90% of copper and 60% of lead was removed. This suggested that lower pH is favour in desorbing the metal ions. The protons in HCl would displace the adsorbed heavy metal by ion exchanging process. Besides, EDTA is functioning as chelating agent, which can simultaneously coordinate to a metal ion between the six donor atoms.

Biomass has shown a greater selectivity towards lead ion based on the result of adsorption. However, the uptake level of metal ion does not reflect the desorption ability. It is because lead has a lower recovery compared with copper in single solution. Lead might have stronger bond with the functional group of the biomass due to its ionic size and electro negativity.

Reference

[1]   Abdullah M.Z., Sites of Metal Deposition on the Biomass of *Schzophyllum commune*, *BEng Dissertation*, USM, Perak, Malaysia, 1999.
[2]   Esposito, A., Pagnanelli, F., Veglio, F., pH-Related Equilibria Models for Biosorpion in Single Metal System, *Chemical Engineering Science*, Vol 57, Elsevier Science Ltd , 307-313, 2001.
[3]   Guangyu, Y., and Viraraghavan, T., Effect of Pretreatment on the Biosorption of Heavy Metals on *Mucor rouxii*, *Water SA*, 26, 119-121, 2000.
[4]   Chu, K.H., Hashim, M.A., Phang, S.M., and Samuel, V.B., Biosorption of Cadnium By Algae Biomass: Adsorption and Desorption Characteristic, *Water Science and Technology*, 35:115-122, 1997.
[5]   Gardea-Torresdey, J.L., Gonzalez, J.H., Tiemann, K.J. and Rodriduez, O., Biosorption of Cadnium, Chronium, Lead, and Zinc By biomass of Medicago Sativa (Alfalfa), *Journal of Hazardous Materials*, 48: 181-190, 1996.
[6]   Chu, K.H., Hashim, M.A., Phang, S.M., and Samuel, V.B., Biosorption of Cadnium By Algae Biomass: Adsorption and Desorption Characteristic, *Water Science and Technology*, 35:115-122, 1997.
[7]   Al-Asheh S., Abdel Jabar N., and Banat F., Packed-Bed Sorption of Copper Using Animal Bones: Factorial Experimental Design, Desorption and Column Regeneration. *Journal of Advance in Environment Research*, 6, 3: 221-227, 1997.
[8]   Waihung, L., Mui-Fong, W. and Hong, C., 2002, <http://www.ct.ornl.gov/symposium/22nd/index_files/poster03.40.htm>
[9]   A. Blanco, B. Sanz, M. J. Llama and J. L. Serra, 2002, <http://bab.portlandpress.com/bab/027/bab0270167.htm>
[10]   Sadowski, Z., Effect of Biosorption of Pb(II), Cu(II), and Cd(II) on the Zeta Potential and Flocculation of *Nocardia Sp.*, *Mineral Engineering*, Vol 14, Elsevier Science Ltd, 547-552, 2001
[11]   Laksanalamai, V and Ilangantileke, I. Comparison of Aroma Compound (2-Acetyl-1-Pyrroline) in Leaves from Pandan (Pandanus amaryllifolius) and Thai Fragrant Rice (Khao Dawk Mali-105), Cereal Chem. 70:381-384, 2002.
[12]   Hashim, M.A., Tan, H.N., Chu, K.H., Immobilized Marine Algal Biomass for Multiple Cycles of Copper Adsorption and Desorption. *Separation and Purification Technology*, 19: 39-42, 2000.

Section 8
Methodologies and practice

Identifying LCA-elements in scrap tire recycling

A. Pehlken[1] & G. Roy[2]
[1]Institute and Chair of Processing and Recycling of Solid Waste Materials, RWTH Aachen University, Germany
[2]Materials Technology Laboratory, Natural Resources Canada, Canada

Abstract

Scrap tire recycling is presented in this paper with LCA-elements identified for further investigation. Life cycle thinking is applied to the life cycle of scrap tires because not much data is available for the recycling of scrap tires. The scrap tires discussed in this paper are provided as raw material input, and the output is shown as different products, such as retreaded tires, crumb rubber, artificial turf energy recovery and many others. The environmental impacts are focussed on the climate change, in particular on CO_2-emissions. The potential to reduce the emissions that support environment-friendly recycling techniques will also be presented. Thermodynamic fundamentals are introduced to set the path for the next phase of the study, where analytical and numerical models based on finite element method (FEM) computer calculations can be performed.
Keywords: LCA, Life cycle thinking, scrap tires, climate change, CO_2-emissions, environment, thermodynamics.

1   Introduction

Life cycle assessment (LCA) is a methodological framework for estimating and assessing the environmental impacts attributable to the life cycle of a product, such as climate change, eutrophication, acidification, depletion of resources, water use, and others. An LCA practitioner tabulates the emissions and the consumption of resources, as well as other environmental exchanges at a very relevant stage (phase) in a product's life cycle, from "cradle to grave", including raw material extractions, energy acquisition, material production, manufacturing, use, recycling and ultimate disposal. The processes within the life cycle and the associated material and energy flows as well as other exchanges are modelled to represent the product system and its total inputs and outputs from and to the natural environment, respectively. This results in a product system model and an inventory of environmental exchanges related to the functional unit [1].

An LCA can be applied to any kind of product and to any decision where the environmental impacts of the complete or part of the life cycle are of interest. In this paper the product scrap tire will be investigated and life cycle-elements will be identified.

2 Scrap tires

Tires that cannot be used for their intended purpose are considered as scrap tires. There are different processing routes for various kinds of tires depending on their size and composition.

The life span of a tire is about 80,000 km depending on the workload. There might be tires, which last for 100,000 km or just 40,000 km. The life span depends on the proper tire pressure as well as on the driving behaviour and the road conditions.

2.1 Composition

The composition can vary from manufacturer to manufacturer. Therefore, the exact composition of a bulk of scrap tire is not always known, and hence it represents an uncertain number in the process. Most countries estimate an average composition for their scrap tires.

Table 1 shows the average composition of a European car tire with carbon black filler and silica filler and an American car tire. As can be seen, the American car tire tends to have more natural rubber content and it weighs more than 2 kg in comparison with the European tire.

Tires intend to loose approximately 15% of their weight during their lifetime, which is mainly rubber loss. The average scrap tire in North America weighs around 9 kg and in Europe around 7.5 kg. Furthermore, only tires with carbon black filler and none with Silica filler are manufactured in North America. In Europe, both silica and carbon black fillers can be found in tires. The exact tire composition is not known because of company secrets. Therefore, the numbers represent an average of the tire composition.

2.2 Scrap tire processing per country

The various possibilities of tire disposition differ from country to country. Most countries are banning or supporting special technologies. In Europe, for example, it is no longer allowed to landfill unprocessed tires (landfill directive 199/31/EC) since the beginning of 2003. The USA and Canada tend to have different legislations in each state or province, respectively. The average dispositions of scrap tires in the USA and Canada and the EU with 15 member countries are listed in Table 2.

Table 1:   Car tire composition [2,3].

Raw Material	European Car Tire (carbon black)	European Car Tire (silica)	American Car Tire
	Amount in wt%	Amount in wt%	Amount in wt%
Synthetic Rubber	24.83	24.17	14.00
Natural Rubber	16.91	18.21	27.00
Carbon Black	26.91	19.00	28.00
Synthetic Silica	0.57	9.65	-
Sulphur	1.35	1.28	Balance
ZnO	1.55	1.58	
Aromatic Oils	7.81	6.12	
Stearic Acid	0.79	0.96	
Accelerators	0.88	1.01	
Antidegradants	1.51	1.47	
Recycled Rubber	0.41	0.50	
Coated Wires	11.70	11.40	15.00
Textile Fabric	4.70	4.70	5.00
Weight (kg)	8.62	8.80	11.00

Table 2:   Disposition of scrap tires in countries [4, 5, 6].

Disposition of scrap tires 2003/2004	USA	Canada	Europe (EU15)
	Amount in wt%	Amount in wt%	Amount in wt%
Tire Derived Fuel	44.7	20.0	30.0
Civil Engineering	19.4	13.0	incl. in Ground Rubber
Unknown	10.3	0.0	-
Ground Rubber / products	9.7	62.0	28.0
Landfill	9.3	0.0	18.0
Export	3.1	0.0	6.0
Punched / Stamped	2.0	incl. in Civil Engin.	0.0
Electric Arc Furnaces	0.2	0.0	0.0
Retreading	only truck tires	only truck tires	12.0
Misc./Agriculture	1.7	5.0	6.0

As can be seen in the table, the USA is processing nearly half of their scrap tires into Tire Derived Fuel, whereas Europe is retreading up to 15% of the passenger and truck tires. The number for Landfill in Europe is changing dramatically because of legislation that bans land filling of scrap tires from 2006.

Canada puts much effort into ground rubber and products, which represents 62% of the processed scrap tires.

3   LCA in Scrap tire processing

Scrap tire processing can be very complex or very simple, depending on the goal. Scrap tires can be retreaded and further used as a tire. But this is limited and not every tire can be retreaded. In order to meet safety requirements, only carefully inspected tire bodies are retreaded. Most scrap tires are used to make other products, or they are used as energy recovery as tire derived fuel (TDF). In the USA nearly the half of all scrap tires are used as TDF, mostly in cement kilns or power plants. Canada is supporting the manufacture of new products from (recycled) rubber obtained by the scrap tire processing. All different technologies implement various input and output. For each process all required energy and additional materials have to be identified.

The life cycle of a scrap tire can be seen in Figure 1.

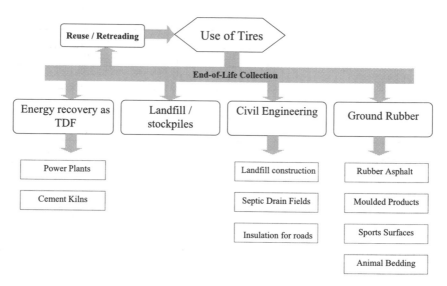

Figure 1:   Life cycle of a scrap tire.

The goal of the assessment can be seen, for example, as product improvement or establishment of the average environmental profile. The average scrap tire composition can be taken as a basis for the calculation or if the recycling plant knows exactly their scrap tire composition this basis is recommended for further investigation. The assessment describes how the environmental exchanges of the system can be expected to change as a result of action taken in the system.

An example of the system boundaries for scrap tire processing is shown in Figure 2. Only the parameters that influence the CO_2-Emissions are taken into account, and they have an impact on the climate change. The impact on eutrophication, acidification, depletion of resources, water use, etc. is not presented in detail in this figure. With any changes within the system boundaries, for example implementing a second or third sorting step, it may have an influence on the input (more energy required) and the output (more air emissions). The induced demand for one unit of product leads to the production and supply of one unit of products with associated emissions and resource consumptions [1].

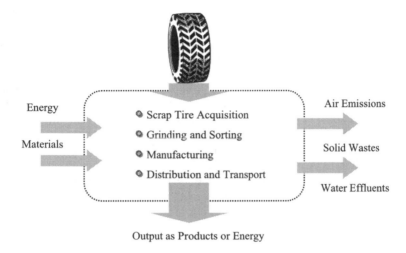

Figure 2:   System boundaries for scrap tire processing.

Very important in regarding the life cycle of a scrap tire is the issue that it (the scrap tire) cannot be separated from the life cycle of a new tire. If anything changes in the composition of a new tire, it will have an influence on the composition of the scrap tire a few years later as well. Therefore, the processors have to overview the tire market as well and make adjustments to their process as needed. The life cycle assessment can help to identify the needed corrections.

4   Life cycle thinking

Often a detailed LCA is not applicable. Because of the lack of data or time simplified LCAs and LCA thinking has to be employed to provide efficient and reliable decision support in a relatively brief period of time. Rebitzer et al. [1] outlines the different methods in detail. There are different strategies for the simplification of the LCA, depending on goal and scope of the study, the required level of detail, the acceptable level of uncertainty, and the available resources. Especially in small companies, like most tire recyclers are, an LCA gives both environmental and economic benefits. They cannot risk it to make an incorrect decision because they might risk the future of the whole company.

4.1 Identified LCA-elements

An LCA-element with major influence on the scrap tire processing is the scrap tire composition. The composition of the scrap tire is never known accurately, because of tire manufacturing companies' secrets and the various different tire models. The tire processor has to deal with the input, which is difficult to predict and can hardly be influenced by the processor. Therefore, an average basis has to be used for the assessment, and it remains uncertain. Table 1 has given the average composition of either European tires or American tires, which can represent the input.

Greenhouse gas (GHG) emissions are a big issue concerning climate change and must be recorded. Most GHG emissions are connected to the energy consumption or the replacement of other virgin materials. If carbon black (100% Carbon) can be recovered from scrap tires through pyrolysis, there is a huge benefit for the GHG emissions, because less virgin carbon black has to be produced. National or international regulations influence the scrap tire process as well because some recycling techniques might be banned or not.

Manufacturing of new products has been identified as another important element within LCA of s scrap tire. Some products require more crushing steps than others, which means that more energy is needed. More energy results in higher greenhouse gas emissions. On the other hand there might be a benefit if other natural resources are replaced (using recycled rubber instead of generating new synthetic rubber).

4.2 Greenhouse gas emissions

The energy input into the recycling plant may be of major interest, depending on how the energy is generated. A power plant based on coal emits far more greenhouse gases than a power plant based on water or wind. Table 3 shows greenhouse gas emission factors for Canadian electricity in 2003 for all Canadian Provinces and Territories.

As can be seen, there are huge differences in greenhouse gas emissions. The province, Quebec, Manitoba and British Columbia have mostly hydroelectric power plants and therefore the emission factor for greenhouse gases is very low.

For the manufacturing of products it makes a difference what kind of energy is used within the process. Therefore, the location of a plant might be of interest because it is related to greenhouse gas emissions. A recycler in the province Alberta has double greenhouse gas emissions than a recycler in New Brunswick, or approximately 125 times higher greenhouse gas emissions than in the province of Quebec. This does not apply to plants, which are using own generated energy, like a pyrolysis or microwave plant.

Another issue is transportation, because it emits CO_2 as well. For example for one litre of diesel approximately 2.7 kg of CO_2 are emitted [8]. This has to be taken into account if scrap tires are transported for long distances. If the energy needed for the transport exceeds the energy recoverable from the scrap tires, the following processing is no longer environmentally beneficial.

Furthermore the recycling technologies must not exceed the potential energy in a scrap tire. The Table 4 gives an overview of the needed energy with regards to (scrap) tires.

Table 3:   Power Plant Emissions in Canada [7].

Greenhouse Gas Emission Factors for Canadian Electricty in 2003 (Factors Expressed in grams per kW-hour of end use)	
Province	**Power Plant Emissions**
Newfoundland & Labrador	152
Prince Edward Island	488
Nova Scotia	642
New Brunswick	468
Quebec	8
Ontario	309
Manitoba	76
Saskatchewan	860
Alberta	1009
British Columbia	24
Yukon	38
North West Territories	399
Nunavut	269
CANADA in total	269

Table 4:   Energy and (scrap) tires [9].

	in kWh/kg
Energy needed to manufacture a tire	32.0
Energy needed to produce tire rubber compound	25.0
Thermal energy released when incinerating scrap tires	9.0
Energy consumed in the process of grinding scrap tires into crumb rubber (0.5 to 1.5 mm)	1.2

Tire manufacturing is an energy intensive process. Therefore, it might be worthwhile to recover the material form tires and take it back into the process. But there are quality issues to deal with, and manufacturing other rubber products from recycled rubber might be a solution as well. Compared to the energy needed to produce tire rubber compound, the energy consumed in the process of grinding scrap tires into crumb rubber is about 20 times less. The

GHG emissions are directly related to the energy consumed by the processing and vary from location to location due to the source of energy supply.

5 Thermodynamics of climate change

To be able to model changes in climate caused by generation and/or influx of CO_2, and to quantify the changes, a definition of a climatic system is introduced. It is understood that the amount of heat content in a certain volume of space, or climatic system, is the total energy content of the volume, known also as enthalpy H, and equal to U + PV, where U is the internal energy of the system, provided by its vibrating molecules, P is the pressure exerted on the system by its environment, and V is the volume of the system. It should be obvious from the definition of the enthalpy, that if any changes inside the system occur unopposed by the external pressure, then the change in enthalpy is equal to the heat transferred during the process. Therefore, when the mass of the volume at a constant pressure gets converted into atoms, the enthalpy will change, and the new state of energy is called enthalpy of atomization. In our case, which is combustion and generation of CO_2, the process is called enthalpy of combustion. The change in enthalpy, ΔH, will cause a change in another thermodynamic variable, entropy, or S. Any small amounts of change of S are denoted ΔS and they are equal to $\Delta Q/T$, where T is the temperature at which the change occurred. The change of entropy is a measure of disorder or randomness of an isolated system, which in our case means the volume of space occupied by the tires and the generated CO_2.

For any testing equipment to measure changes in CO_2, its sensitivity must be gauged against a known standard. For the current study, the standard is assumed to be a change in temperature, $\Delta T_{n \, x \, CO2}$, caused by an n-fold increase of the number of the molecule carbon dioxide, CO_2. To perform such measurements, the number n must be selected before the measurements, and an attempt to calibrate the equipment must be stated.

To predict a global climate change, one can use a heat equation,

$$c_P \, d/dt \, (\Delta T_{n \, x \, CO2}) = \Delta Q - \mu \, \Delta T_{n \, x \, CO2}$$

where c_P denotes the heat capacity of the system, tires and CO_2, under constant pressure, and μ is the thermodynamic parameter, related to the entropy, and expressed as the ratio

$$\Delta Q_{n \, x \, CO2} / \Delta T_{n \, x \, CO2}$$

The heat capacity, c_P, and $\Delta Q_{n \, x \, CO2}$ (to calculate the parameter μ), can be estimated from a variety of measurements on the earth, and typical values for a 2-fold increase of carbon dioxide, i.e. when n = 2, are 1.1 ± 0.5 GJ/mK , [10-11] (Giga Joule per meter Kelvin) and 3.7 W/m^2 (Watt per square m), respectively. The climate sensitivity $\Delta T_{n \, x \, CO2}$, can be determined from the above equations, if μ is well measured for tires.

6   Discussion

The paper shows that there are possibilities for greenhouse gas reduction within scrap tire processing. Life cycle thinking can help to assess the potential. Scrap tires can provide energy in other processes and replace natural resources. They have a higher energy content than coal and can therefore be successful in saving greenhouse gases. But higher potential can be achieved through material recovery and replacing virgin synthetic rubber in other manufacturing processes. Further work is needed to get more reliable data on this issue.

The simple heat equation can be used to predict the change of entropy changes, and thus the amount of randomness if certain scrap tires are used to generate energy, by calculating the change in heat content. The higher energy content in tires than in coal can be quantified through analytical and numerical models based on finite element method (FEM) computer calculations.

References

[1]   Rebitzer et al., Life cycle assessment, Part 1: Framework, goal and scope definition, inventory analysis, and applications, Environment International 30 (2004), pp701-720, 2004.

[2]   BLIC-The European Car Tyre Manufacturers. Life Cycle Assessment of an Average European Car Tyre. Third party report May 2001. www.blic.be.

[3]   RMA, www.rma.org.

[4]   RMA, www.rma.org.

[5]   Pehlken, A., Scrap Tire Recycling in Canada, Report for Natural Resources Canada, Canada, August 2005.

[6]   ETRA, www.etra.eu.com.

[7]   ICF Consulting, Determination of the Impact of Waste Management Activities on Greenhouse Gas Emissions: 2005 Update, Draft report, Environment Canada & Natural Resources Canada, Canada, 2005.

[8]   Atech Group, A National Approach to Waste Tyres, Commonwealth Department of Environment, ISBN 0642 54749 1, Australia, June 2001.

[9]   Reschner, K., Scrap tyre recycling – Market overview and outlook; Waste Management World; July-August 2003.

[10]   Global temperature change and its uncertainties since 1861, by Folland, C.K et al., Geophys Res. Lett, 28, pp 2621- 2624 (2001).

[11]   Warming of the world ocean, by Levitus, S. et al., Science, 287, pp.2225 - 2229, (2000).

Quantification of household waste diversion from landfill disposal by waste management practices

O. Mitaftsi & S. R. Smith

Centre for Environmental Control and Waste Management,
Civil and Environmental Engineering, Imperial College, London, UK

Abstract

The European Landfill Directive requires the phased reduction of biodegradable waste disposal to landfill. One option, with significant potential to divert biodegradable municipal waste from landfill, is to encourage householders to compost their organic waste at home and this approach is supported in principle by the UK Government. However, whilst the benefits of home composting are recognized, there is uncertainty about the actual quantitative effectiveness of this approach to biodegradable waste management and, currently, Local Authorities responsible for municipal waste disposal and recycling do not receive credit or recognition for promoting this activity in the UK. This project aims to quantify the amounts of waste diverted from landfill through home composting, and other household waste recycling methods, by measuring the effects of these practices on the quantity and composition of residual waste produced at the individual household level. An automatic weighing system was fitted to a refuse collection vehicle (RCV) to provide data on the waste arisings from 324 households in the urban Borough of Runnymede, West London. The households were allocated into four groups according to the waste management practices: home composting; home composting + kerbside recycling, kerbside recycling only and a control group who did not compost or participate in the kerbside recycling scheme. The RCV weight data were complemented by a detailed compositional analysis of residual waste collected from the households. This paper presents a summary and overview of the key results and outputs from the research programme.
Keywords: home composting, biodegradable waste, recycling, kerbside collection, landfill diversion, municipal solid waste.

WIT Transactions on Ecology and the Environment, Vol 92, © 2006 WIT Press
www.witpress.com, ISSN 1743-3541 (on-line)
doi:10.2495/WM060361

1 Introduction

One of the major issues associated with waste management in the UK is the reliance upon landfill for disposal of household waste. Landfill has become the dominant waste management route [1] because the geological and hydro-geological conditions have favoured the development of landfill sites at low cost. The continued predominance of landfill in current UK waste management practice reflects the fact that it remains the most adaptable and least expensive option in most geographical areas of the country [2].

However, landfill disposal of household waste is no longer regarded as an environmentally sustainable option [3] due to the environmental impacts arising from the high biodegradable organic content of domestic refuse. Household waste contains approximately 60% of biodegradable material [4], which generates landfill gas and leachate. Approximately 25.4 million tonnes of household waste were generated in 2003/04 in England, of which 4.5 million tonnes(18%) were collected for recycling through kerbside collection schemes or deposited at civic amenity and bring sites, and the residual 20.9 million tonnes(82%) were landfilled [5].

As with other Member States of the European Union, the UK is required to comply with the mandatory targets established by the Landfill Directive (1999/31/EC), to ultimately reduce biodegradable municipal waste (BMW) disposal in landfills to 35% of that produced in 1995 [6]. To assist in complying with these requirements, the Government and National Assembly in England and Wales published the 'Waste Strategy 2000' that established targets to recover at least 30% of household waste through recycling and composting by 2010, increasing to at least 33% by 2015 [7]. The Government's Strategy Unit report emphasized the importance of the role of home composting in diverting household waste from landfill disposal.

Composting is a natural aerobic biochemical process capable of converting biodegradable waste into a beneficial residue termed compost. Home composting is a simple, rewarding way to recycle garden and kitchen waste at home and creates a valuable soil amendment for gardens and lawns. It requires householders to separate and compost their own kitchen and garden waste in compost bins or traditional composting heaps. By composting at home, the cost and environmental risks of managing solid waste materials is reduced. Kitchen and garden waste, such as leaves, grass clippings, garden debris, and small branches, typically make up 27% of household waste [8]. When treated as waste, these materials increase the cost of collection and handling municipal solid waste (MSW). In landfill, they occupy valuable space and create potential pollutants.

Whilst the environmental and economic benefits of home composting are widely recognized, there is uncertainty and insufficient evidence regarding the actual contribution of small scale home composting systems at diverting the biodegradable fraction of household waste from landfill disposal. Under the current recycling credit scheme (RCS), waste collection authorities (WCAs) receive the saving in disposal costs by waste disposal authorities (WDAs) for diverted recyclable materials to motivate recycling of household waste [9].

However, due to the uncertainty about the value of home composting and the problems associated with measuring its positive influence on biodegradable waste reduction, the UK Government has not considered home composting as a practice to be rewarded in the current or revised RCS [10].

Jasim [11] recently monitored the amounts of biodegradable garden, food and paper waste deposited into home compost bins by a group of 64 homeowners for a period of two years in the suburban area of Runnymede Borough Council (RBC), West London. On average, approximately 400 kg of biodegradable waste were deposited annually per household into the compost bins and, if this amount equated directly to waste diversion from landfill disposal, it would correspond to 10% of the total quantity of waste arisings from door-to-door collection (this assumes a participation rate of 20% of households engage in home composting within the community).

This research project aims to further examine the impact of home composting activities on waste disposal. No direct measurements are available of the actual amounts of BMW diverted from landfill disposal by home composting. Therefore, this project adopted an innovative approach to measure waste diversion rates directly using a dynamic, automatic weighing system for individual refuse bins. The effects of kerbside recycling on residual waste disposal were also assessed.

2   Materials and methods

2.1 Quantitative waste diversion assessment

The impacts of home composting and kerbside recycling on landfill disposal of household waste were quantified by a novel research approach using an advanced weighing technology (SULO MGB Ltd, High Wycombe, UK) to directly measure residual waste arisings from individual households in the Borough of Runnymede, West London. The kerbside recycling scheme included the weekly collection of paper, cardboard, glass and steel and aluminium cans in the Borough. The wheeled bins of households participating in the monitoring programme were fitted with a passive read/write microchip to uniquely recognize the characteristics of each property. The weighing system is fitted to the mechanical lifter of the refuse collection vehicle (RCV) and weighs the containers before and after emptying providing an accurate measure of the net weight of the waste. An antenna on the RCV identifies the microchip and transmits the household identification information to the vehicle's on board computer. The weight data are stored on a RAM card, which is transferred to an office computer for data manipulation.

Information regarding the waste measurement practices of individual households in the Study Area on an established waste collection round, that was the focus of the previous research in the Borough by Jasim [11], was obtained from databases of properties that either: (a) had received at least one compost bin distributed by the RBC home composting scheme, and/or (b) were involved in the Council's kerbside recycling scheme. This initial screening identified approximately 450 potential properties in the Study Area for possible inclusion in the residual waste monitoring programme. These properties were further examined to confirm their involvement in home composting and/or kerbside collection or that they did not take part in either activity. Following this process, households that agreed to participate were selected and allocated into four treatment groups as follows:

1. – Recycling bin, – Compost bin ('Control' treatment group): 47 households;
2. + Recycling bin, – Compost bin ('Recycling only' treatment group): 92 households;
3. + Recycling bin, + Compost bin ('Recycling and composting' treatment group): 166 households;
4. – Recycling bin, + Compost bin (Composting only treatment group): 19 households.

Total number of households = 324.

2.2 Waste compositional analysis

Compositional analysis of the residual waste from selected households in the treatment groups was undertaken with assistance of a specialist contractor (Waste Research Ltd, Sheffield, UK), following a standard waste categorization procedure [12]. This involved collecting the entire contents of the wheeled bin from each household and dividing the waste into 13 primary categories and 49 subcategories. Waste samples were collected on two occasions, during the summer (30 June 2004) and autumn (10 – 11 November 2004) periods, to assess seasonal trends in household waste disposal. The compositional analysis determined the total quantity of waste generated weekly by households in each treatment group, the range of materials present in the waste, the amount and relative proportions of these materials, and the quantity of waste that was potentially recyclable or compostable. Table 1 summarises the number of households that were sampled in the waste analysis programme. Households from all treatment groups participated in the summer waste analysis, whereas the 'Composting only' group was excluded in the autumn due to the small number of households in the Study Area that only composted their waste, and because the size of the 'Control' group was increased in the second phase of sampling.

Table 1: Number of households sampled in each treatment group for waste compositional analysis.

Treatment Group	No. of households	
	Summer	Autumn
Control	17	44
Recycling only	50	50
Recycling and composting	37	48
Composting only	12	Not sampled
Total	**116**	**142**

3 Results and discussion

3.1 Quantitative waste diversion assessment

The general trend in waste generation associated with each of the treatment groups was illustrated by weight data collected by the RCV weighing system in two consecutive weeks in October 2005. The average amount of household waste produced by the treatment groups in the Study Area remained below the national average of 23.1 kg/property/wk [5]. Table 2 presents the total number of properties which were serviced on both waste collection days per treatment group, the total number of persons living in the properties and the weight of residual waste deposited in the wheeled bins. The weekly average amounts of waste produced per property and per person are also given. The average occupancy was 3 persons/household. The effect of household occupancy and other socio-economic and demographic factors on waste production will be part of a more detailed statistical examination of the data, but are not considered further in this paper.

Households in the 'Control' group, that do not compost their waste or participate in kerbside collections, produced approximately 18 kg/wk of residual waste. However, properties in the 'Recycling only' group also produced similar or slightly larger amounts of residual waste compared to the 'Control'. This could be explained if, for example, 'Recycling only' households did not actually participate in the kerbside recycling scheme in practice. Alternatively, recyclable materials may be separated by these households, but the spare capacity created by recovering the recyclables is filled with other waste materials, such as surplus bulky garden waste. By comparison, 'Composting only' households produced approximately 1-2 kg (6-9%) less waste than the 'Control' group. Homeowners would presumably dispose of all of their dry (i.e. non-biodegradable) waste first, therefore the observed reduction in the mass of collected waste for this group may reflect a direct consequence of home composting on biodegradable waste diversion. Alternatively, if homeowners in the 'Composting only' group behave in a similar way to the 'Recycling only' group and use the spare capacity in the bin to dispose of bulky garden waste (e.g. for material that may be unsuitable for home composting, for instance), the apparent reduction in residual waste may be explained because denser food waste removed from the residual waste by home composting may be replaced with surplus bulky waste of lower density. These results emphasize the importance of waste compositional analysis data to interpret the effects of recycled and home composting practices on residual waste collection. In contrast to either recycling or composting separately, households that both recycled and composted their waste had a much greater influence on landfill diversion by reducing the average amount of residual waste collected by approximately 3-5 kg (17-25%) compared to the 'Control' group. Monitoring of residual waste collections from the households in the different treatment groups by the automatic RCV system is ongoing and will continue during the winter period 2005/06, and spring and summer periods 2006.

Table 2:   RCV residual weight data collected in two consecutive weeks.

Collection period	Treatment group	Total no. of properties	Total no. of persons	Total waste (kg)	Average waste (kg/property)	Average waste (kg/person)
1st week 11-13 Oct 2005	Control	31	93	580	18.71	6.24
	Recycling only	81	207	1550	19.14	7.38
	Composting only	15	45	255	17.00	5.80
	Recycling&composting	148	400	2069	13.98	5.24
2nd week 18-20 Oct 2005	Control	30	80	525	17.50	6.25
	Recycling only	81	210	1484	18.32	6.72
	Composting only	16	44	263	16.44	6.20
	Recycling&composting	148	399	2141	14.47	5.43
Mean of 1st&2nd weeks	Control	61	173	1105	18.11	6.39
	Recycling only	162	417	3034	18.73	7.28
	Composting only	31	89	518	16.71	5.82
	Recycling&composting	296	799	4210	14.22	5.27

3.2 Waste compositional analysis

Table 3 shows the average amounts of total residual waste, and of ten major waste categories, deposited in the wheeled bins for the sampling periods in June and November 2004. The relatively low total value recorded for the 'Control' in the summer period, equivalent to 12.43 kg/property/wk, may reflect the small number of properties included in the treatment group on this sampling occasion. During the summer, 'Recycling only' households disposed of the greatest quantity of residual waste overall, equivalent to 16.72 kg/property/wk, followed by the 'Composting only' group, which deposited 14.68 kg/property/wk in the wheeled bin for collection. However, in the autumn, the largest amount of residual waste was recorded for the 'Control' group, equivalent to 15.37 kg/property/wk and also a similar amount of residual waste, 15.10 kg/property/wk, was produced by the 'Recycling only' group at that time. At both sampling times, households involved in recycling and composting together disposed of much smaller amounts of waste overall, equivalent to 14.02 kg/property/wk in the summer and 10.78 kg/property/wk in the autumn, compared to the other treatment groups, consistent with the RCV data. Table 4 shows that the patterns in residual waste measured during autumn by the RCV system and from the waste compositional analysis were broadly similar except the amounts had apparently increased by 3-4 kg in October 2005 (RCV) compared to November 2004 (compositional data). This could reflect a general underlying rise in the total amount of residual waste produced between 2004 and 2005 and also differences in the actual properties sampled on the two occasions.

Putrescibles were the predominant waste type in all treatment groups in both the summer and autumn phases of the waste analysis. However, as may be expected, more putrescible waste was collected during the summer compared to the autumn period, reflecting the greater production of garden waste during the active growing season. The results showed that putrescible waste governed the overall waste arisings and, therefore, properties disposing of large quantities of biodegradable waste generated the largest amounts of residual waste. Putrescible waste was sorted into five subcategories and the amounts of these constituents in the residual waste are shown in Table 5.

Table 3:   Average waste arisings (kg/property/wk) for the household treatment groups.

Waste fraction	Control		Recycling only		Composting only	Recycling & composting	
	Summer	Autumn	Summer	Autumn	Summer	Summer	Autumn
Recyclable materials	3.62	5.32	4.20	3.94	3.99	2.87	2.40
Putrescibles	6.31	5.81	8.83	7.27	7.26	6.70	5.01
Paper&card (non recyclable)	0.40	0.65	0.36	0.55	0.44	0.54	0.49
Plastic Film	0.49	0.50	0.59	0.55	0.66	0.55	0.43
Dense Plastic	0.84	0.96	1.14	0.91	0.81	0.94	0.75
Miscellaneous Combustible	0.60	1.40	1.14	1.30	1.15	0.75	1.06
Non-Combustible	0.01	0.08	0.19	0.38	0.18	1.27	0.24
HHW*	0.05	0.24	0.02	0.05	0.01	0.09	0.06
WEEE**	0.02	0.24	0.04	0.01	0.02	0.15	0.17
Fines	0.09	0.17	0.21	0.14	0.16	0.16	0.17
Total kg/property/wk	12.43	15.37	16.72	15.10	14.68	14.02	10.78

Note: Recyclable materials refer to the materials that are currently collected in RBC's kerbside collections. These are recyclable paper and card, glass, ferrous and non-ferrous metals and textiles. *Household Hazardous Waste ** Waste Electrical and Electronic Equipment.

Table 4:   Comparison between total residual waste data measured by the RCV system and from the compositional analysis results.

Treatment group	Total waste generation (kg/property/wk) (no. of properties indicated in brackets)	
	RCV weight data (October 2005)	Waste compositional analysis (November 2004)
Control	18.11 (31)	15.38 (44)
Recycling only	18.73 (81)	15.09 (50)
Recycling and composting	14.22 (148)	10.78 (48)

Garden waste constituted the majority of the putrescible waste in all waste samples. It was also the component which varied to the greatest extent between the summer and autumn sampling periods compared to other types of putrescible waste. Contrary to what might be anticipated, the amount of garden waste collected for disposal was increased by home composting compared to the 'Control' and 'Recycling only' groups. In the autumn, for example, the amount of garden waste disposed by the 'Recycling and composting' group increased by 44% compared to the 'Control'. Nevertheless, home composting reduced the total amount of putrescible waste overall because less kitchen waste was deposited in the wheeled bin for collection. The RCV data indicated that waste substitution could explain the similar overall waste arisings obtained for the 'Control' and 'Recycling only' groups (Table 2). This was confirmed by the compositional analysis, which showed that fewer recyclables were deposited in the residual waste by the 'Recycling only' group, but this was substituted by the increased disposal of garden and other putrescible waste, compared to the 'Control'.

Table 5:   Putrescible waste arisings (kg/property/wk) per treatment group.

Putrescible waste	Control		Recycling only		Composting only	Recycling & composting	
	Summer	Autumn	Summer	Autumn	Summer	Summer	Autumn
Kitchen compostable	1.59	1.77	2.20	2.03	1.28	1.07	1.34
Kitchen non compostable	1.56	2.17	1.99	2.22	1.33	1.10	1.21
Liquids	0.21	0.15	0.17	0.18	0.01	0.46	0.01
Garden waste	2.38	1.60	3.66	2.05	4.59	3.72	2.31
Other putrescibles	0.57	0.12	0.81	0.79	0.05	0.35	0.14
kg/property/wk	6.31	5.81	8.83	7.27	7.26	6.70	5.01

The second dominant type of residual waste was the recyclable material not recovered by recycling. Recyclable materials were reduced in the residual waste from households participating in the kerbside recycling scheme compared to the 'Control' group (Table 6). The 'Recycling and composting' group performed better than the 'Recycling only' group regarding the amount of recyclable materials that were removed from the residual waste and therefore appeared to be the most highly motivated and conscientious recyclers of all the household groups examined. Paper and card is generally the most captured material in kerbside collections at the national level [5]. Nevertheless, recyclable paper and card represented the largest fraction of recyclables in the residual waste for all the household groups. Residual waste from 'Control' households in autumn contained the highest quantity of paper and card, equivalent to 3.52 kg/property/wk, whereas the 'Recycling and composting' households produced the smallest amount, equivalent to 1.64 kg/property/wk. Glass was the next dominant recyclable material and the amount of residual glass disposed of in the wheeled bins was in the range 0.5 to 1 kg/property/wk. The 'Recycling and composting' group were also the most effective recyclers of glass removing 78% and 60% of the glass disposed by the 'Control' and 'Recycling only' groups in the autumn, respectively. Ferrous and non-ferrous metals represented the smallest mass compared to other types of recyclable material and, in all cases, did not exceed 0.3 kg/property/wk in the residual waste. The 'Recycling and composting' properties recycled most of their metal waste and, in autumn 81% and 70% of metal waste was recovered for recycling by this group compared to the amounts disposed by the 'Control' and 'Recycling only' properties, respectively.

According to Parfitt [8] households with large capacity wheeled bins generally take less material to civic amenity (CA) sites for disposal. These results go further and also demonstrate that material recovered from the residual waste collection by home composting and kerbside recycling may be substituted by other surplus garden waste; this type of material is likely to include low density woody clippings and prunings that are unsuitable for home composting.

Table 6:   Recyclable material arisings (kg/property/wk) in residual waste from the treatment groups.

Recyclable material	Control		Recycling only		Composting only	Recycling & composting	
	Summer	Autumn	Summer	Autumn	Summer	Summer	Autumn
Paper & card	2.37	3.52	2.57	2.56	2.61	2.29	1.64
Glass	0.58	0.90	0.70	0.50	0.98	0.16	0.20
Fe metals	0.22	0.27	0.16	0.21	0.17	0.11	0.06
Non-Fe metals	0.08	0.10	0.06	0.02	0.06	0.01	0.01
kg/property/wk	3.25	4.79	3.49	3.29	3.82	2.57	1.91

4 Conclusions

The principal conclusions of this paper are:

- Households that practice both home composting and recycling may reduce the amount of residual waste collected for landfill disposal by approximately 20% compared to households which do neither activity.
- Putrescible matter was the predominant waste type in the residual waste and had an important influence on overall waste arisings. Waste production was greater in the summer compared to the autumn period due to larger quantities of biodegradable garden waste disposed in the summer season.
- Households that composted putrescible waste disposed of more garden waste in the wheeled bin than either 'Recycling only' or 'Control' households. This may be attributed to substitution with non-compostable, bulky garden waste which would otherwise be transferred to a CA site. However, overall putrescible waste arisings were smaller for the 'Recycling and composting' group because of reduced kitchen waste disposal.
- Waste from households involved in kerbside collection contained less recyclable material than households in the 'Control' group, but the total amount of residual waste was similar to the 'Control'. This was because recyclable materials were substituted by surplus putrescible waste in the wheeled bin.
- Households engaged in recycling and composting were the most effective and conscientious recyclers overall compared to the other household groups.

- This research has demonstrated that home composting combined with kerbside collection of recyclable materials is effective in reducing the amount of residual biodegradable waste and the total amount of household waste collected for landfill disposal. Kerbside recycling must be coupled to other initiatives to reduce the amount of putrescible waste deposited in the residual waste bin, otherwise it may have little or no effect on the total quantity of household waste and can increase the amount of biodegradable waste collected for disposal.

References

[1]   Department of Environment, Food and Rural Affairs (DEFRA), *Municipal Waste Management Survey 2002/03*, London, 2004.

[2]   Neil P.A., Waste management–The essential service. *Proc. of 'Options for Urban Waste Management in the 21st century'*. Imperial College London: London, 1997.

[3]   Royal Commission of Environmental Pollution (RCEP), *17th Report, Incineration of Waste*, HMSO, London, 1993.

[4]   Parfitt J. & Flowerdew R., Methodological problems in the generation of household waste statistics: An analysis for the United Kingdom's National Household Waste Analysis Programme. *Applied Geography*, **17(3)**, pp. 231-244, 1997.

[5]   Department of Environment, Food and Rural Affairs (DEFRA), *Municipal Waste Management Statistics 2003/04*, London, 2005.

[6]   European Council (EC), *EU Directives 99/31/EC on the landfill of waste*, Off. J. Eur. Communities L182:1-19, 1999.

[7]   Strategy Unit, *Waste not Want not: a Strategy for dealing with the waste problem in England*, London, 2002.

[8]   Parfitt J., *Analysis of household waste composition and factors driving waste increases*, WRAP, UK, 2002.

[9]   Department of Environment, Food and Rural Affairs (DEFRA), *Municipal Waste Management 1996/97 and 1997/98*, London, 2000.

[10]   Department of Environment, Food and Rural Affairs (DEFRA), *Draft Guidance on the Recycling Credit Scheme*, London, 2006.

[11]   Jasim S., *The practicability of home composting for the management of biodegradable domestic solid waste*, PhD Thesis, Imperial College London, 2003.

[12]   AEA Technology (AEAT), MEL Research Ltd, Waste Research Ltd & WRc, *The Composition of Municipal Solid Waste in Wales, A report produced for the Welsh Assembly Government*, UK, 2003.

Section 9
Landfills, design, construction and monitoring

A methodology for the optimal siting of municipal waste landfills aided by Geographical Information Systems

M. Zamorano[1], A. Grindlay[2], A. Hurtado[1,2], E. Molero[2]
& A. Ramos[1]
[1]Department of Civil Engineering, University of Granada,
Campus de Fuentenueva s/n, 18071 Granada, Spain
[2]Area of Urban and Regional Planning, University of Granada
Campus de Fuentenueva s/n, 18071 Granada, Spain

Abstract

An inappropriate landfill site may have negative environmental, economic or ecological impacts. Landfill siting should therefore consider a wide range of territorial and legal factors to reduce such negative impacts as far as possible. This paper describes the application of an integrated system of landfill siting methodology. The methodology incorporates techniques from various scientific fields as well as GIS (Geographical Information Systems) to generate spatial data for the evaluation of the suitability of an area for optimal landfill siting. The resulting land suitability is reflected on a graded scale with several territorial indexes indicating the risk and probability of contamination for five environmental components: surface water, groundwater, atmosphere, soil and human health. The methodology has been applied to a site in Granada (Southern Spain).
Keywords: landfill siting, municipal waste landfill, Geographical Information Systems, territorial siting criteria, waste management.

1 Introduction

Although authorities are attempting to reduce waste generation and disposal by implementing recycling programs and new facilities, the sanitary landfill remains a necessary part of the municipal waste management system [1, 2, 3].

WIT Transactions on Ecology and the Environment, Vol 92, © 2006 WIT Press
www.witpress.com, ISSN 1743-3541 (on-line)
doi:10.2495/WM060371

Waste disposal in landfills involves a series of complex biochemical and physical processes which lead to the generation of various emissions and environmental hazards. These include ground and surface water contamination, landfill settlement, fires and explosions, vegetation damage, unpleasant odours, air pollution and global warming [4–9].

Siting a landfill is a complex process involving a combination of social, environmental and technical parameters as well as observance of government regulations [2, 10]. Various techniques for landfill siting are described in the literature, including Geographical Information Systems (GIS) [11, 12], a mixed-integer spatial optimization model based on vector-based data [13], multiple criteria analysis [13, 14] and artificial intelligence technology based on fuzzy inference [2, 15].

In recent years a research team from the University of Granada has developed an environmental diagnosis methodology known as EVIAVE. The methodology is designed to facilitate environmental diagnoses of urban waste landfills, providing sufficient information to determine and quantify the set of environmental problems posed by each landfill [16]. This paper presents a new municipal landfill siting methodology based on a combination of EVIAVE and GIS, and describes its application to a landfill in Granada (Southern Spain). Evaluation criteria are based on international landfill siting practice, a review of the relevant literature and Spanish and European Union legislation.

2 Methodology description

The methodology is based on the use of environmental indexes designed to provide quantitative assessment of the possible environmental interaction between a landfill and potentially affected *environmental components* [17]. The original EVIAVE methodology was intended for application to municipal solid waste landfills in countries in the European Union and any other country where similar legislation exists [16].

The decision-making process is conceived as a hierarchical structure consisting of four stages. The first stage concerns the criteria and subcriteria for taking into account spatial aspects of the proposed site. These are used to quantify specific landfill variables and impact indicators, which are in turn used to calculate the different environmental indexes. The second stage represents the Probability of Contamination Indicator for each environmental component (Pbc_i), along with the corresponding Environmental Value (eV_i). The third stage involves the calculation of the Environmental Risk Index (ERI_i) for each environmental component, while the fourth or final stage represents the ultimate objective of the decision-making hierarchy: i.e. the Landfill Suitability Index (LSI). In the following sections we shall analyse each of these stages in turn.

2.1 Stage 1: Landfill variables and impact indicators

2.1.1 Landfill variables
In order to assess contamination probability, *variables* for each environmental component have been identified. These are based on characteristics of the landfill

related to biochemical and physical processes which directly or indirectly affect the environmental components [16, 18]. The variables were established taking into account relevant theoretical and practical studies regarding the siting of landfills, as well as guidelines established in European and Spanish legislation. Examples of landfill variables are: aquifer characteristics, distance from infrastructure, distance from surface water mass, distance from population points, fault lines and rainfall.

Under the EVIAVE methodology, evaluation for each variable (j) may be obtained by means of the *Contamination Risk Index*, whose expression is shown in eqn (1). In this expression, C_j is the *classification of the variable* and provides information about the interaction between disposal processes and environmental characteristics related to the variable, while W_j is the *weighting* of each variable [16]. The range of values of the index may be 1, 2, 3, 4 or 5.

$$CRI_j = C_j \times W_j \tag{1}$$

The weighting of each variable has a value of 1 or 2, determined on the basis of the relationship between the variable and the concept of *structural elements* at the landfill. The *structural elements* considered in the EVIAVE methodology are the existence of organic matter, humidity and density of wastes. These three elements participate in the main biochemical and physical processes produced in the landfill and cause production of gas and leachate, affecting all the variables and providing greater weighting to some of them [16, 18]. W_j reaches a value of 2 when the variable is directly related to the structural elements, or when it affects the environmental components.

Table 1:   Classification of the variable 'Aquifer characteristics'.

Method		GOD	DRASTIC
Classification (Iv=Vulnerability Index)	Very low (C_i=1)	Iv < 0.1	Iv < 28
	Low (C_i=2)	$0.1 \leq Iv < 0.3$	$29 \leq Iv \leq 85$
	Average (C_i=3)	$0.3 \leq Iv < 0.5$	$86 \leq Iv \leq 142$
	High (C_i=4)	$0.5 \leq Iv < 0.7$	$143 \leq Iv \leq 196$
	Very high (C_i=5)	$Iv \geq 0.7$	$Iv \geq 196$

Table 2:   Classification of the variable 'Aquifer characteristics'.

Method		SINTACS	EPIK
Classification (Iv=Vulnerability Index)	Very low (C_i=1)	$Iv \leq 80$	Iv = 2 or 3
	Low (C_i=2)	$81 \leq Iv \leq 105$	Iv = 4 or 5
	Average (C_i=3)	$106 \leq Iv \leq 140$	Iv = 6 or 7
	High (C_i=4)	$141 \leq Iv \leq 186$	Iv = 8 or 9
	Very high (C_i=5)	$Iv \geq 187$	Iv = 10

By way of example, the variable 'Aquifer characteristics' attempts to identify the characteristics of aquifers located near the proposed landfill site and to quantify their vulnerability, taking into account leachate emissions from the waste mass. This variable directly affects the environmental component 'groundwater', and therefore obtains a weighting of 2. Classification of the variable (Tables 1 and 2) is made on the basis of different indexes of vulnerability of the aquifer to pollution, which vary according to its characteristics [19]. Similar justification and quantification are applied to the other variables and environmental components.

2.1.2 Impact indicators

Impact indicators were defined in the 'Environmental Impact Assessment' process in order to measure impact on each environmental component. In the present methodology the indicators are environmental features which could be affected by the landfill project [17]. The features are subsequently used to quantify Environmental Value Indexes for each environmental component. Indicators were selected on the basis of their relevance for impact assessment as viewed by professionals, stakeholders, and the general public. In the case of surface water, for example, impact indicators are: type of surface water mass, use of water and water quality. Each impact indicator may obtain a value of 1, 2, 3, 4 or 5. Table 3 shows justification and quantification in the case of the impact indicator 'Use of water' for the environmental component 'surface water', taking into account Spanish and European Union legislation [20, 21]. Similar justification and quantification are applied to the other characteristics and environmental components.

Table 3:   Impact indicator 'Use of water' for 'surface water'.

A_2	Classification
1	Not for use by humans
2	Hydroelectric, navigation and other uses
3	Industrial
4	Agriculture
5	Human drinking water, aquaculture and recreational uses including beaches suitable for bathing

2.2 Stage 2: Probability Of Contamination Indicator and environmental values

2.2.1 Probability of Contamination Indicator

The *Probability of Contamination Indicator* (Pbc_i) for each environmental component considers possible contamination due to the characteristics of the landfill site. It is expressed by eqn (2), where n is the number of variables affecting each environmental component, CRI_j is the Contamination Risk Index for each variable (j), $CRI_{jminimum}$ is the minimum value obtained by the CRI for each variable in stage 1 and $CRI_{jmaximum}$ is the maximum value obtained by the CRI for each variable in stage 1. The indicator may obtain values between 0 and 1 and generates classifications of 'Improbable' ($0 \leq Pbc_i < 0.2$), 'Seldom probable'

(0.2≤Pbc$_i$<0.4), 'Relatively probable' (0.4≤Pbc$_i$<0.6), 'Probable' (0.6≤Pbc$_i$<0.8) and 'Very probable' (0.8≤Pbc$_i$≤1).

$$Pbc_i = \frac{\sum_{j=1}^{j=n} CRI_j - \sum_{j=1}^{j=n} CRI_{j\,min\,imo}}{\sum_{j=1}^{j=n} CRI_{j\,max\,imo} - \sum_{j=1}^{j=n} CRI_{j\,min\,imo}} \qquad (2)$$

2.2.2 Environmental Value

The concept *Environmental Value* (eV$_i$) is designed to identify and quantify the environmental assessment of each environmental component in the area of the landfill. It is considered as a relative environmental value, since it takes into account the relationship between the landfill environmental and/or social and political characteristics and the *possible* emissions in the landfill [18], as well as the environmental importance for each element in the surroundings of the landfill. Environmental Values for surface water, groundwater, atmosphere and soil are expressed by the mean values for the different impact indicators for each environmental component; for human health Environmental Value is always has maximum value. Values range between 1 and 5 for each environmental component, with classifications of 'Very low' (1≤eV$_i$<1.8), 'Low' (1.8≤eV$_i$<2.6), 'Average' (2.6≤eV$_i$<3.4), 'High' (3.4≤eV$_i$<4.2) and 'Very high' (4.2≤eV$_i$≤5). If an environmental element obtains high or very high values, this indicates that the landfill is located in an area of greater environmental sensitivity for the element in question [16, 18].

2.3 Stage 3: Environmental Risk Index

The *Environmental Risk Index* (ERI$_i$) determines the environmental impact potential for each environmental component, reflecting whether or not interaction exists between the landfill and the characteristics of the environment [16]. For each landfill, the ERI indicates which components are or would be most affected by the presence of wastes, making it possible to determine the extent of possible deterioration in each landfill site.

$$ERI_i = \sum_{i=1}^{i=5} (Pbc_i \times eV_i) \qquad (3)$$

The index is expressed by eqn (3), where Pbi$_i$ is the Probability Indicator and eV$_i$ is the Environmental Value, both for each environmental component (i). The index obtains values between 0 and 5 and is classified as 'Very low' (0≤ERI$_i$<1), 'Low' (1≤ERI$_i$<2), 'Average' (2≤ERI$_i$<3), 'High' (3≤ERI$_i$<4) and 'Very high' (4≤ERI$_i$≤5).

2.4 Stage 4: Landfill Suitability Index

Finally, overall suitability of the landfill site is quantified by a general index known as the *Landfill Suitability Index* (LSI). In the EVIAVE methodology the Environmental Landfill Impact Index (ELI) characterized the overall environmental state of operating landfills [16]. By contrast, the LSI characterizes the overall environmental suitability of the landfill siting. The graded scale used for the Index is from 0 to 25, ranging from the least suitable to the most suitable area. The mathematical expression used to obtain the LSI is eqn (4), where ERI_i is the Environmental Risk Index for each environmental component (i). Classifications generated are 'Unsuitable' (20= LSI = 25), 'Low suitability' (15= LSI <20), 'Average suitability' (10= LSI <15), 'High suitability' (5= LSI <10) and 'Very high suitability' (0= LSI <5).

$$LSI = \sum_{i=1}^{i=5} ERI_i \qquad (4)$$

3 Evaluation of land suitability: an example

This study has described the development of a new methodology to evaluate the process of siting a waste facility. Application of the methodology involves processing a variety of spatial data. In the present case GIS was used to create the digital geodatabase, using the spatial analysis tools provided by GIS. Several algorithms were used to automate the process of determining composite evaluation criteria, perform the multiple criteria analysis (MCA) and perform the spatial clustering process. These algorithms were developed in the Microsoft Visual Basic programming environment, which is compatible with the GIS software ESRI ArcGIS [22]. Various MCA methods have been recommended for the evaluation of the final suitability index, including POPSIS [23] and Compromise Programming [24]. However, in the present study the simple additive weighting (SAW) method was selected as the most appropriate way of solving the multiple criteria problem. The GIS-aided landfill siting methodology presented here combines the spatial analysis tools provided by GIS with MCA to evaluate the entire region, based on specific evaluation criteria (hydrological-hydrogeological, environmental, social, technical/economic).

3.1 Area of study

The area selected has an extension of 300 km² and is situated to the south of the conurbation of Granada (Southern Spain) on the western edge of the Sierra Nevada (fig.1). After Sevilla and Málaga, Granada has the third largest population of the Autonomous Community of Andalusia, with approximately 440,000 inhabitants, of whom two thirds live in the metropolitan area of the city. This figure represents 55% of the total population of the province of Granada

(817,000), concentrated in a surface area of 830 km², i.e. less than 7% of the total surface area. Population density in the area is thus 530 inhabitants per km², as against 32 inhabitants per km² in the rest of the province.

With regard to the treatment, disposal and elimination of solid wastes, there is an urban solid waste treatment plant in the municipal district of Alhendín. This plant handles the waste from the 30 municipal districts which make up the conurbation of Granada, as well as from 36 districts outside the conurbation.

Figure 1:   Map of situation and localization of existing landfill site.

3.2 Modelling the landfill variables

In the field of Geographical Information Systems there are two basic approaches to the question of how to model space. Depending on whether attention is given to properties or localization [24], two different data models may be generated: the vector model and the raster model.

The two models present a series of advantages and disadvantages according to the use for which GIS is intended [25, 26]. However, in the present study the raster model was finally selected due to its speed and efficiency at superpositioning maps, while the vector model was used only to generate the basic cartography and for the initial variable modelling.

An optimal resolution of 10 m was adopted for base cartography at a scale of 1: 10,000 and the following techniques and operations were applied: local analysis (reclassification and map superposition), immediate vicinity analysis (filtrates and slope calculation) and extended vicinity analysis (Euclidean distances and proximity or 'buffer' analysis).

3.3 Model implementation

'Cartographic modelling' is a more ample term than the set of steps described above. The method involves the arrangement of a series of data layers in logical sequence, including topological and thematic operations, information external to GIS and value judgements, with an aim to finding solutions to specific spatial problems [27]. Tomlin [28] describes cartographic modelling as a 'general methodology for the analysis and synthesis of geographical data', and defines it as 'the use of the basic operations of a GIS in a logical sequence to resolve complex spatial problems'. These are the phases of the applied model:

1. Cartographs of the Contamination Risk Index (CRI_j). Each localization variable is modelled and reclassified and subsequently each W_j is measured using map calculator algorithms [22] and the product operator. Each localization variable landfill generates a cartograph for each impact on the environmental components, and the value for the Contamination Risk Index is indicated on each pixel.

2. Cartographs of the Probability of Contamination Indicators (Pbc_i). In a subsequent step results are grouped using arithmetic superposition in order to obtain cartographs of the Probability of Contaminations Indicators, with one image corresponding to each environmental component.

3. Calculation of Environmental Values (eV_i) and cartographs of the Environmental Risk Index (ERI_i). The values obtained are used to determine the Environmental Risk Index (ERI_i), by means of arithmetic superposition of the Environmental Value (eV) for each environmental component.

4. Cartographs of the Landfill Suitability Index (LSI). To conclude the model, the cartograph of the LSI is obtained by means of multi-criteria analysis techniques (MCA), taking as factors the different Environmental Risk Indexes for each environmental component (surface water, ground water, atmosphere, soil and human health). The value associated with each pixel of the map gives a final indication of the suitability of the site.

Results obtained in the application of the methodology gave a Landfill Suitability Index of 6.48 for the siting of the landfill, equivalent to 'high' according to the classification described above. Figure 2 shows results for the model applied with GIS.

The Probability of Contamination Indicator for the environmental components obtained classifications of 'Improbable' for ground water and 'Seldom probable' for surface water, soil and human health. The component 'atmosphere' obtained a rather higher value of 'Probable'. These results indicate that the siting of the landfill does not present general characteristics which might contribute to the contamination of the different environmental components, except in the case of atmosphere. In this case, the landfill could present an impact influenced by high rainfall, seismic risk in the area and wind characteristics.

Results for the Environmental Risk Index (ERI_i) show final contamination risk values for each environmental component. In this case the Environmental

Value for each component is also taken into account, with classifications of 'very high' for health and atmosphere, 'high' for groundwater, 'low' for soil and 'very low' for surface water. With the addition of these data, results for ERI_i are finally 'very low' for surface water, groundwater and soil; 'low' for human health and 'high' for atmosphere. Again, atmosphere presents a higher risk of being affected, due not only to the higher probability value but also to the very high Environmental Value.

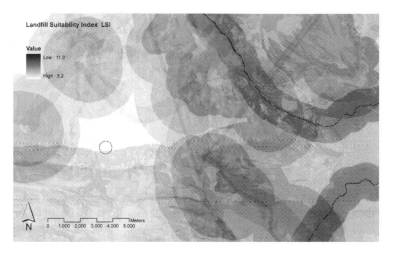

Figure 2:   LSI for the area under study.

3.4 Analysis of model sensitivity

Sensitivity analyses are directly related to modelling in any scientific field. A model is always a simplified version of reality which enables us to describe a specific problem and reach a better understanding of it, through the representation of essential elements and mechanisms of systems in the real world, whether physical, social, economic or environmental. In order to demonstrate that a model is a reliable representation of such a real system, it is necessary to carry out certain validation processes to lend sufficient credibility to the model. In the present study a verification of the methodology was undertaken, together with a results validation test and model stability analysis.

4 Conclusions

The methodology has been shown to be applicable to analysing the suitability of a site for landfill localization, due chiefly to the generation of a final suitability index. However, analysis of results for the other indexes may also facilitate the study of potential problems related to the different environmental components, thus contributing to the decision of whether or not to locate the landfill in a particular setting. On the basis of the results of the practical application of the model in Granada as well as the sensitivity analysis, we may conclude that Geographical Information Systems may be usefully applied to the optimal siting of facilities with such exceptional characteristics as landfills. It seems reasonable, moreover, to predict that this instrument has the potential to assist planners, decision-makers and other agents involved in the process of selecting suitable sites for municipal landfills, by extending their knowledge of the physical terrain and facilitating the analysis and execution of plans of action.

Acknowledgement

This research is part of an R+D Project funded by the Spanish Ministry of Science and Technology entitled "Design and implementation of methodologies for the environmental diagnosis of urban waste landfills and waste dumps".

References

[1]   Komilis, D.P., Ham, R.K. & Stegmann, R., The effect of municipal solid waste pretreatment on landfill behavior: a literature review. *Waste Management and Research*, **17**, pp. 10-19, 1999.
[2]   Al-Jarrah, O. & Abu-Qdais, H., Municipal solid waste landfill siting using intelligent system. *Waste Management*, **26(3)**, pp. 299-306, 2005.
[3]   Vaillancourt, K. & Waaub, J-P. Environmental site evaluation of waste management facilities embedded into EUGÈNE model: A multicriteria approach. *European Journal of Operation Research*, **139**, pp. 436-448, 2002.
[4]   Abu-Rukah, Y. & Al-Kofahi, O., The assessment of the effect of landfill leachate on ground-water quality – a case study: El-Akader landfill site, North Jordan. *Journal of Arid Environments*, **49**, pp. 615-630, 2001.
[5]   Edgers, L., Noble, J.J. & Williams, E., A biologic model for long-term settlement in landfills. *Proceedings of the Mediterranean Conference on Environmental Geotechnology,* (M.A. Usmen and Y.B. Acat., eds), Cesme, Turkey, pp. 177-184, 1992.
[6]   El-Fadel, M., Findikakis, A.N. & Leckie, J.O., Environmental impacts of solid waste landfilling. *Journal of Environmental Management*, **50**, pp. 1-25, 1997.
[7]   Raybould, J.G. & Anderson, D.J., Migration of landfill gas and its control – a case history. *Journal of Engineering Geology,* **20**, pp. 75-83, 1987.
[8]   Gilman, E.F., Flower, F.B. & Leone, I.A., Standardized procedures for planting vegetation on completed sanitary landfills. *Waste Management & Research*, **3**, pp. 65-80, 1985.
[9]   Sarkar, U., Hobbs, S.E. & Longhurst, P., Dispersion of odour: a case study with a municipal solid waste landfill site in North London, United Kingdom. *Journal of Environmental Management*, **68**, pp. 153-160, 2003.
[10]   McBean, E., Rovers, F. & Farquhar, G., *Solid Waste Landfill Engineering and Design*. Prentice-Hall PTR, Englewood Cliffs, New Jersey, USA, 1995.

[11]   Siddiqui, M.Z., Landfill siting using Geographical Information Systems: a demonstration. *Journal of Environmental Engineering*, **122(6)**, pp. 515-523, 1996.

[12]   Kontos, Th.D., Komilis, D.P. & Halvadakis, C.P., Siting MSW landfills in Lesvos Island with a GIS-based methodology. *Waste Management and Research*, **21(3)**, pp. 262-27, 2003.

[13]   Lin H-Y, & Kado, J-J., A vector-based spatial model for landfill siting. *Journal of Hazardous Materials*, **58**, pp. 3-14, 1998.

[14]   Kao, J-J., Multifactor spatial analysis for landfill siting. *Journal of Environmental Engineering*, **122(10)**, pp. 902-908, 1996.

[15]   Gupta, R., Kewalramani, M.A. & Ralegaonkar, R.V., Environmental impact analysis using Fuzzy Relation for landfill siting. *Journal of Urban Planning and Development*, **129(3)**, pp. 121-139, 2003.

[16]   Zamorano, M., Garrido, E., Moreno, B., Paolini, A. & Ramos, A., Environmental diagnosis methodology for municipal waste landfills as tool for planning and decision-making process. *Sustainable Development and Planning*, Wessex Institute of Technology, 1, pp. 545-554, 2005.

[17]   Antunes, P., Santos, R. & Jordao, L. The application of Geographical Information Systems to determine environmental impact significance. *Environmental Impact Assessment Review*, **21**, pp. 511-535, 2001.

[18]   Calvo, F., Moreno, B., Zamorano, M. & Szanto, M., Environmental diagnosis methodology for municipal waste landfills. *Waste Management*, **25**, pp. 68-79, 2005.

[19]   Foster, S. & Hirata, R. Determinación del riesgo de contaminación de aguas subterráneas. Una metodología basada en datos existentes. [Determining the risk of contamination in groundwater. A methodology based on existing data.] *CEPIS,* pp.1-81, 1991.

[20]   Directive 2000/60/EC of the European Parliament and of the Council of 23 October 2000 establishing a framework for Community action in the field of water policy.

[21]   Real Decreto 927/1988, de 29 de julio, por el que se aprueba el Reglamento de la Administración Pública del Agua y de la Planificación Hidrológica, en desarrollo de los Títulos II y III de la Ley de Aguas. [Royal Decree 927/1988 of 29 July, approving the Regulation proposed by the Public Administration for Water and Hydrological Planning, in application of Chapters II and III of the Water Act.]

[22]   McCoy, J. & Johnston, K., Using *ArcGis Spatial Analyst*. Environmental Systems Research Institute. Redlands, California, 2002.

[23]   Yoon, K. & Hwang, C.L., *Multiple Attribute Decision Making: an Introduction*. Sage Publications Inc., London, UK, 1995.

[24]   Zaleny, M., Fuzzy Sets. *Information and control*, **8**, pp. 338-353, 1982.

[25]   Bosque, J. *Sistemas de información geográfica*. [Geographical Information Systems.] Madrid, Rialp, 1997.

[26]   Bosque, J., Díaz, M.A., Gómez, M., Rodríguez, V.M., Rodríguez, A.E. & Vela, A., Un procedimiento, basado en un SIG, para localizar centros de tratamiento de residuos. [A GIS-based procedure for siting waste

treatment centres.] *Anales de Geografía de la Universidad Complutense*, **19**, pp. 295-323, 1999.

[27] Barredo, J.I. & Gómez, M., *Sistemas de Información Geográfica y Evaluación Multicriterio en la Ordenación del Territorio*. [Geographical Information systems and multicriteria evaluation in the organization of territory] Ed. Ra-Ma. Madrid, 2005.

[28] Tomlin, D., *Cartographic modelling*. Maguire, D., Goodchild, M. and Rhind, D. (eds). Geographical Information Systems. Volume. 1, New York, Longman, 1991.

Windbreaks for odour dispersion

S. Barrington[1], L. Xingjun[1], D. Choinière[2] & S. Prasher[1]
[1]Faculty of Agricultural and Environmental Sciences,
Macdonald Campus of McGill University, Québec, Canada
[2]Consumaj inc., Québec, Canada

Abstract

Windbreaks are believed to help disperse odours emitted by livestock facilities. This project compared odour dispersion for a site without and another with a windbreak consisting of a single row of deciduous trees with an optical porosity of 35%. Both sites, without and with a windbreak were subjected to the emission of swine manure odours produced by an odour generator and the resulting odour plume was measured by three groups of four trained panellists calibrated using a forced choice dynamic olfactometer. This calibration was used to relate the group's field odour intensity perception to an actual odour concentration in odour units (OU). In September 2004, the windbreak site was evaluated 15 times while the site without a windbreak was evaluated 4 times. The odour plume which developed on both sites was found to be quite erratic with zones of high odour concentration intercepted and others of low and even non-detectable odour concentration. Nevertheless, a regression method using classification for the data on each site indicated that the windbreak could reduce the odour dispersion distance by 20%, on the average.
Keywords: windbreak, odour dispersion, separation distance.

1 Introduction

The development of the livestock industry in North America and in Europe is jeopardized by their emissions of odours which are a nuisance for neighbours. These odours are produced mainly during the handling of manures, but can also originate from feed storage and dust emissions. Since there are no technologies which can completely eliminate these odours, livestock operations and environmental authorities use air dilution as remediation measure. The concept consists in building the livestock operation far enough that ambient climatic and atmospheric conditions can dilute the emitted odours below their threshold before reaching neighbouring sites. Such distances separating the livestock shelters from the neighbouring buildings are called "set back" or "separation" distances. Most province in Canada, state in the USA and country in Europe have adopted a method of calculating setback distances based on practices adopted by local livestock operations, climatic conditions and topography [1]. Among other options, this method allows for the introduction of correction factors accounting for the use of odour attenuation technologies.

Natural windbreaks are known to affect air currents and improve air mixing, thus be able to reduce separation or setback distances between livestock operations and their neighbours [2, 3]. Accordingly, the objective of this paper was to compare the length and strength of the odour plume developing in the presence and absence of a natural windbreak.

2 Method

2.1 Site and equipment

The sites without and with a windbreak were located on relatively flat farm land with a slope of 0.1% and in the absence of other trees or fences, in the region South West of Montreal, Canada. The experimental windbreak consisted of a single row of deciduous trees offering an optical density of 35%. A mobile odour generator was used to emit a controlled level of swine manure odour and to carry out the test away from any infrastructure capable of interfering with the results. During the tests, the odour generator was positioned 15m upwind from the windbreak and liquid swine manure was used to produce the odour source. A weather tower was installed 200m upwind from the site to measure, every minute, wind velocity and direction and ambient air temperature.

2.2 Method

The wind direction was measured before starting each test to establish the monitoring path of the three groups of four trained panellists. Five minutes after starting the odour generator, the panellists started to walk their designated path, stopping at random locations, taking off their face mask and evaluating the odour for 1 minute using a scale of 1 to 10. During this time, odour samples were collected from the outlet of the odour generator every 30 minutes. Once the odour plume was evaluated, up to 500m downwind from the odour generator, the panellists travelled to the olfactory laboratory, where they were rated using n-butanol, and were used to evaluate the odour samples collected at the generator. At the same time, their field odour intensity perception was correlated to an actual odour concentration in terms of odour units (OU). A forced choice dynamic olfactometer was used for this laboratory work. The sites without and with a windbreak were evaluated 4 and 15 times, respectively in September 2004, when the ambient air temperature ranged between 18 and 26°C (Table 1).

Table 1:   Experimental conditions for windbreak evaluation.

site	test number	Average OU generated	Average wind		Average temperature (°C)
			Velocity (m/s)	direction	
without windbreak	1	621	6.4	90	19
	2	760	6.0	90	20
	3	859	2.5	50	17
	4	551	2.5	50	20
	average	698			
with windbreak	1	1373	3.9	90	21
	2	492	4.4	90	23
	3	578	4.7	40	18
	4	585	4.2	40	19
	5	214	1.0	60	22
	6	218	1.1	70	20
	7	5360	1.2	30	22
	8	1096	2.7	20	27
	9	559	1.2	50	23
	10	294	1.0	40	26
	11	744	5.1	90	28
	12	745	1.5	90	23
	13	1879	1.5	40	24
	14	1352	1.4	50	21
	15	846	2.2	60	26
	average	997			
average		934			

Note: a wind direction of 90° is perpendicular to the windbreak.

2.3 Analysis of the data

The odour observations and evaluations of all four panellists in each group where averaged before analyzing the data. All odour plumes were standardized for comparison. The odour generator produced a different odour level because the swine manure had to be changed before each test and, also, during each test, the odour intensity would decrease over time, but no more than 50%. Thus, the odour concentration measured at every point by each group of panellists, at a given time, was divided by the odour concentration of the generator at that time, and multiplied by the average odour level of 934 OU (odour unit) calculated from all 19 tests. Furthermore, the wind direction with respect to the windbreak was readjusted to 90° (perpendicular to the windbreak) for purposes of comparison. For each 10 minute of wind direction and speed measurement, the windbreak was assumed to stand perpendicular to the average wind direction and new x and y coordinates were computed for each odour point observed by defining them as perpendicular and parallel to the windbreak, respectively, with the odour generator standing at the origin.

To compare the odour plumes on both sites, without and with windbreak, a regression analysis was performed with classification, using SAS [4, 5] and all the data collected. The least square method was used to establish the significance of the treatment (absence or presence of a windbreak) at a 95% confidence level.

3 Results and discussion

3.1 The typical odour plume shape

The respective and typical odour plumes which developed on both sites, without and with a windbreak, are illustrated in Figure 1a and b. Although the odour intensity was found to drop with distance away from the odour generator, the odour plumes were far from being uniform. Rather, zones of peak odour concentration were surrounded by other zones of low and even non-detectable odour concentration.

For the site without windbreak, the first zone of high odour concentration of 10 OU was observed at 75m downwind from the generator and its intensity dropped with distance, but still remained over 2 OU even at a distance of 500m. Also, the drop in odour intensity was progressive with distance away from the source.

For the site with the windbreak, the zones of highest odour concentration occurred immediately on the leeward side of the trees, reached a value of 30 OU which was much higher than that of the no windbreak site, and spanned a distance of 100m. Nevertheless, for distances greater than 100m, the odour concentration dropped drastically, to become undetectable at a distance of 500m. The windbreak therefore appeared to trap the odours on its leeward side, and then to release these odours with a dilution effect.

3.2 Comparison of odour dispersion efficiency

Figures 2 a) and b) illustrate the regression equations obtained from all the data collected, for the site without and with a windbreak, respectively. The least square analysis indicates that the windbreak did have a significant effect on odour level with distance from the source ($P < 0.5$). Whereas the odour concentration was higher at the source, for the site with the windbreak, the odour concentration decreases faster downwind, compared to the site without the windbreak. This phenomenon results from the fact that the windbreak appears to hold the odours immediately on its leeward side and then release them with a dilution effect. Two regression lines were produced for each site. The higher line is computed from the zones of high odour intensity while the other line is computed from all the data collected on the individual sites. If a separation distance is computed from the maximum and overall lines, for both sites, 2 OU is reached without the windbreak at a distance of 717 and 561m from the source, while with the windbreak the distance is of 564 and 355m. Thus, the windbreak did reduce the separation distance by 21 and 37%, respectively, for the maximum and overall data, as compared to the site without windbreak. In determining the separation distance, the maximum odour levels are preferred as they are more conservative than the overall data.

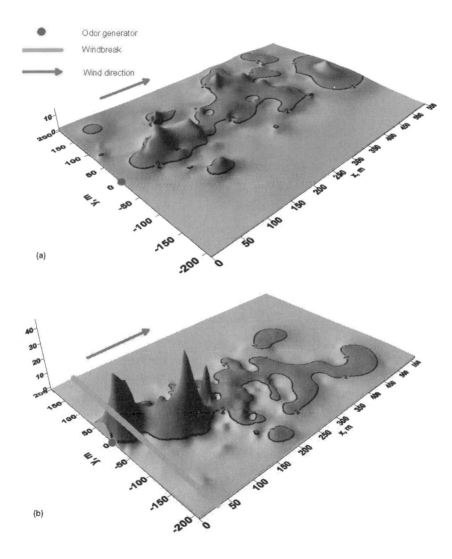

Figure 1:   Typical odour plumes which developed in the absence (a) and presence (b) of a windbreak. An odour concentration of 2 Ou/m^3 is used to draw the final contour of the odorous zones.

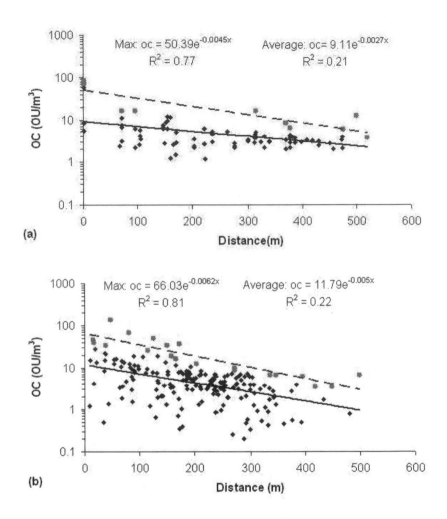

Figure 2: Odour concentration (OC) with distance from the source and regression lines for the site without (a) and with windbreak (b), for all the data collected and standardized. The top line is associated with the zones of peak odour concentration while the bottom line of each graph is associated with all the data.

4 Conclusion

This project observed the odour plumes which developed on a site without a windbreak and on another with a natural windbreak consisting of a single row of deciduous trees with an optical density of 35%. The results indicate that natural windbreaks can effectively help dilute odours, by holding them immediately on their leeward site and then releasing them with a dilution effect. A regression analysis with classification indicated that the effect was significant at a 95% confidence level.

Therefore, windbreaks warrant further investigation to obtain the optimal value of parameters establishing their effectiveness in diluting odours. Parameters such as windbreak porosity and type of tree, wind direction and speed, ambient air temperature with height and therefore air stability or convective effects can affect the performance of windbreaks in diluting odours. Because field trials are expensive, time consuming and not perfectly controllable, it is advisable to work with a model simulating the dilution effect to windbreak.

Acknowledgement

The authors wish to acknowledge the financial contribution of Consumaj inc., CDAQ, The Livestock Initiative Program, Agriculture and Agro-Food Canada and the Natural Sciences and Engineering Research Council of Canada.

References

[1]   Guo, H., Jacobson, L.D., Schmidt, D.R., Nicolai, R.E. & Janni, K.A., Comparison of five models for setback distance determination from livestock sites. *Canadian Biosystems Engineering*, **46**, pp. 6.17-6.25, 2004.
[2]   Heisler, G.M. & Dewalle, D.R., Effects of windbreak structure on wind flow. *Agriculture, Ecosystems and Environment*, **22-23**, pp. 41-69, 1988.
[3]   McNaughton, K.G., Effects of windbreaks on turbulent transport and microclimate. *Agri. Ecosystem & Environment* **22-23**, pp.17-39, 1988.
[4]   SAS Institute. 1990. SAS/STAT, *User's Guide*. Volume 2, 4th ed. Cary, NC, USA. SAS Institute Inc.
[5]   Miller, I., Miller, M. & Freund, J.E., *John E. Freund's mathematical statistic,* Prentice Hall: Upper Saddle River, New Jersey, USA, pp. 624, 1999.

Chemical hazard exposure as a result of waste land filling: a review

M. E. El Megrahi, G. Karani & K. Morris
Cardiff School of Health Sciences, University of Wales Institute Cardiff, Wales, UK

Abstract

Waste is any substance that is discarded, emitted or deposited in the environment in such volume, constituencies or manner as to cause damage in the environment. Waste can include domestic waste, commercial waste, industrial, medical and radioactive material. All types of wastes are deposited in the environmental ecosystem frequently by the land filling process. Land filling is one of the methods used to reduce and manage the waste as well as being the most common method by which waste is disposed of in the UK. After dumping the waste in landfill sites, several chemical hazard materials such as volatile organic compounds, (VOCs), methane, heavy metals, dioxin, furan, hydrogen sulphide and natural occurrence radioactive material (NORM) are vented to environment as result of several biological and chemical processes occurring in the landfill site. This paper is a review of the current knowledge on the importance of monitoring of chemical hazards and its effects on environment and health.
Keywords: land filling, sources, monitoring, chemical hazard, exposure.

1   Introduction

The quantity of waste generated has considerably increased for several reasons, but major factor is lifestyle improvement and the increase in consumption of food and goods associated with it. However, another factor associated with increased waste production is the extension in industries throughout the world. While this expansion in industrial capacity offers more diversity in production, the increased waste production brings with it considerable associated problems. In addition, waste disposal is subject to legislation governing treatment before

WIT Transactions on Ecology and the Environment, Vol 92, © 2006 WIT Press
www.witpress.com, ISSN 1743-3541 (on-line)
doi:10.2495/WM060391

land filling. However, in many countries such legislation either does not exist or is not sufficiently implemented. There has emerged increasingly the necessity of environmental waste management systems. Landfill is a place used to dispose waste. These are normally in the form of cells which can be lined, then filled with waste materials which are progressively compressed and enclosed with further soil, and finally stored with a permanent cap.

Land filling is one of the most frequently used options for safe disposal of domestic, commercial and industrial waste. Landfill is generally regarded as the simplest and most cost-effective method of disposing of waste, Barrett and Lawlor [1].

In most low-to medium-income developing nations, almost 100% of generated waste goes into landfill. Even in many developed countries, most solid waste is land filled In the UK landfill continues to be the most common method for waste disposal. In 2004 Wales generated an estimated 14.5 million tones of waste EPW [2]. Consequently, Wales in 2007 may face a critical crisis as result of waste land filling. DEFRA [3]. Furthermore although the proportion of waste to landfill may in future decrease, the total volumes of municipal solid waste being produced are still increasing at a rate in excess of 3% per year, Douglas [4].

Land filling is likely to remain a relevant source of ground water contamination for the foreseeable future, Allen [5] and as several unlined landfill sites in UK are situated on major aquifers, the potential impact on the water table needs to be investigated, Bloor *et al.* [6].

2 Source of chemical hazard exposure from land filling

Waste is a mixture of packaging materials, food cans, biodegradable waste and garden waste, commercial and industrial waste. Solid household waste contains between 40-90% moisture and contains levels of soluble organic compound which can play major role to the growth of unnecessary bacteria. These may lead to the generation of toxic substances through chemical and biological processes such as, hydrolysis, fermentation, cytogenesis, and methanogenesis, oxidation, Satheropouls [7].

2.1 Volatile organic compounds (VOCs)

VOCs are compounds that have a high vapour pressure greater than 0.0013atm at standard temperature and low water solubility. VOCs can be classified as several types; Halogenated Volatile organic compounds, oxygenated VOCs, sulphanted VOCs, and aliphatic, aromatic VOCs, Hamideh [8].

2.2 Dioxins and furans

Dioxins are chlorinated dibenzo-paradioxine and are formed during the composition of material containing chlorine. Dioxins and furans are released into the atmosphere as   by products of many combustion processes. These includes the incineration of municipal waste, the burning of fuels including coal, oil, and also some industrial processes such as bleaching pulp, paper and combustion of landfill gas, ASTDR [9]

2.3 Heavy metals

Heavy metals are defined as metals that have specific density of more than $5g/m^3$. Metals account for a approximately one-quarter of the earths mass. The majority of metals occur in nature as natural ores, which are compounds of the metal combined with typically oxygen and sulphur. Common heavy metals found in landfill are cadmium, mercury, arsenic and lead, Jarup [10].

2.4 Acid gas

Acid gases such as nitrogen oxide, sulphur dioxide, carbon monoxide can be released as result of decomposition of waste and gas flaring landfill site. Tennessee Division of Solid Waste Management Programme [11].

2.5 Odour

Odour is an oraganoleptic attribute Perceptible by the olfactory organ on sniffing certain volatile substances, CEN: [12]

Most landfill sites release nuisance odour especially when a lot of putrescible wastes are dumped in the site. In some cases landfill odours have been traced over 1km. There is suggestion that odour is result of hydrogen sulphide, dimethyl sulphide, and mercaptans found in the landfill site and may produce strong rotten-egg sell even in very low concentration, El-Fadel et al. [13].

2.6 Methane

Methane is the most common gas emitted from landfill as result of waste decomposition in the absences of oxygen. Landfills are the single largest sources of U.S. man-made methane emissions, ASTDR [14].

2.7 Natural radioactivity

Radioactivity is spontaneous nuclear transformation that results in the formation of new elements. These transformations are accomplished by emission of particles from the nucleus or by the capture of orbital electron.

Each of these reactions may or may not be accompanied by a gamma photon, WHO [15]. Around 90% of human radiation exposure comes from natural sources such as cosmic radiation, exposure to radon and terrestrial radiation. However some industrial process such as coal, peat, mineral, sands by- product may enhance radionuclide to such a degree that they may pose risk to human and environment if they are not controlled, Organo and Lee [16].

3   Environmental problems caused by land filling emissions

The deposition and burning of waste cause a profound strain on the environment with its potential to contaminate ground water resources, emit gases cause air and soil contamination, Poulsen *et al.* [17]. Air quality has become a major interest in area where several waste facilities are located. This interest is based on the increased odour and emission vented from landfill. On global scale, landfills are responsible for about 10% of human intrusion methane.

Landfill gas collection systems do not prevent significant emission of the harmful greenhouse gas, methane, from landfills. In fact, less than 50% and possibly only 10-20% of methane produced is likely to be captured as result of collection systems. In order to prevent an explosive mixture oxygen and methane gas is drawn from the landfill with vacuum pumps. The vertical collection pipes cannot be properly perforated at the top third of the pipe; moreover, the pipes cannot extend to the very bottom of the waste, United Nations Environment Programme [18].

The magnitude of annual global emission of methane from municipal solid waste landfills without landfill gas control system implies that these landfills are significant contributors to the atmospheric load of green house gases Fourie and Morris [19].

Study on ambient air monitoring by, Zu *et al.* [20] at Datianshan landfill, Guangzhou, South China in 1998 to investigate the seasonal and horizontal variations of trace volatile organic compounds (VOCs) identified thirty eight vocs in the winter, whereas 60 were detected in the summer.

Study by Tolvanen *et al.* [21] to investigate the abundance of volatile organic compounds within landfill site; found that the common VOCs were found are aldehydes, carboxylic acids, ketones. Landfill leachate contains wide range of toxic substances arising from the decomposition of waste, and causes contaminates of domestic groundwater sources and excess of nitrogen in watercourses. Substances found in leachate from household waste such as toluene and mercury, are toxic to living organisms that come into contact with leachate-contaminated water or soil (Slack *et al.* [22]).

Research on the impact of peat by- product land filling in Ireland has been undertaken. Although the study confirmed that there was low exposure due to natural occurrence radioactivity material in the peat used in producing electricity in Ireland, the study did not expand on the inhalation of peat ash dust on the landfill sites arising from the generation of windborne ash on the ash pond, Organo [23] Furthermore the study did not investigate the possibility of leaching of Ra- 226 from ash pond to underground water underneath the ash pond and the transport of dry ash by wind to different residential and agricultural area. Moreover internal and external radiation exposure to the public may occur when some of the ash may be used as building material.

4 Health problems caused by land filling emissions

Health is defined as a state of complete physical, mental and social well-being and not totally the absences of disease or infirmity, WHO [24].

Concern about possible adverse health effects for population living nearby land fill sites and have been highlighted, especially in relation to those sites where hazardous waste is dumped Vrijheid [25].

Several studies on occupational exposure or health problems in waste facilities carried out in the USA, Mandsdorf *et al.*, [26] Great Britain,

The studies of Crooks *et al.*, [27] and Rahkonen, [28] have indicated that workers in this industry have an excess risk of work- related health problems such as musculoskeletal problems, pulmonary diseases, organic dust toxic syndrome (ODTS) symptoms (cough, chest-tightness, dyspnoea, influenza-like symptoms such as chills, fever, muscle ache, joint pain, fatigue and headache), gastrointestinal problems and irritation of the eye, skin and mucous membranes.

Hertzman *et al.*, [29] stated that people who lived or worked near an industrial dump revealed significantly elevated the following symptoms bronchitis; difficulty berating; cough; skin rash; hearth problems; muscle weaknesses. Paigen and Goldman [30] stated children living near the chemical waste dump in love canal Niagara Falls showed significantly reduced height compared with a control group of children living further from the dump site.

Berry and Bove [31] studied birth weight at the Lipari Landfill in New Jersey, a site for municipal and industrial waste. Leachate from the site migrated into nearby streams and a lake adjacent to a residential area. Inhalation of volatile chemicals emitted from the landfill and contaminated waters was thought to be the most important exposure pathway. The site closed in 1971 after complaints of residents, but the heaviest pollution was estimated to have occurred during the late 1960s to the mid-1970s. The study found a convincing increase in proportion of low birth weight babies ($<$ 2500 g) and a lower average birth weight in the population living closest (within a radius of 1 km) to the landfill in the time period when potential for exposure was thought to be greatest (1971-1975) compared to these factors in a control population.

Study by Kharrazi *et al.* [32] at BBK landfill California sites detected waste of a significant association disposal waste and the weight foetal, with a negative correlation between mortality rate and waste disposed.

Dolk *et al.* [33] noticed the risk of congenital anomaly in babies whose mothers lived close to landfill sites was in both the study and control areas. The authors felt that direct measure of exposures and birth defects would better establish a causal relationship.

Study on the health impact of waste on workers and domestic waste collectors showed signs of increased prevalence of respiratory symptoms and also increased concentrations of variables indicative of upper respiratory inflammation in nasal lavage compared with controls Wouters [34].

Although several researchers have addressed the risk of waste facilities, for instance congenital anomalies and malignancies in relation to living in close-proximity to landfill sites, the health impacts of new waste management technologies and the increasing use of recycling and composting will require further assessment and monitoring, Rushton [35].

Hamer [36] stated the effects of the waste management practices on public health and environment safety remain unanswered.

5 Conclusion

In order to develop environmentally sound and sustainable waste management, it is necessary to establish research that focuses on modern analytical technique and use the latest sensory technology for the measurement and evolution of chemical hazard exposure.

Studies on natural occurrence radioactive material specific activity from waste land filling process limited and the potential sources of exposure is not characterised. In this context, natural radioactivity comprises vital subject, which need to be considered in updated landfill regulation and standards. This will help and enhance the limitation of radiation exposure risks emitted from NORM.

6 Future research

For the future planning and regulation of land filling, it is important to understand which types of sites are most likely poses risks. Research in to the environmental and health effects of land filling is relatively rare, and further research may improve our current understanding. This only through multidisciplinary approach drawing from the fields of environmental Sciences, toxicology and epidemiology.

Biological monitoring plays important role in industrial hygiene practice to evaluate the dose-response relationship between internal exposure and adverse health effects of exposure of chemical, there fore research would be worth doing it using biological marker as technique as tool to see the effect of waste on public and environmental species. More research into effects of chemical mixtures and possible interaction between single chemicals is needed to improve understanding of effects of multiple chemical exposures.

References

[1] Barrett, A., Lawlor, J. The economics of waste management in Ireland. Economic and Social Research Institute, Dublin, pp 129, 1995.
[2] EPW. Environment Protection Agency Wales, 2005
[3] Department for Environment food and Rural Affairs, Review of Environmental and Health Effects of Waste Management, London, UK, 2004.
[4] Douglas, T, Patterns of land, Water and air pollution by waste. In: Newson, M. (ed.) Managing the human impact on Natural Environment John Wiley &Sons, 150-171,1992.
[5] Allen, a contaminated landfills: The myth of sustainability. *J. Eng. Geol*, **60**, pp3-19, 2001.
[6] Bloor M.C, *et al.*, Acute and sub lethal toxicity tests to monitor the impact of leachate on an aquatic environment, *Environmental International*, **31**, pp.269-273, 2005.

[7]   Satheropouls M a study of volatile organic compounds evolved in urban waste disposal bins, *Journal of AtmosphericEnvironment,* **39**,pp.4639-4645,2005.

[8]   Hamideh Soltani Ahmadi, *A Review of the literature Regarding Non-Methane and Volatile Organic Compounds In Municipal Solid Waste Landfill Gas.,* SWANA/Hickman Intern, Department of Civil and Environmental Engineering, University of Delaware, Near, Delaware 19716, 2000.

[9]   ASTDR Agency For Toxic Substances and Disease Registry, Landfill Gas Primer, An overview for Environmental Health Professionals, USA, 2001.

[10] Jarup Lars, Hazards of heavy metal contamination, *British Medical Bulletin,* **68** pp.167-182, 2003.

[11] Tennessee Division of Solid Waste Management Programme. Landfill Gas Monitoring and Mitigation. The technical Section/Solid Waste Management Programme, 1999.

[12] CEN: Air quality- Determination of Odour concentration by dynamic olfactmetry. CEN, European committee for standardization, Draft prEn13725, Brussels, 1999.

[13] El-Fadel, *et al, M,* Environmental impacts of solid waste land filling. *Journal of* Environmental *Management,* 50(**1**), 1997.

[14] ASTDR, *op. cit.*

[15] WHO, World Health Organization, Depleted Uranium Sources, exposure and health effect, Department of protection of the human environment, World Health Organization, Geneva, 2001.

[16] Organo C, E M, Lee Investigation of occupational radiation exposures to NORM at an Irish peat-fired power station and potential use of peat fly ash by construction industry, *Journal of Radiological protection,* 25, pp.461-474, 2005.

[17] Poulsen M. Otto, *et al,* Sorting and recycling of domestic waste. Review of occupational health problems and their possible causes, *the sciences of the total Environment* **168**, pp.33-56, 1995.

[18] Agency, United Nation Environment Programme, Global Warming: National Emission. P table es, 2000.

[19] Fourie AB, Morris JW Measured gas emissions from four landfills in south Africa and some implications for landfill design and methane recovery in semi-arid climates, *Waste Manag Res* ;**22**(**6**) pp. 440-53, 2004.

[20] Zu C, *et al,* Characterization of ambient volatile organic compounds at a landfill site in Guangzhou, South China, *Chemosphere.* **51**(**9**) pp. 1015-22, 2003.

[21] Tolvanen O, Nykanen J, Nivukoski U, *et al,* Occupational hygiene in a Finnish drum composting plant, *Waste Manag;***25**(**4**) pp.427-33, 2005.

[22] Slack R J, *et al,* Household hazardous waste in municipal landfill: contaminates in leachate. *Sci Total Environ* 20; 337(1-3) pp.119-37, 2005.

[23] Organo C, *op. cit.*

[24] World Health Organization, (WHO) Definition of Health, 2005.

[25]   Vrijheid Martine, Health effects of residence Near hazardous waste Landfill sites: A Review of Epidemiologic Literature, Environmental health Perspective Supplements Volume 108, Number **S1**, March, 2000.

[26]   Mandsdorf *et al.,* Industrial Hygiene characterization and Aerobiology of Resource Recovery Systems. NIOSH, Morgantown, 1982.

[27]   Crooks *et al* Airborne Microorganisms Associated with Domestic Waste Disposal. Crop Protection Division, AFRCA Institute of Arable Crops Research, Rothamasted Experimental Station, Harpenden, Herts, AL5 2JQ, pp.1-119. 1987.

[28]   Rahkonen, P Airborne contaminates at waste treatment plants. *Waste Management. Research*, 10 pp. 411-421, 1992.

[29]   Hertzman, *et al*, Upper Ottawa Street landfill site Health study" *Environmental Health Perspectives.* (**75**), pp581-97, 1987.

[30]   Paigen B, Goldman LRR. Growth of Children living near the hazardous waste site, Love Canal. *Hum Boil* **59**pp.489-508, 1987.

[31]   Berry M & Bove; Birth weight reduction associated with residence near a hazardous waste landfills, *Environ Health Perspective.* **105(8)**, pp856-861. 1997.

[32]   Kharrazi, Von Behren, J., Smith, M., A community based study of adverse pregnancy outcomes near a large hazardous waste landfill in California, *Toxicol.Ind Health*, **13** pp.229-310, 1997.

[33]   Dolk H, *et al*, Risk of congenital anomalies near hazardous waste landfill sites in Europe: The EUROHAZCON Study. *Lancet.* **352**, pp, 423-27, 1998.

[34]   Wouters, I .M. Upper Air inflammation and respiratory symptoms in domestic waste collectors, *Journal Occupational and Environmental Medicine.* **59**, pp 106-112. , 2002.

[35]   Rushton L. Health hazards and waste   management. *Br Med Bull.* **68**, pp183-97. 2003.

[36]   Hamer Geoffrey Solid waste treatment and disposal: effects on public and environment, *Biotechnology advance*, **22**, 1-2, 77-79, 2003.

Section 10
Soil and groundwater cleanup

Soil flushing by surfactant solution: pilot-scale tests of complete technology

M. Šváb[1], M. Kubal[2] & M. Kuraš[2]
[1]Dekonta, a.s., Ustí nad Labem, Czech Republic
[2]Institute of Chemical Technology Prague, Czech Republic

Abstract

This paper relates to PCBs contaminated soil flushing process, in which the aqueous solution of anionic surfactant Spolapon AOS 146 was passed through the sandy soil having an average PCBs concentration of 34.3 mg/kg of dry matter.

The laboratory part was focused in particular on the selection of a suitable surfactant for PCBs solubilisation from the soil and to the development of the soil extract processing technique which took place before the realisation of the pilot-scale demonstration of the soil flushing technology.

The experimental pilot-scale facility used consisted of a steel column (3 m length, 1.5 m diameter) containing 1.7 m^3 of polluted soil and a liquid circulation system, by which an aqueous solution of the surfactant was supplied to the soil. Spolapon solution (40 g/l) was passing through the soil column for 2.5 months. The concentration of surfactant and PCBs in the final aqueous extract was monitored during this time period. The final PCBs concentration profile in the soil was determined after stopping the liquid flow. After passing through the soil the PCBs containing aqueous extract was pumped out from the steel column bottom to a treatment unit, where it was processed by the adsorptive micellar flocculation followed by carbon black adsorption. The degree of PCBs removal from aqueous extract to coagulation sludge of 99.99% was observed. In terms of the mass balance all PCBs removed from the soil were concentrated in 14 kg of sludge with moisture content of 66%. A decrease in PCBs concentration to 15 mg/kg of dry soil was achieved, but it is certain that a decrease to level of units mg/kg could be achieved in the case of a longer duration of flushing.
Keywords: soil, PCBs, surfactant, flushing.

WIT Transactions on Ecology and the Environment, Vol 92, © 2006 WIT Press
www.witpress.com, ISSN 1743-3541 (on-line)
doi:10.2495/WM060401

1 Introduction

Proposals for remediation of soils polluted by polychlorinated biphenyl (PCB) compounds have included incineration, solidification/vitrification, and electrokinetic approaches. However, mainly because of costs, environmental constrains and efficacy many of these approaches have never been applied in the field-scale system. That is why other more efficient methods for treating PCBs contaminated soils continue to be proposed, optimized and evaluated [1].

Numerous advanced remediation technologies have been developed recently for the clean up of soils polluted by PCBs. Significant attention is devoted to the process in which the PCBs are leached out from the soil by using surfactants solutions. This technique is mostly called soil flushing when used as in-situ process or soil washing for ex-situ batch arrangement [2, 3, 4].

The positive effect of surfactant presence in aqueous phase in contact with PCBs contaminated soil to PCBs solubility has been known for many years [5]. Treatment technique based on surfactant solution flushing/washing has been found to be able to reduce hydrophobic hydrocarbons content in solid contaminated materials [6]. Especially, leaching with surfactant can effectively and cheaply substitute very expensive thermal methods (the only methods able to reduce the PCBs and other persistent species from soils), at least for less-contaminated soils.

2 Background

Many papers describing background of surfactant effect on NAPL (non aqueous phase liquid) aqueous solubilisation from soil have been published [7]. Main important property of the surfactants discussed in many studies is a critical micelle concentration (CMC). Above this concentration the surfactants become to be efficient for solubilisation of non-polar species in aqueous solution. This solubilisation takes place inside of the micelles formed. Due to the wide range of variable parameters (available surfactants, type of the soil and contaminant, contaminant-soil bond type etc), it is difficult to provide some general methodology for suitable surfactant selection in case of particular soil and contaminant. Usually, only some general principles are suggested for surfactant selection [8]. For the NAPL contaminants solubilisation from soils, only surfactants of anionic and non-ionic type are mostly considered [7]. Cationic surfactants are not suitable because of their high sorption onto mostly negatively charged sorption surfaces in the soils. Finally, each project focused on application of the surfactant solution leaching of NAPL contaminants from soil should consist of the laboratory research (in which the suitable surfactant is selected based on some experiments) followed by the pilot-scale demonstration of the technology to verify the laboratory-obtained data.

The project presented deals both with the laboratory part and with the pilot-scale demonstration of the soil flushing. The laboratory part containing also suggestion of the simple mathematical model of the studied process has been presented on ConSoil conference 2005 [9]. This study was focused mainly to the results of the laboratory research (suitable surfactant selection, determination of the CMC value, suggestion of the mathematical model) and their comparison with the model calculations of flushing in the pilot-scale demonstration.

The study presented here describes complete pilot-scale demonstration including processing of soil leachate, mass balance and evaluation of the technology.

3 Pilot-scale demonstration description

The aim of the pilot-scale demonstration was to verify efficacy, time-demands and technology aspects, which are not available in laboratory scale. Experiment was designed based on knowledge of results of preliminary laboratory part, in which Spolapon AOS 146 was selected as suitable anionic surfactant for solubilisation of PCBs from soil to the aqueous solution. Its CMC was estimated as 1.34 g/L [9]. Methods for processing of soil extract have been also tested in laboratory part involving flocculation, absorption into organic solvent and adsorption by carbon black. The best method proved to be flocculation. The soil used for pilot-scale demonstration was sampled on the same site (and same place) in Czech Republic like the soil for the laboratory experiments.

3.1 Chemicals, solvents and surfactants

Sample of anionic surfactant with trade name Spolapon AOS 146 was provided by Enaspol Velvety Co. (Czech Republic). The product contains 38 weight% of active surfactant component. Structure of the surfactant is based on linear sodium alkane-sulfonates (C14-C16).

3.2 Analytical methods used

The PCBs content in the soil was analyzed in the following way: known mass of the soil sample between 0.8 and 1.5 g (precisely measured on the analytical balances) was extracted for 4 hours by hexane at the temperature of its boiling point (Soil was placed between the boiling test-tube and back flow condenser of hexane, in this way, the soil extraction by the fresh solvent had been achieved.). The soil extract in the boiling test-tube containing whole PCBs content of the soil was then passed through a column with length of 10 cm and diameter of 3 mm filled with an activated Florisil. The extract was filled up to the volume of 10 ml by the hexane and PCBs concentration was measured on the gas chromatograph with the ECD detector. Only six congeners (number 28, 52, 101, 153, 138, 180) of the PCBs were quantitatively determined in accordance with the Czech legislation. PCBs content in the sample represent sum of those six congeners concentrations.

The PCBs concentration in the liquid aqueous samples was determined after extraction with 10 ml of hexane for 2 hours with shaking, separation of the hexane phase sample and purification of it by use of the same column of Florisil as in the case of the soil samples extracts.

Analysis of anionic surfactant content in the liquid aqueous samples was done by the volumetric method in two-phase system chloroform-water. Methylene blue solution was used as indicator while solution of cationic substance Septonex was a volumetric agent. At the beginning of the titration in the presence of anionic surfactant blue color remains in the chloroform phase. As titration (in term of addition of the volumetric solution) continues, blue color passes into the water phase. Point of equivalence was determined as decolorizing of the chloroform phase.

3.3 Experimental facility

Technological arrangement shown schematically in the Figure 1 was constructed for purpose of the soil flushing demonstration. Most important part of the technology was the flushing tank with diameter of 1.46 m and height of 3 m which has been equipped with the filtration bed on the bottom part. It consisted of a layer of the gravel covered by layer of sand and geotextile. The flushing tank was placed in the retaining tank for collection of soil leachate equipped by pump. The pump was able to pump the soil leachate either to the tank for waste water (not displayed in the Figure 1) or to the tank for collection of soil extract. Bottom of the retaining tank was equipped with valve enabling to discharge tank completely and then to measure flushing solution flow through the flushing tank (in assumption that flow through valve and flushing tank should be equivalent). In the upper part of the flushing tank was installed an electrode system connected with control unit and pump of fresh flushing solution installed near the tank with flushing solution. This system was able to control layer of flushing solution in the flushing tank on constant level.

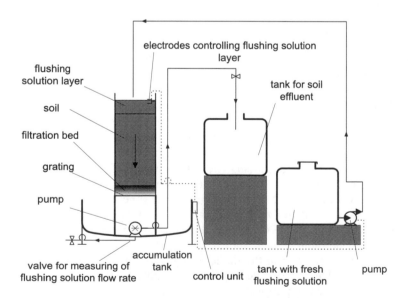

Figure 1: Pilot-scale demonstration facility scheme.

3.4 Process description

Soil used was sandy soil with organic carbon content of 1.45%, bulk density of 1.63 g/cm3 and estimated effective porosity of 40%. At the beginning of the pilot-scale demonstration, about 1.7 cubic meter of the soil was mixed with water to form a dense suspension, which was then introduced into the flushing column by use of the barrels and forklift. After settling down the soil formed 1 m high layer in the column. Rest of the water set-off from the soil was removed from the flushing tank by the pipe (by use of the effect of layers levels difference). Samples of the initial soil were taken from the homogenized suspension before introducing it to the flushing column to know an input PCBs concentration. Initial PCBs concentration founded was about 34 mg/kg of dry soil.

Flushing solution concentration of anionic surfactant Spolapon AOS 146 was 40 g/L. Relatively high concentration was chosen due to the results of laboratory experiments and with respect to expected duration of the pilot-scale demonstration needed to reach some demonstrable changes in the PCBs concentrations in soil. The flushing solution was pumped up to the soil in the flushing column to make a layer of 30 cm. It was kept constant during the flushing process by electrodes connected with control unit. In fact, by operation, the flushing pilot-scale demonstration was started.

During the process, one or twice per week, flow of flushing solution through the soil layer was measured after pumping out the soil leachate from retaining tank either to the tank for waste water or to the tank for processing of it. Content of the retaining tank was pumped to the wastewater tank during the beginning phase of the soil flushing. Soil sorption capacity was not yet saturated and because of it, concentrations of both the surfactant and PCBs were very low. Moment, in which it was needed to pump leachate into the tank for processing was observed only visually – from leachate color and foam, what gives evidence that concentrations of both the surfactant and PCBs became to increase. When the retaining tank was empty, valve in its bottom was opened for at least 1 hour to get the same flow through valve and flushing column. After that, flushing solution was measured by use of volumetric cylinder. The soil leachate collected in the cylinder was then taken as sample. In this way, the flushing process was simulated for approximately 2 months.

Flushing process was terminated by removal of the flushing solution layer from the top of the flushing tank. After the predominant part of the flushing solution drained out from the soil layer, the final sampling of the flushed soil has been carried out. To this purpose, three holes were made into the soil layer in the flushing tank by use of the equipment, which allows cutting the sample of the whole soil profile. Three sampled profiles were divided into 8 sections and then, mixed sample has been prepared from the same sections of each profile. In these 8 samples, final residual PCBs concentration was determined.

Soil leachate in the tank for processing of it was processed by patented method based on coagulation [10]. Dose of ferric trichloride was 0.017 mol/L. After addition of this agent and neutralisation by addition of calcium hydroxide to reach suspension of pH 7, period of fast mixing (about 600 rpm) for 5 minutes was followed by period of slow mixing (about 150 rpm) for approx. 10 minutes. Simple stirrer (usually used for stirring of paints) powered by handy electric drill was used for mixing of the suspension. After sedimentation for 2 days, filtrate was separated from the sludge by filtration through textile bags, type PM10MY (Czech producer). Before filtration, it was possible to directly remove approx. 70-80% of liquid above the settled sludge.

4 Results

First important information provided by pilot-scale demonstration was flushing solution flow rate through the soil layer during the demonstration period. This flow rate was expected to be decreasing with time according to the theoretical assumption. Flow rate in beginning phase of the flushing was 2 L/hour and it decreased to 0.6 L/hour during the period of 2 months. It was possible to interpolate the flow rate lowering by powered function. Further flow rate decreasing seemed to be slow, so we could expect the limiting constant flow rate in case of longer flushing should range between 0.5–0.55 L/hour. This flow rate should be constant for potential further long duration of flushing.

Figure 2 shows both the PCBs and the surfactant concentration in the soil leachate during the flushing. There is evident relation between concentration of the surfactant and of the PCBs in the solution. It is clear, that presence of the surfactant directly determines higher concentration of the PCBs in the solution in comparison with the pure water. It is expectable, that concentration of the PCBs would rise above to the final value measured in the pilot-scale demonstration. It is evident from this fact, that decontamination is faster and more efficient (for example in term of flushing solution consumption needed to remove unit of PCBs from soil) with process time because of the increasing of the PCBs concentration in leachate. It remains to comment why the concentrations became to rise before one pore volume of the flushing solution flowed through the soil, especially when sorption of the surfactant onto soil particles has been expected from laboratory experiments (approx. 8 g/Kg). Probably, it was caused by combination of two facts. In first, Figure 2 was constructed for estimated soil porosity of 40%, but even little change in the porosity estimate strongly affects the pore volumes in which the concentrations became to rise. We can also expect that the flow of solution through the soil in the pilot-scale experiment was not piston-flow, e.g. some preference pathways could occurred in the soil layer. Those two effects probably predominates the sorption, from which we expected increase of both (PCBs and surfactant) concentration later. Finally, it was positive observation, that PCBs became to be removed from the soil sooner than it could be expected from surfactant sorption behavior.

The Figure 3 shows final PCBs concentration profile after the flushing was terminated. It is obvious from the profile that PCBs were concentrated in the bottom part of the soil layer, what has been caused by the surfactant adsorption onto soil in the top part of the layer during the beginning phase of the flushing. Concentration of PCBs in the top part of the layer was probably affected by gravitation separation of the soil particles in the column that occurred during introducing the soil suspension into the column. Finest particles, which are usually most contaminated due to their larger sorption surface, remained on the top of the layer. It was the reason why the residual PCBs concentration in the soil (top part of the column) seemed to be unexpectedly high (about 15 mg/kg). This observation was confirmed by sieving analysis of the soil used. There is a relevant reason to await the residual PCBs soil concentration lower than the observed value at the top part of the column. It can be expected that in case of sufficiently long flushing, concentration in whole profile of the soil should decrease on level equal to the top part of the soil layer. In this way, decontamination efficacy observed was about 56%, but it can be expected that efficiency could be substantially higher (>90%). Time needed to reach this efficiency in whole profile can be estimated between 6–12 months (estimated simply from mass balance of PCBs).

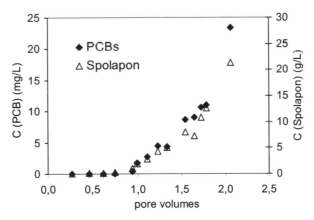

Figure 2: Concentration of both PCBs and surfactant in soil leachate during pilot-scale demonstration of soil flushing.

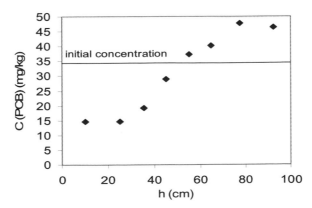

Figure 3: Residual PCBs concentration profile in the soil layer after the pilot-scale flushing was terminated.

Mass balance of the flushing showed the recovery value of the PCBs of 98.8% what was significantly better than it could be expected before. This high recovery was surprisingly better than it has been expected. Total volume of the soil leachate was 1.4 cubic meters. Of this volume, about 500 L was directly pumped to the tank for wastewater during the beginning phase of the experiment (see Figure 2), because both concentrations were very low. Rest of soil leachate, 900 L, was collected in the plastic tank with volume of 1 cubic meter. Concentrations of PCBs and of Surfactant in the soil leachate (solution in tank) were 8.1 mg/L and 8 g/L, respectively. The soil leachate was processed by above described patented coagulation method. Complete information about the processing of soil leachate are summarised in Table 1.

Table 1:   Summarisation of soil leachate processing.

volume of leachate processed	~900 L
initial concentration of PCBs	8.1 mg/L
initial concentration of surfactant	8 g/L
residual PCBs concentration in filtrate	24 ng/L
residual surfactant concentration in filtrate	2.2 g/L
mass of produced sludge	14.2 kg
moisture of sludge	66%
concentration of PCBs in sludge	1145 mg/kg (dry sludge)

It is obvious from Table 1, that PCBs were removed from the soil leachate into the sludge with very high efficiency exceeding 99.99%. It remains to refer about the analytical complication caused by presence of phthalates in the filtrate (from plastic tank). Therefore, it was necessary to eliminate this disturbing effect by modified analytical process in specialised analytical laboratory. From Table 1 also results that PCBs removed from 2.7 tons of soil were concentrated into 14 kg of the sludge. The sludge was then burned in the incinerator (in Ostrava, Czech Republic).

5   Conclusions

The pilot-scale demonstration study confirmed that it is possible to remove the PCBs from the real contaminated soil by the flushing with surfactant solution. Decontamination efficiency proved was 56%, but it could be higher than 90% in case of longer duration of the flushing process. Time needed to reach this efficiency can be estimated as 6–12 months. Water consumption can be estimated to be approximately 2–3 cubic meters of water per 1 cubic meter of soil. However, it should be possible to recycle water in the process because the residual PCBs concentration in processed soil leachate was very low (24 ng/L). So, total fresh water consumption should be relatively low due to its recirculation. Costs of decontamination of 1 cubic meter of soil can be estimated between 200–250 Euro/t (including wastes utilisation, costs of investments are not included). In principle, it could be possible to operate the technology both in-situ and ex-situ. We can expect, that ex-situ method should be more efficient, faster and safer towards surrounding environment, but more cost demanding (due to investments). Decision about the technological design of flushing process has to be done with respect to the particular conditions – soil permeability, hydraulic conditions on locality, local legislation etc.

Technology of soil flushing by surfactant aqueous solution could be serious competition to expensive thermal methods especially for less-contaminated soils (up to approximately hundreds mg/kg).

References

[1] Wu O., Marshall W.D.: Approaches to the Remediation of Polychlorinated Biphenyl (PCB) Contaminated Soil – a Laboratory Study, Journal of Environmental Monitoring, 3 (2001), 281-287.

[2] Otten A., Alphenaar A., Pijls CH., Spuij F., Wit H.: In Situ Soil Remediation, Kluwer Academic Publishers, Dordrecht 1997.

[3] Roote D. S.: In Situ Flushing, GWRTAC Technology Overview Report, TO-97-02, 1997.

[4] Mann J.M.: Full-scale and pilot-scale soil washing, Journal of Hazardous Materials 66 (1999), 119-136.

[5] Haigh S.D.: A Review of the Interaction of Surfactants With Organic Contaminants in Soil, The Science of the Total Environment 185 (1996), 161-170.

[6] Mulligan C.N., Yong R.N., Gibbs B.F.: Surfactant-enhanced Remediation of Contaminated Soil: a Review, Engineering Geology 60 (2001), 371-380.

[7] Haigh S.D.: A review of the interaction of surfactants with organic contaminants in soil, The Science of the Total Environment 185 (1996), 161-170.

[8] Jafvert Ch.T.: Sufactants/Cosolvents, Technology Evaluation Report TE-96-02, 1996 Purdue University.

[9] Svab M., Raschman R., Kubal M.: PCBs Removal from Contaminated Soil by means of Surfactants: Pilot-scale Tests, ConSoil 2005, 3.-7. 10.2005, Bordeaux, France.

[10] Svab M., Raschman R.: Method for Processing of Soil Leachate Containing PCBs and Anionic Surfactant, patent nr. 294812, 21.1.2005, Czech Republic.

Optimization of *in situ* emplacement of nano-sized FeS for permeable reactive barrier construction

T. M. Olson & J.-H. Lee
Department of Civil and Environmental Engineering,
University of Michigan, Ann Arbor, Michigan, USA

Abstract

Construction techniques for permeable reactive barriers (PRBs) have commonly involved trench-and-fill methods. These strategies, however, are limited in use to shallow groundwater remediation applications. To extend the technology to deeper, less accessible zones of contamination, *in situ* emplacement methods of the reactive media are required. In this paper, development studies of an *in situ* PRB construction method are presented that rely on the injection and deposition of nano-sized iron sulphide particles. Optimal chemical conditions for depositing FeS particles in sand bed media were ascertained that maximize surface coatings and minimize reductions in hydraulic conductivity. Our studies indicate that moderately alkaline injection conditions are needed to establish optimal coverage of FeS on sand. The constraints necessitating these chemical conditions are discussed in terms of FeS surface charge characteristics and particle filtration theory.
Keywords: permeable reactive barriers, iron minerals, construction methods, trace metals, particle deposition, metal sulphides, zero valent iron, groundwater remediation.

1   Introduction

Permeable reactive barriers (PRBs) have been applied with increasing frequency as a technology to clean up or manage groundwater contamination. Such passive, *in situ* methods of treatment and/or immobilization of contaminants offer a potentially economically attractive option for this type of waste management. Among the many reactive media which have been used, granular forms of zero

WIT Transactions on Ecology and the Environment, Vol 92, © 2006 WIT Press
www.witpress.com, ISSN 1743-3541 (on-line)
doi:10.2495/WM060411

valent iron (ZVI) are the most common, comprising at least 45% of the media in PRB applications [1]. The use of economical sources of ZVI, however, is limited to trench-and-fill PRB construction methods and relatively shallow installations. Deeper aquifer environments, however, are more likely to present the desired anoxic conditions that are required for an effective PRB, but methods to construct PRBs for these locations are lacking. An *in situ* method of PRB construction involving the injection and deposition of colloidal reduced iron minerals on native subsurface materials has been examined in this study. Iron sulphide was selected as the reactive media since it can be readily synthesized as a nano-sized colloidal suspension and because it is a versatile sequestering agent for many contaminants.

2 FeS media properties

2.1 Reactivity

Iron sulphide minerals have a high capacity to sequester 1) non-redox active metals by forming insoluble metal sulphides, and 2) redox-active metals, by reduction first, followed by precipitation of the reduced metal as mixed metal sulphide solid. Redox active metals such as As(V) for example, can be reduced by FeS to As(III) and precipitated as the solid orpiment (As_2S_3) [2]. Although the reactivity of a FeS barrier would ultimately degrade as FeS is converted to other sulphide minerals, opportunities to microbially regenerate FeS surfaces are possible. Sulfate reducing bacteria can oxidize organic matter and reduce sulfate to sulphide and produce FeS if sufficient reduced iron is present. Such microbially mediated FeS production has been shown to produce mackinawite (FeS) [3]. Based on the versatile metal ion sequestration potential and the novel biogenic opportunities for FeS regeneration, the media was selected as the reactive barrier media for development of *in situ* emplacement methods.

2.2 Nano-particle synthesis and characterization

Nano-particulate FeS suspensions were prepared for use in this study in an anoxic chamber according to a method outlined by Butler and Hayes [4]. The resulting particles were harvested by centrifugation in tightly sealed bottles, sequential rinsing with deoxygenated distilled water, freeze-drying under vacuum, and storage of the dry powder in a nitrogen atmosphere. Powder X-ray diffraction analysis established that the mineral was a poorly crystalline mackinawite phase, with slightly greater-than-stoichiometric sulphur content. Photon correlation spectroscopy measurements of the re-suspended FeS, performed in an anoxic glovebox, revealed that the particles had a mean diameter of 3 nm that remained stable for over two hours.

Based on the synthesis method, the ambient pH of the resuspended nano-particles in distilled water was relatively alkaline. Preparation of a 1 g/L FeS suspension in distilled water, for example, yielded a solution pH of 10.3. Borate buffer and/or HCl were required to establish more circumneutral pH conditions in column experiments. From literature studies the FeS point of zero charge, pH_{pzc}, i.e., the pH at which positive and negatively charged surface groups are equal in concentration, appears to be relatively uncertain. Some reports indicate its pH_{pzc} may be as low as 3 [5]. The nanoparticulate FeS used in this study may be significantly less acidic, however. Unpublished electroacoustic characterizations of FeS surface charge by another research group, using the same synthesis method used herein, indicate its pH_{pzc} is about 5 [6]. Over the pH range of most groundwater environments, the FeS, therefore, would have an overall negative surface charge.

3 FeS deposition experiments in sand columns

3.1 Column description and apparatus

The deposition of FeS particles was studied as a function of solution chemistry in packed columns of clean quartz sand. Prior to use, the sand (mean diameter approximately 175 μM) was treated with sequential rinses of sodium dithionite and hydrogen peroxide solutions to remove metals and organics. The sand packed columns, 2.6 cm in diameter and 25 cm in length, had an average porosity of 0.34. A schematic diagram of the column experiment apparatus is shown in Figure 1.

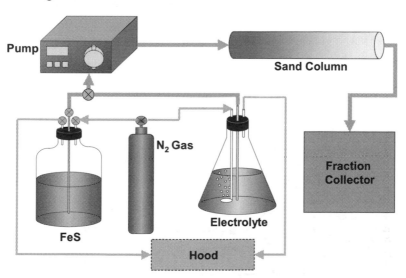

Figure 1:   Schematic diagram of column apparatus.

3.2 FeS injection experiment and analysis procedures

Prior to injection of the FeS, the sand media was initially equilibrated with two pore volumes of the background electrolyte. To minimize oxidation, the influent

1 g/L FeS suspension was maintained under a nitrogen atmosphere while it was pumped through the sand column. Ten pore volumes of the suspension were processed in each experiment. Borate buffers and HCl were used to adjust and fix the pH of the suspension over the range of 8.2 to 10.3. Sodium chloride was added to adjust the ionic strength.

During the FeS deposition experiments, column effluent samples were collected in a fraction collector and analyzed by ICP-MS for their iron content. After the experiment, two pore volumes of the background electrolyte were applied to remove any suspended particles, the sand was sectioned by hydraulic extrusion, and the deposited FeS concentrations on the sand were determined by extraction of the iron in strong acid, and iron analysis by ICP-MS.

3.3 Results and data analysis

Chemical electrolyte conditions were selected on the basis of particle stability and pH buffering requirements. The solution pH range considered, pH 8.2 to 10.3, was determined at the acidic end of the range by the stability of a 1 g/L FeS suspension. At pH conditions below 8.2, the suspensions were observed to aggregate significantly. The ionic strength range used in the column experiments was similarly determined by suspension stability and pH buffer requirements. The minimum ionic strength when buffers were added was approximately 0.025 M. The maximum ionic strength used was 0.05 M, due to suspension instability at higher salt contents.

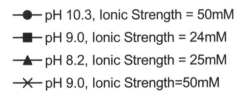

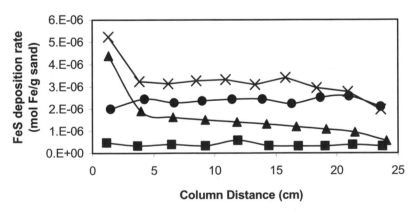

Figure 2: Effect of pH and ionic strength on deposited iron concentrations as a function of column depth after injecting 10 pore volumes of a 1 g/L FeS suspension.

Profiles of the deposited FeS concentrations as a function of column depth are presented for a selected set of pH and ionic strength conditions in Figure 2. As these results indicate, optimum surface coverage was obtained at pH 9.0, with 0.05 M ionic strength. They furthermore suggest that deposited concentrations of the FeS throughout the column were reasonably uniform, although somewhat higher concentrations near the inlet were observed in some cases. Mean sand coverages and the fractional efficiency of FeS are reported in Table 1.

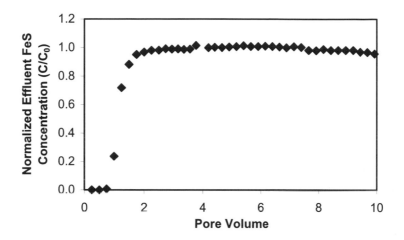

Figure 3:   Effluent breakthrough curve deposition experiment conducted at pH 8.2, ionic strength = 0.025 M.

Table 1:   FeS coating concentrations and fractional deposition in sand column experiments.

Coverage parameter	pH 10.3, I=0.05M	pH 9.0, I=0.025M	pH 9.0, I=0.05M	pH 8.2, I=0.027M
Mean deposited Fe conc. (mole Fe/g sand)	2.34E-06	3.83E-07	3.24E-06	1.60E-06
% of injected FeS deposited	11.0%	1.8%	15.3%	6.7%

In each of the experiments in Figure 2, particle breakthrough concentrations were also observed to be complete (i.e., effluent concentrations achieved a steady state concentration nearly equal to the influent concentration). A sample breakthrough curve illustrating this behaviour is shown in Figure 3 at pH 8.2,

I = 0.025M.   The achievement of complete breakthrough suggests that interactions of depositing particles with the *coated* sand media are sufficiently unfavourable (electrostatically repulsive) and that saturation FeS coverage of surface is obtained.

Attempts to emplace FeS particles at pH 8.2 at a higher ionic strength of 0.05 M, however, resulted in incomplete particle breakthrough and eventual plugging of the column inlet region. Influent suspensions at these electrolyte conditions were also relatively unstable. Given the instability of the suspensions at pH < 8.2 in the presence of modest concentrations of background electrolyte, the FeS surface pH_{pzc} is likely to be closer to the more weakly acidic values discussed above.

4   Conclusions and implications

To achieve relatively uniform FeS sand coverage across a PRB zone and avoid inlet plugging during emplacement, requires chemical conditions that are moderately alkaline (pH > 8.2) and relatively low in ionic strength ($I_{max} = 0.05$ M). Under these conditions, FeS—FeS interactions are sufficiently unfavourable while FeS-sand interactions are sufficiently favourable. In many groundwater systems these alkaline conditions may not be practical, especially where an adjustment to an alkaline pH causes the precipitation of new solid phases. Additional strategies are needed therefore to extend the chemically feasible 'window' for FeS emplacement to lower pH. On-going research in our laboratory is focused on the development of such strategies.

References

[1] Scherer, M.M., Richter, S., Valentine, R.L. & Alvarez, P.J.J., Chemistry and microbiology of permeable reactive barriers for groundwater clean up, *Crit. Reviews in Environ. Sci. and Technol.*, **30(3)**, pp. 363-411, 2000.
[2] Rochette, E.A., Bostick, B.C., Li, G., & Fendorf, S., Kinetics of arsenate reduction by dissolved sulphide, *Environ. Sci. Technol.*, **34(22)**, pp. 4714-4710, 2000.
[3] Vaughan, D.J., & Lennie, A.R., The iron sulphide minerals – Their chemistry and role in nature, *Science Progress*, **75(298)**, pp. 371-388, Part 3-4, 1991.
[4] Butler, E.C., & Hayes, K.F. Effects of solution composition and pH on the reductive dechlorination of hexachlorethane by iron, *Environ. Sci., Technol.*, **32(9)**, pp. 1276-1284, 1998.
[5] Bebie, J., Schoonen, A.A.M., Furhrmann, M., & Strongin, D.R., Surface charge development on transition metal sulphides: An electrokinetic study, *Geochimica Cosmochimica Acta*, **62**, pp. 633-642, 1998.
[6] Gallegos, T., Personal communication, 19 January 2006, Doctoral student, Civil & Environ. Engineering, Univ. of Michigan, USA.

Improved DRASTIC method for assessment of groundwater vulnerability to generic aqueous-phase contaminants

N. Z. Jovanovic[1], S. Adams[1], A. Thomas[1], M. Fey[2],
H. E. Beekman[3], R. Campbell[3], I. Saayman[3] & J. Conrad[4]
[1]*Department of Earth Sciences,*
University of the Western Cape, South Africa
[2]*Department of Soil Science, University of Stellenbosch, South Africa*
[3]*CSIR, Stellenbosch, South Africa*
[4]*GEOSS, Stellenbosch, South Africa*

Abstract

Increased use and protection of groundwater resources are seen as possible solutions to mitigating water scarcity in arid and semi-arid regions. The DRASTIC index method is one of the most commonly used approaches to assess groundwater vulnerability to pollution. However, this method has been criticized in the past due to its subjectivity as well as the failure to account for some important hydrogeological characteristics (e.g. multi-layer vadose zone and preferential flow) and specific properties of contaminants (e.g. sorption and decay). These problems were addressed in this study with the objective of improving the DRASTIC method for assessment of groundwater vulnerability to generic aqueous-phase contaminants. Literature data and laboratory measurements were used in order to define categories and weighing factors for hydrogeological characteristics and specific contaminant properties. The new DRASTIC method developed in this study provides an improved categorization of the impact of the vadose zone, which accounts for the following factors: multi-layer vadose zone, based on site-specific conceptual models; hydraulic properties of the unsaturated zone (flow mechanism, drainage and travel time); and specific chemical properties (sorption and decay). The information was packaged in a user-friendly format for rating groundwater vulnerability. The method can be used for applications in site-specific environmental impact assessments for new developments, for regional groundwater vulnerability assessment as well as in integrated water resources management.
Keywords: decay, DRASTIC, groundwater vulnerability, hydraulic conductivity, preferential flow, recharge, sorption, travel time, vadose zone.

WIT Transactions on Ecology and the Environment, Vol 92, © 2006 WIT Press
www.witpress.com, ISSN 1743-3541 (on-line)
doi:10.2495/WM060421

1 Introduction

Increased use and protection of groundwater resources are seen as possible solutions to mitigating water scarcity in arid and semi-arid regions. Groundwater all over the world is becoming a natural resource of strategic importance due to its limited availability, quality deterioration, increasing demand and limited replenishment. In South Africa, water resources are generally scarce and unevenly distributed, and a large number of towns and rural settlements depend on groundwater for their drinking water supply and development. Previous research aimed at establishing a groundwater protection strategy (Sililo et al. [1]). One of the main outcomes of this research was that the vadose (unsaturated) zone, in particular the soil, should be seen as the "first line of defense" to transport of pollutants from overlying land-based sources to groundwater.

Groundwater vulnerability to contamination is defined as the tendency or likelihood for contaminants to reach a specified position in the groundwater system after introduction at some location above the uppermost aquifer (National Research Council [2]). The degree of groundwater contamination depends on the intrinsic hydrogeological characteristics and the physio-chemical properties of specific contaminants. Different types of pollutants are attenuated to a different degree depending on the characteristics of the site and speciation. Knowledge is therefore required on the properties of the porous medium through which the pollutant travels, the properties of the pollutant as well as the physical, chemical and biological processes (Sililo et al. [1]).

Several methods are available for assessing groundwater vulnerability. These were classified as index and overlay methods, process-based models and statistical methods (National Research Council [2]). Overlay and index methods are based on combining maps of various physiographic attributes (e.g. geology, soils, depth to water table) of the region and assigning a numerical index or score to each attribute. Process-based simulation models include analytical or numerical solutions to mathematical equations that represent processes governing contaminant transport. Statistical methods incorporate a probability of contamination as the dependent variable. Burkart et al. [3] presented examples of application of statistical, overlay and index, as well as process-based modeling methods for groundwater vulnerability assessment to a variety of data from the Midwest United States. Burkart and Feher [4] developed a strategy for regional groundwater vulnerability assessment that integrates elements of overlay, process-based and statistical methods.

Amongst the methods for assessment of groundwater vulnerability, the DRASTIC index (Aller et al. [5]) is the most commonly used in South Africa. This method makes use of the hydrogeological factors of an area in order to determine the relative groundwater vulnerability to contaminants. These hydrogeological factors, making up the acronym DRASTIC, are depth to water table, net recharge, material of the aquifer, soil properties, topography, properties of the vadose zone and hydraulic conductivity. Each factor is assigned a weight based on its relative importance to groundwater contamination potential, as well as a rating for different ranges of values. The DRASTIC index is then computed as the sum of the products of rating and weight for each factor. The DRASTIC index is generally built into Geographic Information Systems (GIS) – based maps to facilitate planning and management of groundwater protection, where each hydrogeological setting is a mappable unit with common hydrogeological factors (Thirumalaivasan et al. [6]). This practice, aimed at indicating areas at low, moderate or high risk from groundwater contamination, was applied in Israel (Secunda et al. [7]), Japan (Babiker et al. [8]), Jordan (Al-Adamat et al. [9]) and New Zealand (McLay et al. [10]).

The DRASTIC method has been criticized in the past due to its subjectivity as well as the lack of some important hydrogeological characteristics and specific properties of contaminants. For example, Dixon [11] indicated that groundwater contamination potential maps were more consistent with field data when soil structure was taken into consideration in a study conducted in Woodruff County in the Mississippi Delta region of Arkansas. Sandersen and Jorgensen [12] explained that, even when aquifers are deep-seated and appear to be well protected, preferential flow paths for downward transport of contaminated water from shallow aquifers may occur. Melloul and Collin [13] suggested that both vertical and lateral flow play an important role in groundwater contamination. Worrall et al. [14] found that interaction between site and chemical factors represents the most important process in the occurrence of pesticides in groundwater, based on multi-annual monitoring datasets from the United Kingdom and Mid-Western United States. Worrall and Kolpin [15] indicated that the best-fit model to predict the occurrence of herbicides in groundwater of the Mid-West United States combined organic carbon content, percentage sand content and depth to the water table with molecular descriptors representing molecular size, molecular branching and functional group composition of the herbicides.

In this study, the aim was to modify the original DRASTIC method (Aller et al. [5]) to account for some important hydrogeological characteristics (e.g. multi-layer vadose zone and preferential flow) and specific properties of contaminants (e.g. sorption and decay) in order to improve, in particular, the reliability of the rating "I" (impact of the vadose zone). An additional aim was to provide a more detailed (less subjective) description of the ratings in order to adapt the DRASTIC method for assessment of groundwater vulnerability to generic aqueous-phase contaminants under South African environmental conditions.

2   Improved DRASTIC method

The rating of the vadose zone (high, medium and low) is based on a combination of factors that contribute to the likelihood of contaminants reaching the saturated zone following the path of aquifer recharge. The ability of the vadose zone to attenuate and/or prevent any contaminant from reaching groundwater depends on the following factors: thickness of the unsaturated zone, hydraulic properties and flow mechanism, recharge, travel time, sorption and decay. The approach used in order to improve DRASTIC, involved the description and quantification of these factors, as well as the provision of guidelines to assist in quantifying their relative importance. The factors are discussed individually below.

The thickness of the unsaturated zone depends on the nature of both aquifer (confined, unconfined, leaky and semi-confined) and regolith material (unconsolidated, consolidated, weathered, consolidated fractured or a combination). Table 1 summarizes the impact of the thickness of the unsaturated zone and the type of media on groundwater vulnerability. The ranges of values reported in Table 1 were estimated based on experience.

Table 1: Unsaturated zone thicknesses, type of media and resulting impact on groundwater vulnerability.

Unsaturated zone medium		Thickness (m)		
Unconsolidated material	Gravel	> 50	30-50	0-30
	Clean sand	> 50	30-50	0-30
	Silty sand	> 30	15-30	0-15
	Silt	> 15	5-15	0-5
	Clay	> 5	2.5-5	0-2.5
Consolidated fractured medium		> 30	5-30	0-5
Leaky aquifers		> 30	5-30	0-5
		Vulnerability impact		
		Low	Medium	High

The dominant mechanism of aqueous-phase contaminants transport in the unsaturated zone is by advection along wetting front edges (with water content close to saturation) during the infiltration process. Due to this transport mechanism, the inaccessibility, high cost of measurement and spatial variability, saturated hydraulic conductivity is often used as a substitute for unsaturated hydraulic conductivity. Typical values of saturated hydraulic conductivity for most porous media are easily accessible (Freeze and Cherry [16]). Unsaturated hydraulic conductivity can be estimated if additional hydraulic properties of the medium are known, e.g. the water retention curve (Van Genuchten et al. [17]). The vertical hydraulic conductance is then calculated as a function of hydraulic conductivity (Eimers et al. [18]). The ranges of values for the impact of vertical hydraulic conductance on groundwater vulnerability are defined in Table 2.

Preferential flow is a complicating factor in estimating travel times of contaminants in dual porosity environments. Dual porosity consists of two interacting pore regions with different hydraulic properties, the one associated with the macro-pore or fracture network, and the other with micro-pores inside soil aggregates or rock matrix blocks. It is generally assumed that flow occurring through preferential paths is fast (Table 2), resulting in lower contaminant attenuation through, for example, sorption and decay.

The definition of recharge used in this study is the amount of rainfall that reaches the saturated zone, either by direct contact in the riparian zone or by downward percolation through the unsaturated zone (Rushton and Ward [19]).

Aquifer recharge depends on factors such as groundwater depth, climate, geology (lithology and structures), geomorphology, vegetation, soil conditions and antecedent soil moisture. A variety of methods can be applied for the estimation of recharge in semi-arid conditions, e.g. CMB (Chloride Mass Balance), CRD (Cumulative Rainfall Departures), EARTH model (Xu and Beekman [20]), as well as the "Qualified Guess" method based on information on soil, vegetation, geology as well as South African maps of groundwater recharge, recharge of soil water into the vadose zone and harvest potential. The rating of recharge rate is defined in Table 3 for various unsaturated zone thicknesses.

Table 2:   Unsaturated zone flow mechanism, hydraulic properties and resulting impact on groundwater vulnerability.

Flow mechanism	Vertical hydraulic conductance ($m^2\,d^{-1}$)		
Matrix	< 45	45-9000	> 9000
Preferential			> 1
	Vulnerability impact		
	Low	Medium	High

Table 3:   Unsaturated zone thickness, recharge and resulting impact on groundwater vulnerability.

Thickness (m)	Recharge ($mm\,a^{-1}$)		
0-5	0-1	1-5	> 5
5-30	0-5	5-10	> 10
> 30	0-10	10-100	> 100
	Vulnerability impact		
	Low	Medium	High

Table 4:   Unsaturated zone media type, travel time to water table and resulting impact on groundwater vulnerability for a 5 m thick unsaturated zone.

Unsaturated zone medium		Travel time		
Unconsolidated material	Gravel			< 1 h
	Clean sand		1 mo	1 d
	Silty sand	> 1a	< 1 a	< 1 mo
	Silt	> 0.5 a	< 0.5 a	< 1 mo
	Clay	>> 1 a		
Consolidated fractured medium		> 1 a	1 mo	1 h – 1 d
		Vulnerability impact		
		Low	Medium	High

Travel time is the time it takes a contaminant to move from the soil surface to the groundwater. This can be calculated as the ratio of the travel distance divided by flow velocity, where flow velocity is the volumetric flux divided by the volumetric water content. Alternatively, Foster and Hirata [21] suggested simple equations to calculate travel time as a function of unsaturated zone thickness, effective porosity and saturated hydraulic conductivity (gross surcharging conditions), or specific retention and annual infiltration (natural infiltration conditions). The travel time rating for various vadose zone media is defined in Table 4. The database can be further expanded to include different thicknesses of the vadose zone.

Extensive laboratory experiments were undertaken in order to correlate sorption of different groups of contaminants to soil properties. The experiments made use of an extensive database of soil samples collected in South Africa over several decades. Sorption of representative cationic metals (Cu and Zn) and anions (SO_4 and PO_4) was measured on a large number of soil horizons. The main outcome of the laboratory experiments was that the diagnostic horizons, as defined by the Soil Classification Working Group [22], are not good predictors of sorption. However, six soil properties were found to somewhat correlate to sorption, namely clay content, organic matter content, pH, exchangeable basic cations, extractable Fe and Al. It was observed that sorption is variable within certain ranges or maximum values of these properties, and unlikely beyond these limiting values. The impact of sorption on groundwater vulnerability was therefore defined as in Table 5.

Table 5: Contaminant species, sorption in different soil types and resulting impact on groundwater vulnerability.

Contaminants	Soil properties		
Cationic (inorganic and polar organic)	Thick, clayey profiles, margalitic soils; strongly calcareous clays; eutrophic peats	All other soils	Dystrophic sands low in humus
Anionic (inorganic and polar organic)	Deep, dystrophic, ferrallic clays	All other soils	Eutrophic sands
Organic (non-polar)	Deep humic clays and peats	All other soils	Pure sands low in humus
	Vulnerability impact		
	Low	Variable	High

Table 6: Unsaturated zone sorption capacity, contaminant persistence and resulting impact on groundwater vulnerability.

Sorption capacity	Half-life		
Low	< 1 h	1-24 h	> 24 h
Variable	< 1 d	1-15 d	> 30 d
High	< 15 d	15-50 d	> 50 d
	Vulnerability impact		
	Low	Medium	High

Decay, expressed in terms of half-life, is related to organic contaminants subjected to physical (e.g. photolysis), chemical (e.g. hydrolysis) and biological (e.g. microbial) degradation. Half-lives of specific contaminants may vary by orders of magnitude depending on environmental factors (e.g. microbiological activity, pH, moisture, temperature etc.). A database of half-lives and other properties of contaminants was compiled by Usher et al. [23]. The impact of decay on groundwater vulnerability was defined in Table 6.

3 Software application

The rating of the factors relevant to the attenuation of contaminants in the vadose zone was incorporated into a user-friendly Excel-based calculator of the improved DRASTIC index. An example printout of the main menu of the improved DRASTIC calculator is shown in Figure 1. The purpose of the main menu is to summarize scores for each factor based on rating and weighing. The ratings for each factor are determined in sub-menus that include guidelines and theoretical description. The improved DRASTIC calculator also allows for a multi-layer vadose zone.

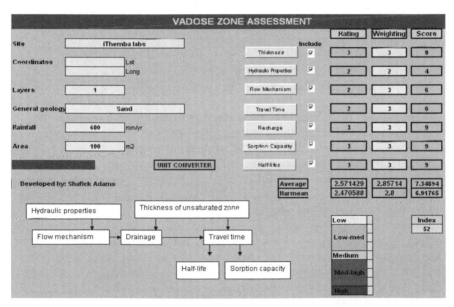

Figure 1: Example printout of the main menu of the Excel-based improved DRASTIC index calculator.

The vulnerability of an aquifer to contamination originating at the soil surface can be assessed using data that are readily available, that can be calculated or that can be estimated using the information provided in the database of the improved DRASTIC calculator, based on a step-by-step procedure and conceptual models for specific sites. The first step in the procedure is to define the unsaturated zone thicknesses using water level depths and national groundwater databases, as well as the nature of the unsaturated zone using borehole logs. The rating for the unsaturated zone thickness is then selected (Table 1). The second step involves the calculation of the vertical hydraulic conductance using available data (Freeze and Cherry [16]; Van Genuchten et al. [17]; Eimers et al. [18]), as well as the estimation of dominant flow mechanisms from available borehole logs and geological information. The rating for hydraulic properties and flow mechanism is then selected (Table 2). The next step is the estimation of recharge using available information and any of the methods included in the software ("Qualified Guess", CMB, CRD, EARTH). The rating for recharge is selected from Table 3. Travel time is calculated using available data and any of the methods included in the software (flow velocity; Foster and Hirata [21]). The rating for travel time is selected using Table 4. Sorption is defined from available information on soils and types of contaminants, whilst decay is determined based on the contaminant species and the database of their properties (Usher et al. [23]). The rating for sorption is selected from Table 5, whilst the rating for half-lives is selected from Table 6. The rating scores for each factor are finally combined in the main menu to yield the rating of the impact of vadose zone on groundwater vulnerability.

4   Conclusions

An improved DRASTIC index was developed to account for hydrogeological factors like multi-layer vadose zone and preferential flow, as well as specific properties of contaminants like sorption and decay. In addition, a database of values and descriptors related to the rating "I" (impact of the vadose zone) was included in a user-friendly, Excel-based, improved DRASTIC index calculator. The algorithms presented in this study can be easily included into a GIS system for assessment of groundwater vulnerability at different scales.

The method can be used for applications in site-specific environmental impact assessments for new developments, for regional groundwater vulnerability assessment as well as in integrated water resource management. This has, however, implications with uncertainties of groundwater vulnerability assessment as well as data availability and scarcity. The level of uncertainty increases with the coarsening of input data. Similarly, the level of uncertainty and accuracy is associated with the coarsest dataset. National data sets can be used in the absence of field data. However, the recommended approach is to assess vulnerability at a site-specific scale. If finer resolution data sets are available for the area being studied, then improved levels of data certainty will be achieved.

Acknowledgement

The authors acknowledge the funding of the Water Research Commission (Pretoria, South Africa) through the research project on "Improved methods for aquifer vulnerability assessment and protocols for producing vulnerability maps, taking into account information on soils".

References

[1] Sililo, O.T.N, Saayman, I.C. & Fey, M.V., Groundwater Vulnerability to Pollution in Urban Catchments, Water Research Commission Report No. 1008/1/01: Pretoria, South Africa, 2001.

[2] National Research Council, Groundwater Vulnerability Assessment: Predicting Relative Contamination Potential under Conditions of Uncertainty, Committee for Assessing Groundwater Vulnerability, National Academy Press: Washington, D.C., 1993.

[3] Burkart, M.R., Kolpin, D.W. & James, D.E., Assessing groundwater vulnerability to agrichemical contamination in the Midwest US. Water Science and Technology, 39(3), pp. 103-112, 1999.

[4] Burkart, M.R. & Feher, J., Regional estimation of ground water vulnerability to nonpoint sources of agricultural chemicals. Water Science and Technology, 33(4-5), pp. 241-247, 1996.

[5] Aller, L., Bennet, T., Lehr, J.H. & Petty, R.J., DRASTIC – A Standardized System for Evaluating Groundwater Pollution Potential Using Hydrogeological Settings, EPA Report 600/2-87/035: US Environmental Protection Agency, 1987.

[6] Thirumalaivasan, D., Karmegam, M. & Venugopal, K., AHP-DRASTIC: software for specific aquifer vulnerability assessment using DRASTIC model and GIS. Environmental Modelling & Software, 18(7), pp. 645-656, 2003.

[7] Secunda, S., Collin, M.L. & Melloul, A.J., Groundwater vulnerability assessment using a composite model combining DRASTIC with extensive agricultural land use in Israel's Sharon region. Journal of Environmental Management, 54(1), pp. 39-57, 1998.

[8] Babiker, I.S., Mohamed, A.A., Hiyama, T. & Kato, K., A GIS-based DRASTIC model for assessing aquifer vulnerability in Kakamigahara Heights, Gifu Prefecture, central Japan. Science of the Total Environment, 345(1-3), pp. 127-140, 2005.

[9] Al-Adamat, R.A.N., Foster, I.D.L. & Baban, S.M.J., Groundwater vulnerability and risk mapping for the Basaltic aquifer of the Azraq basin of Jordan using GIS, Remote sensing and DRASTIC. Applied Geography, 23(4), pp. 303-324, 2003.

[10] McLay, C.D.A., Dragten, R., Sparling, R. & Selvarajah, N., Predicting groundwater nitrate concentrations in a region of mixed agricultural land use: a comparison of three approaches. Environmental Pollution, 115(2), pp. 191-204, 2001.

[11] Dixon, B., Groundwater vulnerability mapping: A GIS and fuzzy rule based integrated tool. Applied Geography, 25(4), pp. 327-347, 2005.

[12] Sandersen, P.B.E. & Jorgensen, F., Buried Quaternary valleys in western Denmark—occurrence and inferred implications for groundwater

resources and vulnerability. Journal of Applied Geophysics, 53(4), pp. 229-248, 2003.

[13] Melloul, A.J. & Collin, M., A proposed index for aquifer water-quality assessment: the case of Israel's Sharon region. Journal of Environmental Management, 54(2), pp. 131-142, 1998.

[14] Worrall, F., Besien, T. & Kolpin, D.W., Groundwater vulnerability: interactions of chemical and site properties. The Science of The Total Environment, 299(1-3), pp. 131-143, 2002.

[15] Worrall, F. & Kolpin, D.W., Aquifer vulnerability to pesticide pollution—combining soil, land-use and aquifer properties with molecular descriptors.
Journal of Hydrology, 293(1-4), pp. 191-204, 2004.

[16] Freeze, R. & Cherry, J.A., Groundwater, Prentice-Hall Inc.: New Jersey, 1979.

[17] Van Genuchten, M. Th., Leij, F.J. & Yates, S.R., The RETC Code for Quantifying the Hydraulic Functions of Unsaturated Soils, EPA Report 600/2-91/065: US Environmental Protection Agency, 1991.

[18] Eimers, J.L., Weaver, J.C., Terziotti, S. & Midgette, R.W., Methods of Rating the Unsaturated Zone and Watershed Characteristics of Public Water Supplies in North Carolina, Water Resources Investigations Report 99-4283: US Geological Survey, 2000.

[19] Rushton, K.R. & Ward, C., The estimation of groundwater recharge. Journal of Hydrology, 41, pp. 345-361, 1979.

[20] Xu, Y. & Beekman, H.E., (eds). Groundwater Recharge Estimation in Southern Africa. UNESCO IHP Series No. 64: UNESCO, Paris, 2003.

[21] Foster, S.S.D. & Hirata, R., Groundwater Contaminant Risk Assessment: A Methodology Using Available Data, Pan American Centre for Sanitary Engineering and Environmental Sciences (CEPIS): Lima, Peru, 1995.

[22] Soil Classification Working Group, Soil Classification. A Taxonomic System for South Africa, Dept. of Agricultural Development: Pretoria, South Africa, 1991.

[23] Usher, B., Pretorius, J.A., Dennis, I., Jovanovic, N., Clarke, S., Cave, L., Titus, R. & Xu, Y., Identification and Prioritization of Groundwater Contaminants and Sources in South Africa's Urban Catchments, Water Research Commission Report No. 1326/1/04: Pretoria, South Africa, 2004.

Section 11
Community involvement
and education

Attitudes of consumers on E-waste management in Greece

K. Abeliotis, D. Christodoulou & K. Lasaridi
Harokopio University, Athens, Greece

Abstract

Waste from electric and electronic equipment (WEEE) is the fastest growing waste stream in the European Union. WEEE may cause environmental problems due to its hazardous materials contents. On the other hand, recycling of WEEE is an important issue, not only from the point of waste treatment, but also from the aspect of recovering valuable materials. This paper presents data on WEEE generation and the status of WEEE recycling in Greece, in its current early stage. It also presents the attitudes and behaviour of Greek consumers regarding WEEE management practices. A field survey based on a closed-type questionnaire was conducted. The survey took place in the city of Athens in the first months of 2005. The main results of the study indicate that WEEE generation in Greece will continue to grow and that people are willing to get involved provided that the needed infrastructure is established.
Keywords: WEEE, E-waste, Greece, public opinion.

1   Introduction

The production of electric and electronic equipment is increasing worldwide. Both technological innovation and market expansion continue to accelerate the replacement of equipment leading to a significant increase of waste electric and electronic equipment (WEEE) [1]. WEEE is recognised as the fastest growing waste stream in the European Union (EU), with estimates of 14-20 kg per person per annum, and it is increasing at about three times higher rate than the average for municipal waste [2]. Currently in the EU-15, WEEE accounts for 8% of all municipal waste [3]. WEEE consists of a wide range of electronic devices that can be classified into three groups, namely white goods, brown goods and information technology (IT) scrap [4].

WIT Transactions on Ecology and the Environment, Vol 92, © 2006 WIT Press
www.witpress.com, ISSN 1743-3541 (on-line)
doi:10.2495/WM060431

E-waste contains more than 1,000 different substances, many of which are toxic, such as lead, mercury, arsenic, cadmium, selenium, hexavalent chromium, and flame retardants that create dioxins emissions when burned [3]. WEEE may cause serious environmental problems during the waste management phase if it is not properly pretreated.

On the other hand, recycling of WEEE is an important issue not only from the point of waste treatment but also from the aspect of recovering valuable materials. E-waste contains valuable substances such as gold and copper [3]. For example, early generation PCs used to contain up to 4 g of gold each; however this has decreased to about 1 g today [3]. One ton of e-waste contains up to 0.2 tons of copper, which can be sold for about €500 at the 2004 world price [3]. Recyclable material from WEEE typically consists of ferrous metals (47%), plastics (22%), glass (6%) and non-ferrous metals (4%) [4]. Well established take-back and recycling systems for WEEE, such as the one in Switzerland, have proven clear environmental advantages, compared to the complete incineration of all WEEE [5].

The European Commission, recognising the need for legislation to address the escalating problem of WEEE, introduced Directive 2002/96/EC on WEEE. The central theme of the Directive is extended producer responsibility and aims to increase, rates of reuse, refurbishment and recycling. The Directive sets a target: by December 31, 2006 at the latest, a minimum rate of separate collection of 4 kg on average per inhabitant per year of WEEE must be achieved [6]. This target is already met in the UK [2] while in well organised WEEE management systems, such as the one in Switzerland, 11 kg per capita per year are reported [5]. However, for EU members like Greece, that are starting its WEEE recycling from scratch, this target is very ambitious. The Directive also requires producers to set up systems for recovery and recycling of WEEE, either collectively or individually. It is the responsibility of the producers that the specific recovery and recycling targets are met (see Table 1).

The treatment of the hazardous material as well as its precious metal recovery depends on the collection rate of WEEE. White goods, i.e. household appliances such as refrigerators and cookers, have a high metal content and therefore a high collection rate. For example, the collection rate in the U.K. in 2000 was reported at 90% [4]. Brown goods, i.e. household electrical entertainment appliances (DVD players, TVs, radios, etc.), easily fit into the municipal solid waste (MSW) collection bins. Therefore, its collection rate in U.K. was reported for 2000 at 50% [4]. Large household appliances, small household appliances, IT and telecommunication equipment, and consumer equipment account for almost 95% of the WEEE generated [3].

2 The situation in Greece

The electrical and electronic goods trade sector is one of the most active economic sectors in Greece. It is reported that in 2003, 1.2 billion € was spent on electric appliances trade, excluding the part of IT and telecommunications [7]. In 2005, 540,000 TV sets were sold in Greece, 140,000 Hi-Fi systems, 600,000

DVD players, 180,000 DVD recorders, 270,000 digital photo cameras, 100,000 videocameras, 500,000 PC monitors and 220,000 laptops [7]. The latest data for white appliances are from 2003: 330,000 refrigerators were sold, 30,000-50,000 freezers, 320,000 cookers, 250,000 washing machines and 80,000 dishwashers [7]. In addition 2.8-3.0 million mobile phones were sold in Greece in 2005 [7]. The annual WEEE production in Greece is currently estimated at 170,000 tons. It is projected that, by the end of 2008, it will reach 185,000 tons, which corresponds to 14.4 kg per capita annually.

Table 1:   Targets of 2002/96/EC Directive on WEEE [1].

	Minimum recovery rate (%)	Minimum rate of component material and substance reuse and recycling (%)
Large household appliances and automatic dispensers, by an average weight per appliance	80	75
IT and telecommunications equipment and consumer equipment, by weight of the appliances	75	65
Small household appliances; lighting equipment; electrical and electronic tools; toys, leisure and sport equipment; monitoring and control instruments, by an average weight per appliance	70	50
Gas discharge lamps, by weight of the lamps	-	80

The WEEE Directive was incorporated into the Greek law as the Presidential Decree 117/2004. In order to fulfil the requirements of the aforementioned Directive, a joint venture company has been established, by the major players of electrical and electronic equipment trade in Greece, called "Equipment Recycling S.A." which is responsible for the organisation and the operation of a collective WEEE recovery system. The target of the system is, by 31 December 2006, coverage of 90% of Greek households and the collection of at least 44,000 tons annually, which is the national target for Greece. As of today, 400 companies have already joined the collective system and contracts have been signed with 40 municipalities for WEEE collection. The calculation of the recycling charges per unit weight is based on the break-even principle, i.e. the aim of "Equipment Recycling S.A." is to balance its operational costs with the recycling charges and not make a profit out of it. The calculation of the recycling charges is based on the following parameters [8]:

- The financial viability and the management costs of the system,
- The principle of no interference to the competition among similar products,
- The quantities of WEEE produced per product category,
- The difficulty of collection and disassembly per product category,
- The cost of removing hazardous materials,
- The income from reselling the recovered valuable materials.

The charge for each item of WEEE is presented in Table 2. The classification of WEEE in this table is based on the ordinances of 2002/96/EC Directive. The prices of this catalogue are valid from February 1, 2005. For electrical and electronic equipment sales from 1/7/2004-31/1/2005 there is a single cover charge for all categories at €59.5 /ton (including VAT).

Table 2:   Recycling charge for participation in the WEEE management scheme in Greece [8].

Product category	Recycling charge per unit weight including 19% VAT (€/ton)
Large household appliances	85.72
Small household appliances	95.81
IT and telecommunications equipment	302.55
Consumer equipment	302.55
Lighting equipment	148.75
Gas discharge lamps	0.120 (per piece)
Electrical and electronic tools	121.02
Toys, leisure and sport equipment	181.52
Medical technology equipment	59.50
Monitoring and control instruments	181.52
Automatic dispensers	90.76

In terms of management practices in Greece, before the enforcement of the WEEE Directive, large white appliances at the end of their life were handled as bulky solid waste. All the municipalities have collection services for bulky items. People leave their large electrical items next to the MSW collection bins and then call the pick-up service. However, for large white electrical goods with high metal content, there is also an informal collection mechanism consisting of individual garbage collectors who pick up the appliances left for collection. These people then sell the collected appliances as metal scrap, without any form of recovering the valuable components. Until the enforcement of the Directive, this informal collection and scrap recycling was the only action taken to divert WEEE from legal or illegal landfilling. This informal mechanism accounts for

80% of the collected WEEE, even after the introduction of the collective system. Small appliances were placed directly to the MSW bin.

Today, the first WEEE recycling facility is operating 60 km west of Athens with an annual operating capacity of 20,000 tons, i.e. almost half of the required capacity by the WEEE Directive for the whole country. For this reason, a second recycling facility is underway in northern Greece.

In addition, after the appearance in the Greek market of low price eastern Asian appliances, the second hand market of electrical and electronic goods is very limited. Old PCs are given to charities and friends or stored at home while older appliances are given to friends or family members that, for example, study in a different city. This trend was also reported in a recent telephone survey regarding the use of PCs in Greece [9]. Up to the middle of 2004, all WEEE was landfilled in Greece. The only pre-treatment taking place was crashing them in order to create smaller parts and avoid the operational problems at the landfill.

3   Methodology

A field survey was conducted in the first three months of 2005 in the city of Athens using a questionnaire, consisting of 18 closed-type questions. The sample size of the survey was 100 persons. SPSS 10 was used for data analysis. The first part of the questionnaire dealt with the gender, age, education level and family income of the sample. The demographics of the sample are presented in Table 3. The sample is, therefore, mainly consisting of young people. However, each person was asked to represent its household.

Table 3:   The demographics of the sample.

Gender			
	Male	31%	
	Female	69%	
Age (y)			
	18-25	50%	
	26-45	32%	
	46-65	16%	
	> 65	2%	
Education			
	Primary	9%	
	Secondary	57%	
	University	34%	
Family Income (€)			
	< 15,000		39%
	15,001-30,000		42%
	> 30,000		19%

In the second part of the questionnaire, questions relating to the specific goals of the survey were included. More specifically, this study aims to:

- report the way that consumers manage their WEEE,
- estimate the rate of WEEE production,
- examine what people know regarding WEEE management,
- survey the willingness of the people to participate in WEEE recycling.

4 Key findings

The first question was "How often do you read the operational instructions of new electric and electronic equipment?" The aim of this question was to identify if people realise that WEEE has to be handled apart from the rest of MSW. This piece of information is included in the instruction manual of each appliance by the use of a certain logotype. 60% replied that they always read the manual, 36% replied sometimes while 4% never read their manual.

The next question was "How do you handle the large white appliances (refrigerators, washing machines, etc.) at the end of their life?" 56% replied that they put it on the road, next to the MSW collection bin. 21% replied that they call the local municipality service for bulky waste. The rest (23%) replied that they handled their large white goods differently, meaning that they call on purpose a garbage collector, or they keep them at home. Even though large white goods make up the majority of WEEE by weight, small and medium sized items are the vast majority by number [2]. Therefore, it seems reasonable to expect different management practices from the people regarding small appliances. In the question "How do you handle the small appliances at the end of their life?" 82% replied that they trash it alongside the rest of MSW.

The next set of questions aimed at identifying the rate of WEEE generation. 78% of the respondents replied that they bought at least one electrical or electronic appliance during the past year. From the 78 buyers of the sample, 60.3% bought an IT or telecommunications appliance, 42.3% bought an entertainment appliance, 35.9% bought a small household appliance while 23.1% bought a large white appliance.

The next question, addressed to the buyers of at least one appliance, was "What happened to the respective old appliance?" In this question, 39% replied that they still own the old appliance and have stored in their home, 22% claimed that they didn't have a similar appliance or that they bought it for the needs of a new household, 21% replied that they throw it away in the MSW bin, 14% gave it as a gift, while only 4% gave it to an individual collector for scrap recovery. In order to have a better view on the fate of appliances at the end of their life, the appliances that were bought for the very first time are taken out of the calculations. Breaking down the remaining replies into the four major appliances categories, Figure 1 is generated. From there it is evident that old large appliances are more likely (46.2%) to be disposed off or given to a third party (30.8%). For old small appliances it is easier to keep them at home (42.9%) or again give it to a third party (28.6%). Old entertainment appliances are mainly kept at home (53.6%) or disposed off (28.6%). Old IT and telecommunications devices are kept at home (62.2%) or given to a third party (21.6%).

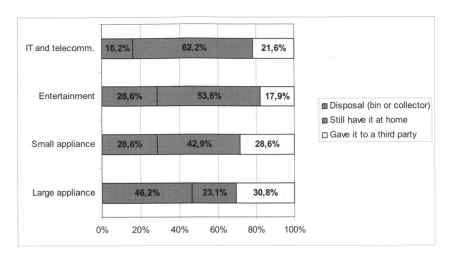

Figure 1:   Old appliance management per appliance category.

The following question was "What is the single main reason for buying a new electric or electronic device?" 48% replied that they buy a new appliance when the old appliance is broken, 40% replied that they buy a new appliance in order to replace it with a technologically advanced one, while 12% replied that they buy a new appliance when they don't have a similar old one. The next question was "Identify the single most important criterion when buying a new electrical or electronic appliance?" 67% replied to be technologically advanced, 14% replied to have lower price, 18% replied to be more energy efficient or environmentally friendly while only 1% replied that they look for better appearance. Breaking down the replies in this question into the four appliances categories, Figure 2 is generated. It is evident that the main reason for buying a new appliance of any kind is to be technologically advanced. For IT and telecommunication devices this statement is also confirmed by the χ^2 independence test ($\chi^2=14.094$; $p=0.001$). For the large and small household appliances category, energy efficient and environmentally friendly appliances are preferred by more than 20% of the respondents while low price comes into play only for small appliances.

The next question was "Did you buy a second-hand electric or electronic appliance at least once in your life?" Only 14% of the respondents gave a positive answer to this question. Out of these 14 people, 5 bought a used large appliance, 8 bought an IT or telecommunications appliance (personal computer or mobile phone) while 1 person bought both a large appliance and a telecommunications device.

The final set of questions aimed at identifying which are the preferences of the sample regarding issues relating to the collection of WEEE. The first question in this set was "Are you willing to pay a fee when you buy a new appliance which will be refunded at the end of its life?" 54% replied positively, 14% replied negatively while the rest (34%) replied "maybe". In the question

"Which is the most preferable way for you to get rid of a large white appliance?" 57% of the respondents replied that the most preferable way for them is the retailer from whom they buy the new to collect the old one. 38% replied that they prefer to put the old appliance next to the MSW bin so that the municipality can pick them up while 5% asked for a different form of management. The next question was "What is the strongest motive for you to take the old small appliances in pre-specified collection centre?" 36% replied that a refundable amount of money would be the strongest motive, 36% that more information on the adverse effects of WEEE on the environment and human health would be the strongest reason, 17% replied that strong law enforcement would be the strongest motive while 11% replied that there is no strong enough motive for them to take their old WEEE to a collection centre.

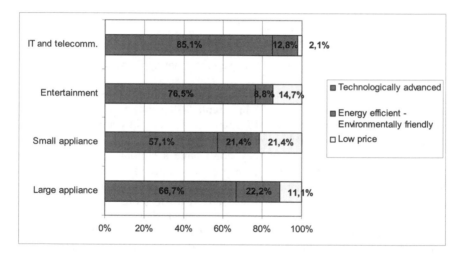

Figure 2:   Main reason for buying a new appliance per appliance category.

5 Discussion

Prior to the enforcement of the WEEE Directive by the EU, the dominant management practice in Greece was landfilling. Greece is currently building the necessary infrastructure for dealing with its WEEE. The necessary legislative background is already set, allowing the establishment of the first collective WEEE recovery system. In addition to the collective system, there are three more collection ways for domestic WEEE: the infrastructure of the municipalities, the retailers that are required to receive the old appliances on a one to one basis, and the individual collectors. In all four alternatives, the participation of the public is of paramount importance. In terms of WEEE recycling, a single facility operates in Greece, capable of dealing with half of the WEEE amount required by the Directive. Therefore, in order to meet the challenge of the EU Directive, Greece has to speed up.

The aim of this paper was to present the current status of WEEE management in Greece and to outline, by means of a questionnaire, the opinion and practices of people living in the area of Athens regarding WEEE management. The sample size of 100 persons is small and can only allow for the identification of some trends. From the analysis of the responses to the questionnaires the following trends are identified:

The majority of the people read the instruction manual from the appliances that they buy. This is important because in the manuals it is clearly stated that WEEE must be managed differently and not along with the rest of the MSW.

Since small electric and electronic appliances fit easily in the MSW bin, people tend to get rid of them more easily. Large appliances, due to their size, have to be handled differently. Therefore, for the success of any small appliances recovery program it is essential that this is easily accessible by the public. For large white appliances the easiest way, for the consumers, is to have them picked up by the retailers that bring home the new appliance.

The present survey reveals that consumers in Greece tend to keep home the majority of old electric and electronic appliances, with the exception of large appliances. This is also reported by other authors [10] in a similar survey of 126 households in the urban area of Thessaloniki (see Table 4). This means that the useful life of each appliance is extended; however, eventually, all appliances end up as WEEE.

Table 4:   Fate of old appliances in Greece per appliance category.

Appliance category	% disposal (landfill&collection)		% given to third party		% stored at home	
	This study	[10]	This study	[10]	This study	[10]
Large	46.2	50	30.8	30	23.1	20
Small	28.6	80	28.6	-	42.9	-
Entertainment	28.6	35	17.9	-	53.6	-
IT and telecom.	16.2	20	21.6	40	62.2	40

The main reason that people buy new appliances is the technological advancements. This is especially true for IT and telecommunications appliances. Moreover, the trade of second hand electrical and electronic appliances in Greece is very limited. Both these results indicate that WEEE generation will continue to grow in Greece.

In terms of people's involvement in the recovery process, people think than a refundable amount of money and more information are the strongest motives for getting involved.

Closing this paper, it is important to state that the involvement of consumers is of paramount importance in the WEEE recovery process, provided that the infrastructure is there. In order to get people involved, more information and education is needed regarding the adverse environmental impacts that WEEE may cause. Information sources may be the municipalities, the collective system, the retailers of the electrical and electronics goods, and the mass media.

Acknowledgement

This research was supported in part by Pythagoras II programme (MIS: 97456, subproject 1) co-funded by the European Social Fund (75%) and National Resources (25%).

References

[1]   Cui, J. & Forssberg, E., Mechanical recycling of waste electric and electronic equipment: a review. Journal of Hazardous Materials, B99, pp. 243-263, 2003.
[2]   Darby, L. & Obara, L., Household recycling behaviour and attitudes towards the disposal of small electrical and electronic equipment. Resources, Conservation and Recycling, 44, pp. 17-35, 2005.
[3]   Widmer, R., Oswald-Krapf, H., Sinha-Khetriwal, D., Schnellmann, M., & Böni, H., Global perspectives on e-waste. Environmental Impact Assessment Review, 25, pp. 436-458, 2005.
[4]   Mohabuth, N., & Miles, N., The recovery of recyclable materials from Waste Electrical and Electronic Equipment (WEEE) by using vertical vibration separation. Resources, Conservation and Recycling, 45, pp. 60-69, 2005.
[5]   Hischier, R., Wager, P., & Gauglhofer, J., Does WEEE recycling make sense from an environmental perspective? The environmental impacts of the Swiss take-back and recycling systems for waste electrical and electronic equipment (WEEE), Environmental Impact Assessment Review, 25, pp. 529-539, 2005
[6]   Directive 2002/96/EC of the European Parliament and of the Council of 27 January 2003 on waste electrical and electronic equipment. Official Journal L037:0024-39 [13/02/2003], europa.eu.int/eur-lex/en/
[7]   STATBANK, www.statbank.gr/sbstudies.asp
[8]   Equipment Recycling S.A., www.electrocycle.gr
[9]   Valta, A.A., Menegaki, M.E. & Kalampakos, D.C., Electronic Waste Management in Greece: Are people ready to get involved in the process? Proc. of the 9th International Conference on Environmental Science and Technology, ed. T.D. Lekkas, Rhodes, Greece, pp. A1560-A1565, 2005.
[10]   Karagiannidis, A., Papadopoulos, A., Moussiopoulos, N., Perkoulidis, G., Tsatsarelis, Th., & Michalopoulos, A., Characteristics of wastes from electric and electronic equipment in Greece: Results of a field survey. Proc. of the 8th Int. Conf. on Environmental Science and Technology, ed. T.D. Lekkas, Lemnos, Greece, pp. 353-360, 2003.

A concept of education in sustainable electronic design

V. B. Litovski, J. Milojković & S. Jovanović
Faculty of Electronic Engineering,
University of Niš, Serbia and Montenegro

Abstract

A postgraduate curriculum on sustainable electronic design is briefly described. That is, to our knowledge, the only postgraduate curriculum on sustainable electronic design in South-East Europe. The motivation of this new educational effort comes from the fact that longer tradition in electronic production exists in the local region and, in the same time, long tradition in higher education in electronics exists.

There is no general awareness on the interrelation of electronic-product's life cycle (e.g. electronic production and electronic waste management) and the environment in the local communities. The main goal of this curriculum is to establish this awareness, to emphasize the role of the electronic designer in the life cycle of an electronic product, and to introduce sustainability-aware design methods. Of course, special attention is paid to end of life and especially to collection and management of electronic waist.

After the overview of the curriculum, contents of the subjects will be listed. Every content will be accompanied by some comments trying to stress some specifics of the subject, some specific goals to be achieved, or some characteristic data or method to be stressed. These are tailored for graduate electronic engineers that are expected to enter the field of sustainable design and management of the end of life of electronic products.

Part of our published research and results that are available for this curriculum will be described too. It is related mostly to information infrastructure necessary for implementation of sustainability concepts.
Keywords: sustainable electronic design, curriculum, postgraduate studies, ecological education of electronic engineers.

1 Introduction

The production of electrical and electronic equipment is one of the fastest growing domains of manufacturing industry in the Western world [1]. Electronics as a human activity becomes more and more influential. There is hardly any part of life where electrical and electronic equipment are not used. The number and weight of electronic components and equipment in use becomes so high that makes electronics comparable with other much "heavier" industries. In the year of 1988 the amount of electronic equipment reaching end-of-life measures 6 million tones and is expected to double in 2010 [1]. This means that about 20% of the municipal waste stream will be related to WEEE (Waste Electrical and Electronic Equipment).

In addition, the growth of the production of electronic equipment is directly related to the rise of use of virgin materials giving no opportunity to the nature to regenerate this kind of resource. This is especially related to rare elements being used in modern electronic components.

Finally, electronic equipment is using energy for its work. The amount of energy spent by electronic appliances both active and stand-by is enormous and any optimisation leads to significant savings. This is why the electronic engineers and especially electronic designers are to accept that they are playing an exceptionally important role in the world economy and the eco-system. Slowly but surely they become part of the most influential people in the world. This is why regular education of sustainable electronic design becomes more and more important.

In this paper we will try to give an overview of the very first curriculum for postgraduate studies of sustainable electronic design in our country. It started in the year 2002. two main orientations are expected to be given. One is related to the design of electronic components, devices and equipment, while the other is expected to prepare electronic engineers to develop subsystems within the collective social effort for sustainable development. The specifics of the sustainability of electronic design are mainly thought based on publications having as a subtitle synonyms of "green book" such as [2–7].

2 Overview of the curriculum

It is a two years curriculum where students are though in six subject, three per year. The subjects are listed in table 1. Number one and two are used at the beginning of the subject code (column 1) to denote first and second year of studies, respectively. Two of the subjects are compulsory while students may choose among the rest according their personal interest. Small projects are part of the examining process in any subject and an MS thesis compulsory, too. At least, two publications of papers, the content being related to the projects or the MS thesis, are expected from every student.

The compulsory subjects noted by the letter A in column one of table 1 are introductory in both general subject of sustainable development, and specific relation of electronics to the sustainable development.

The rest (marked by B) of the subject may be combined in many ways. For example, a combination containing the subjects: 1.SD.03B, 2.SD.06B, 2.SD.07B, and 2.SD.10B should prepare an expert, in the electronic design team, who may be responsible for the aspect of sustainability of the whole project.

Table 1:   The curriculum.

CODE	SUBJECT
1.SD.01A	Sustainable development
1.SD.02A	Sustainable electronic design
1.SD.03B	Electronic circuit design
1.SD.04B	Electronic measurement and data acquisition systems
2.SD.05B	Method of standardization and standards of sustainable design
2.SD.06B	Physics and technology of electronic circuits and systems
2.SD.07B	Simulation and optimisation of electronic circuits and systems
2.SD.08B	End-of-life management
2.SD.09B	Ecological information systems for monitoring and control
2.SD.10B	Methods and tools of sustainable design

In the next the contents of the subjects will be listed together with some comments stressing some specific aspects of the teaching and/or relation to other general subjects or subjects within this curriculum.

3 Contents and comments

The contents listed below are expected to be the ones that, we consider, are necessary for the student at the moment. We intend to monitor the developments in any subject and to accommodate according to any important change. We are aware, of course, that any teacher will give a specific taste to the subject he or she is teaching.

3.1 1.SD.01A Sustainable development

Content: Human-nature interaction. Pollution of natural resources such as water, air, and soil. Energy and ecology. Electronics and Ecology. Economical aspects. Relation of the overpopulation, over-consumption etc. to the ecological principles. Ecological problems in cities. Concepts of sustainable design. International organizations dealing with sustainable development. The electronic designer and the eco-tasks. Problems of ecological education of electronic engineers.
Comment:
The point here is to establish awareness that electronics has relation with sustainability at a global level. Examples of how electronics is influencing pollution are given exemplifying all aspects: material (frequently poisonous) and energy consumption, electromagnetic radiation, and air, water, and soil pollution.

3.2 1.SD.02A sustainable electronic design

Content: The life-cycle. Sustainable pricing of products. Exploitation phase. End-of-life. Reuse. Dismantling. Disposal. Take-back. Concepts of sustainable design. Design phase. The role of the designer. Effects of sustainable design. Methods of sustainable design. Sustainable design of electronic products.
Comments:
-Design for dismantling and similar aspects is stressed here.
-In addition, design methods are discussed that allow for monitoring of the life cycle of critical parts within an electronic equipment so enabling decision making about its reuse at the end-of-life of the equipment.
-Special attention is paid to maintenance. Namely, in the first life of the product maintenance means making it longer (among other) by use of spear parts. These spear parts, however, may be taken from a used system so prolonging its second life. This becomes especially important for maintenance of industrial facilities that, when operating, perform well even if old electronic technology is applied.

3.3 1.SD.03B Electronic circuit design

Content: Electronic circuits CAD. Systems for CAD. Hardware description languages. Application on nonelectrical systems design. Integrated circuits design. Technology design, simulations, physical design, symbolic topology design, DRC, compaction. Techniques for ASIC design. System and systematic design aspects. IC testing. Specifics of printed circuit board design.

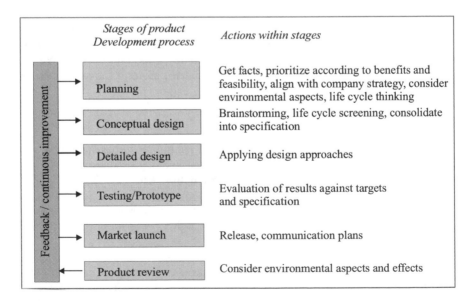

Figure 1:   Generic model of integrating environmental aspects into product development process.

Comments:

-Fig. 1 represents the generic model advocated for the integration of environmental aspects into the product development process [8]. This, of course, should be considered as one among many concepts (such as [3]) and should not exclude any further improvement.

-Choice of digital alternatives is recommended while electronic circuit synthesis is considered.

-Bare board design is twofold optimisation problem. First, one should think on placement and routing on the board what is usual electronic design problem. The goals here is to achieve minimum area, short connections (high frequencies), low voltage drops etc. On the other hand, having in mind the production process and waste generated during production, one is supposed to optimise: the copper fraction on a single layer, the number of boards per panel, and the total number of signal layers [9].

3.4 1.SD.04B Electronic measurement and data acquisition systems

Content: Measurement of physical, chemical and biological quantities. A/D and D/A conversion. Analogue and digital processing of measured data. Advanced digital signal processing. Structure of a data acquisition system. Communication interfaces and standards.

Comment:

-Among environmentally important quantities to be measured one should recognize: temperature, humidity, wind velocity, toxic chemical (CO_2, SO_2 etc.) and biological agents in air, water, and soil, personal contamination vibration, noise etc. Special attention is paid to detection/diagnosis of harmful chemical and biological agents including explosives, nerve gases, mycotoxins, viruses, pathogenic bacteria and biological toxins.

3.5 1.SD.05B Method of standardization and standards of sustainable design

Content: Complexity of the standardization process. Procedures for standardization. Standardization organizations. Access to standards. Standards related to industrial environment. European standardization bodies. Home standardization institutions. Anti-trust low. Concepts and contents of ISO14000.

Comments:

-When speaking on ISO 14000, a short overview of ISO9000 is thought and its relation to ISO14000 is stresses,

-Environmental management system (14001) gives specification with guidance for use,

-Guidelines for environmental auditing (14010). Gives general principles of environmental auditing,

-General principles of environmental labelling, (14020)

-Environmental management (1404x, x=0,…,9) includes principle and guidelines for LCA,

-Terms and definitions (14050),

-Guide for inclusion of environmental aspects in product standard (14060),
-The important thing to introduce, however, is that every organization that wants to adopt the ISO 14000 standards is supposed to obey not only national but international regulations, too. In addition when the certificate is issued the organization will be expected to have an effective system of internal management that is accommodated to the application of the ISO 14000.

3.6 2.SD.06B Physics and technology of electronic circuits and systems

Content: Materials used in electronic production. Process modelling. Reliability. Yield. Health hazard assessment. Reuse of materials and components. Passive and active discrete electronic component technologies. Integrated circuits. Thick film technology. Bipolar technologies. MOS technologies. PCBs. Assemblies.
Comments:
-Lists of poisonous materials and health hazards during production are introduced first.
-Theory of computation of yield reliability is related to the waste generated during production and to the prolongation of the product's life. In that sense these two categories become important criteria for sustainability evaluation of a design.

3.7 2.SD.07B Simulation and optimisation of electronic circuits and systems

Content: Electronic circuit analysis in time and frequency domain. Methods of analog optimisation. Constrained optimisation. Evolutionary algorithms for optimisation. Deterministic and statistical tolerance analysis and design. Power and delay estimation methods. Logic simulation. Discrete event simulation. Test pattern generation. Basics of diagnostics.

Table 2:   Logic versus discrete event simulation.

LOGIC SIMULATION	DISCRETE-EVENT SIMULATION
deterministic event generation	stochastic event generation
logic models	analog and/or logic models
deterministic delay models	stochastic delay models
next event algorithm	
selective trace algorithm	

Comment:
-As an illustration of how sustainability may be related to ordinary knowledge of electronic design a comparison is given in table 2 relating logic and discrete event simulation. The former is used for design verification and test signal evaluation in logic design, while the letter is supposed to predict the timing in the life-cycle of a product [10].

3.8 2.SD.08B End-of-life management

Content: Resell, upgrade, and recycle options. Evaluating the environmental effectiveness of reselling, upgrading and recycling. Re-manufacturing. Product returns systems. Logistics. Producer responsibility. Recycling processes and technologies. Demanufacturing. Disposal.
Comment:
-Awareness of the amount of electronic waste is one of the important aspects of teaching this subject.
-Establishment of an effective product return system accommodated to the way of functioning the society both locally and on the state level is of prime importance.
-Specific recycling processes such as PCB recycling or monitor recycling are studied.
-Concepts of (computer) upgrading are studied, too.

3.9 2.SD.09B Ecological information systems for monitoring and control

Content: Eco-parameters that may be measured. Concepts of conversion of natural quantities into electrical ones. Environmental transducers. Eco-data acquisition at the level of industrial objects, at the level of municipality, and at the level of a country. Ecological information systems. Monitoring, statistical evaluation, and control of ecological parameters. Avoiding hazardous situation by monitoring ecological parameters.
Comment:
-This is a subject that is most convenient for practical evaluation of projects. Namely the students are directed to different university partners having their own, more or less complete, information systems related to sustainability.

3.10 2.SD.10B Methods and tools of sustainable design

Content: Classification of the procedures for sustainable design. Full Life Cycle Analysis (LCA). Life cycle costing (LCC). Partial LCA: Ecoindicator, Product improvement method (PIM), Method of assessment of ecological data. Optimisation of logistic systems. Integrated system for eco documentation. Databases for sustainable design.
Comments:
-We suggest the classification done by Digital Equipment Corporation and the MIT Program on Technology, Business and Environment to be accepted. They classify DfE tools according to three primary attributes. First, a tool's applicability to product development stages will indicate whether a tool may be used for simply establishing project goals or actually developing detailed designs. Second, its degree of applicability to product life cycle stages determines how appropriately life-cycle stages are applied to a product's design and how many life cycle stages the tool supports. Finally, a tool's degree of decision support accounts for the type of data output generated (inventory tools

generate raw data, impact tools aggregate the data, and improvement tools generate design alternatives that minimize environmental impact [11].

-The standardized [12, 13]. LCA is a complex tool for assessment of the environmental impact during the whole life-cycle of the product. Main difficulties in the implementation of LCA are related to some system modelling due to the data uncertainty for product's end of life because it is nearly impossible to collect real data on end-of-life processing at the design stage. End-of-life effects are difficult to anticipate for complex products with long lifetimes.

-The detailed study being time consuming, expensive and, in some cases difficult to realize, one use simplified methods. These methods mainly take into account only part of the whole life-cycle and consider only some impacts to the environment.

4 Toward an integrated life-cycle design tool

Within our research, partly by help of students but mostly in collaboration with some local small high-tech enterprises, we started a development of a complex informational infrastructure that is supposed to be used as a support to sustainable product design [14]. It is expected to be used by students preparing projects in most of the subjects described, and as a production company resource, as well.

Here the tasks, the goals, concepts and preliminary result will be described for such an attempt. The objectives were to build an integrated information system that will be open both vertically and horizontally. It will include XML Web Services and software tools for documenting services together with software tools for life-cycle evaluation services. The system under development is depicted in fig. 2. A front-end substructure was designed in order to enable communication between the project manager and production, customers, market partners and research and development departments. Several data-bases (Db) were to be designed and created, such as material, components, modules, products technologies (production, recycling) standards, legislation, professional and scientific literature, life cycle, important research institutions, and similar. The middle layer of the system consists of a set of development tools (CAD/CAM/CAE, Products and Technology Documenting Tools, Life Cycle and Cost/Benefit Evaluation) and SW tools for working with Databases. The system is intended to be organized as company products development system, which can be distributed as multi location system, or it can be organized as service for products development, which could be used on rental base.

Among other, SW tools for Life cycle and Cost/Benefit will be introduced in order to provide life-cycle simulation and Cost/Benefit evaluation as well as to define recycling logistics and technology. This part of the system has a subsystem to document all this, based on solution already described earlier. To allow Life cycle data acquisition separate B^2 service and B-C service will be provided, so that product sales and service network as well as customers can submit appropriate data.

Project management system is going to be built to provide tool for product design teams management. Messaging within the system will be done using standard products.

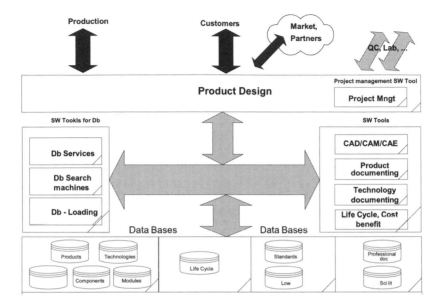

Figure 2:   Structure of an information system supporting sustainable design.

Integrated Information Infrastructure for Sustainable Product Design System described here can be implemented as a company resource or as service resource available for public use or to be used as commercial service.

When used as company resource, based on described functionality it can serve development activities on the company Intranet. Its functionality can be extended to Extranet to allow business partners to be included in. Db-s could be both centralized and distributed, but to the users they will be exposed as unique Db-s, provided that they will be loaded according to the procedures which will be built as admin tools. Built in XML Web services will provide efficient.

The system can be organized and used as Internet service resource, either public or commercial one. For this purpose administration of the system is customized to allow security administration for users of the system.

5   Conclusion

A postgraduate MS level curriculum is described that started in the year 2002 at the Faculty of Electronic Engineering, University of Nis, Serbia and Montenegro. Contents of subjects are given together with comments related to some aspects of the realization of the educational process. Some research results are mentioned that are, among other, intended to become available to students in the future.

References

[1]   Appelbaum, A., Europe Cracks Down on E-Waste, *IEEE Spectrum*, **39(5)**, pp. 46-51, 2002.
[2]   *Eureka E! 2009, Strategic Comprehensive Approach for electronics REcycling and Re-use, SCARE*, Green Book, International CARE "VISION 2000" Office, Wiener Neustadt, 2000.
[3]   Ferrendier, S. et all., *Environmentally improved product design case studies of the European electrical and electronics industry*, ECOLIFE Thematic Network, Eco-design subgroup, 2002.
[4]   Goldberg, L., *Green electronics/Green bottom line*, Newnes, Boston, 1999.
[5]   Milojkovic, J., and Litovski V., Eco-design in Electronics - The State of the Art, *Facta Universitatis, Series: Working and living Environment Protection*, University of Nis, **2(2)**, pp. 87-100, 2002.
[6]   Wimmer, W., Zust, R., & Lee, K., *Ecodesign Implementation*, Springer, The Netherlands, 2004.
[7]   Kuehr, R., & Williams, E., *Computers and the Environment*, Kluwer Academic Publishers, Dordrecht/Boston/London, 2004.
[8]   Stevels, A., Ecodesign for competitive advantage, *Proc. of the 1st regional conference for manufacturing companies: Eco-design for Competitive Advantage*, Inst. of Mechanical Engineers, London, pp. 1.1-1.5, 2001.
[9]   Balakrishnan, S., and Pecht, M., *Placement and routing of electronic modules*, Marcel Dekker, New York, pp. 95-96, 1995.
[10]   Litovski, V, and Zwolinski, M., *VLSI circuit simulation and optimisation*, Chapma and Hall, London, 1997.
[11]   Roll, M., and Ehrenfeld, J., *Implementing design for environment: A primer*, Digital Equipment Corporation, Maynard, MA, pp. 15-16, 1997.
[12]   *ISO 14040, Environmental Management - Life Cycle Assessment - Principles and Framework*, Berlin, Deuth, 1997.
[13]   Weidema, B.P. Improving the performance of LCA, *Proc. of the 2. International Conference on EcoBalance*, Tsukuba, pp. 247-252, 1996.
[14]   Jovanovic, S., Milojkovic, J., and Litovski, V, Information Infrastructure for Sustainable Product Design System, *Proceedings of the Symposium: e-ecological manufacturing*, Berlin, pp. 75-77, 2003.

Section 12
Biological treatment of water

Biological transformation of PCBs in hazardous site waste sludge

J. Tharakan
Department of Chemical Engineering, Howard University, Washington, DC, USA

Abstract

In this paper we review the use of biological technologies for the treatment of PCB contaminated hazardous wastes and report on some results that have been obtained from our laboratory on the utilization of two specific biological technologies for PCB biotransformation, including a microbial amended anaerobic-aerobic cycling bioreactor and an earthworm-inoculated vermicompost bioreactor. Our review of PCB biological transformation research suggests a large potential for biological transformation of PCBs using anaerobic reductive dechlorination, with several mediating organisms and effective pathways identified. The bulk of the research demonstrates reductive dechlorination of higher chlorinated PCB congeners accompanied by the appearance of increased levels of lower chlorinated congeners. Aerobic biodegradation using cosubstrates demonstrated degradation of lower chlorinated PCBs, mostly as a cometabolic function of cosubstrate oxidation. In our laboratories, actual hazardous waste sludge contaminated with PCBs demonstrated reductions of as much as 75% in an anaerobic-aerobic bioreactor amended with PCB-dechlorinating anaerobic sediments. The addition of PCB cometabolizing aerobic microbes did not reveal significant additional PCB reductions. Our research with earthworms demonstrated PCB removals, mainly through bioaccumulation in earthworm biomass, with little evidence of further PCB biotransformation. However, these results do suggest high potential for microbial or vermicompost systems and further research is warranted to establish mechanisms and elucidate procedures for biological transformation of PCBs in hazardous sludges.
Keywords: polychlorinated biphenyls (PCBs), biotransformation, microbial, hazardous waste, earthworms.

WIT Transactions on Ecology and the Environment, Vol 92, © 2006 WIT Press
www.witpress.com, ISSN 1743-3541 (on-line)
doi:10.2495/WM060451

1　Introduction

Polychlorinated biphenyls (PCBs) are a family of compounds that were produced commercially by the direct chlorination of biphenyl, and there are 209 possible PCB congeners. PCBs were widely used in electrical transformers and a multitude of other industrial applications because of their excellent insulating properties, extremely low volatility and combustibility, as well as their stability and resistance to decomposition [1]. The extensive use of PCBs led to discharges that have resulted in an almost ubiquitous distribution of PCBs in the environment. The production of PCBs was banned in 1978 after accumulating evidence of their potential toxicity and carcinogenicity [2]. In the U.S. they are regulated under the Toxic Substances Control Act (TSCA) and the Comprehensive Environmental Response Compensation and Liability Act (CERCLA). Remediation of PCB contaminated sites is a national priority, but current technologies are expensive and often lead to byproducts that are environmentally problematic. Biological transformation of PCBs in hazardous waste site sludges is an attractive option because of its economic benefits and the potential for complete transformation of the waste components.

The objective if this paper is two-fold: to review the research status of biological treatment technologies for PCB contaminated sludges and to report on two biological treatment technologies that have been utilized for the remediation of PCBs in a hazardous site sludge, including an anaerobic-aerobic cycling bioreactor and a vermicompost bioreactor utilizing earthworms for the reduction of PCBs in the target sludge.

2　Traditional treatment of PCB contaminated wastes

The most common approved treatment technology for PCB contaminated wastes are land filling or incineration. Complete destruction of PCBs requires incineration at a temperature of 1200°C. Combustion at temperatures below this often leads to the generation of dioxins and other hazardous emissions. Hence, the degree of acceptance by the public for this technology is low; in addition, the high cost of incineration of around $2,300 per ton [3] renders it economically non-viable in most instances. Land filling of PCB contaminated sludges, soils and sediments is also an approved method of disposal [3], although there is opposition to this procedure due to the potential for PCB volatilization as well as escape of PCB through leaching and contamination of groundwater. Land filling also only contains the PCB contaminated wastes – actual elimination of PCBs does not occur. Finally, both these technologies are strictly *ex situ* technologies that require the excavation and transport of the PCB contaminated wastes. In the case of PCB contaminated sediments, dredging is especially troublesome as it results in the re-suspension of PCBs in the water and in making them more bioavailable. Excavation and dredging are also expensive technologies and project costs can be prohibitive.

3 Biological treatment of PCBs

Extensive research to date has demonstrated that under anaerobic conditions, reductive dechlorination of PCBs occurs resulting in the reduction of levels of higher chlorinated PCB congeners and the appearance of lower chlorinated PCB congeners [4]. Under the influence of anaerobic microcosms, reductive dechlorination results in the removal of chlorines from the *meta* and *para* positions of the biphenyl molecule and a concomitant increase in the levels of mono-, di-, tri- and tetra-chlorobiphenyls [2]. Sediments from the Hudson River have demonstrated this type of dechlorination, resulting in the predominance of *ortho* substituted PCB congeners, and there are eight dechlorination pathways that have been elucidated [5]. It is generally believed that methanogenic and sulphate reducing bacterial populations are responsible for other dechlorination pathways, although there are indications that dehalorespiring bacteria that utilize the halogenated compound itself as the terminal electron acceptor may be responsible for some of the reductive dechlorination that has been observed in the field [1, 6]. Several studies have identified individual microbial species that catalyze the reductive dechlorination of PCBs [7, 8].

In addition to anaerobic reductive dechlorination, aerobic biodegradation of PCBs has also been extensively investigated, beginning with the early work of Ahmed and Focht [9], who demonstrated that lightly chlorinated PCBs were degraded through the cometabolic oxidation of biphenyl through the catabolic biphenyl pathway. In this pathway, the biphenyl molecule is converted to a benzoic acid, eventually resulting in acetyl-CoA that is used in the tricarboxylic acid cycle [4]; however, the aerobic degradation of PCBs has generally been limited to trichlorobiphenyls, although more recent research has demonstrated the aerobic degradation of PCBs with as many as four or five chlorines using different cosubstrates including naphthalene and terpenes [10]. Although there are many microorganisms that code for the biphenyl pathway and hence are capable of cometabolically biodegrading PCBs, the actual mineralization of PCBs is rare. To get around this, there has been much work on the genetic engineering of microbes that can completely mineralize PCBs, and great potential has been demonstrated by some microorganisms that have demonstrated capability of attacking the resistant *ortho* positioned chlorine on the PCB molecule [11].

The capability of adapted anaerobic microcosms to reductively dechlorinate higher chlorinated PCB congeners and the ability of selected and/or engineered microbes to aerobically biodegrade lower chlorinated PCB congeners through cometabolic oxidation of suitable cosubstrates has lead to the investigation of cycling anaerobic-aerobic biotreatment schemes. Such a scheme in our laboratories has demonstrated the potential to reduce total PCB levels in soils spiked with a mixture of PCB congeners including Aroclor 1248 [12].

The investigation reported here utilized an anaerobic-aerobic cycling bioreactor that was amended in sequence with anaerobic sediments and then with aerobic microbes. We also report on the results of earlier investigations that utilized earthworms in a vermicomposting bioreactor to examine the potential for biotransformation of PCB in hazardous sludge. Both sets of investigation utilized sludge directly from a TSCA hazardous waste site and extends our earlier investigations that utilized laboratory soils spiked with PCB mixtures.

4 Materials and methods

4.1 PCB Contaminated hazardous sludge

The PCB contaminated hazardous sludge samples were from the Gary Sanitary District's in Gary, Indiana from their Ralston Street Lagoon. PCB concentrations (reported as Aroclor 1248) averaged around 180 ppm (dry weight basis), ranging as high as 1200 ppm. In the anaerobic/aerobic microbially amended bioreactor study, two separate bioreactors were established with sludge samples from two lagoon locations, Midwest and Southwest, with initial total PCB levels of 1000 and 220 ppm, respectively. In the vermicompost bioreactors the Midwest sludge sample was used.

4.2 Anaerobic-aerobic cycling bioreactor set-up

The reactors were constructed from modified desiccators incorporating fluid delivery plumbing with 4.5 L total reaction volume. An initial 1.5 L layer of gravel was covered with 3.0 L of the sludge-soil mixture. Sludge was air dried, ground and mixed with 1.5 L of sand. The bioreactors were operated in an anaerobic glove box; initially bioreacors were amended with 200 ml of anaerobic sediment. Reduced anaerobic mineral media (RAMM) was circulated through the sludge mixture which was maintained under nitrogen for the duration of the anaerobic stage. Bioreactors were then switched to aerobic conditions and inoculated with *Rhodococcus erythropolis* (strain NY05), a gram positive aerobic cocci isolated from the Hudson River (NY). This strain has earlier demonstrated cometabolic biotransformation of several different PCBs [12]. Aerated minimal salt media (MSM) was then circulated through the bioreactor for the rest of the aerobic phase.

4.3 Vermicompost experimental setup

Vermicomposting bioreactors (VBs) were established with sludge volume fractions ranging from 0%, 10%, 25%, 50% and 75% (v/v). Control VBs without earthworms were simultaneously initiated. Sterile potting soil was mixed with dried sludge in appropriate fractions and layered over gravel, followed by inoculation with 9 gms of earthworms. VBs were supplemented with 3 gms of cornstarch on a weekly basis. VBs were covered with mesh to prevent earthworms escaping; VB walls were covered with aluminium foil to block light and minimize PCB photodegradation. At the end total earthworm biomass in different VBs was measured. The total sludge in the VBs was also weighed and dried for extraction and analysis. The sludge-bedding matrix was dried, extracted and analyzed for PCBs. Total earthworm biomass was dried, extracted and analysed. This permitted the completion of the mass balance on fate and distribution of PCBs.

4.4 PCB measurement

Samples were air dried and then Soxhlet extracted prior to injection into gas chromatograph (HP 5890 Series II GC) utilizing a 0.32 mm internal diameter, 30 m fused silica column with a 0.5 μm film ECD. Calibration curves were generated with Aroclor 1248. Protocols followed Quality Assurance Procedures established by the site regulators [13].

5 Results and discussion

Total PCB levels decreased over time in all VBs regardless of the sludge volume fraction. Overall reductions of around 80% were achieved [14]. The worm-free control VBs revealed that PCB levels were being reduced here as well, albeit to different levels. A mass balance model was developed in order to discriminate between PCB biotransformation, bioaccumulation and abiotic PCB removal, including irreversible adsorption to surfaces, volatilization, photodegradation, as well as PCB loss in sampling. In the calculations used in this model, volatilization and photodegradation were assumed to be negligible, due to the extremely low volatilization rates for PCBs and because VBs were light-protected. Irreversible adsorption was assumed negligible due to rigorous extraction protocols. The total PCB mass (in mg) at the outset of the vermicompost experiments and that at the end are shown in Tables 1 and 2, for VBs with and without earthworms, respectively, along with the percentage reduction of PCB. The percentage reduction is calculated after accounting for PCB loss due to sampling and, in the case of the VBs with earthworms, bioaccumulation in the earthworm biomass. With the earthworms, there is a slight demonstrated increase in percentage PCB reduction with the earthworms present, with the largest increase in percentage reduction (14%) seen with the VB established at a sludge fraction of 50% on a volume basis. Interestingly, the VBs with both the highest and lowest initial PCB level demonstrate no appreciable increase in PCB reduction percentages with the addition of the earthworms.

The bioaccumulation of PCBs was also measured and the data demonstrated that there was an increase in bioaccumulation as the PCB concentrations in the surrounding sludge-soil matrix increased. This data is presented in Table 3, which also includes data on the increase in earthworm biomass. This data suggests that there is some enhancement in PCB reduction with the addition of earthworms, although it is not clear how effective a strategy this may be, given the variability seen in percentage PCB reduction with different initial levels of PCB. The data on biomass increase demonstrates that the PCBs are detrimental to earthworm viability and health, contributing to significant increases in biomass reduction and viability as the PCB concentration is increased.

Table 1:   PCB concentrations and reductions in VBs with earthworms.

Sludge Mass Fraction % (V/V)	Initial Total PCB Mass in VB (mg)	Final Total PCB mass in VB (mg)	% PCB Reduction*
10	10.3 + 0.7	1.2 + 0.1	80.9
25	30.6 + 1.2	3.8 + 0.4	81.6
50	60.3 + 2.5	4.3 + 0.01	90.3
75	95.8 + 3.9	12.3 + 0.7	82.9

*After accounting for PCB mass lost to sampling and in earthworms.

Table 2:   PCB concentrations and reductions in control VBs without earthworms.

Sludge Mass Fraction % (V/V)	Initial Total PCB Mass in VB (mg)	Final Total PCB mass in VB (mg)	% PCB Reduction*
10	9.8 + 0.1	1.3 + 0.2	78.7
25	29.5 + 1.2	2.9 + 0.1	84.5
50	59.8 + 2.4	10.0 + 0.7	77.0
75	93.9 + 3.8	10.9 + 0.2	85.6

*After accounting for PCB mass lost to sampling.

Table 3:   PCB bioaccumulation and percentage earthworm biomass increase.

Sludge Mass Fraction % (v/v)	Final PCB Mass Bioaccumulated in Earthworms	Final Earthworm Biomass Increase (%)
0*	0	104
10	148 + 18	86
25	213 + 17	75
50	189 + 9	52
75	313 + 9	4

*Control VB with earthworms and no sludge.

In the anaerobic-aerobic cycling bioreactors, the results of these investigations have demonstrated that the bioreactor with the higher initial PCB level demonstrated a 75% reduction in PCB level at the end of the anaerobic phase. The percentage PCB reduction that was achieved following the anaerobic phase and at the end of the aerobic phase was not significantly increased. The bioreactor with the lower initial PCB level only revealed an initial reduction of 25%. Subsequent to this initial reduction in the anaerobic phase, there were no significant further reductions, even after switching the bioreactor to aerobic conditions, as has already been reported [15].

The extensive research on biological treatment of PCB contaminated soils has mostly focused on the use of clean soils and systems that have been spiked with varying levels and compositions of different PCB congeners. Anaerobic reductive dechlorination is the mechanism by which higher chlorinated PCB congeners get reduced in sludges and in aquatic sediments [1, 4], which occurs through the replacement of chlorine by other substituents, usually hydrogen [4]. The different dechlorination pathways that have been identified each have specificity for different congeners and different chlorination positions [2]. Aerobic degradation and ring cleavage has also been demonstrated, beginning with the investigations conducted in the early 1970s and the catabolic biphenyl pathway has been well characterized [16], but the degree of chlorination of the PCB congeners susceptible to aerobic breakdown is generally three or less.

The cycling anaerobic-aerobic bioreactors did, however, reveal changes in the composition of the PCB congeners at the end of each phase. Table 4 shows the presence or absence of particular congeners present in the sludge in the initial sludge sample, at the end of the anaerobic phase and at the end of the aerobic phase of the bioreactor study. As the data demonstrate, reductive anaerobic dechlorination does occur, while demonstrating aerobic reduction of certain congeners and recalcitrance of others.

Table 4:   Presence or absence of specific PCB congeners.

PCB Congener	Present in Initial Sludge	Present at End of Anaerobic Phase	Present at End of Aerobic Phase
Biphenyl	+	--	--
2,2'-diCB	+	--	--
4,4'-diCB	+	+	+
2,4,4'-triCB	+	+	+
2.2',5-triCB	--	+	--
2,3,4,5-tetraCB	+	+	--
2,3',4',5-tetraCB	+	+	--
2,3',5,5'-tetraCB	+	+	+
2,2',4,6-tetraCB	+	--	--
2,2',5,6'-tetraCB	+	+	+
2,2',4,5'-tetraCB	--	--	+

The results of these investigations confirm that anaerobic dechlorination and aerobic biodegradation, when conducted in sequence, affect the PCB congener composition in the sludge. Vermicomposting results demonstrate bioaccumulation with possible biodegradation, likely due to the combined effects of earthworm gut microbial ecology and earthworm digestive actions [17–19].

6   Conclusion

The results from the anaerobic phase of the anaerobic/aerobic experiment showed evidence of biotransformation, likely due to amendment of sludge with anaerobic Hudson River sediments. The results suggest suitable amendments would support anaerobic dechlorination for biotransformation of sludge contaminated with PCBs; this may be a potentially viable option for on-site remediation.

The vermicomposting study results demonstrate earthworms can bioaccumulate PCBs from surrounding sludge in proportion to sludge PCB levels. Only minimal biotransformation of PCBs was demonstrated. However, since PCB concentrations in sludge decrease after addition of earthworms, this may be a methodology to reduce total mass of sludge requiring final disposal, especially if all the PCBs can be transported into the earthworms. As a treatment and elimination technology, vermicomposting requires further research before it can be a viable alternative for PCB removal.

References

[1]   Abraham, W.R., B. Nogales, P. Golyshin, D. Pieper and K. Timmis, (2002) "PCB-Degrading Communities in Soils and Sediments, "*Current Opinion Microbiology*, **5** (3) 246 – 253.
[2]   Wiegel, J. and Q. Wu, (2000) "Microbial Reductive Dehalogenation of PCBs" *FEMS Microbial Ecology,* **32** (1) 1- 15.
[3]   US EPA (2003) "Management of PCBs in the USA" Office of Pollution Prevention and Toxics, Washington, DC.
[4]   Bedard, D. (2003) "PCBs in Aquatic Sediments: Environmental fate and Outlook for Biological Treatment," in Haggblom and Bossert (eds) *Dehalogenation: Microbial Processes and Environmental Applications*, Kluwer Press, 443 – 465.
[5]   Zwiernik, M., J. Quensen, and S. Boyd (1998) "FeSO4 Amendments Stimulate Extensive Anaerobic PCB Dechlorination" *Environmental Science and Technology,* **32** (21) 3360 – 3365.
[6]   Kim, J., and G-Y Rhee (1997) "Population Dynamics of PCB Dechlorinating Microorganisms in Contaminated Sediments," *Appl. Env. Microbiol.* **63** (5) 1771 – 1776.
[7]   Cutter, L., J. Watts, K. Sowers and H. May (2001) "Identification of a Microorganism that Links its Growth to the Reductive Dechlorination of 2,3,5,6-Chlorobiphenyl" *Env. Microbiol.* **3** (11) 699 – 709.
[8]   Wu. Q., *et al* (2002) "Identification of a Bacterium that Specifically Catalyzes the Reductive Dechlorination of PCBs with Double Flanked Chlorines" *Appl. Env. Microbiol.* **68** (2) 807 – 812.
[9]   Ahmed, D., and D. Focht (1973) "Degradation of PCBs by two Species of *Achromobacter" Can.J.Microbiol.* **19**, 47 – 52.
[10]   Chawla, R.C., R. Liou, J.H. Johnson, Jr., J.P. Tharakan, "Biodegradation of PCB's in Aqueous and Soil Systems," in Wise, Trantolo, Cichon, Inyang, and Stottmeister (eds) Bioremediation of Contaminated Soils: New York, Marcel Dekker, Inc., 2000.
[11]   Tsoi, T., E. Plotnikova, J. Cole, W. Guerin, M. Bagdasarian and J. Tiedje "Cloning, Expression and Nucleotide Sequence of the *Pseudomonas*

aeruginosa 142 *ohb* Genes Coding for Oxygenolytic *ortho* Dehalogenation of Halobenzoates" (1999) *Appl. Env. Microbiol.* **65** (5) 2151 – 2162.

[12] Tharakan J.P., E. Sada, R. Liou and R. C. Chawla, "Transformation of Aroclor by Indigenous and Inoculated Microbes In Slurry Reactors," in Alleman and Leeson (eds) *In Situ* and On-Site Bioremediation: The fifth International Symposium Proceedings, Batelle, Columbus, OH, 1999.

[13] Gary Sanitary District, Indiana. Standard Operating Procedure for the Determination of PCB's As Aroclors by GC Capillary Column Technique for Gary Sanitary District. Southwest Lab of Oklahoma, Inc., 1996.

[14] Tharakan, J., A. Addagada, D. Tomlinson, and A Shafagati, "Vermicomposting for the bioremediation of PCB Congeners in SUPERFUND site media," in Popov et al (eds) *Waste Management and the Environment II,* WIT Press, Southampton, UK, pp.117 – 124 (2004).

[15] Tharakan, J., D. Tomlinson, A. Addagada and S. Shafagati (2006) "Biotransformation of PCBs in Contaminated Sludge: Potential for New Biological Technologies," *Engineering Life Sciences,* In Press.

[16] Sylvestre, M. (2004) "Genetically Modified Organisms to Remediate PCBs: Where Do We Stand" *Intl. Biodeterioration and Biodegradation,* **54** 153 – 162.

[17] Edwards, C. A., Neuhauser, E. F., Earthworms in waste and environmental management, pp: 321-328 (1988).

[18] Fanelli et al., (1980a). Routine analysis of 2,3,7,8-Tetrachlorodibenzenzo-p-dioxin in biological samples from the contaminatcd arca of Scvcso, Italy. *Bull Environ Contam Toxicol* **24:**818-823.

[19] Kreis et al., (1987). The dynamics of PCB's between earthworm populations and agricultural soils. *Pedobiologia* **30:**379-388.

Aerobic biological treatment of landfill leachate

R. J. Matthews, M. K. Winson & J. Scullion
*University of Wales Aberystwyth, Institute of Biological Sciences,
Penglais, Aberystwyth, UK*

Abstract

There is an ongoing need to treat leachates from landfills using approaches that
avoid expensive installation and operating costs. Faced with such a problem,
Powys County Council (Wales, UK) developed a treatment system based on
practical experience. Leachate was re-circulated through aeration towers
containing a biofilm supported on plastic media before being polished in
reed/filter beds.

Investigations were undertaken to evaluate the performance of these
processes. Replicated model aeration towers (1/300 site scale) were used to
establish treatment rates for pollutants in leachates of varying strengths; the
effects of temperature and variations in pollutant concentrations were also
evaluated. Data from these model experiments were corroborated with findings
from pilot-scale plants (1/10 site scale) operating on landfill sites. Assessment of
treatment performance was based on the degree of amelioration of standard
chemical (ammoniacal-N and total organic carbon) parameters. A range of
related parameters including nitrate, pH and redox potential were measured in
support of these assessments.

Model system experiments indicated treatment rates at 15 °C (883-1895 mg
NH_3-N m^{-3} h^{-1}; 347-1600 mg TOC m^{-3} h^{-1}) that were similar across a wide range
of leachate concentrations (37-1880 mgl^{-1} ammoniacal-N; 130-5315 mgl^{-1} COD).
Marked changes in the concentration of leachates did not affect treatment
efficiency following a short lag effect. Process rates roughly halved with each
5°C decrease in temperature below 15°C, significant nitrification was maintained
at temperatures of 0.5-1 °C. High variability in treatment capacity was observed
for individual plastic media. Treatment rates in pilot scale plants were slightly
higher but broadly consistent with those obtained in model systems.
Keywords: leachate treatment, aerobic, temperature, leachate strength, biofilm.

WIT Transactions on Ecology and the Environment, Vol 92, © 2006 WIT Press
www.witpress.com, ISSN 1743-3541 (on-line)
doi:10.2495/WM060461

1 Introduction

Long-term, cost effective treatment of polluting emissions from landfills is important [1]. As water percolates through wastes emplaced in a landfill it dissolves organic and inorganic components and decomposition products to produce a potentially polluting liquid leachate. It is only relatively recently that the treatment of leachate has started to receive the type of thorough investigation which has been normal practice for sewage and other industrial effluents [2].

The putrescible components in municipal solid waste (MSW) have the potential to generate highly polluting and toxic leachates as they degrade. This degradation progresses through a series of more or less well defined stages towards maturity and eventual stability [3]. Leachate composition varies between stages, being highly enriched with organic breakdown products initially, then acidic with elevated concentrations of metal, chloride, ammonium and phosphate ions [4], then with higher pH and gradually declining concentrations of ammoniacal-N and organic concentrations through the prolonged methanogenic phase. In practice, leachates will be generated by wastes at various stages of degradation on all but the older closed sites. The timescale for each of the degradation stages has been shown to vary considerably according to; the type and nature of the emplaced wastes [5], landfill management practices [6] and local environmental conditions [7].

The composition of aged MSW leachates is such that they may be both polluting [8] and toxic [9] capable of producing severe environmental impacts especially in vulnerable recipients such as aquifers and surface waters. These effects may include eutrophication or toxic effects on aquatic organisms resulting from ammonia, heavy metals or organic compounds. Ammoniacal-N concentrations often present more of a long-term problem, than the leaching of degradable organic substances such as volatile fatty acids [10].

Remediation of leachate has been achieved primarily by aerobic biological treatment [11]. Although these technologies have developed in operational complexity to enhance the rate or extent of the biological conversions taking place [12], the fundamental biological reactions have remained essentially the same [13]. For leachates containing high concentrations of BOD, treatment processes (aerated lagoons, sequencing batch reactors, activated sludge) operating with flocculated heterotrophic micro-organisms in suspension and the assimilation of ammoniacal-N into biomass, have been widely reported [14]. For methanogenic leachates, nitrification utilising autotrophic, fixed microbial film processes (biofilm towers, trickling biofilters, rotating biological contractors) have been shown to be generally effective in reducing ammoniacal-N concentrations [15].

This study examines the effectiveness of a leachate treatment technology developed for former MSW landfill sites managed by Powys CC in Mid Wales, UK. Previous on-site investigations [16] indicated that leachate treatment was achieved primarily within aeration towers, so this treatment stage became the focus of more detailed study using replicated 'model' systems and pilot scale systems for on-site validation of findings.

2   Materials and methods

Replicated model systems (1/300 of typical site scale) of aeration towers filled with plastic media (Mass Transfer Cascade Filterpak ™ Stoke, UK) were used to establish treatment rates for leachates of varying strengths (see Figure 1 for schematic) operating under different temperature regimes.

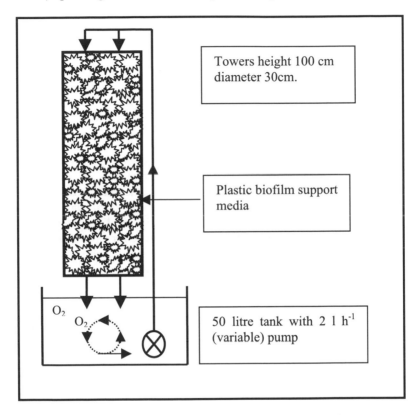

Towers height 100 cm
diameter 30cm.

Plastic biofilm support
media

O_2

O_2

50 litre tank with 2 l h^{-1}
(variable) pump

Figure 1:   Schematic of model system experimental set-up.

Data reported are for a series of experiments using leachates from three landfill sites with widely varying characteristics (Table 1); Nantycaws 2 and Trecatti were operational. For practical reasons, experiments involved treating leachate in batches to completion. Experiments typically included 3-4 replicates depending on the number of treatment comparisons. Findings from these model experiments were corroborated with data from pilot-scale plants (1/10 of typical site scale).

Leachates were treated in temperature controlled growth rooms (4 replicates) at temperatures of 0.5-1, 10, 15 and 30 °C; the first three temperatures were chosen to represent the predominant range under UK conditions whilst the highest temperature may represent close to 'optimum' conditions for nitrification

[17]. Data presented here are for one site (Borth) only; data comparing treatment rates for different leachates are for Borth, Nantycaws 1 & 2 and Trecatti leachate at 15 °C. Air and leachate temperatures were routinely monitored during pilot plant trials; for the data presented, mean leachate temperatures were 11, 16.5, 13 and 16.1 °C for Borth, Nantycaws 1, Nantycaws 2 and Trecatti respectively. Repeat runs were treated as replicates for the pilot plant.

Table 1: Typical characteristics of landfill leachates included in studies.

Site	pH	NH$_4$-N mgl^{-1}	Total Fe mgl^{-1}	COD mgl^{-1}	BOD mgl^{-1}
Borth	7.5	177	10.3	137	42
Nantycaws 1	7.7	37	0.5	131	22
Nantycaws 2	8.1	1880	2.1	5315	639
Trecatti	7.9	1195	3.4	1575	124

Treatment rates were routinely recorded following changes in leachate concentration or composition to identify any differences in treatment rate resulting from variations in the prior history of individual model towers. Data are presented for one occasion where intensive monitoring was undertaken. Replicate towers were exposed to full strength (130 mg l^{-1} ammoniacal-N) or fully treated (< 1 mg l^{-1} ammoniacal-N) Borth leachate for several days, then all towers were switched to full strength Borth leachate and changes in ammoniacal-N recorded over the subsequent days treatment completion.

Individual media elements were investigated to determine the degree of heterogeneity in nitrification potentials within the model systems. These were submerged in solutions of 20 mgl^{-1} ammoniacal-N (pH 6.3) and placed on orbital shakers (20 °C). Ammoniacal-N concentrations were measured initially and twice daily until treatment completion.

Ammoniacal-N was determined [18] using a Cecil Series 2 spectrophotometer measuring absorbance at 630 nm. Nitrate was determined by anion exchange chromatography (Dionex QIC Ion Chromatograph); samples were filtered (< 0.2 μm) prior to injection. Total organic carbon was measured using a Shimadzu analyser (TOC-5050). The pH and Eh of samples were determined routinely using Hanna (301) meters. Dissolved oxygen concentrations were recorded using a Jenway (970) oxygen meter.

Treatment rates were determined by fitting linear regressions to changes in concentrations with time for individual replicate towers. Rates were then re-calculated as mg m^{-3} media h^{-1}. Most experiments, involved in excess of 10 sampling occasions with sampling intervals varied according to the period of the experimental run as affected by leachate strength. There was no evidence of these rates being affected by substrate concentrations except when treatment approached completion as illustrated in Figure 2; regression R^2 values were consistently > 90 and usually > 95%. Mean process rates were then compared by one-way analysis of variance.

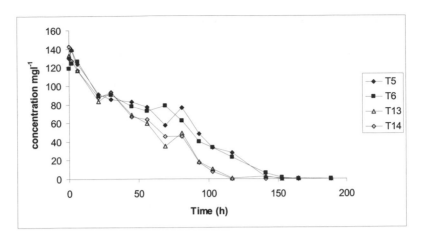

Figure 2: Linear reductions in ammoniacal-N concentration (Borth leachate) for 4 replicate model tower systems.

3 Results

Effects of varying temperature on treatment of the Borth leachate are described in Table 2. Nitrification rates and nitrate-N production rates broadly corresponded when ammonification of organic-N is taken into account. Between 0.5 °C and 15 °C process rates roughly doubled for each 5 °C temperature increment, with rates at 15 °C roughly a third of the optimum 30 °C rates. Significant amelioration of leachate was maintained even at temperatures close to freezing point. For this leachate, pH remained above 7.5 throughout the treatment period; apart from the initial 2 hour period when dissolved oxygen concentrations and redox potentials were low, these parameters stabilised in excess of 7 mg l^{-1} and 170 mV respectively for most of the remaining period.

Table 2: Effects of varying temperature on process rates (mg $m^{-3}h^{-1}$) for Borth leachate; a) ammoniacal-N, b) nitrate-N and c) organic carbon.

a)

Temperature °C	0.5	10	15	30
Mean	200	469	861	3109
Standard deviation	30.3	119.8	133.9	94.7

b)

Temperature °C	0.5	10	15	30
Mean	169	505	765	3353
Standard deviation	42.0	142.8	119.0	236.6

c)

Temperature °C	0.5	10	15	30
Mean	408	1097	1600	4343
Standard deviation	18.9	290	123.5	611.8

Comparisons of process rates for different leachates at 15 °C are presented in Table 3. Ammoniacal-N concentrations decreased more rapidly for the stronger leachates, particularly Nantycaws 2. This difference was not reflected in nitrate production rates. Ammonia volatilisation detected for the Nantycaws 2 leachate could not fully explain this discrepancy; denitrification was detected and this occurrence was consistent with low Eh values (c. 100 mV) for much of the early treatment of this leachate. Organic C treatment rates for the Nantycaws 1 leachate, which had low initial OC, were notably lower than for other leachates. Values for the two operational sites were intermediate and highly variable

Table 3:   Process rates at 15 °C constant temperature (mg m^{-3} h^{-1}) for a) ammoniacal-N, b) nitrate-N and c) organic carbon for model systems.

a)

Leachate	Borth	Nantycaws 1	Nantycaws 2	Trecatti
Mean	968	883	1895	1335
Standard deviation	123.5	327.9	242.7	97.7

b)

Leachate	Borth	Nantycaws 1	Nantycaws 2	Trecatti
Mean	861	1191	982	1095
Standard deviation	133.9	266.6	617.1	529.7

c)

Leachate	Borth	Nantycaws 1	Nantycaws 2	Trecatti
Mean	1600	347	1304	1024
Standard deviation	123.5	136.0	635.5	415.8

Table 4:   Process rates at variable temperature (mg m^{-3}h^{-1}) for a) ammoniacal-N, b) nitrate-N and c) organic carbon for pilot plant systems.

a)

Leachate	Borth	Nantycaws 1	Nantycaws 2	Trecatti
Mean	2428	2184	2333	2181
Standard deviation	4.8	422.7	57.7	508.6

b)

Leachate	Borth	Nantycaws 1	Nantycaws 2	Trecatti
Mean	1636	2651	1656	1050
Standard deviation	221.2	72.7	7.8	378.2

c)

Leachate	Borth	Nantycaws 1	Nantycaws 2	Trecatti
Mean	2582	3404	1254	354
Standard deviation	455.2	1318	619.3	137.7

Batches of leachate were treated in pilot plants on each of the landfill sites (Table 4). In making comparisons between model and pilot plant systems differences in temperature and general environmental conditions must be considered. Rates of ammoniacal-N reduction and nitrate-N production were generally comparable and broadly consistent with data from model systems with the exception of the very high OC values for Nantycaws 1.

Nitrification capacity (Figure 3) was highly variable between individual media and there was some evidence of a trend towards a reduction in this capacity with depth in the model system towers.

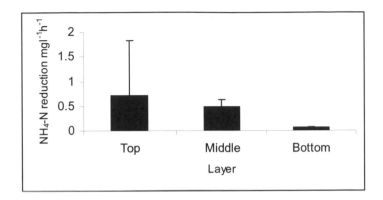

Figure 3: Reductions in ammoniacal-N concentrations for individual media (6 reps) in standard solutions for 3 days (error bars = standard deviations).

Treatment rates were affected (Table 5) by the nature of the preceding leachate to which model systems were exposed. Towers with low strength leachate treated ammoniacal-N at rates about two thirds those of towers treating stronger leachate; this difference was not apparent beyond the first day of the trial.

Table 5: Reduction in ammoniacal-N after 1 and 15 h as affected by previously treated leachate (H = 130 and L = <1 mg l^{-1} ammoniacal-N).

Previous leachate	H	L	H	L
Time (h)	1	1	15	15
Mean mg l^{-1}	9.8	6.3	96.1	66.9
Standard deviation	5.26	5.12	5.70	24.14

5 Discussion

Work undertaken by Tschui *et al.* [19] has shown the specific growth rates of nitrifying bacteria are extremely slow compared with those of heterotrophic bacteria, necessitating the lowering of organic loading rates to reduce the competition between heterotrophic and autotrophic bacteria [20]. Our results indicate that heterotrophic and nitrifying bacteria can co-exist within this treatment system at least across the range of leachates tested.

Nitrification has been demonstrated in soil close to 0 °C [21]. Kettunen *et al.* [22] have reported ammoniacal-N removal rates of between 65-99% respectively at process residence times of 2.7 and 4.2 days, for aerobic leachate biotreatments at temperatures within the range 1-5 °C. Nitrification occurs optimally within the range 25-35 °C [17]. Findings reported here indicate process rates at 0.5 °C 7, 5 and 9% those of optimum values for ammoniacal-N, nitrate-N and organic carbon respectively. In the UK where the majority of leachate production is concentrated during the winter months, data for optimum temperatures are often unrealistically high [11]. Based on laboratory scale experiments the effects of temperature on the rate of nitrification have been estimated at 12% $°C^{-1}$ within the temperature range 7-20 °C [23]. Our results are again broadly consistent with these general findings.

Nitrification requires a significant amount of oxygen (4.33g O_2 per gram of NH_4-N) [24]. The natural passage of air through voids present in the biofilm support media [25] has been shown to avoid oxygen limitation and in our trials dissolved oxygen concentrations in bulk leachate were maintained close to saturation even for the strongest leachates.

The oxidation of free ammonia to nitrite consumes two moles of alkalinity per mole of ammonia oxidised [26] and may result in acidification. Acidic pH in poorly buffered leachates reduces the concentration of free ammonia and may also inhibit oxidation by reducing the specific growth rate of the bacteria [27]. Although the optimal pH required for nitrification has been estimated at 8.5 [28], our experiments indicate sustained rates of nitrification at pH's < 7.00 although treatment stalling was observed at variable pH's below this value.

It has also been suggested that ammonia may inhibit nitrite oxidation by *Nitrobacter* bacteria. However, more recent studies [27] have however shown that significant variability exists in the degree of inhibition. Ammonia present in high concentrations has also been reported to inhibit its own oxidation by *Nitrosomonas* bacteria but again the degree of inhibition is variable [14] and may be influenced by temperature and pH [29]. There was no evidence of any such inhibition across the range of ammoniacal-N concentrations and pH ranges tested in the trials reported here.

Although there was some evidence of delayed response to marked changes in leachate concentration, the treatment system evaluated proved to be robust both at model and pilot scales across a very wide range of leachate characteristics encompassing long closed to fully operational sites. Clearly there is high spatial and perhaps temporal variability in process capacity within the tower media. Some degree of variability is inevitable but there may be scope for further improvements in treatment rates if some of this variability could be avoided.

Acknowledgements

This work was jointly funded by Biffaward and Biffa Waste.

References

[1]   Haarstad, K. & Mæhlum, T., Important aspects of long-term production and treatment of municipal solid waste leachate. *Waste Management and Research* **17,** pp. 470-477, 1999.

[2]   Williams, P.T., *Waste Treatment and Disposal.* John Wiley & Sons Ltd, West Sussex, England, 1999.

[3]   Heyer, K.U. & Stegmann, R. 1997. The long-term behaviour of landfills: Results of the joint research project Landfill Body. Sardinia, 6th International Landfill Symposium, Conference Proceedings, CISA, Caliari, Italy, Vol. **1,** pp. 73-87.

[4]   Zweifel, H.R., Johnson, C.A. & Holm, E., Long-term analyses of the main leachate constituents from old disposal sites. *Müll und Abfall,* **31**, pp. 727-732, 1999.

[5]   Department of the Environment (DoE), A review of the composition of leachates from domestic wastes in landfill sites. Report No. CWM 072/94. In: Waste Management Paper 26B. *Landfill Design, Construction and Operational Practice.* HMSO, London, 1995.

[6]   Hjelmar, O., Johannessen, L.M., Knox, K., Ehrig, H.J., Flyvbjetg, J., Winther, P. & Christensen, T.J., *The composition and management of leachate from landfills within the EU.* In: Christensen, T.J., Cossu, R. and Stegmann, R. (Eds.) The 5th International Landfill Symposium. CISA, Environmental Sanitary Engineering Centre, Cagliari, 1995

[7]   Blight, G.E. & Fourie, A.B. 1999 Leachate generation in landfills in semi arid climates. *Geotechnical Engineering* **137,** pp. 181-188.

[8]   Gray, D.A., Mather, J.D. & Harrison, I.B., A review of groundwater pollution from waste disposal sites in England and Wales, with provisional guidelines for future site selection. *Quarterly Journal of Engineering Geology* 7, pp. 181-186, 1974.

[9]   Osborn, D., Malcom, H.M., Wright, J., Freestone, P., Wyatt, C. & French, M.C., *Assessment of the role of landfill operations in the contamination of wildlife by organic and inorganic substances.* National Groundwater and Contaminated Land Centre (EA). Report CWM 102/94. 1994.

[10]   Harrington, D.W. & Maris, P.J., The treatment of leachate: A UK perspective. *Water Pollution and Control* 20, pp.45-55, 1986.

[11]   Robinson, H.D. & Barr, M.J., Aerobic biological treatment of landfill leachates. *Waste Management and Research* 17, pp. 478-486, 1999.

[12]   Robinson, H.D. The UK: leading the way in leachate treatment. *Proceedings of the Institute of Waste Management (IWM)* 12, pp 4-10, 1996.

[13]   Britz, T.J. Landfill leachate treatment. In: Senior, E. (Ed.) *The Microbiology of Landfill Sites,* 2nd Ed. CRC Press, Boca Raton, Florida, USA, 1995.

[14]   Butler, P.R., Kitchin, J.E., Parry, C. & Lane, P.G. Characterisation of a sequencing batch activated sludge process for treatment of high ammonia landfill leachates. *Proceedings Institute of Wastes Management, March,* pp. 9-18, 1999.

[15]   Knox, K., Leachate treatment with nitrification of ammonia. *Water Research,* **19**, pp. 895-904, 1985.
[16]   Matthews, R.J. Field and laboratory investigations of biological landfill leachate schemes in Wales. MSc thesis, University of Wales, Aberystwyth, 2001.
[17]   Bock, B.R. Efficient use of nitrogen in cropping systems. In: Nitrogen in Crop Production. Hauck, R.D. (Ed.) American Society of Agronomy, pp. 273-294, 1984.
[18]   Harwood J.E & Kühn, A.L. A colorimetric method for ammonia in natural waters, Water Research 4, pp. 805-811, 1970.
[19]   Tschui, M., Boller, M., Gujer, W., Eugster, J., Mader, C. & Stengel, C., Tertiary nitrifcation in aerated pilot biofilters. *Water Science and Technology* 29, pp. 109-116, 1994.
[20]   Boller, M., Gujer, W. & Tschui, M., Parameters affecting nitrifying biofilm reactors. *Water Science and Technology* **29**, pp. 1-11, 1994.
[21]   Schmidt, E.L. Nitrification in soil, In: *Nitrogen in Agricultural Soils.* Stevenson, F.J. (Ed.) American Society of Agronomy, Crop Science of America, Soil Science Society America, Madison, Wisconsin, USA, pp. 253-288, 1982.
[22]   Kettunen, R.H., Keskitalo, P., Hoilijoki, T.H. & Rintala, J.A., Biological treatment for removal of organic material and ammonium from leachate at low temperatures. *Waste Management and Research* 17, pp. 487-49, 2000.
[23]   Pelkonen, M., Kotro, M. & Rintala, J., Biological nitrogen removal from landfill leachate: a pilot scale study. *Waste Management and Research* **17**, pp. 493-497, 1999.
[24]   Gujer, W. & Boller, M., Operating experience with plastic media tertiary trickling filters for nitrification. *Water Science and Technology* **16**, pp. 201-213, 1986.
[25]   Daigger, G.T., Heinemann, T.A., Land, G. & Watson, R.S., Practical experience with combined carbon oxidation and nitrification in plastic media trickling filters. *Water Science and Technology* **29**, pp. 189-196, 1994.
[26]   Fisher, M. & Fell, C., Ammonia removal from high strength leachate using a sequencing batch reactor. *Wastes Management,* March, pp 25-26, 1999.
[27]   Hunik, J.H., Meijer, H.J.G. & Tramper, J., Kinetics of Nitrosomonas europea at extreme substrate product and salt concentrations. *Applied and Environmental Microbiology* 37, pp. 802-807, 1992.
[28]   U.S. Environmental Protection Agency (EPA) Technology Transfer. *Process Design Manual for Nitrogen Control,* Washington D.C., USA, 1975.
[29]   Kim, D.J., Lee, D.I. & Keller, J., Effect of temperature and free ammonia on nitrification and nitrite accumulation in landfill leachate and analysis of its nitrifying bacterial community by FISH. *Bioresource Technology* **97**, pp 459-468, 2006.

Section 13
Air pollution control

Air pollution impacts from open air burning

B. Sivertsen
NILU, Kjeller, Norway

Abstract

As part of the air quality monitoring and assessment, emissions from the open air burning of waste and biomass have proven important sources of impact on the population exposure. The emissions of a variety of air pollutants from different types of open-air burning have been evaluated. Simple model estimates are used to estimate the impact in areas downwind from backyard burning. Different models have been used to estimate the importance of impact and the reasons for odours detected downwind from waste dumping areas.

Data from Cairo and other major cities have been used to estimate the impact of emissions from agricultural waste burning. In several urban areas especially in developing countries the contribution to the PM exposure has been demonstrated to range between 30 and 50% dependent upon weather conditions. The design and application of combined monitoring programmes and models may be part of the systematic evaluation of impacts and optimal control of the sources.

Keywords: air pollution, waste burning.

1 Introduction

Waste is a growing environmental, social and economic issue for all modern economies. Waste volumes are increasing at rates equalling and sometimes outpacing economic growth. The way that waste is generated and handled has an impact on everyone, from individual citizens and small businesses to public authorities and international trade.

The generation of more and more waste is a symptom of inefficient use of resources, which should actually have been utilised in a better way as a resource for energy and materials. The air pollution generated from waste causes impacts felt by a large part of the population living in and around major urban areas. The pollutants are partly emitted due to leakages and evaporation of gases causing

WIT Transactions on Ecology and the Environment, Vol 92, © 2006 WIT Press
www.witpress.com, ISSN 1743-3541 (on-line)
doi:10.2495/WM060471

annoying smells and partly generated by self-burning or active fires onset by humans at dumpsites.

In the following we will discuss emissions to the atmosphere and exposure to humans caused by dump sites, burning of household-, garden and agricultural waste as well as pollutants due to the use of waste in small industries.

2 Estimated emissions to the atmosphere

Air pollution emissions from waste deposits can be generated through leakages of gases or from combustion when waste is being burned. One of the major concerns at dumpsites is the leakage of methane and other green house gases. This subject will, however, not be covered in this paper.

The complete combustion and open air burning of waste require sufficient heat flux, adequate oxygen supply, and sufficient burning time. The size and quantity of the emissions of different compounds from this type of burning depends on a number of factors such as density, moisture and wetness, topography (slope and profile of underlying ground) and meteorological conditions (wind, precipitation etc.)

2.1 Waste deposit areas and dumps

From an environmental perspective, dumping or 'land filling' is the worst waste management option. It uses up space and creates a future environmental liability. It represents a waste of resources. European legislation has introduced high standards for landfills to prevent soil and groundwater pollution, and to reduce emissions to air; for example of methane and other air pollutants. In some cases the legislation is poorly implemented and there are still thousands of mismanaged and unauthorised landfills across the member states.

Because of their high vapour pressures and low solubility, many toxic VOCs are observed in landfill gas. In a report by the State of California Air Resources Board, the average surface emission rate of hazardous chemicals was estimated to be 35 kg per million kg of refuse (Bennett [3]).

Numerous investigations have been conducted with the objective of characterizing landfill gas emissions. Significant variation in landfill gas composition has been observed. A number of environmental parameters influence the production rate and the composition of landfill gas. The main factors are waste composition, time since placement, pH, moisture content, availability of nutrients, and soil type.

Temperature is also a key variable. Within a few days of placement, the temperature in the dump will rise due to aerobic decomposition. Intensive reduction of organic matter by putrefaction is usually accompanied by disagreeable odours of hydrogen sulphide and reduced organic compounds which contain sulphur, such as mercaptans. After the oxygen is depleted, the temperature drops back again because anaerobic processes are slower and less energetic. The optimal temperature for the production of methane and other organic pollutants is between 32 and 38°C (Cooper et al. [7]).

NILU measured gases released from a landfill outside Oslo in 1989 (Braathen and Schmidbauer [5]). Samples were taken at 7 locations of VOC, Volatile aldehydes, H_2S and methanthiol. Concentrations measured at the side of the landfill were much higher than on top of the dump. It was thus concluded that most of the odours identified in the surrounding areas was due to old waste.

In other studies ethyl benzene was found at concentrations more than 20 times the detection threshold (OU_{50D}). The H_2S concentrations on top of the deposit were between 120 and 450 $\mu g/m^3$.

Tønnessen [16] performed a specific study on a landfill to identify the reasons for complaints about odours in the surrounding areas. Calculations of gas emissions and dispersion of odorous gases were based upon measurements and models. At this specific site measurements showed that H_2S was the odorous component with highest release rate compared to the odour threshold.

At this relatively small deposit area, where 60,000 tons were deposited every year, the total amount of gases released was estimated at 590 m^3/h. Measurements indicated average concentrations of H_2S at 0.9 $\pm$ 0.47 $\mu g/m^3$. For local adverse dispersion conditions, that occurred 6% of the time, odour from the landfill occurred in residential areas around the landfill. At these conditions the dispersion models gave half hourly concentrations of about 14 $\mu g/m^3$ in one of the living areas. In conclusion it was stated that H_2S also could be used as an indicator for any odour problem occurring around landfills of this kind.

Figure 1:   The experimental layout of measuring emissions from bonfires and burning of rubbish.

2.2 Burning of garden waste

Emissions to the atmosphere from ground level open burning are affected by many variables, including wind, ambient temperature, composition and moisture content of the debris burned, as well as compactness of the pile. In general, the relatively low temperatures associated with open burning increase emissions of particulate matter, carbon monoxide, and hydrocarbons and suppress emissions of nitrogen oxides. Emissions of sulphur oxides are a direct function of the sulphur content of the material burned.

Field experiments were undertaken at NILU to study the amount of particles and polycyclic aromatic hydrocarbons (PAH) emitted when typical garden debris was burned in Norway (NILU [14]). PAH may be formed by the thermal decomposition of any organic material containing carbon and hydrogen. More than 30 PAH compounds and several hundred PAH derivatives have been identified to have carcinogenic and mutagenic effects, making them the largest single class of chemical carcinogens known (Bjørseth and Ramdahl [4]).

In smoke from garden deposits it was found that the emission rates of CO could be as much as 3 to 4 times the emission rate from an idling car. Ten kilos of garden debris gave the following emissions to the atmosphere:

- 1 kg Carbon monoxide (CO)
- 1 kg PM (soot and ash)
- 25 g Aldehydes
- 5 g Polycyclic Aromatic Hydrocarbons (PAH)

PAH compounds have been studied in many countries. Fang et al. [8] found that the total PAH concentration in smoke from biomass burning ranged from 7 to 46 ng/m^3. The highest value of B(a)P was measured at 3 ng/m^3 while the lowest was at the same site 0.1 ng/m^3. These values are comparable to those measured at street level in urban centres such as Hong Kong.

Emissions of suspended particles have been investigated and reported over a number of years. Darley et al. [1] carried out both laboratory and field studies and concluded that moisture content of fine fuel residues was the most significant factor influencing emission levels.

At good burning conditions the particulate emission factors for some crops were:

- Orchard pruning, citrus, almond, grape   2-4 g/kg
- Rice, barley, wheat   3-5 g/kg
- Russian thistle   10-13 g/kg
- Asparagus   14-20 g/kg

2.3 Agricultural waste

Organic refuse burning consists of field crops, wood, and leaves, and the emissions are again dependent mainly on the moisture content of the refuse. In the case of the field crops the emission rates depends upon whether the refuse is burned in a head fire or a backfire. Head fires are started at the upwind side of a field and allow to progress in the direction of the wind, whereas backfires are started at the downwind edge and forced to progress in a direction opposing the wind.

Emission factors for open air burning of agricultural waste are presented in Table 1 as a function of the type of waste. The emission rates may also be a function of burning techniques and/or moisture content.

Hays et al. (2005) indicated that the $PM_{2.5}$ mass emission factor from burning of wheat and rice were 4.7 ± 0.04 and 13.9 ± 0.3 g/kg of dry mass respectively.

Table 1:   Emission factors (g/kg) for some selected compounds when burning different type of agricultural waste.

Type of waste	PM	PM₁₀	PM₂.₅	SOx	NOx	CO	VOC
Municipal refuse 1)	8			0.5	3	42	6.5
Open air refuse 3)	37	37		0.5	3		
Agricultural waste 3)	10	5					
Rice1) 2)	4	3.7	2.9	0.6	2.6	29	3
Alfalfa 2)		14.5	13.6	0.3	2.3	60	12
Canopy & shrubs2)			10.7				
Grassland 2)		8.0	7.6	0.3	2.3	57	5.5
Forest 2)		10-15	8-14	0.05	1.8	75-150	10

1) Jenkins, [12] AP42 2) Gaffney [9] 3) Larssen et al. [13].

Figure 2.

2.4 Municipal waste burning

In contrast to municipal combustors, which operate under highly controlled conditions designed to reduce formation and emission of air pollutants, backyard trash burning is uncontrolled. The low temperature burning and smouldering conditions typical of backyard trash fires promote the formation of air pollutants including polychlorinated dibenzodioxins and dibenzofurans, sometimes collectively called "dioxins", fine particulate matter and PAHs. These pollutants form during backyard trash burning regardless of the composition of the material being burned.

In 2001 a typical US County processes approximately 400,000 tons of waste/yr with an estimated emission rate of about 200 ng dioxins per kg waste. This is equivalent to 72.6 g/yr of dioxins from backyard trash fires if all trash were domestically burned (Chlorine Chemistry Council [6]).

3   Concentrations and impacts

Burning of trash and biomass such as leaves and wood is common practices throughout most of the developing countries. The impact to people in these areas is mainly due to high concentrations of $PM_{2.5}$ and PM_{10} which also contain toxic and carcinogenic elements. Biomass burning often takes place after sunset, when the atmospheric dispersion conditions are poor. Very high concentrations of PM and other pollutants are measured in the smoke during these conditions.

The relative contributions to the total PM exposure have been estimated in several cities around the world. In Mumbai the contribution from refuse burning to the total PM_{10} urban exposure has been estimated to about 26% (Larssen et al. [13]). Samples of PM were collected in a semi-residential area of Dhaka and in an urban area of Rajshahi, a city in northwestern region of Bangladesh. In Rajshah it was found based on receptor modelling that biomass burning contributed about 50% of the $PM_{2.5}$ mass (Begum et al. [2]). A number of source apportion studies have indicated that biomass burning often represents 30 to 50% of the total PM load.

3.1 Concentrations downwind from back yard burning

Model estimated downwind concentrations of PM_{10} are presented in Figure 3 based on emissions of particles as a result of open air burning.

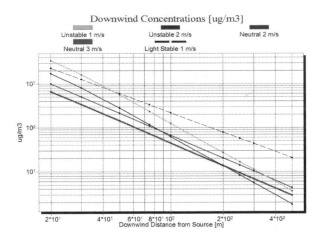

Figure 3:   Estimated concentrations of PM_{10} downwind from open-air burning of backyard waste.

Assuming that a small waste dump is set on fire and that about 10 kg of waste is being burned during one hour the estimated emission rate of PM_{10} at around 0.1 g/sec may lead to concentrations of PM_{10} in the smoke plume around 100 m away from the fire of between 80 and 200 $\mu g/m^3$ depending on meteorological conditions.

At a distance of 200 m from a small bonfire of this kind the one-hour average concentration of PM_{10} may still be 20-30 $\mu g/m^3$ during neutral atmospheric conditions and up to 80 $\mu g/m^3$ in light stable conditions. The latter case is typical for the effect of open-air burning that is taking place in developing countries at sunset or in the early evening.

3.2 High PM_{10} concentrations from agricultural waste burning

Measurements of suspended particles in Cairo were undertaken during the season when rise waste and other agricultural waste were burned in the Nile Delta. Measurements in Cairo was taken downwind from these sources, and it has been shown that the burning of waste contributed considerably to the general air pollution load in Cairo, several tens of kilometres downwind (Sivertsen and Dreiem [15]).

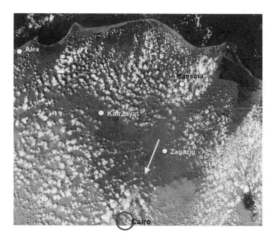

Figure 4:   A satellite photo taken on 11 October 2004 shows plumes from burning of agricultural waste in the Nile Delta. The wind transported the plumes towards the city of Cairo.

On 11 to 12 October 2004 very high concentrations of PM_{10} were recorded at the monitoring stations at Abbaseya and at Fum-AlKhalig in Cairo as seen in Figure 5. The winds were from the north at 2 m/s at night and in the morning hours. Plumes of dust were observed moving into Cairo in the morning of 11 October 2004. The site Fum AlKhalig is located in the northern part of Cairo (closest to the Nile Delta) and hourly average PM_{10} concentrations exceeded 400 $\mu g/m^3$ at this site

Gertler et al. [10] performed a source attribution study (SAS) to determine contributions from various sources in Cairo. Samples of PM_{10}, $PM_{2.5}$, PAH, VOC were collected on a 24-hour basis to identify the importance of the burning of agricultural waste in the Delta. At a site in northern Cairo, close to the Delta, PM_{10} contribution from burning was estimated at 40%, for $PM_{2.5}$ the contribution was 30%. Inside the city centre the open-air burning contribution was estimated at 42% both for $PM_{2.5}$ and PM_{10}.

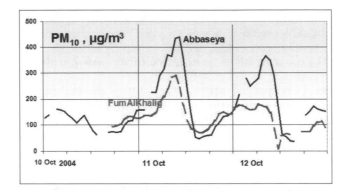

Figure 5: PM_{10} concentrations measured at Abasseya (north Cairo) and Fum AlKhalig (city centre) from 10 to 12 October 2004.

4   Conclusions

Leakages of gases from waste dumps have been shown to include a number of odorous compounds. Measurements undertaken in Norway indicated that H_2S and Ethyl Benzene had contributed to bad smell.

Burning of waste and biomass produces PM, PAH, dioxins and a number of toxic air pollutants. Concentrations downwind may frequently exceed limit values and air quality standards.

Air pollutants from biomass and agricultural burning have been demonstrated to contribute 30 to 50% to the general PM exposure in urban areas dependent upon weather conditions.

References

[1]   Darley, E.F., Miller, G.E., Coss, J.R. and Sisswell, H.H. (1974) Air pollution from forest and agricultural burning. Berkeley, University of California (Project report 2-017-1).

[2]   Begum, B.A., Kim, E., Biswas, S.K. and Hopke, P.K. (2004) Investigation of sources of atmospheric aerosol at urban and semi-urban areas in Bangladesh. *Atmos. Environ., 38*, pp. 3025-3038.

[3]   Bennett, G.F., (1987) Air quality aspects of hazardous waste landfills. *Hazard. Waste Hazard. Mat., 4*, 119-138.

[4]   Bjørseth, A. and Ramdahl, T., (1985) Sources of emissions of PAH. In: Bjørseth A., Ramdahl T. (Eds.), *Handbook of polycyclic aromatic hydrocarbons, Vol. 2*, New York, Marcel Dekker, pp. 1-20.

[5]   Braathen, O.A. and Schmidbauer, N. (1989) Mapping emissions to air from Grønmo landfill. Lillestrøm (NILU OR 55/89) (In Norwegian).

[6]   Chlorine Chemistry Council (2004) Backyard trash burning: the wrong answer.
URL:
http://c3.org/chlorine_issues/understanding_dioxin/trash_burning.html.

[7]   Cooper, C.D., Reinhart, D.R., Rash, F., Seligman, D. and Keely, D. (1992) Landfill gas emissions. Gainesville, Florida center for solid and hazardous waste management (Report no. 92-2).

[8]   Fang, M., Zheng, M., Wang, F., To, K.L., Jaafar, A.B. and Tong, S.L. (1999) The solvent-extractable organic compounds in the Indonesia biomass burning aerosols characterization studies. *Atmos. Environ.*, *33*, pp. 783-795.

[9]   Gaffney P. (2000) Emission factors compiled by California Air Resources Board. URL: http://www.arb.ca.gov/smp/techtool/arbef.pdf.

[10]   Gertler, A.W., Abu-Allaban, M. and Lowenthal, D.H. (2004) Measurements and sources of PM_{10}, $PM_{2.5}$, and VOC compounds in greater Cairo. Presented at the 13th World Clean Air & Environmental Protection Congress and Exhibition, 22-27 August 2004, London, U.K.

[11]   Hays, M.D., Fine, P.M., Geron, C.D., Kleeman, M.J. and Gullett, B.K. (2005) Open burning of agricultural biomass: Physical and chemical properties of particle-phase emissions. *Atmos. Environ., 39,* pp. 6747-6764.

[12]   Jenkins, B. (1996) Atmospheric pollutant emission factors from open burning of agricultural and forest biomass by wind tunnel simulations, April 1996, Davis, University of California (CARB Project A932-126).

[13]   Larssen, S., Gram, F., Hagen, L.O., Jansen, H., Olsthoorn, X., Aundhe, R.V., Joglekar, U., Mehta, K.H. and Mahashur, A.A. (1995) URBAIR. Urban air quality management strategy in Asia. Greater Bombay. City specific report. Kjeller (NILU OR 56/95).

[14]   NILU (2003) Effects of burning bonfire and burning of rubbish. NILU Fact sheet (NILU 3/2003 N). URL: http://www.nilu.no/index.cfm?ac=topics&text_id=7560&folder_id=4580 &view=text (In Norwegian).

[15]   Sivertsen, B. and Dreiem, R. (2004) DANIDA. EIMP phasing-out phase, 2003-2004. End of mission report, air quality monitoring, mission 05, October 2004. Kjeller (NILU OR 76/2004).

[16]   Tønnesen, D. (2001) Estimated emissions and dispersion of odours from a waste disposal area. Kjeller (NILU OR 70/2001) (In Norwegian).

[17]   U.S. EPA (1996) AP42 Emission Factor Database. U.S. Environmental Protection Agency.

[18]   Young, P.S. and Parker, A. (1983) The identification and possible environmental impact of trace gases and vapours in landfill gas. *Waste Man. Res., 1,* 213-226.

Performance standards and residential energy efficiency in Egypt

K. Tiedemann
BC Hydro and Simon Fraser University, Canada

Abstract

The Egypt Organization for Energy Planning (OEP) has put in place energy efficiency labels and standards to increase energy efficiency in the residential sector. In the first phase, energy efficiency standards were put in place for three household appliances – refrigerators, washing machines and air conditioners. In the second phase, energy efficiency labels and standards are being considered for three more end use technologies – electric water heaters, electric lighting and electric motors. The purpose of this paper is to report on an impact evaluation of recent Egyptian minimum energy performance standards for residential appliances.

Keywords: energy conservation, program evaluation, energy efficiency, greenhouse gas emissions.

1 Introduction

Egypt has considerable endowments of energy resources including natural gas, petroleum and hydro-electric power. As a result of investments in the energy sector and strong and increasing demand, the rate of growth of primary energy production has more than kept pace with the growth of population over the past twenty years. The Government of Egypt has historically kept energy prices at below world levels to foster economic development and to provide economic benefits to residential energy consumers. As result, energy use is at a relatively high level compared to many other countries with similar levels of income or GDP per capita.

With respect to electricity prices, tariffs have been quite low although the Government of Egypt has made efforts to increase electricity prices to reduce the gap between selling prices and long-term supply costs. Various estimates suggest that the ratio of average price paid for electricity across all customer groups as a proportion of long term marginal cost is from 80% to over 100%. Not withstanding the overall price-cost situation, many residential electricity customers are still heavily subsidized.

There are significant opportunities to increase energy efficiency in the residential, commercial and industrial sectors in Egypt. The Government of Egypt has recognized the importance of these opportunities and passed enabling legislation in the form of the National Environmental Law of 1993. Subsequent activities included preparation of a National Climate Change Plan and National Strategy for Improving Energy Efficiency in Egypt. The objectives of these policies include: improving energy efficiency, reducing energy consumption and reducing greenhouse gas emissions.

A number of market barriers have been identified which limit the adoption of energy efficient technologies in Egypt. These include:

- Low residential, commercial and industrial customer awareness and knowledge of the nature and benefits of specific energy efficient techniques and practices;

- Relatively low residential, commercial and industrial customer interest in purchasing and adopting these technologies and practices;

- Reluctance to invest in capital projects even with paybacks as rapid as two or three years;

- Risk aversion towards the adoption of new technologies; and

- Inadequate market infrastructure in terms of the availability of efficient equipment and appropriate skills and knowledge for the support of installation and maintenance and use of high efficiency equipment.

To help overcome these barriers, the Organization for Energy Planning (OEP) has been established with a mandate to put in place energy efficiency labels and standards. In the first phase, energy efficiency standards were put in place for three household appliances – refrigerators, washing machines and air conditioners. In the second phase, energy efficiency labels and standards are being considered for three more end use technologies – electric water heaters, electric lighting and electric motors.

The purpose of this paper is to report on an impact evaluation of recent Egyptian minimum energy performance standards (MEPS) for residential appliances. Engineering analysis was used to estimate the impact of the program on incremental sales of efficient appliances, energy savings, and reductions in carbon dioxide emissions.

2 Overview of residential electricity use

Table 1 summarizes the key results of several surveys of residential electricity use that have been undertaken in Egypt over the past ten years. Average annual consumption per household varies from about 1,800 kWh in Alexandria Governorate in 1994 to about 3,200 kWh in Luxor Governorate in 2002.

Table 1:   Average residential energy consumption.

Governorate	Year of survey	Monthly consumption (kWh)	Annual consumption (kWh)
Alexandria	1994	149	1788
Suez	1995	208	2496
Port Said	1996	214	2568
Cairo	2000	239	2868
Luxor	2002	267	3204
Assuit	2002	225	2700
Average		217	2604

Source: Egypt Organization for Energy Planning.

Table 2 provides information on residential electricity consumption by end use, and this information may be viewed as broadly representative of the residential sector in Egypt. Important end uses by consumption level are lighting, refrigerators, water heating, television, and air conditioners.

Table 2:   Residential end use consumption, Cairo, 2000.

Appliance	Unit energy consumption (kWh)	Saturation rate (%)	Average consumption (kWh)	Consumption share (%)
Television	303	96	290	10.1
Radio/cassette	172	92	158	5.5
Fan	137	90	123	4.3
Refrigerator	555	93	516	18.0
Washer	157	97	152	5.3
Water heater	1055	31	327	11.4
Air conditioner	1980	12	198	6.9
Iron	88	91	80	2.8
Lighting	872	100	872	30.4
Other	-	-	155	5.4
Total	-	-	2868	100.0

Source: Egypt Organization for Energy Planning.

3   Refrigerators

The basis for regulation of labeling and standards for refrigerators is Egypt (2003) Ministerial Decree 180/2003, Egyptian Energy Standards for Refrigerators, Refrigerator-Freezers and Freezers. The label allows the consumer to compare energy efficiency among refrigerators or freezers with the same capacity. The labels include information on the manufacturer's name or trade mark; adjusted volume in liters; maximum energy consumption for similar products; minimum energy consumption for similar products; energy consumption of the specific unit; and efficiency grade on a five-point scale.

Several technologies can be used to improve the energy performance of refrigerators and freezers to meet the minimum energy performance standards. These include more efficient motors, improved compressors, more accurate temperature controls, higher quality insulation and improved gaskets and seals.

A key concept used in the minimum energy performance standards is that of adjusted volume. The adjusted volume for refrigerators is given by equation (1), and the adjusted volume for freezers is given by equation (3). Equation (2) provides the adjustment factor.

$$AV_R = V_R + A_F * V_F \tag{1}$$
$$A_F = (T_A - T_F) / (T_A - T_R) \tag{2}$$
$$AV_F = A_F * V_F \tag{3}$$

where,

AV_R = adjusted volume of the refrigerator in liters
A_F = adjustment factor
V_R = net volume of the fresh food compartment
V_F = net volume of the freezer
T_A = ambient temperature in the test area of 32 degrees Celsius
T_F = standard temperature of the freezer, typically –6 degrees.

Table 3:   Refrigerator minimum energy performance standards.

Product Type	Maximum consumption in kWh per year	
	Year 2003	Year 2005
1-door refrigerator, manual defrost	0.48AV + 784	0.48AV + 627
2-door refrigerator, partial automatic defrost	0.37AV + 721	0.37AV + 577
2-door refrigerator, automatic defrost	0.57AV + 1130	0.57AV + 904
Upright freezer, manual defrost	0.36AV + 330	0.36AV + 264
Upright freezer, automatic defrost	0.53AV + 469	0.53AV + 375
Chest freezer	0.39AV + 979	0.39AV + 784
Compact refrigerator, less than 14 liters	0.48AV + 408	0.48AV + 326
Refrigerator without freezer	0.48AV + 300	0.48AV + 270

Source: Egypt Organization for Energy Planning.

Table 3 summarizes the maximum limit of energy consumption for various models of refrigerators and freezers. Initial performance requirements were established for 2003 and these were made more stringent for 2005. Table 4 estimates customer energy savings due to minimum energy performance standards for refrigerators. The above MEPS share is the estimated share of product not meeting the minimum energy performance standard before program launch (that is, consumption is above MEPS). Annual added compliant units are the product of sales in millions of units (mn), the above MEPS share and the assumed compliance rate. Annual customer savings is the product of added sales and unit savings and this is cumulated to estimate the level of cumulative annual savings over time due to the gradual switch over in the capital stock to the efficient product.

Table 5 estimates generation and carbon dioxide savings. Annual customer savings times system loss factor (one plus the eight percent of electricity generated lost in transmission) gives cumulative annual system savings at the generation level. Cumulative annual system savings times the emissions factor provides the annual reduction in carbon dioxide.

Table 4:   Customer refrigerator savings analysis.

Year	Sales (mn)	Above MEPS share	Compliance rate	Annual added compliant units (mn)	Unit saving (kWh per year)	Annual customer saving (GWh)
2003	1.068	0.60	0.75	0.481	64	30.8
2004	1.098	0.60	0.75	0.494	64	31.6
2005	1.135	0.80	0.75	0.681	120	81.7

Table 5:   System refrigerator savings analysis.

	Annual customer saving (GWh)	Cumulative annual customer saving (GWh)	Loss	Cumulative annual system saving (GWh)	Emissions factor (kt per GWh)	Annual CO$_2$ reduction (ktonne)
2003	30.8	30.8	1.08	33.3	0.506	16.8
2004	31.6	62.4	1.08	67.4	0.506	34.1
2005	81.7	144.1	1.08	155.6	0.506	78.7
Total						129.6

4   Clothes washers

The basis for regulation of labeling and standards for washing machines is Egypt (2003) Ministerial Decree 4100/2003, Measurement and Calculation Methods of Household Clothes Washing Machines. The label includes the following information: manufacturer name or trademark; model; capacity in kilograms; maximum energy consumption for this type of product; minimum energy consumption for this type of product; energy consumption of the unit; efficiency grade. The minimum energy performance standard is 0.26 kWh per kilogram per load. The energy efficiency of clothes washers can be improved by using more efficient motors, better sizing of the motor to the load, and more efficient transmissions.

Table 6 estimates customer energy savings for clothes washers due to minimum energy performance standards. Annual customer savings is the product of added sales and unit savings and this is cumulated to estimate the level of cumulative annual savings over time due to the gradual switch over in the capital stock to the efficient product. Cumulative annual savings times the system loss factor gives cumulative annual system savings (these are savings at the generation level). Finally cumulative annual system savings times the emissions factor provides the annual reduction in carbon dioxide.

Table 6:   Customer clothes washer savings analysis.

Year	Sales (mn)	Above MEPS share	Compliance rate	Annual added compliant (mn)	Annual Unit saving (kWh)	Annual customer saving (GWh)
2003	0.368	0.70	0.75	0.193	50	9.7
2004	0.378	0.70	0.75	0.198	50	9.9
2005	0.391	0.70	0.75	0.205	50	10.3

As shown in table 7, cumulative annual savings times the system loss factor gives cumulative annual system savings or savings at the generation level. Cumulative annual system savings times the emissions factor provides the annual reduction in carbon dioxide.

Table 7:   System clothes washer savings analysis.

Year	Annual customer saving (GWh)	Cumulative annual Customer saving (GWh)	Loss	Cumulative annual system saving (GWh)	Emissions factor (kt per GWh)	Annual CO_2 reduction (ktonne)
2003	9.7	9.7	1.08	10.5	0.506	5.3
2004	9.9	19.6	1.08	21.2	0.506	10.7
2005	10.3	29.9	1.08	32.3	0.506	16.3
Total						32.3

5   Air conditioners

The basis for regulation of labeling and standards for air conditioners is Egypt (2003) Ministerial Decree 4100/2003, Measurement and Calculation Methods of Household Clothes Washing Machines. The label includes the following information: manufacturer name or trademark; model; capacity in kilograms; maximum energy consumption for this type of product; minimum energy consumption for this type of product; energy consumption of the unit; efficiency grade. The minimum energy efficiency ratio (EER) for window type air conditioners is 8.5 and 9.0 for split type air conditioners. The energy efficiency of air conditioners can be increased by using more efficient motors, using better compressors, and improving thermostats and air conditioner controls.

Table 8 estimates customer energy savings for air conditioners due to minimum energy performance standards. Annual customer savings is the product of added sales and unit savings and this is cumulated to estimate the level of cumulative annual savings over time due to the gradual switch over in the capital stock to the efficient product.

Table 8:   Customer air conditioner savings analysis.

Year	Sales (mn)	Above MEPS Share	Compliance rate	Annual added compliant units (mn)	Unit saving (kWh per year)	Annual customer saving (GWh)
2003	0.577	0.60	0.75	0.260	165	42.8
2004	0.592	0.60	0.75	0.266	165	44.0
2005	0.610	0.60	0.75	0.275	165	45.3

As shown in table 9, cumulative annual savings times the system loss factor gives cumulative annual system savings or savings at the generation level. Cumulative annual system savings times the emissions factor provides the annual reduction in carbon dioxide.

Table 9:   System air conditioner savings analysis.

Year	Annual customer saving (GWh)	Cumulative annual Customer saving (GWh)	Loss	Cumulative annual system saving (GWh)	Emissions factor (kt per GWh)	Annual CO_2 reduction (ktonne)
2003	42.8	42.8	1.08	46.2	0.506	23.4
2004	44.0	86.8	1.08	93.7	0.506	47.4
2005	45.3	132.1	1.08	142.7	0.506	72.2
Total						143.0

6   Conclusions

In this study, we have analysed the impact of the Government of Egypt's minimum energy performance standards for residential appliances. Key findings are as follows.

(1) Residential Energy Use. Average residential electricity use in Egypt is about 2,600 kWh per household per year. Unit energy consumption per year for major uses are 1980 kWh for air conditioning, 1055 kWh for water heating, 872 kWh for lighting, 555 kWh for refrigerators, 303 kWh for televisions, 172 kWh for radio/cassettes, 157 kWh for washers, 137 kWh for fans and 88 kWh for irons.

(2) Refrigerators. Refrigerator sales are about 1.1 million units per year, with about 60 percent of units in the base year 2001 not meeting the 2003 performance standards. Electricity saving for refrigerators was about 30.8 GWh in 2003, 62.4 GWh in 2004 and 144.1 GWh in 2005. Carbon dioxide savings were about 16.8 ktonnes in 3003, 34.1 ktonnes in 2004 and 78.7 ktonnes in 2005.

(3) Clothes Washers. Clothes washer sales are about 0.4 million units per year, with about 70 percent of units in the base year 2001 not meeting the 2003 performance standards. Electricity saving for clothes washers was about 9.7 GWh in 2003, 9.9 GWh in 2004 and 10.3 GWh in 2005. Carbon dioxide savings were about 5.3 ktonnes in 2003, 10.7 ktonnes in 2004 and 16.3 ktonnes in 2005.

(4) Air Conditioners. Air conditioners sales are about 0.6 million units per year, with about 60 percent of units in the base year 2001 not meeting the 2003 performance standards. Electricity saving for air conditioners was about 42.8 GWh in 2003, 44.0 GWh in 2004 and 45.3GWh in 2005. Carbon dioxide savings were 23.4 ktonnes in 2003, 47.4 ktonnes in 2004 and 72.2 ktonnes in 2005.

References

[1]   Economist Intelligence Unit Egypt *Survey*, 2005.
[2]   S J. Huang, J. Deringer, M. Krarti and J. Masud, Development of Residential and Commercial Building Energy Standards for Egypt, *Proceedings of the Energy Conservation in Buildings Workshop*, Kuwait, 2003.
[3]   United Nations Development Program. United Nations Program of Technical Cooperation. *Energy Efficiency Improvements. Project No. EGY97G31*, 2005.
[4]   World Energy Council, *Energy Information: Egypt. Extract from the Survey of Energy*, 2005.
[5]   Egypt Organization for Energy Planning, *Energy Efficiency Improvements and Greenhouse Gas Reduction, 1st Interim Report, Residential Customers Survey*, 1998.
[6]   Egypt Organization for Energy Planning, *Energy Efficiency Improvements and Greenhouse Gas Reduction. 6th Interim Report. Measuring and Testing Method for Refrigerators and Refrigerator-Freezers*, 2004.
[7]   Egypt Organization for Energy Planning, *Energy Efficiency Improvements and Greenhouse Gas Reduction. 7th Interim Report. Measuring and Testing Method for Air Conditioners*, 2004.
[8]   Egypt Organization for Energy Planning, *Energy Efficiency Improvements and Greenhouse Gas Reduction. 8th Interim Report. Measuring and Testing Method for Clothes Washers*, 2004.

Section 14
Hazardous waste, disposal and incineration

Numerical modelling of a bubbling fluidized bed combustor

S. Ravelli & A. Perdichizzi
Department of Industrial Engineering, Bergamo University, Italy

Abstract

A numerical model of a bubbling fluidized bed combustor fed by refuse derived fuel is presented. The combustor is divided into two regions: the bed and the freeboard. The calculation of mass and energy fluxes entering from the bed into the freeboard provides the boundary conditions for the subsequent CFD analysis. The three-dimensional freeboard model, implemented by means of the commercial code FLUENT 6.1, is mainly concerned with employing the two-mixture fraction-pdf approach to track both the flue gas coming from the bed and the solid fuel particles that do not burn in the bed, but above it. The excess air is injected through four series of nozzles; the heat exchange between the combustion gases and the boiler tubes is also taken into account in the model. The comparison between the predicted and the experimental data shows a good agreement about chemical species concentrations, velocity and temperature profiles along the freeboard height. The reliability of the simulation results proves that CFD modelling is a valid instrument to analyse the behaviour of non conventional furnaces.

Keywords: CFD modelling, fluidized bed combustion, freeboard, refuse derived fuel.

1 Introduction

The fluidized bed combustor (FBC) is a flexible and reliable technology that carries out combustion at lower operating temperatures (700-950°C) in comparison with conventional boilers. So, it's possible to reduce considerably the pollutants emitted with the flue gas; besides, limestone or dolomite can be brought into the bed to adsorb sulphur and halogen compounds. Carbon monoxide and unburned hydrocarbon discharged in atmosphere are very low since fluidization guarantees high combustion efficiency and high gas-solid mixing. These features make the FBC suitable not only for coal but also for alternative fuels, such as RDF or biomass.

Investigations on solid fuel combustion have been carried out not only with one-dimensional but also with multi-dimensional models: the former has been diffusely employed in chemically reactive fluidized bed simulation (Sriramulu *et al.* [1], Marias *et al.* [2], Scala and Salatino [3]); the latter is currently being applied only to conventional combustion processes (Stopford [4], Eastwick *et al* [5], Eaton *et al.* [6]). Attention has been drawn to multi-dimensional numerical models focused on the combustor simulation. While once confined to specialized researches, these models are gradually becoming more accessible as features in commercially available CFD codes. Time requirement for their running is also becoming much more acceptable due to improved numerical methods and advanced hardware technology. These positive circumstances has made the three-dimensional, comprehensive model of a real furnace an attainable goal. The present investigation is limited to the CFD analysis of the freeboard since numerical method can't deal with the 3D modelling of reactive fluidized beds at the moment. As a consequence, the bed behaviour has been estimated by means of simple mass and energy balances that provide the boundary conditions for the freeboard model. This has been implemented by FLUENT version 6.1 [7]. Continuity, momentum, turbulence, energy and mixture fraction equations are solved sequentially in an iterative way until the converged criteria are satisfied. Predictions have been validated against a set of experimental data measured by the FBC monitoring system. The great potential offered by CFD stands out in the post-processing: qualitative and quantitative information about the flow field can be easily obtained throughout the whole domain.

2　The bubbling FBC

The present investigation has been carried out for the furnace of the waste-to-energy plant owned by BAS-ASM, a municipal utility operating in North Italy. The FBC (fig.1) is fed by RDF by means of fluff or pellets. Two stokers throw the solid particles into the combustor; their trajectory varies according to their shape and their weight: the heaviest ones quickly fall in the bed while the finest ones are carried by the gas phase throughout the freeboard.

In the bed, the RDF particles burn thanks to the primary air, which is injected through the bottom of the furnace by means of tuyeres, and provide the necessary heat to keep the bed temperature constant (750-800°C). In the freeboard, the fine particles are yielded to combustion by the secondary air, injected on two different levels.

The boiler tubes, completely embedded in the furnace, are vertically set: they are held together in 4 bundles and each bundle is composed of 11 tubes. In the bottom part, approaching the bed, they assume an "U"; the horizontal lines of the "U" are sloped of 10° in order to promote water circulation.

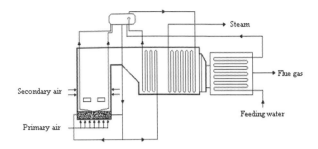

Figure 1: Schematic diagram of the FBC, part of the heat recovery steam generator.

3 Mass and energy balances in the fluidized bed

Since it is impracticable to measure the fraction of fuel that falls into the bed, an hypothesis has been made according to which a percentage varying from 60% to 75% of the fuel fed to furnace burns in the bed. This percentage is related to the temperature of the bed, that is supposed to be isothermal.

The ultimate and proximate analysis of the RDF are shown in table 1. The assumption about the complete conversion of the fuel fixed carbon to CO_2 allows to compute the flue gas flow rate and composition, in terms of CO_2, H_2O, N_2 and O_2. The bed temperature, resulting from the energy balance, increases consistently to the fuel percentage burning in the bed (fig. 2). The combustion heat is partly taken out by the boiler tubes located inside the bed (33%) and partly carried by the combustion gases entering the freeboard (66%). Since the case in which 60% of the RDF burns in the bed describes the most recurrent combustor configuration, it has been selected as the reference for the subsequent freeboard modelling.

Table 1: RDF data.

Ultimate analysis (%wt)		Proximate analysis (%wt)	
C	51.244	Moisture	8
H	4.232	Fixed C	3.9
N	0.644	Volatiles	75.6
O	16.376	Ash	12.5
S	0.46	LHV (kJ/kg)	19374

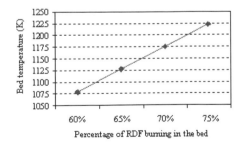

Figure 2: Bed temperature as a function of the RDF percentage burning in the bed.

4 Freeboard modelling

4.1 The grid

The three-dimensional model represents accurately the full scale freeboard with its internals (fig. 3). From the base (4.58 m by 5.928 m) the flue gas enters the freeboard; the secondary air is injected into the furnace through four series of nozzles, whose diameter is 0.051 m. Each series is composed of 12 nozzles. The distance between the base and the two secondary air injection planes is 2.96 m and 3.96 m, respectively. The freeboard is 12.36 m high and is fed by two RDF entry ports (0.9 m by 0.2 m) equally spaced on the front. The combustion products, before leaving the freeboard through the side exit (5.928 m by 1.6 m), transfer heat to 44 evaporative tubes. Their diameter is not constant: it equals 0.076 m at the tube bend and it decreases to 0.064 m at the straight stretch. The diameter reduction along the freeboard height favours the natural circulation of the two-phase fluid flowing in the tubes.

An unstructured grid of 5.344.716 cells has been applied to the freeboard volume; the resultant total number of nodes is 1.111.183. The desire to thicken the mesh just to achieve high calculation accuracy faced up with the need to limit computation time. Consequently, the mesh had to be thickened only in the critical regions. Since heat transfer modelling may be severely affected by grid features, the highest concentration of cells has been restricted to the areas close to the boiler tubes.

4.2 The numerical model

The framework of the CFD model, composed of equations for conservation of mass, energy and momentum combines sub-models accounting for gaseous species mixing and chemical reactions, fuel particles devolatilization and char oxidation. The temperature field is mainly governed by radiant energy transport as well as the heat exchange between the combustion gases and the boiler tubes. The incompressible steady state form of the conservation equations was solved with the segregated SIMPLE-based approach. Additional transport equations for the turbulent kinetic energy k and its dissipation rate ε were considered: the realizable k-ε model was chosen for modelling the turbulence. The solid fuel combustion has been modelled using the two-mixture fraction-pdf formulation. In this approach, the reacting system can include three streams at most:
- one "fuel", i.e. the RDF flow rate that doesn't burn in the bed;
- one "secondary stream", i.e. the flue gas coming from the bed;
- one "oxidiser", i.e. the secondary air.

Transport equations for conserved scalars quantities, the mixture fractions, are solved instead of the individual species transport equations. The mixture fractions are simply the mass fractions of each stream, therefore the sum of all three mixture fractions is always equal to 1. Species mole fractions and other instantaneous fluid properties, such as density and temperature, derive from the predicted mixture fractions distribution.

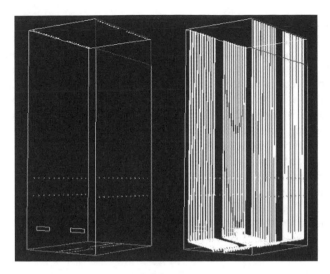

Figure 3:   Three-dimensional freeboard model with its internals. The mass flow inlet faces are coloured blue; the outlet face is coloured red; the wall faces are coloured white.

The reaction system is computed according to the infinite rate chemistry: mole fraction of every species is predicted in accordance with the chemical equilibrium. The species included in the model were the following: C_3H_8, CO, CO_2, $H_2O(L)$, H_2O, H_2, N_2, C(S), O_2, C, H, N, O, according to the kinetic models of solid organic fuels combustion (Boiko and Pachkovskii [8]).

The RDF particles were tracked in a Lagrangian frame of reference by the discrete phase model. The impact of the discrete phase on the continuous phase is accounted by the inter-phase exchange of momentum, heat and mass transfer: all these terms derive from the sub-models describing the combustion process, such as heating, devolatilization and char burnout.

Since the optical thickness is greater than unity, the P1 model has been considered the most suitable to simulate the radiative energy transport. Although it's easy to solve with little CPU demand, it accounts for scattering and radiation exchange between gas and fuel particles. The gas absorption coefficient was set to be a function of local concentration of water vapour and carbon dioxide whereas fuel particles were characterized by constant emission and scattering coefficients.

4.3 Input data and boundary conditions

The fuel features were defined through a series of chemical and physical properties (table 2), including not only the composition but also the particles shape and size. The cumulative size distribution of the RDF particles burning in the freeboard is shown in fig. 4. The subsequent step towards the complete model was the definition of the boundary conditions (table 3): the flue gas entering the freeboard from its base and the secondary air injection through the nozzles were modelled by the "mass flow inlet" condition; the freeboard exit was specified by the "pressure outlet" condition; finally, the condition of "wall" was applied to the freeboard walls and the boiler tubes, whose temperature was supposed to be constant and equal to 1175 K and 600 K, respectively.

The turbulence boundary condition was indicated as turbulence intensity: 10% is enough to represent fully developed turbulence. The gauge pressure is zero: it means that the freeboard pressure was assumed to be constant and equal to the atmospheric pressure. The boundary conditions required by the discrete phase model (DPM) on each surface determine the fuel particle trajectory on the physical boundary. The available conditions are the following: "trap" means that the trajectory calculation terminates and the remaining volatile mass is passed into the vapour phase; "escape" denotes that the trajectory calculation is terminated; "reflect" signifies that the particle is rebounded off the boundary. The first condition was applied to the freeboard base, as the fuel particles falling on this surface volatilise.

Table 2:   RDF data.

Density (kg/m³)	1230
Specific heat (J/kgK)	1000
Thermal conductivity (W/mK)	0.0454
Devolatilization temperature (K)	400
Binary diffusivity (m²/s)	0.0005
Swelling coefficient	2
Emissivity	0.8
Scattering factor	0.6

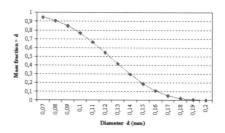

Figure 4:   Cumulative size distribution of RDF particles burning in the freeboard.

Table 3:   CFD freeboard model boundary conditions.

Surface	Boundary Condition	Input	
Freeboard base	Mass flow inlet	T = 1078.5 K	DPM: trap
		m = 18.23 kg/s	Emissivity = 1
Nozzles for secondary air injection	Mass flow inlet	T = 298 K	DPM: reflect
		m = 2.97 kg/s	Emissivity = 1
Freeboard outlet	Pressure outlet	Gauge pressure = 0	DPM: escape
RDF entry ports	Wall	T = 298 K	DPM: fuel injection
		m = 0.833 kg/s	Emissivity = 1
Freeboard walls	Wall	T = 1175 K	DPM: reflect
			Emissivity = 0.75
Boiler tubes	Wall	T = 600 K	DPM: reflect
			Emissivity = 0.5

4.4 Results and validation

The big CPU demand has imposed the parallel processing using a network of UNIX workstations. The solution has met the physical convergence criteria in about 1600 iterations. The analysis of the results calls attention to the combustion process, the velocity and temperature field and the chemical species concentration. The final purpose is the validation of the freeboard model.

The history of each single RDF particle can be tracked in a qualitative and quantitative way: the trajectory has been displayed in the domain (fig. 5) by selecting the desired number of streams. Besides, the DPM report gives information about the average residence time of the fuel particles in the freeboard (i.e. 4 s) and the combustion efficiency. Volatiles are completely released and burnt in the gas phase and the char conversion, which is obtained through heterogeneous reaction with oxygen, is 100%, in accordance with experimental data.

The velocity field shows that the gas velocity varies from zero to 30 m/s; the maximum value occurs at the secondary air injection through the nozzles. The velocity contours are shown in fig. 6 for four different planes:
- $z = 1.6$ m, at the top of the combustion chamber;
- $z = 8.4$ m and $z = 9.4$ m, where the secondary air is injected into the freeboard;
- $z = 11.15$ m, where the RDF is thrown into the furnace.

The interaction between continuous and discrete phases is shown by fuel particles drag around them: as a result, the gas velocity increases from 1.5-3 m/s to 16-18 m/s. It is important to notice the fluid dynamic interaction between the air flows injected through the two series of nozzles, on each freeboard side (fig. 7): the penetration of the freeboard by the air jet increases with the series of nozzles along the z axis. Near the outlet section the average velocity rises: the cross section reduces from 27.15 m^2 to 9.48 m^2 therefore the gas velocity increases to about 10 m/s.

In the freeboard, the coexistence of the combustion process and the heat transfer between the flue gas and the evaporative tubes makes the temperature field examination quite significant. The radiation transport assures a homogeneous temperature distribution along the x axis (fig. 8). The maximum temperature reaches 1550 K: this value is much lower than the peak flame temperature (i.e. 2070 K) because of the excess air.

As far as the chemical species concentration is concerned, carbon dioxide and oxygen are taken into consideration. The contours of CO_2 molar fraction (fig. 9) are considerably affected by the presence of the secondary air that supplies the necessary oxygen to combustion; the lower values (about 7%) in the base of the freeboard are due to the flue gas coming from the bed. On the contrary, the O_2 concentration (fig. 10) rapidly decreases where the greater fuel amount is burnt.

The results here shown needed to be validated by means of comparison between the predicted and the experimental data. The experimental data available in the freeboard are the following:
- the temperature in proximity to the secondary air injection and to the exit (each value is measured by three thermocouples);

- the chemical species concentration at the outlet.

In addition, the plant monitoring system measures the steam flow rate produced in the heat recovery steam generator and the one at the turbine inlet. The difference between these values equals the steam flow rate generated in the furnace: this datum is necessary for the estimation of the heat removed by boiler tubes.

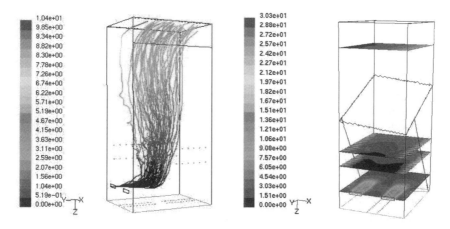

Figure 5: RDF particles traces coloured by particle residence time (s).

Figure 6: Velocity magnitude (m/s) at z = 1.6 m, z = 8.4 m, z = 9.4 m and z = 11.15 m.

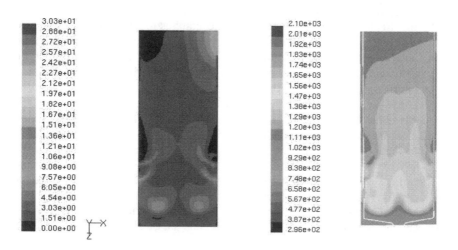

Figure 7: Velocity magnitude (m/s) at y = 2.964 m.

Figure 8: Static temperature (K) at y = 2.5 m.

The differences of the computed data with respect to the experimental values are shown in the fourth column of table 4. It can be observed that, in general, there is a good agreement: the largest error (6.7%) is on the O_2 concentration. The model overestimates the heat transfer to the steam of about 4-5% mainly because the P1 model tends to overvalue the radiative fluxes from localized heat source. Consequently, the temperature of the gases leaving the freeboard is slightly undervalued. Nevertheless these differences, it can be concluded that the accuracy of the model predictions is more than satisfactory, because one has to take into account the big complexity of the physical process taking place in a real furnace and modelled by the computation. As regards the velocity field, no measured data are available for comparison; anyway, the prediction seems reasonable.

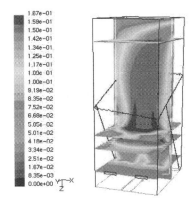

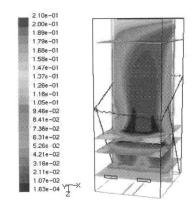

Figure 9: CO_2 mole fraction at $z = .6$ m, $z = 8.4$ m, $z = .4$ m, $z = 11.15$ m and $y = 2.5$ m.

Figure 10: O_2 mole fraction at $z = 1.6$ m, $z = 8.4$ m, $z = 9.4$ m, $z = 11.15$ m and $y = 2.5$ m.

Table 4:   Comparison between model outputs and experimental data.

Variable	Model output	Experimental data	Δ%
Flue gas temperature (K) at the freeboard outlet	1176.4	1239.5	-5%
Flue gas temperature (K) at $z = 8.4$ m	1220	1218	+0.16%
Gas-boiler tubes heat exchange (kW)	8065	7718	+4.3%
Species molar fraction at the freeboard outlet:			
C(S)	2.4e-5	-	
CO	4.23e-7	-	
CO_2	0.110	0.112	-1.8%
H_2O	0.084	0.086	-2.3%
O_2	0.074	0.069	+6.7%
N_2	0.732	0.733	-0.1%

5 Conclusions

In theory, a FBC has a number of highly useful properties such as low combustion temperature, absence of temperature gradients in the bed, high combustion efficiency and easy pollutants reduction. In practice, BAS-ASM experience has shown that it is difficult to get a regular operation of a FBC fed by RDF, because of the high variability of the fuel heat content, the combustive air constraints imposed by bed fluidisation and the slag clustering. It followed that the combustor behaviour had to be analysed in depth and CFD modelling was regarded as the best way to pursue this aim.

At the moment, the development of the numerical techniques can't stand up to the multi-dimensional modelling of reactive fluidized beds so CFD analysis was restricted to the freeboard. The 3D freeboard modelling presented here has shown to include all the physico-chemical phenomena of main importance in solid fuels combustion processes: turbulence coupled with infinite rate chemistry, radiation and RDF reaction schemes are taken into account. The results have been compared with the experimental data: as the differences have been found in the range 4%÷6%, it can be concluded that the accuracy of the computed results is satisfactory. The high complexity of this case has seriously tested the numerical method: the reliability of the simulation outputs has proved that CFD is a valid instrument to analyse, retrofit and optimise the performance of a full scale industrial furnace fed by a non conventional fuel.

References

[1] Sriramulu, S., Sane, S., Agarwal, P. & Mathews, T., Mathematical modelling of fluidized bed combustion. *Fuel*, **75**, pp. 1351-1362, 1996.

[2] Marias, F., Puiggali, J. R. & Flamant, G., Modeling for simulation of fluidized-bed incineration process. *AIChE Journal*, **47**, pp. 1438-1460, 2001.

[3] Scala, F. & Salatino, P., Modelling fluidized bed combustion of high-volatile solid fuels. *Chemical Engineering Science*, **57**, pp. 1175-1196, 2002.

[4] Stopford, P. J., Recent applications of CFD modelling in the power generation and combustion industries. *Applied Mathematical Modelling*, **26**, pp. 351-374, 2002.

[5] Eastwick, C. N., Pickering, S. J. & Aroussi, A., Comparisons of two commercial computational fluid dynamics codes in modelling pulverised coal combustion for a 2.5 MW burner. *Applied Mathematical Modelling*, **23**, pp. 437-446, 1999.

[6] Eaton, A. M., Smoot, L. D., Hill, S. C. & Eatough C. N., Components, formulations, solutions, evaluation, and application of comprehensive combustion models. *Progress in Energy and Combustion Science*, **25**, pp. 387-436, 1999.

[7] FLUENT user's guide, vols. 1-4, Lebanon, 2000.

[8] Boiko, E. A. & Pachkovskii S. V., A kinetic model of thermochemical transformation of solid organic fuels. *Russian Journal of Applied Chemistry*, **77**, pp. 1547-1555, 2004.

Hazardous waste management: educating industrial communities in Egypt

A. R. Ramadan[1] & A. H. Nadim[2]
[1]*Chemistry Department, American University in Cairo, Egypt*
[2]*Hazardous Waste Management Unit, Environics, Cairo, Egypt*

Abstract

Hazardous waste management is in its initial stages in Egypt. In this regard, the need to develop awareness and capacities has been identified as a priority, particularly for the industrial establishments, constituting a major contributor to HW generation nationally. Delivery of capacity building programs for HW management to these communities has led to the development of a methodology suited to the particularities of this area in Egypt, ensuring both the effectiveness and success of these programs. Based on accumulated experience in this regard, an educational CD has also been developed. The elements of this methodology are presented, followed by an overall description of the CD.
Keywords: hazardous waste management, capacity building, Egypt.

1 Introduction

Environmental protection became a priority in Egypt over the last decade. One of the primary aims of the Egyptian Law 4 for the Environment, issued in 1994, is the reduction of pollution from industrial sources. The industrial sector in Egypt contributes to about 20% of the Egyptian GDP and employs approximately 15% of the workforce [1]. Moreover, it is the primary generator of hazardous waste (HW) [2]. Promoting and ensuring compliance to the legal stipulations in this regard is a necessity. As the implementation of proper HW management practices is relatively new in Egypt, there has been a need to build the capacity of all concerned parties with a strong focus on HW generators. In this regard, the development and delivery of targeted awareness and capacity building programs for the industrial communities in Egypt came to be recognized as an effective tool in educating such communities about their responsibilities as well as best practices to meet such responsibilities. This has led to the recognition of capacity building programs as a priority on the national level [3]. This paper presents a methodology developed to this end, taking into consideration the particularities and challenges of HW management in Egypt.

2 Hazardous waste management in Egypt

The implementation of proper HW management practices is relatively new in Egypt. The issuance of the Egyptian Law 4 in 1994, and its Executive Regulations (ER) in 1995, marked the recognition, for the first time, of HW as a source of environmental degradation [4]. However, initial enforcement activities of the stipulations of Law 4 and its ER did not place a high priority on environmental performance with regards to HW management. This was due to a number of reasons. Most significantly were the more pressing and obvious issues of air and water pollution, as well as the novelty of HW identification as a source of pollution and the practices necessary for its proper management [5].

This has led to particularities and challenges for developing and delivering capacity building programs to concerned parties. Such particularities and challenges primarily encompassed a strong sense of ambiguity and danger regarding HW, emphasized by the less specific, yet more stringent legal stipulations particularly with regards to penalties. Furthermore, proper HW management was considered as not only necessitating thorough and detailed technical knowledge not widely available in Egypt, but also management practices forming an overall integrated system extending beyond the boundaries of the generating entity, also considered a novelty in Egypt. In this respect, an underlying overall objective of all capacity building programs for HW management in Egypt has been the elucidation of the "mystery" of HW, and the demonstration that HW identification and proper management is not only feasible, but relatively straightforward, once the necessary basic information and skills for it have been acquired.

3 Design of programs: factors taken into account

In general, capacity building programs are designed to increase awareness, inform, change behaviour and/or improve performance of the target audience. In this respect, the type of audience, its knowledge base, and the improvements expected as a result of the programs, are all significant factors to be taken into consideration for a successful program design and implementation. They have an impact on program content, level of details, presentation tools and the assessment and evaluation methods used.

3.1 Type of target audience

The target audience for HW capacity building programs varies significantly: it includes middle and top management, technical staff as well as workers of industrial establishments and HW management facilities. The variation in educational background, technical knowledge and skills, and the roles played within establishments are the principle factors contributing to this diversity. This variation is to be taken into account when designing and delivering programs, particularly with regards to the delivery tools: whereas lectures and presentations would form the main constituents of programs for participants from management and technical departments, little text, enhanced by visual tools such as posters and drawings as well as practical drills, are to be used for workers who, in many cases, might include people with limited literacy.

3.2 Needs identification

The knowledge base of the target audience and improvements expected as a result of the program (i.e. the program objectives) form the basis of needs identification for a program.

$$Needs = Expected\ outcomes - Knowledge\ base$$

This in turn determines the contents of the program. For a successful program such contents necessitate that knowledge and skill gaps between what the participants know, and what is expected of them to know, be addressed so that improvements could be achieved. This affects the level of complexity of the topics and skills addressed by the program, as well as the degree of details covered.

4   Methodology

A number of factors were found necessary for an effective HW capacity building program in Egypt taking into account the particularities and challenges in this regard. These factors encompass the need for closely tailoring the contents to the target audience, flexibility with regard to the level of details covered during delivery, the need for audience maximum engagement, the need for enhancing the audience's realization of its "learning" progress, and the need to ensure that the increased capacities of the audience is sustained. In this respect, five components represent key features for a successful program.

4.1 Contents

Capacity building programs on HW management in industries must incorporate a number of fundamental topics in order to ensure their effectiveness. These include what HW is, what renders it hazardous, as well as the best practices necessary to handle such waste, i.e. the components and alternatives of a HW management system within an industrial establishment and their relation to HW management practices and services outside the establishment. However, the contents of the program must take into account two issues: the type of participants, with the identified needs for capacity building in light of their roles and responsibilities, as well as any actual specificities of the participants, such as their association to a particular industrial sector, a given geographical location or a certain size of industrial establishments.

The type of participants would reflect on the degree of detail covered for each of the fundamental topics, but more significantly, on the focus of the program. A program targeting workers would typically focus on best practices and safety issues for handling HW containers, whereas one targeting top management would focus on the necessary steps for the implementation of proper HW management, or how to identify HW minimization options. Both programs would address the fundamental topics, but each within its own focus, resulting in somewhat different contents.

Geographical, sectorial as well as size-of-establishment specificities also impact program content. This is typically manifested not only in examples used, but also in details of HW management components and alternatives within establishments, and their relation to practices and services outside the establishments. For example, program content for small and medium textile-dying establishments located within an old urban agglomeration would be different from one targeting a wide variety of industries located in a more recently developed industrial estate. Both programs would cover the fundamental topics, but the first would make use of specific examples and present collection, transportation, treatment and disposal alternatives different from what the second program would portray.

4.2 Structure of presentations

Flexibility with regards to level of detail covered during delivery has been found to have a number of significant advantages. Not only does it allow optimum use of allocated time, but also ensure the engagement of the participants throughout the presentation time. This flexibility is achieved by having the information covered during the presentations (power point format) structured in layers accessible through hyperlinks. In this regard, the top layer, composed of the power point slides forming the main body of the presentation, would contain the fundamental information to be covered. Links imbedded in these slides would lead to the other slides, offering clarifying details, diagrams, pictures, examples, etc., constituting the second layer of information. Within the slides of this second layer, other hyperlinks would lead to a third layer with further information deemed not crucial for clarifying the topics at hand, but useful for the knowledge base of these topics. The third layer could then possibly lead to a fourth with reference material. In this respect, the more the layers, the more added details.

For example, a presentation about storage of HW would include in its main slides the conditions for storage containers, one of which would be the presence of the necessary labels. This would lead to a second level where these labels are presented graphically, and in turn this would lead to a third level giving a descriptive account of each label and its specifications, possibly leading a fourth level addressing specific examples of HW for which these labels would be used. This structure offers the flexibility to the presenter to decide about the level of details/information covered while delivering the presentation, depending on the interest and engagement of the participants. This ensures that they would neither feel frustrated from the absence of information they consider necessary, nor bored with an overflow of details they consider irrelevant. It is important to note that all information of a presentation is given to the audience as handouts (electronic and/or a hardcopy) even if not dwelt on during the delivery. This is discussed further in section 4.5 below.

4.3 Self assessment

The effectiveness of a capacity building program is primarily based on the success of knowledge assimilation by the participants. Moreover, for HW programs, where topics are interdependent, there is a need to ensure that a clear understanding of the basics has been achieved prior to addressing more complex issues. For example a clear understating of what HW is must be achieved for an understanding of best practices for handling. In this respect, it was found that feedback throughout the program is necessary from capacity building recipients. This achieved through exercises carried out at a number of identified milestones at which it is deemed essential to verify participants' assimilation of information. Such exercises are usually short, typically entailing multiple choice, right and wrong questions, etc., and are carried out individually or as group work depending on the topic and the number of participants. During the exercises, several members of the program delivery team rotate between the participants, providing comments, answering queries, and taking part in any discussions. The exercises are not marked, but corrected collectively at the end the session. This was found to be more effective in allowing interactions between participants, as well as between participants and the program delivery team. The exercises are viable for technical, non-technical and less-educated participants.

These exercises were found to offer several advantages. With regards to delivery, it gives an insight to its effectiveness, allowing changes in pace, focus, use of examples and level of details to be carried out for the following sessions. More significantly, they allow the participants to self-assess their assimilation of information and gain in their knowledge base. Issues they realize still unclear can be discussed or enquired about before the end of the program, and topics they realize having been assimilated usually give them an increased enthusiasm and ensure their continued engagement in following sessions. In addition, the exercises allow a break where participants and program delivery teams can informally interact while still focusing on the topics at hand.

4.4 Skill development

Capacity building programs should not be limited to provide information. They must also contribute to the development of new skills for participants. A primary objective of capacity building programs is to ensure that participants are able to utilize gained knowledge in analysing situations, solving problems, and justifying their opinions. To this end, exercises, case studies, and actual site visits are used.

Exercises, different in nature from self assessment ones, primarily focus on developing skills for using existing tools aiding in HW identification and management. Examples include the ability of effectively using databases for identifying HW. Case studies on the other hand, simulate situations in which participants are likely to find themselves, requiring decisions relying on assessment and analysis of different conditions and variables. Site visits develop this further, requiring participants to note, analyse, comment on, and offer solutions for real-life HW issues in actual establishments. These different tools are carried out at different milestones within the capacity building program, and would usually be less in number than the self-assessment exercises. Typically, self assessment exercises would be carried out at the end of every session, whereas the above tools would be carried out at the end of a day (encompassing two to three sessions), or of a module (typically encompassing two days). They are typically carried out as group work, with outcomes discussed collectively at their end. These activities have been found not only useful to skill development, but also to the participants' realization of their own "learning" progress. It is noteworthy to mention that for programs addressing workers, case studies and site visits are usually replaced by practical demonstrations and drills.

4.5 Material

In Egypt, for HW management which is perceived as vague and confusing, material and presentation tools used for capacity building play a significant role in conveying the necessary information effectively. In this respect, these must be adapted to the type of audience, their educational background and capabilities. For top and middle management, as well as the technical staff, material and presentation tools include power point slides, printed legal and technical handouts, and other computer-aided tools such as electronic databases and flow diagrams for decision support. On the other hand, for programs targeting workers or drivers the focus is on visual material and tools. Typically for such cases, the material encompass drawings, photos and/or sketches conveying the necessary messages with minimum text. Handouts would then consist of posters and drawing of various sizes which participants can keep at hand for reference and/or hang up in the work place.

Material handed out to participants would include all information incorporated in the presentations, even if not covered in detail during the delivery of the program. In addition, reference legal and technical material would also be included. All such handouts (whether hard or electronic copies) would be compiled and clearly indexed in a manner to facilitate their later use. This is believed to sustain the developed capacities of the participants over longer spans of time. Participants would thereafter have the programs contents at hand, with the necessary basic skills to make use of them.

5 Responses to the methodology

The above methodology can be considered effective if it succeeds in a better development of participants' capacities, in comparison to the more widely used approach of direct lectures and presentations. In addition, its success in rendering the participants more conscious of their newly gained knowledge base and skills is also important. In this respect, both these factors have been measured for different capacity building programs. Typically, results of formal examinations carried out for some of the programs (as a requirement from the funding entities, or in order to obtain a certificate) would be expected to reflect a better performance for modules following this methodology. Indeed, figure 1 below demonstrates that participants to capacity building programs composed of different modules using different approaches faired better on the HW module using the above methodology. For the "Technical Training" program addressing decontamination of industrial sites, the average score of the 48 participants was about 10% higher for the HW management module than the average scores of other modules not using the above methodology. For the "Health & Safety" program, addressing health and safety issues within industrial establishments with contamination problems, the average score of the 42 participants was about 5% higher for the HW handling module than the average scores of other modules not using the above methodology.

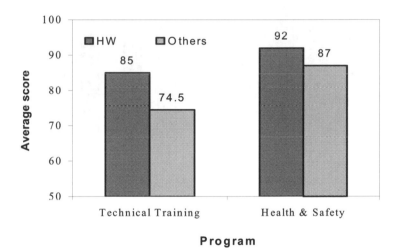

Figure 1:   Examination results for the programs of "Technical Training" and "Health & Safety" for contaminated industrial sites [6,7].

As for making participants more conscious of their newly gained knowledge base and skills, responses of program evaluation are indicative. In this respect, figure 2 summarizes responses of participants with regards to program contents, relevance, methodology, and acquired skills, for three programs addressing "HW Classification", "HW permitting", and "HW Storage". For the evaluation, the participants were requested to rate each of these parameters (among others) following a five-step scale (very satisfied, satisfied, neutral, dissatisfied, and very dissatisfied). The evaluation responses are overtly positive, with 96% to 100% of the 60 participants being satisfied or very satisfied with program content, 85% to 97% being satisfied or very satisfied with program relevance to their job responsibilities, 97% to 98% being satisfied or very satisfied with the methodology used for program delivery, and 84% to 96% being satisfied or very satisfied with gained skills. The lowest ratings of 84% for gained skills in the "HW Permitting" program mirrors the perception of some participants of the somewhat limited significance of the procedural skills for permitting HW activities gained during this program. Similarly, the low of 85% for relevance in the "HW Classification" program mirrors the perception of some participants of the somewhat limited relevance of classifying generated HW to their job responsibilities. Figure 3 summarizes responses of participants regarding program content, delivery methodology, and relevance to their job responsibilities, for an overall "HW Management" program, using a three-step scale (good, adequate, not suitable). Responses are overtly positive, and the rating of 100% for the delivery methodology is noteworthy.

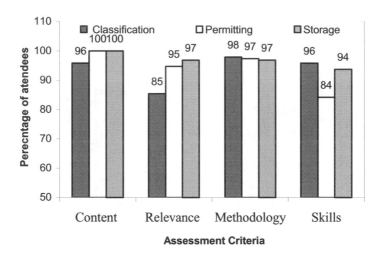

Figure 2:   Participants feedback for three HW programs [8-10].

6 Hazardous waste management educational CD

With the objective of increasing awareness with regards to HW management issues in Egypt, and making use of accumulated experience in the development and delivery of capacity building programs in this regard, an educational CD was developed compiling all relevant information for HW management in Egypt. This encompasses legislation, operational guidelines, as well as references and databases. Based on experience and feedback from capacity building programs, a decision was carried out to structure the information in layers with multi access. Drop down menus, and sub-menus would allow direct access to information from an indexed list of choices. Moreover, hyperlinks imbedded within the information would allow the access to further details and interrelated topics without having to go back to the menus. This has been found to greatly facilitate navigation within the CD. The contents encompass the topics necessary for promoting the operationalization of proper HW management on the national level in Egypt. In this respect, information is grouped under eight main categories: legislation, the national HW classification system, operational guidelines for HW storage, the operational guidelines for HW transportation, alternatives for HW treatment and disposal, HW minimization, HW registers, self monitoring of HW management activities in industrial establishments. This CD, made available free of charge, was first launched in 2003, and was met with significant demand. So far, it has been widely distributed not only to industrial establishments, but also to regulating entities, and industrial non-governmental organizations and business associations.

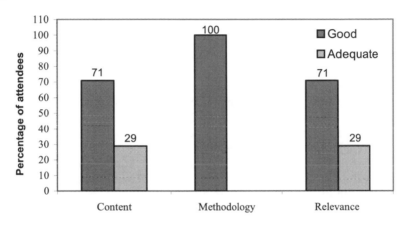

Figure 3:   Participants' feedback for an overall "HW Management" program [11].

7 Conclusion

Capacity building is a key component to the operationalization of HW management in Egypt. Building the capacities of industrial communities, constituting a major contributor to HW generation, is faced with a number of challenges. These primarily encompass a sense of ambiguity and danger regarding HW, with the practices for its proper management being perceived with vagueness and confusion. To develop and deliver effective programs addressing these challenges, the type of audience together with its specific needs in this regard, must be taken into account. In this respect, a methodology has been developed where program contents, structure of presentations, handout material together with exercises for self assessment and skill development ensure maximum tailoring of delivered programs to participants' needs, participants' continuous engagement, the enhancement of their realization of their "learning" progress, as well as the sustainability of their developed capacities. To further support these efforts, an educational CD on HW management was developed and widely distributed.

References

[1]   Egyptian Pollution Abatement Project, *Strategy and Action Plan for Cleaner Production in Egyptian Industry*, Egyptian Environmental Affairs Agency, 2004.

[2]   Egyptian Pollution Abatement Project, *Hazardous Waste Management Manual for Industry*, Egyptian Environmental Affairs Agency, 2003.

[3]   *National Environmental Action Plan (NEAP)*, Egyptian Environmental Affairs Agency, 2002.

[4]   Executive Regulations of Law 4/1994 for the Environment, Chapter 2, 1995.

[5]   Ramadan, A.R, Nadim A.H, Hazardous Waste Management in Egypt: Status and Challenges, *Proc. of Waste Management and the Environment II*, eds. V. Popov, H. Itoh, C.A. Brebbia and S. Kungolos, WIT Press: Southampton and Boston, pp. 125-135, 2004.

[6]   LIFE Lead Pollution Clean-up in Qalyoubia, *Technical Remediation Training*, Egyptian Environmental Affairs Agency, 2005.

[7]   LIFE Lead Pollution Clean-up in Qalyoubia, *Health & Safety Training*, Egyptian Environmental Affairs Agency, 2005.

[8]   Development Training II (DT2), *How to Characterize/ Classify Hazardous Waste*, Egyptian Environmental Affairs Agency, 2002.

[9]   Development Training II (DT2), *Permitting of Hazardous Waste,* Egyptian Environmental Affairs Agency, 2002.

[10] Development Training II (DT2), *On-site Storage of Hazardous Waste*, Egyptian Environmental Affairs Agency, 2002.

[11] Hazardous Waste Management Project in Alexandria, *Hazardous Waste Management*, Egyptian Environmental Affairs Agency, Alexandria Governorate, 2003.

Section 15
Waste management

Waste management policy and citizen participation from the aspect of waste management planning theory

T. Okayama[1] & M. Yagishita[2]
[1]Division of Integrated Research Projects, EcoTopia Science Institute,
Nagoya University, Japan
[2]Graduate School of Global Environment Studies, Sophia University,
Japan

Abstract

In worldwide measures against environmental problems, the 'participation of all actors and stakeholders' has become of paramount importance. The system of law concerning Japan's environment has also emphasized citizen participation as a long-term target. In addition, laws concerning Japan's waste, which aim towards the construction of a Sound Material-Cycle Society (SMCS) has decreased the use of natural resources as far as possible and reduced the impact on the environment. Waste management laws in Japan have been converted from their status of 1R, to 3R, that of a sustainable society, for which the participation of citizens and individual stakeholders is essential.

However, there is no legal obligation in the municipality, which could instruct an area how to execute the policy. It is not easy for a society to achieve 'citizen participation and a sound material-cycle society'.

A forum that aimed at 'creating a sound material-cycle society based on citizen participation" was held as a social experiment in Nagoya from 2002 to 2005.

By comparing conventional policy-making with participatory policy-making, it was possible to see whether or not participatory policy-making can overcome the weak points of conventional policy-making in terms of policy effectiveness.

Keywords: waste management planning, sound material-cycle society, citizens' participation, forum, policy effectiveness.

WIT Transactions on Ecology and the Environment, Vol 92, © 2006 WIT Press
www.witpress.com, ISSN 1743-3541 (on-line)
doi:10.2495/WM060511

1 Purpose of the study and definition of "participation" and "partnership"

'Participation' and 'partnership' are, in principle, what is necessary to manage the worldwide environmental problem. In order to cope with these problems, the general population needs to reform their individual lifestyles countries will be required to reform their national policy. In order to implement such a reform, participation and partnership of all citizens and sectors of society are necessary.

The 'importance of participation of the all stakeholders and citizens' is highlighted even in the Rio declaration of the United Nations Conference for Environmental Development (UNCED) in 1992, under the 10th principle, as follows:

"...Environmental issues are best handled with **participation of all concerned citizens**, at the relevant level."

Participation of the citizen is addressed as one of the long-term targets in the Japanese Basic Environment Plan, which will be decided in the next years of The Basic Environment Law (1993) that was enforced following the Rio declaration and Agenda 21.

These participation concepts are varied and extensive, and include participation in activities, which preserve the environment and participation in a decision-making and administrative plan. The participation in public decision-making is of extreme importance.

Therefore, "the citizen's participation in environmental policy" in this manuscript means participation in the environmental policy decision-making process. Many cases of the citizen participation technique and method have been done at so-called forums in Europe and America, and there are many reviews of the cases that have so far been developed in detail, for example by Renn et al. in 1995 [1], and more recently by Nishizawa in 2003 [2].

The purpose of this study is to make it clear how and what the participation in conventional policy is, and what the issue of conventional policy is, especially from the viewpoint of policy effectiveness, at the same time as comparing conventional policy and participatory policy. The municipal waste management plan in Japan is taken as a specific example case, and it is shown whether or not participatory planning can overcome the weak points of conventional planning from the viewpoint of policy effectiveness. Furthermore, an ideal, method of citizen participation in a waste management plan is proposed by this study.

2 The Japanese modulo system regarding the waste management

In Japan, the Basic Environment Law is thought of as the most significant environmental policy. The Fundamental Law for Establishing a Sound Material-Cycle Society (FLESMCS) was formed in 2000, at the same time the Basic Environment plan and Waste Management and Public Cleansing Laws (WMPCL) were amended (Figure 1).

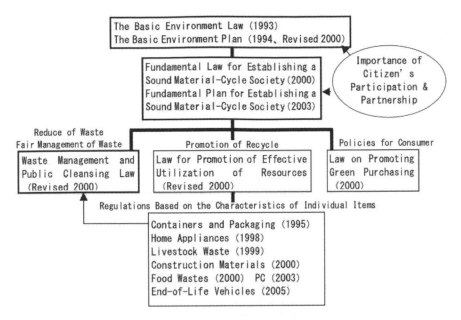

The Basic Environment Law (1993)
The Basic Environment Plan (1994, Revised 2000)

Fundamental Law for Establishing a
Sound Material-Cycle Society (2000)
Fundamental Plan for Establishing a
Sound Material-Cycle Society (2003)

Importance of
Citizen's
Participation &
Partnership

Reduce of Waste
Fair Management of Waste

Waste Management and
Public Cleansing Law
(Revised 2000)

Promotion of Recycle

Law for Promotion of Effective
Utilization of Resources
(Revised 2000)

Policies for Consumer

Law on Promoting
Green Purchasing
(2000)

Regulations Based on the Characteristics of Individual Items

Containers and Packaging (1995)
Home Appliances (1998)
Livestock Waste (1999)
Construction Materials (2000)
Food Wastes (2000) PC (2003)
End-of-Life Vehicles (2005)

Figure 1:   Legislative framework to establish a sound material-cycle society.

The Japanese waste policy was converted from a 1R (Recycle) policy to 3R (Reduce, Reuse and Recycle). In Japan, the part of society that this 3R applies to is called a "Sound Material-Cycle Society (SMCS)". For SMCS formation, all sectors of society have to take a second look at their lifestyle and business in order to convert to 3R. The natural resource consumption is suppressed as much as possible, and environmental load should decrease.

Regarding the Tripartite Environment Ministers Meeting between Japan, China and Korea (TEMM), presently "constructing the SMCS" is advanced through 3R initiatives.

The conduct and partnership of citizens is indispensable in the implementation of 3R, because it is needed for the promotion and procurement of eco-friendly goods and services, for example: mending or reusing old goods, and separating and recycling to reduce the waste. Individual citizens execute these activities. In order to make the 3R policy more effective, the partnership-type policy should encourage the cooperation and partnership of the citizen. Partnership-type policy, rather than a top-down administrative plan, is an action plan that all stakeholders, including the administration, the citizen and the enterprise make together. With such a partnership, SMCS can effectively be implemented.

Regarding Figure 2, each municipality has a responsibility to dispose of municipal waste (not industrial waste) and therefore has the responsibility of choose the municipal waste management plan to be used. Article 6 of WMPCL requires a decision to be made about a plan.

National	Waste Management and Public Cleansing Law	Fundamental Plan for ESMCS
Local Government (Prefecture)	Waste Management Plan	Waste & SMCS Infrastructure Service Plan
Municipality (City, Town)	Municipality Waste Management Plan	Waste Management Facility Improvement Plan

Figure 2: Waste management policy system from national level to local level.

It can be interpreted from The Basic Environmental Plan that WMPCL should be enforced on the basis of being a fundamental plan for establishing an SMCS. However, the only condition is that the waste is processed on the basis of WMPCL, and that the resource should be recycled on the basis of FLESMCS, so in other words, although the target of WMPCL is the correct processing of waste, there is little stipulation regarding SMCS formation. This legal technicality in the municipal waste management plan, which aims to form an SMCS and to keep partnership of all sectors is considered, strictly speaking, as going beyond the necessary function of a municipal waste management plan. Because participation cannot be legally forced, 3R is not stipulated sufficiently in the municipal waste management plan, and the plan has not become a partnership. In addition, traditionally in the Japanese indirect democracy system, the administration has the main decision-making powers. The opportunity for citizens and enterprises to participate in decision-making is basically limited to elections and in forming public opinion. Recently, the administration has requested opinions from citizens regarding policy-making and has encouraged public comment, which was institutionalized in 1999, questionnaires and candidacy to conference members are sometimes used. However, participation is very limited in the decision-making process.

3 Cases of participatory planning in Japan

Current details regarding the participation in international environmental policy were shown in Section 1. Regarding Japanese waste policy, as shown in Section 2, it is thought that in order to convert the conventional waste policy into the much more effective SMCS formation policy (i.e. 3R policy), participation and partnership of stakeholders are necessary. In addition, deciding on a location for the final disposal place and its construction becomes difficult, local finance becomes tight; agreement formation and citizen participation are encouraged. The citizen's understanding and participation are indispensable in such a difficult situation. When experiencing this kind of status, there seem to be some cases where citizens mainly participate in the process of waste management planning. In Aichi prefecture in Tsushima city (2003), Nisshin city (2004) and Mie prefecture Kuwana city (2005), through NGO coordination, citizens decided on a waste management plan. In this case, it was possible for the administrative plan to be agreed with the citizens (Hirose et al. [3]), but this plan did not come about as a result of legal obligation, which is stipulated in article 6 of WMPCL, the citizens made the decision instead. The plan was already in place with the main administrative leadership, but the citizens' plan followed the former plan, which basically concerned the decision-making process and the planned contents.

As a pilot case, policy-making was trialled at a forum in Aichi prefecture in Nagoya city from 2002 to 2005. The purpose of this social experiment was to see whether citizens would positively grasp the concept of an SMCS and be able to make SMCS policy themselves without being restricted by legalities. If it were to work, it was expected that this policy would be more effective than conventional policy (Yagishita et al. [4]). In this paper, this experimental case is called 'participatory planning'.

4 The difference between the conventional planning and participatory planning

Figure 3 shows the processes of these two types of planning.

If a conventional type plan is selected, the process will take place as outlined on the left in Figure 3. The foundation of the plan is conceptual planning, the creation of a framework, and the basic planning process. The conceptual planning of municipal waste management is the point at which it is possible to incorporate some specific concepts of general civil planning, which sets the target, purpose and method for carrying out the policy. This is very closely connected to the basic plan, which decides on the location of facilities at each stage of development in order to further the concept of the plan.

At the basic planning phase, the quantity of waste, the quantity of resource material for collection and the final disposal quantity are concretely identified as the targets. These targets are set on the basis of objective data, including evaluations of the present condition, current land use, social status and the amount of waste, resource amount and amount within a year of Business as Usual (BAU). The targets and goals of plans are usually set safely inside the municipality's waste processing ability, even if it is decided from a political aspect first (which is rare). Furthermore, at roughly the same time as the target is set, concrete measures such as compilation of a list of items to separate; collection of resources and distribution of a list of facilities and services for recycling is put together.

Sometimes the municipality requests citizen opinion and public comment concerning the output of the basic plan. During the conceptual planning and the basic planning phases, the planning process is conducted and advanced through conferences. Specialist forms mainly depend upon the municipality. Recently, this conference has included some citizens who have responded as a result of a desire to voice their opinion. For example, in the case of Tsushima city, all members of the conference were citizens offering their opinions. The above is an explanation of conventional planning.

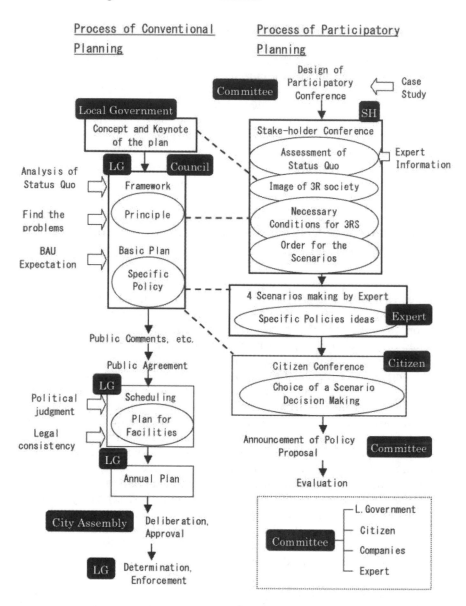

Figure 3:   Process of model planning, conventional plan and Participatory plan.

On the other hand, on the right hand side of Figure 3, participatory planning is shown. When planning in this way, 'the Committee for the Forum for Creating a Sound Material-Cycle Society based on Citizen participation' (below, committee) was organized first, and it became the management parent for this forum. Committee members are the stakeholders, the specialists in waste management, Nagoya City, and the citizens. On this committee, first of all the method of participatory conference was examined.

Figure 4 shows the concept of participatory planning. It first became clear what the stakeholder image of Nagoya becoming SMCS was, and what the necessary conditions and composition requirements of SMCS would be. Secondly, the plans to transform Nagoya into an SMCS were examined. Before making scenarios, the 'order', which would influence the design of the scenarios was extracted. Finally, one scenario was selected. The committee designed the hybrid-type forums: the stakeholder meeting first and the citizen meeting second.

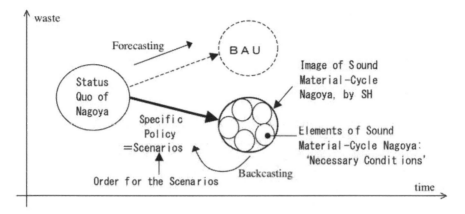

Figure 4:   The idea of backcasting planning.

At the stakeholder meeting, the stakeholders firstly evaluated Nagoya city's waste reduction policy from the point of view of the status quo of waste management. They also discussed and decided upon the index for evaluation. This index was set by the necessary conditions and element requirements of SMCS, for example: 'the mechanism of 3R', 'everyone carries out responsibility', 'the person who endeavours is rewarded' and 'communication and consensus' etc. These requirements attach importance to a person's state and sense rather than to physical flow and the environmental load. Next, the stakeholders decided upon the order for creating the scenario. On receiving this order, the specialists created four scenarios.

The committee continuously redesigned the citizens' meeting. The participants of this meeting were extracted at random from the Nagoya City electoral roll, and 16 people gathered. The citizens' meeting discussed the scenario six times and voted twice. One scenario was finally selected, partly revised, and announced as a 'proposal'. The above is an explanation of the method of participatory planning.

Table 1 shows the basic philosophy of the waste management plan and its title.

Nagoya City's plan is used as an example of a conventional plan. The plans of other cities do not greatly differ from this plan. The basic philosophy of Nagoya City's plan is 'administration, citizens and enterprise partnerships'. The behaviour of the administration, the citizens and the entrepreneur is being written in a concrete measure. For instance, Nagoya City is requesting its citizens to be greener. Yokohama City has decided that the role of the city is to coordinate citizens and enterprise partnerships. On the other hand, 'participation of all sectors' and 'impartial load' are assumed to be the most important philosophies because of the SMCS construction.

Table 1: Basic philosophy of waste management plan and title of plan.

Nagoya (Participatory)
[Title] SMCS that is created with effective separation and eco-friendly goods -Done based on the partnerships and impartial loads of all people who act in Nagoya- [Basic philosophy] • Continuous action is most important for the achievement of an SMCS, so that all people who live in Nagoya share equal responsibility. • For the achievement of the SMCS, "The producer must make the effort to produce eco-friendly goods and to use the recycling system, to reduce waste as far as possible" and "thoroughness in effective waste separation by the citizens" should be set as a prop.
Nagoya City (Conventional) [Title] Challenge for SMCS [Basic philosophy] Challenges for SMCS and Sustainable Society, Eco-Partnership for SMCS

Table 2 shows the target of the plan. The plan based on WMPCL has been decided upon to show the amount of the target of the waste processing from the data like the situation of waste discharge, BAU and the target fiscal year.

Table 2: Amount in standard year and target amount of plan (1,000ton).

Type of planning	Standard Year		BAU	Target amount	
	Discharge	Land-fill	Discharge	Discharge	Land-fill
Participatory	–	—	–	–	–
Conventional	1084	150	1200	1080	20

However, the participatory plan has no numerical target amount. Reading the "proposal", the target of a participatory plan is written below. These targets are: partnership among all sectors of Nagoya and mechanisms, which aim at achieving a new society. The environmental impact and the cost expected as a result of this policy are just fixed for referral. The differences mentioned above are summarized in Table 3.

Table 3:   Features of two types of planning.

	Conventional Planning	Participatory Planning
Process	Planning process is fixed.	Designed by participants (committee) →process conducted by citizen
	The plan (output) partially reflects citizens' opinions.	The output from the forum becomes the plan.
Contents	Forecasting The target is set forecasting the future from the current state. Using BAU	Backcasting The target is not expressed by the fixed quantity index. Moreover, it is not a target led by BAU either.
	The target is expressed by the fixed quantity index.	The quantification index of the amount waste etc. is a value of the result of the policy. It was fixed as verification data.
	"Partnership" and "fairness"etc. are just notes. ▼	"Fairness" is the most important. Fairness and partnership becomes the target of plan. ▼
	The administrative body systematizes a necessary measure for itself. It is not easy to become partnership type.	All of actors' behavior is committed. This is agreed on among actors. Partnership type

5 Conclusion

In order to increase the effectiveness of the plan, 'understanding and agreeing on the policy' and 'commitment to acting to the policy' are indispensable. However, there is no opportunity for citizens to participate in conventional waste management planning because the waste management plan is an administrative plan, and an administrative plan must be one that the administration decides upon. However, the administration has requested to go beyond the limit of an administrative plan to a waste management plan and an SMCS policy, which require participation and partnership from its citizens. This creates a dilemma for the administration, which creates a weakness and directly impacts upon policy effectiveness. If it is assumed that conventional planning in Table 3 has its weak points, the new participatory method should be able to overcome them.

Therefore, even if it cannot obtain legal proof to make plans and policies by forum in the present system, it has the means for each municipality to hold a participatory forum and make the partnership-type policy.

The part of administration in the plan is assumed to be a conventional waste management plan, then concerning the citizen and the enterprise, the plan also become their action plan which they committed.

References

[1] Renn, O. Webler, T. and Wiedermann, P., Fairness and Competence in Citizen Participation; Evaluating Models for Environmental Discourse, Dordrecht, The Netherlands: Kluwer Academic Publishers, 1995.

[2] Nishizawa, M., The "Stick-in" Nature of Citizen Deliberation for Technology-Environmental-and Health-Risk Management in Social Environment: Case from Germany, Dissertation collection of Social Technology Study, Social Technology Study Society, Vol.1, pp.133-140, 2003.

[3] Hirose, Y. Shoji, T. and Asai, N., What Participatory Waste Management Plan is needed to be accepted by Citizens –Case Study from Tsushima City-, Proceedings of the 16th Annual Conference of the Japan Society of Waste Management Experts, pp.186-187, 2005.

[4] Yagishita, M. et al, Final Report of the Study for Creating a Sound Material-Cycle Society Based on Citizens' Participation, (temporary), 2006.

Sustainable supply chain networks – a new approach for effective waste management

H. Winkler & B. Kaluza
Department of Production/Operations Management/Business Logistics and Environmental Management, University of Klagenfurt, Austria

Abstract

In order to realise sustainable development, sufficient waste management is rather important. It is inappropriate for the future only to reduce waste and save energy. Sustainable economic growth is necessary for prosperity but this also in turn causes waste. Based on the literature and formal empirical research, we can state that a movement towards sustainability is only possible if we manage to develop concepts that integrate economic and ecological goals. These challenges can be realised best with the application of certain Sustainable Supply Chain Networks and supply chain management methods. The set up of Sustainable Supply Chain Networks strongly supports the realisation of a circular economy by way of closing process chains between enterprises within an industry and by implementing joint environmental protection measures. This significantly enables the avoidance and reduction of waste and saves costs as well as improves the competitiveness of the members.
Keywords: waste management, waste avoidance, waste reduction, sustainability, Supply Chain, Sustainable Supply Chain Network.

1   Introduction

Reducing and/or avoiding waste was already an important issue since the early 1970s. Initially, waste was reduced by end-of-pipe-technologies, followed by process- and product-integrated environmental technology developments in the 1980s [1]. These concepts were isolated applications that focussed on a single enterprise.

Today, waste management has to be considered in the context of sustainable development. Sustainable development claims the simultaneous improvement of

WIT Transactions on Ecology and the Environment, Vol 92, © 2006 WIT Press
www.witpress.com, ISSN 1743-3541 (on-line)
doi:10.2495/WM060521

economic, environmental, and social conditions. This means in turn for the business to construct a certain circular economy, where it is that used products, scrap, residuals, and other waste materials are collected, conditioned, and reused or recycled. A single enterprise cannot implement and run a closed circular economy effectively for itself. Holistic waste management can be realised best with the application of certain Sustainable Supply Chain Networks (SSCN). With the methods from supply chain management, a sustainable circular economy is created by closing process chains between companies in an industry and by taking joint economical and ecological useful measures.

In our contribution, we will present an approach for the implementation of a SSCN. Therefore, we will characterise different forms of waste and discuss the reasons for waste emergence. In the next step, some general proposals for waste management will be set forth. Subsequently, we introduce the concept of SSCN and demonstrate the economic and ecological benefits for participating enterprises.

2   Types and emergence of waste along the product life cycle

Producing, distributing, and consuming products [2] cause an enormous amount of various types of waste. Within production, various types of waste emerge. Initially, there are waste materials and by-products as a direct result of the transformation processes [3]. In addition, scrap arises as a result of not achieving the required output specifications. Finished and semi-finished goods as well as raw materials require certain packaging that fulfils different functions, such as protection, transport, storage, selling, or easier handling and usage [4]. The lack of the adoption of not re-usable packaging is responsible for most of the generated waste and counts for approximately half of the emerging waste volume in Western Europe [5, 6, 7]. At the end of the life cycle, used products are mostly disposed of by the original user. Hence, they have to be reused, recovered, disposed of, or incinerated.

The emergence of waste is assignable to the product life cycle, which can be divided into three stages: the development-, production- and disposal-stages [8]. From the strategic perspective, the development stage has significant implications for the subsequent life cycle stages. Decisions made during development determine the amount and quality of the generated waste. The configuration of the product properties results in determining most of the used processes, materials, and possibilities to recover and to reuse [9, 10]. Therefore, it is the most effective way for avoiding waste. The waste that appears in the production stage is highly influenced by the specifications of a product, production processes, and production technologies.

Frequently, waste in the production stage emerges as a result of not having effective processes and/or technologies. The use of hard tooling, e.g. milling, drilling, grinding, or lathing manufacturing processes determines the emergence of waste. Soft tooling processes, e.g. laser sintering or stereo lithographic processes, can be used in order to minimise waste materials. In addition, apart from the used technologies and processes, the efficiency of the transformation processes specifies the amount of generated waste. The appearance of scrap emerges due to a mismanaged process or inappropriate processing methods.

A lot of waste along the product life cycle is derived from bad coordinated production and logistics processes, as well as from information deficits between the supply chain members. Bad coordinated logistic processes and a lack of information all lead to the so-called "bull-whip-effect" [11]. This causes overproduction and unnecessary inventory. Overproduction means that production has commenced excessively early and/or too many units have been produced. This generates longer lead times and spoilage. The use of suitable packaging systems can avoid almost half of the waste volume as mentioned above. Therefore, it is tremendously important to use re-usable packaging systems or at least minimise packaging along the entire supply chain. This also includes the consumer packaging that should be considered from the very beginning of the product life cycle to ensure a materially efficient approach.

The disposal stage is characterised through the transition of a good to waste. This stage normally begins when the original user no longer needs the product and chooses to dispose of it. Hence, it should be solicited to prolong the life cycle or intensify the service units provided to minimise the waste impact. Therefore, the emergent waste should be reusable or recyclable in order to achieve a high degree of sustainability [10, 12].

3   Waste management in the context of a circular economy

There are three main reasons for enterprises to make efforts in improving their waste position: 1) If the waste producers are charged for the full environmental cost, it in turn becomes more cost-effective to reduce the amount of waste they produce, 2) it can be profitable to reduce waste because the raw material charge also decreases and 3) an improved environmental position can strengthen customer loyalty, which is beneficial in a favourable market position.

For effective waste management the types and reasons of emerging waste have to be detected. The amount of produced waste is a consequence of how effectively and efficiently we use resources, and once waste has been produced, how it is dealt with has a strong impact on the environment. Most of the generated waste can be avoided and/or recovered by constructing a circular economy.

In order to realise a circular economy, every type of waste has to be a part of a closed loop process chain. Therefore, companies have to provide a framework for realising a circular system. In detail, an integrated reuse and recovery system must be organised and implemented to ensure adequate waste treatment. Consequently, waste management has considerable impacts on the competitiveness and profitability of an enterprise [13].

The principles of waste management are elimination, reduction, reuse, recovery, and disposal of waste, whereas elimination has the highest priority, and disposal has the lowest priority [12, 14, 15, 16]. For realisation, the principles of waste management enterprises are forced to close open process chains. This means that most of the generated waste has to be returned to the manufacturing or consuming processes. In this context, the best solution is reusing waste because there is no need for additional material input. When reusing is impossible, waste should be recovered in a recycling process to guarantee a circular flow. This advances the material efficiency as well as the positive impact on the environment because only a minimum of the produced goods and waste must be assimilated by the environment. Incineration or disposal should be avoided because the waste materials have to be replaced by new material inputs.

4   Application of Sustainable Supply Chain Networks for the avoidance and reduction of waste

In business, waste management is often seen as a duty of a single company [17]. We reject this limited view and suggest, due to the aforementioned facts, to configure a SSCN to realise a circular economy and construct a closed loop waste management. We think that a SSCN configuration is a holistic approach to avoid and reduce waste and gain economic advantages for members.

4.1 Organisation and structure of a Sustainable Supply Chain Network

A SSCN is a set of different types of companies that work together to realise a sustainable circular economy considering the potentials for waste reduction and avoidance from the development stage to the end-of-life of the life cycle of a product.

With the forming of a circular economy, the various process chains of all members of the network should be closed to avoid and reduce waste. With closing the process chains, the share of recycled materials can be increased up to 80 per cent, instead of 1 per cent with unclosed process-chains [8].

The main principles of the SSCN refer to a sustainable configuration of the network. Therefore, it is necessary to set up sustainable goals and strategies, formulate and introduce sustainable win-win relationships, implement intelligent systems and resources, as well as arrange cooperative R&D with focus on sustainability. These principles influence and interact with each other and must be coordinated to establish an effective and efficient SSCN.

The participants of a SSCN are not only companies of an existing supply chain, but also service providers specialised in collecting, exchanging, conditioning, recycling, or eliminating used products and waste [18]. These companies adopt duties for remanufacturing/reuse, recycling and disposal/incineration to enable a circular economy in the SSCN. In addition, special logistics providers are integrated to manage the physical distribution and redistribution of the material flows between the participants. Furthermore, together with the service providers they are also responsible for collecting, saving, and analysing waste specific information and providing them to the participants of the network. In Figure 1 shows exemplarily the relationships between the enterprises of a SSCN. These relationships include the information-, material-, residual-, used product-, waste- and financial exchange processes among all of the companies.

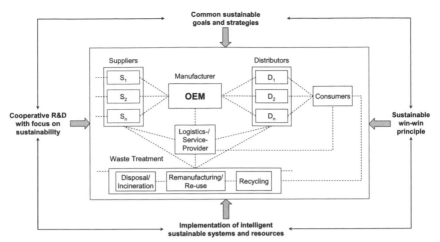

Figure 1:   Generic structure and principles of a Sustainable Supply Chain Network.

Within the SSCN, two main planning levels have to be considered. The first planning level corresponds to the network-level, the second to the enterprise-level. To run a circular economy the companies formulate environmental goals, strategies, measures and measurement principals for co-operation on the network-level in common.

Preventing and reducing ecological mismanagements such as wasting energy and resources by organising closed process chains and gaining economical benefits are goals of the SSCN. Strategies to accomplish the determined goals among all of the companies have to be formulated. These strategies refer to recycling, reusing, remanufacturing, as well as on product development. Deriving from these strategies, the appropriate measures and metrics have to be set on the enterprise-level. Purchasing, marketing, distribution, logistics, and operational measures have to be set within the enterprises, involving all corporate divisions and value creating activities [19, 20]. It is here that the main restriction on the enterprise-level, which is the sustainable goals of the network, must not be in conflict with the concrete strategic and operative goals of a single enterprise. In fact, the companies remain legally and economically independent, making their own strategic and operational decisions influenced by the network strategies.

The implementation of a SSCN leads to the strong improvement of the competitiveness of all participants of the network as well as to an increase of all the companies' values. This is based on the higher efficiency of the used resources as well as on the value of the remaining waste. An improvement of most of the ecological and economical results is a realistic consequence. Firstly, this is evident because the production of waste and residuals is avoided at the development stage of the product life cycle where it is economically and ecologically useful. Secondly, when waste is a valuable good for one or more members of the SSCN, this waste is no longer only garbage, but a sellable good.

4.2 Possibilities for waste avoidance and waste reduction

One target of implementing a SSCN is to avoid waste along the entire product life cycle. Identifying all of the factors that influence the existence and the emergence of waste at every member of the supply chain is very important to reach this objective. In fact, proactive SSCN management begins in the stage of R&D, in the conceptual design-phase of the product. Decisions that are made in this life cycle stage have the most significant influence on the product life cycle costs and on preventing environmental costs at the disposal stage of the product [9, 10].

Integrating the SSCN partners in this early stage of product development prevents the conception of a product that contains hazardous (toxic), non-reusable, or non-recyclable materials [9]. With this measure, bad economic and ecological results for various members of the network are avoided. The major objective of cooperative R&D is to find the possibilities to prevent waste and save resources. Not only materials and other factors for production are determined at the developing stage. Furthermore, this stage has effects on the product design (shape) and the packaging. Designing a product that is constructed for an easy dismantling makes it easier to avoid waste that is not reusable. Connecting this with developing a packaging-free product and/or reusable packaging systems where it is technologically, economically, and ecologically possible, avoids packaging waste. The usage of reusable packaging-systems has economic and ecological advantages for the entire supply chain network because of standardising the logistical interactions between the SSCN participants. Reusable packaging systems can be implemented in the distribution and redistribution of the materials and goods. For developing these packaging systems, the knowledge and expertise of the recycling-companies and the logistic service providers has to be used. Specialised in managing the transportation, handling, storekeeping, collecting, and packaging of materials and goods, logistic-service-providers know various options to avoid waste in an effective way. The recycling companies are acquainted with the possibilities to disassemble all of the products or semi-products of the supply chain. Consequently, they can provide the essential information of how the producers can design the products and their components free of waste.

Cooperative developing of ecological and waste free products is not the only reason for joining a SSCN. The possibility to use resources of the network-partners is another reason for participating. It is here that the focus is on the available tangible and intangible resources of the partners. In co-operation with the partners, previous production technologies and techniques could be improved to eliminate waste and raise productivity. Similar technologies that have already successfully been implemented in enterprises could be transferred to the other companies.

Apart from the proposed methods to prevent the production of waste, the enterprises have to analyse their processes, methods, and products to reduce existing waste significantly. Analysing the production and logistics processes within a company to find potential for ecological improvements is an option for reducing waste. The main objective here is to increase the efficiency of processes, for example, by improving the use of materials and other production factors in the manufacturing process. Especially packaging waste can be reduced by using larger or standardised packaging-units.

Apart from the processes, the used resources for procurement, manufacturing, distribution, and redistribution also have to be dissected. With a change in the handling of some waste intensive production and logistics devices scrap can be decreased [19, 21]. Therefore, manufacturing technologies in the SSCN should operate in a way that no wastefulness of materials and semi-finished goods occur at any stage. Another opportunity to reduce waste is to prolong the product life. The exchange of spare parts/consumable parts of the products as well as preventive maintenance are services that extend the life of products significantly.

Additionally, improving the planning systems and activities are opportunities for cutting waste. With coordinated planning methods, it is possible to predetermine the quantities of materials needed for manufacturing more precisely.

5   An example for the reduction of waste by using a kind of Sustainable Supply Chain Network

For illustrating the benefits of a SSCN, it is suitable to choose an industry that works diligently to achieve the goal of sustainability. Within the automotive industry, many enterprises exist that have introduced environmental initiatives for gaining a competitive advantage towards sustainability. Consequently, the automobile is the consumer article with the highest rate of recycling, which is even higher than paper [22].

From 2006 at least 85% and from 2015 at least 95% of the average weight of an end-of-life vehicle must be recovered whilst the remaining materials may be disposed of. Over 80% accordingly 85% of the total recovery quota is taken up by reuse and recycling whilst the proportion of incineration is restricted to a maximum of 5% accordingly 10% of the average weight per vehicle [23].

For achieving these targets, enterprises have to implement a closed-loop waste management system, which must be organised by integrating all of the participants along the supply chain. This means that all of the possible waste flows during the product-life-cycle must be considered. With this framework, it is possible to generate a synergetic result, which means that all of the waste flows are planned by a central logistics-/ service-provider.

Figure 2 shows a possible flowchart of a disassembly and utilisation system in connection with a SSCN in the automotive industry. Many different processes have to be regarded and planned. Therefore, the service-provider has to coordinate the different material and waste flows to secure an optimum in recovering the end-of-life products, in this case used cars. In addition to the material flow activated by the origin supply chain, this process results in high material efficiency.

Running the recovery facilities at full capacity, and cutting the overall recovery costs; different manufacturers should consider aligning their recovery processes [23]. For this reason, it should be possible to create a SSCN where all of the participants are able to reach a high degree of competitiveness regardless of the examined industry. Therefore, the manufacturers are able to decrease their material inputs and to realise the required recovery quota. The enterprises involved in waste treatment can optimise their processes because of having a high equipment utilisation, which results from the coordinated waste flows.

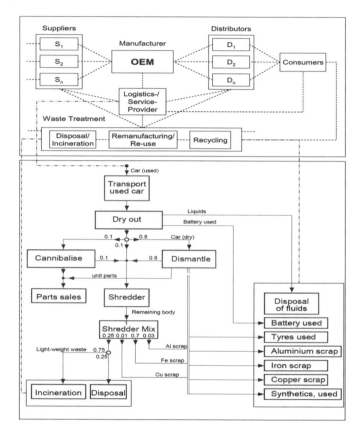

Figure 2:   A part of SSCN in the automotive industry.

6   Conclusion

In our contribution, we highlighted the actual deficits of waste management and claimed the possibility to construct a SSCN in order to implement an integrated waste management. With a SSCN, it is possible to turn from a flow economy to a closed loop economy. This benefits ecological and economic development within the SSCN, which leads to the improvement of the competitive position for all participants.

We have shown that four principals are essential to reach a sustainable configuration of the SSCN. These are the setting up of sustainable goals and strategies, formulating and introducing sustainable win-win relationships, implementing intelligent systems and resources, as well as arranging cooperative R&D with a focus on sustainability. These strategic principals shape the SSCN, its specific configuration, and the relationships between the participating enterprises. As we have demonstrated, the members of the SSCN are not only companies of the origin supply chain. Remanufacturing/Re-use, Recycling, and Disposal/Incineration enterprises, but also logistics and service providers are important partners in a SSCN.

In example, we demonstrated that in the automotive industry legal changes occurred that encourage automotive manufacturers to introduce a circular economy. Meeting the legal requirements and achieving economic advantages is only possible by constructing a SSCN.

Future research in this area must focus on the practical implementation of the concept of SSCN. Measurement-systems must be developed and verified for steering the SSCN effectively and efficiently. Within empirical projects, the validity of the stated arguments has to be tested.

References

[1]   Nagel, M. H., Managing the environmental performance of production facilities in the electronics industry: more than application of the concept of cleaner production, *Journal of Cleaner Production*, **11(1)**, pp. 11-26, 2003.

[2]   Feser, H.-D., Vom End-of-Pipe zum Integrierten Umweltschutz? Die volkswirtschaftliche Perspektive. *Integrierter Umweltschutz. Umwelt und Ressourcenschonung in der Industriegesellschaft,* eds. H.-D. Feser, W. Flieger & M. v. Hauff, Transfer Verlag: Regensburg, pp. 41-56, 1996.

[3]   Dijkema, G. P. J., Reuter, M. A. & Verhoef, E. V., A new paradigm for waste management. *Waste Management*, **20(8)**, pp. 633-638, 2000.

[4]   Souren, R., *Konsumgüterverpackungen in der Kreislaufwirtschaft. Stoffströme-Transformationsprozesse-Transaktionsbeziehungen,*   neue betriebswirtschaftliche forschung 293, Deutscher Universitäts-Verlag: Wiesbaden, pp. 40-43, 2002.

[5]   Europäische Kommission, *Im Visier der EU: Abfallwirtschaft*, Amt für amtliche Veröffentlichungen der Europäischen Gemeinschaften: Luxemburg, p. 11, 2000.

[6]   Kogg, B., Power and Incentives in Environmental Supply Chain Management. *Strategy and Organization in Supply Chains,* eds. St. Seuring, M. Müller, M. Goldbach & U. Schneidewind, Physica-Verlag: Heidelberg New York, pp. 65-82, 2003.

[7]   Cardinali, R., Waste management: a missing element in strategic planning. *Work Study,***50(5)**, pp. 197-201, 2001.

[8]   Blecker, Th., Logistische Aspekte der Kreislaufwirtschaft. *Kreislaufwirtschaft und Umweltmanagement. Duisburger Betriebswirtschaftliche Schriften Band 17*, ed. B. Kaluza, S + W Steuer- und Wirtschaftsverlag: Hamburg, pp. 97-134, 1998.

[9]   Waage, S. A., Reconsidering product design: a practical "road-map" for integration of sustainability issues. *Journal of Cleaner Production,* Article in Press, Corrected Proof, pp. 1-12, 2006.

[10]   Maxwell, D. & v. d. Vorst, R., Developing sustainable products and services. *Journal of Cleaner Production,* **11(8)**, pp. 883-895, 2003.

[11]   Wildemann, H., *Supply Chain Management. Effizienzsteigerung in der unternehmensübergreifenden Wertschöpfungskette.* TCW-report 39, TCW Transfer-Centrum: München, pp. 14-19, 2003.

[12]   Bates, M. P. & Phillips, P. S., Sustainable waste management in the food and drink industry. *British Food Journal,* **101(8)**, pp. 580-589, 1999.

[13]   V.Hoek, R. I., From reversed logistics to green supply chains. Supply Chain Management, **4(3)**, pp. 129-134, 1999.

[14]   Kumar, S. & Malegeant, P., Strategic alliance in a closed-loop supply chain, a case of manufacturer and eco-non-profit organization. *Technovation,* Article in Press, Corrected Proof, 2005.

[15]   Ravi, V., Shankar, R. & Tiwari M. K., Productivity improvement of a computer hardware supply chain. *International Journal of Productivity and Performance Management,* **54(4)**, pp. 239-255, 2005.

[16]   McDougall, F. R., Life Cycle Inventory Tools: Supporting the Development of Sustainable Solid Waste Management Systems, *Corporate Environmental Strategy,* **8(2)**, pp. 142-147, 2001.

[17]   Shen, J.-B., Chou, Y.-H. & Hu, Ch.-Ch., An integrated logistics operational model for green-supply chain management. *Transportation Research Part E: Logistics and Transportation Review,* **41(4)**, pp. 287-313, 2005.

[18]   Prahinski, C. & Kocabasoglu, C., Empirical research opportunities in reverse supply chains. *Omega,* **34(6)**, pp. 519-532, 2006.

[19]   Rao, P. & Holt, D., Do green supply chains lead to competitiveness and economic performance?. *International Journal of Operations & Production Management,* **25(9)**, pp. 898-916, 2005.

[20]   Hervani, A. A., Helms, M. M. & Sarkis, J., Performance measurement for green supply chain management. *Benchmarking: An International Journal,* **12(4)**, pp. 330-353, 2005.

[21]   Rao, P., Greening the supply chain: a new initiative in South East Asia. *International Journal of Operations & Production Management,* **22(6)**, pp. 632-655, 2002.

[22]   Schweimer, G. W. & Levin, M., Life Cycle Inventory for the Golf A4, www.volkswagensustainability.com/nhk/nhk_folder/en/download.Par.0022.Download.pdf

[23]   Krinke, S., Boßdorf-Zimmer, B. & Goldmann. D., Executive Summary Life Cycle Assessment of End-of-Life Vehicle Treatment, www.volkswagen-sustainability.com/nhk/nhk_folder/en/leistungen/umwelt/recycling/sicon-verfahren.Par.0002.Download.pdf

Integrated waste management of special and municipal waste – a territorial case study

L. Morselli[1], S. Cavaggion[1], C. Maglio[1], F. Passarini[1], G. Galeazzi[2]
& G. Poltronieri[2]
[1]Department of Industrial Chemistry and Material,
Bologna University, Italy
[2]Mantova Province, Italy

Abstract

An Integrated Waste Management System must join flows, methods, and techniques of collection, treatment and disposal, in order to achieve environmental benefits, economic optimisation and social acceptability. The aim is to define, by means of the quantification of waste produced, collected, reused and disposed within the studied territory, a practical system of management of these materials. Results obtained in the Province of Mantova (Northern Italy) revealed an overall production of special waste (year 2003) of about 1 million and 137 thousands tons, of which about 74,500 tons of hazardous waste. The most important material flows are represented by the different typologies of wooden rejects, agricultural residues, construction and demolition inert debris, refuses containing asbestos, medical waste, biosolids from sewage sludge. The quantification of municipal solid waste was, on the contrary, performed directly by the local Governments (Municipalities), which must forward these data to the Provincial Waste Observatory. In 2004, the production of urban waste amounted to about 216,000 tons, about 81,000 tons of which were selectively collected. In the second step of the work, plants for waste recovery, treatment, storing and disposal were examined to determine their capacity (according to the authorisations released by the Province); these authorised capacities were compared to the waste quantities to be managed. From this comparison, a good plant capacity resulted, so that the Province can face the problem of waste produced in its territory, in a position to balance supply and demand for the present and future.
Keywords: waste management system, planning, treatment plants, recovery, landfill's useful life.

WIT Transactions on Ecology and the Environment, Vol 92, © 2006 WIT Press
www.witpress.com, ISSN 1743-3541 (on-line)
doi:10.2495/WM060531

1 Introduction

The governmental decree currently in force in Italy (Ministry of Environment Decree 22/1997), which implemented three European Community Directives (91/156/EEC on waste, 91/689/EEC on hazardous waste and 94/62/EC on packaging and packaging waste), fixes Regional and Provincial competence, concerning waste management. In particular, each Region must prepare a plan of waste management, in which the guidelines and the direction strategies of the single Provinces are presented. Provinces, according to the specific territorial features (demographic, geographic, productive…), elaborate provincial plans of waste management, which are in force for 10 years and undergo a revision every 5 years. Within these plans, indications concerning the management of both municipal and industrial waste are contained. The following study consists in the revision of the provincial plan of waste management in the Province of Mantova. This province extend over about 2,340 km^2 and is mainly constituted (for about 92% of its surface) of plan areas, typical of Po valley. The territory is divided into 70 Municipalities, generally of low demographic density (165 inhab./km^2). The productive fabric is distinguished by the small sized enterprises (93% of the total), but some large industrial facilities are also present. The main business fields represented are food, textile, clothing, engineering sectors.

2 Urban waste

The governmental decree 22/97 defines urban waste as "any substance or object included in the categories reported in the Annex A, and which the holder get rid of (or has decided to or must get rid of)". This means all waste deriving from household and productive consumers (which are comparable for quantity and typology). In order to know physico-chemical and commodity characteristics of wastes, sorting of waste according to IPLA [1] methodology were performed in 5 Municipalities: waste sample was sorted in 18 categories (fines < 20 mm, glass, other incombustibles, metals, aluminium, batteries, drugs, packaging with toxic and/or flammable substances, other hazardous waste, textiles, leather and hide, plastic films, plastic packaging, other plastics, fermentable organics, paper and cellulosic materials, cardboard, wood). Fines are sorted according to the size (10-20 mm, 5-10 mm, 3-5 mm and < 3 mm). Coarse wastes were separated, weighted and sorted in the different categories.

As can be seen in Figure 1, about one third of the undifferentiated waste is constituted of organic matter, i.e. the waste coming from consumers' kitchens.
An integrated management system provides that the organic component is distinguished by a separated collection and afterwards treated in composting plants. In this way, it is possible to obtain a high quality compost (competitive in the market of amendments) and to take away the humid fraction of municipal waste from landfill disposal (significant reduction of leachate and biogas production). Waste from Electrical and Electronic Equipment (WEEE) represents a further critical component in municipal waste management, because they contain some hazardous substances for human and environment, and require

careful precautions in all the steps of disposal, from collection to treatment. Another Italian Decree (n. 151, 25 July 2005), which implemented three European Community directives (2002/95/EC, 2002/96/EC and 2003/108/EC), introduces in this wide class all the Equipments whose correct operation depends on electric currents or electromagnetic fields and those which operate in generating, transferring and measuring these currents and fields. Within the Decree, the quantitative and financial goals of separated collection (4 kg/inhab.*y) and recovery (about 60-80%, depending on the class) are defined.

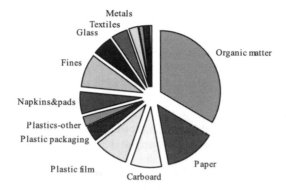

Figure 1:   Percentage of waste categories analysed by sorting.

2.1 Production

In 2004 in Province of Mantova about 216,000 thousands of municipal waste were managed; the amount pro capita is of 550 kg/inhab*y, basically in line with the regional and national pro capita production (510 and 533 kg/inhab*y).

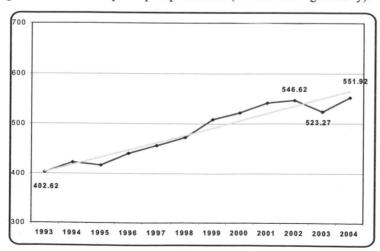

Figure 2:   Pro capita production of municipal waste from 1998 to 2004.

Within the study performed on the Province of Mantova, many interventions aimed to counter the linear growth of waste have been proposed. In particular, the practice of domestic composting has been promoted, being easily diffusing in an agricultural territory, like Mantova. Furthermore, in the next years, prevention actions will be proposed, as the development of Last Minute Market: it consists in the building of relationships between supply and demand, and in the appreciation of products in the last step of their life cycle. Finally, financial tools will be adopted to stimulate the reduction of waste production, as the tariff of environmental hygiene. In the case it was possible to calculate that tariff in the single cases, it could be possible to measure the direct production of each consumer.

2.2 Separated collection

The above mentioned Ministerial Decree 22/97 defines separated collection as "the collection suitable to sort municipal waste in homogenous fractions, from a commodity point of view".

Lombardy Region calculates the percentage of separated collection as the ratio between the amount of waste separately collected and the total extent of waste produced; at the numerator the bulky waste reclaimed in the proper selection plants are included. In 2004, the Province of Mantova reached a percentage of separated collection of 37.5%; this value is in line with the result obtained in North-Italian Regions (35.5%). However, Regional average is slightly higher (40.9%), because in many Lombard Provinces more efficient collection systems have been implemented. In the Province of Mantova bring scheme of collection is the most common system. In the next years the introduction of kerbside collection system is going to be introduced, in order to reach higher percentage of separated collection and an improved product quality. Furthermore, it is fundamental to divide organic fraction, which now is separately collected only in 10 Municipalities out of 70. In order to encourage Municipalities to adopt efficient systems of separated collection and reach high qualitative standards, a financial incentive, named Project Tribute, has been proposed: each Municipality is classified depending on the percentage of separated collection, the pro-capita production and other management parameters; Municipalities which show the most efficient results will benefit from the relief of provincial tax, due to the activities of environmental safeguard, protection and hygiene. This incentive has been acknowledged by the European Environmental Agency.

2.3 Recovery-disposal plants

Province of Mantova is served by two mechanical-biological treatment plants, a landfill disposal plant, some composting plants and a recovery plant for bulky waste. The first ones treat the undifferentiated waste which is suitably sieved, deferrized and stabilized. In output, stabilized organic and inert material, metals are recovered and, most of all, a Refuse Derived Fuel (RDF) is produced.

From the analysis of the current flows and of those expected for the next decade, a management overcapacity (160,000 tons per year are permitted) and the necessity of a re-vamping of treatment technologies resulted (comparison with Best Available Techniques), in order to obtain quality RDF (according to UNI 9903 standard).

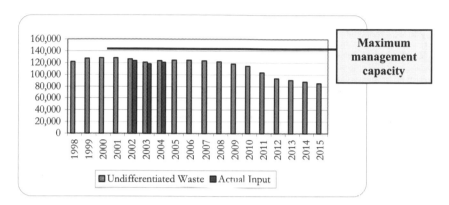

Figure 3:   Prevision of undifferentiated waste flows destined to mechanical-biological treatment plants

Existing composting plants are of different type: some of them treat only the "green" waste, coming from the maintenance of public parks and gardens, others treat a mix of green, organic material and sludge. All in all, the expected flows for the Province of Mantova appear fulfilled; in the next decade an intervention on composting process variables will be performed, in order to improve the quality, and thus the marketability of the compost obtained (according to the law n. 748/1984)

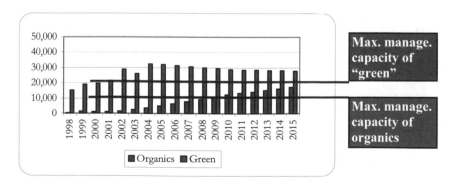

Figure 4:   Amount of waste treated in composting plants.

The plant devoted to the recovery of bulky waste is authorized to treat about 6,000 tons/y of waste. From the analysis of historical fluxes of waste flown into this plant (in 2004 about 6,150 tons were treated) and from the prediction of the future flows a certain plant shortage results; indeed, at present, about 5,000 tons of bulky wastes are conveyed directly into the disposal plant. This practice appears in contrast with Italian law (Decree 36/2003) which, since 2007, allows conveyance of waste not further exploitable in terms of matter and quality, and not further treatable.

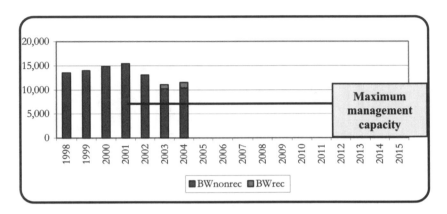

Figure 5: Bulky waste (BW) collection, years 1998-2004.

Table 1: Landfill's life, depending on annual input and degree of compaction.

Capacity of the second lot (m³)	950,000		
Degrees of compaction (t/m³)	1.3	1.0	0.8
Capacity of the second lot (t)	1,235,000	950,000	760,000
Used capacity (t)	140,478		
Residual Capacity (t)	1,094,522	809,522	619,522
Hypothesis of input quantities (t/y)	**Landfill's useful life**		
30,000	36.5	27.0	20.7
40,000	**27.4**	20.2	**15.5**
60,000	18.2	13.5	10.3
80,000	**13.7**	10.1	**7.7**
100,000	10.9	8.1	6.2
120,000	**9.1**	6.7	**5.2**
150,000	7.3	5.4	4.1

The Province of Mantova is not equipped with waste-to-energy plants, and with landfills for inert and hazardous waste. At present, residual municipal wastes are conveyed into one landfill for non-hazardous waste, made up by two lots, measuring about 464,000 and 950,000 m³. The estimation of landfill life has been performed considering different hypothesis, according to the coefficient of compression and to the annual amount accepted by the plant in the next decade. The degree of compaction is an index of "density", calculated as ratio between the total tons input in the plant and the projected volume capacity of it; it depends on material in input's characteristics (e.g., fermentable substance content), and on the compression techniques applied (natural or mechanical volumetric reduction). To this parameter, the used capacity and thus the residual capacity of landfill are strictly linked. In Table 1 it can be seen that residual capacity ranges from 620,000 t, in the hypothesis of coefficient equal to 0.8 t/m^3, and 1,100,00 t, in the best management case (compression of 1.3 t/m^3). The other decisive parameter is the amount of waste annually conveyed to the landfill; this value is highly unpredictable, thus in this study a range between 30,000 and 150,000 t/y is suggested. As can be observed in Table 1, landfill's useful life is strongly related to input quantities: keeping the same amount accepted in the previous years (about 80,000 t/y), landfill would be active until 2012 (compaction of 0.8), or 2018 (optimal situation, compaction of 1.3). The cases of 50% less or more (40,000 or 120,000 t/y) are also highlighted in the table.

3 Special waste

The Decree 22/97 states that, as for Industrial Waste, there is no obligation of planning, because this waste can be transported, swapped and traded as commodities or consumer goods. The Province, in this case, must only monitor waste flows produced, imported, exported and managed within its territory, to check if the treatment and disposal capacity of its plants is adequate.

The same decree defines an inventory system of the different material flows (M.U.D. Data Base, consisting in the compulsory declarations of the producers, managers and carriers of special and hazardous waste) and makes the Province responsible for authorisation of plant construction and working. The decree requires annual declarations by waste producers and managers. These declarations must contain the waste quantities produced and managed in their industries. After a correction of mistakes in declarations at first, we initially calculated the quantities of industrial waste produced in the Province, adding the single quantities declared by every activity; then, we identified the fraction of the hazardous waste. Values are shown in Tab. 2.

Table 2: Total Industrial waste products in the Province at 2002.

Industial waste (t)	Hazardous waste (t)	Total Industrial Waste produced in the Province, 2002 (t)
1,062,902.54	74,580.08	1,137,505.60

The most important material flows are waste from wooden applications and panels, furniture, pulp, paper, cardboard production. This category represents the 38% of the whole quantity of industrial waste produced. This trend of production is due to the presence on this territory of wood manufactures with a volume of business prevailing compared to other kinds of territorial activities. The second category, as for production, is "19" (from waste treatment plants, sewage sludge, water purification and its preparation for industrial use): wastes deriving from treatment plants are considered in their turn as industrial wastes that must be properly disposed. The third category considered is construction and demolition waste (11% of the total product). As for hazardous waste, the most important categories are "07" (waste from organic chemical experiments) and "16" (e.g., discarded cars, oil filters, lead batteries, electric and electronics equipments).

The second step was aimed to determine different kinds of waste management, practised in Mantova Province; divided into recovery and disposal activities. The processed data have been inferred from the declarations of treatment plants managers, in 2002. The total quantity of industrial waste, managed in the Province of Mantova, was 3,171,996.57 tons (recovery and disposal). The difference between the produced and managed quantity is due to a great amount of waste imported inside the territory. Province is the institution responsible for authorization release and for specific plant capacities. The total plant capacity of recovery and disposal is essentially given by the sum of all capacities of the single authorized plants, in tons.

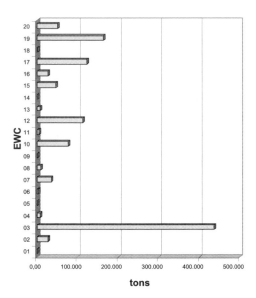

Figure 6: Total quantity of industrial waste products in Mantova Province, classified according to EWC codes (tons).

The authorized quantity, recovery and disposal, in Mantova territory, is 8,832,685.70 tons. Authorized plants for hazardous waste treatment are few and their treatment capacity is not so high; thus, the greatest part of hazardous waste produced in the Province is exported out of the territory. In this way, Mantova Province is not self-sufficient for recovery and disposal of this kind of waste; even if the small produced quantities, often, do not justify construction and starting of specific plants.

Table 3:   Total capacity of all the different kinds of plants authorized to
             waste management, in Mantova Province (tons)

Authorized operations	Total capacity, in tons
Recovery/treatment/disposal	1,418,540.88*
Own disposal	131,234.87
Simplified recovery	7,282,352.50
TOTAL (t)	**8,832,685.70**

A first comparison between produced and managed quantities (produced +
imported), on the territory, is showed in Tab. 4.

Finally, the study's results have demonstrated a good total potentiality of
waste management on the whole provincial territory and a clear prevalence of
plants capacities in comparison to the quantities conferred at 2002.

Table 4:   Comparison between capacities of recovery/disposal and waste
             quantities, in 2002.

Total Authorized Recovery Plant Capacity (t)	Total waste recovered in 2002 (t)	Total Authorized Disposal Plant Capacity (t)	Total waste disposed at 2002 (t)
8,023,416.50	2,884,863.74	307,513.48	173,309.56

4   Conclusion

This study was aimed to the implementation of an integrated waste management
system in a local territory. A detailed survey of the current management system
in the Province of Mantova was performed: waste analysis, technical-economical
solutions adopted, separated collection systems. Furthermore, a comparison
between the present treatment and disposal capacity authorized and the
requirements expected for the decade 2005-2015 was performed. Waste
Management system resulted sufficiently integrated; however, some critical
points emerged for particular kinds of waste (bulky, biodegradable, WEEE, etc.).
The current status appear adequate to meet the needs expected for the next
decade. However, it is necessary to improve technological processes to increase
the percentage of recovered matter and secondary raw material of quality (RDF
and compost). Finally, a prediction of the useful life of the only landfill present
in the territory was performed; this shortage could represent a critical point for
the next decade.

References

[1] DI.VA.P.R.A. (Dipartimento di valorizzazione e protezione delle risorse
     agro-forestali, Univ. Turin) & I.P.L.A. (Istituto per le piante da legno e

l'ambiente), Metodi di analisi dei compost - Determinazioni chimiche, fisiche, biologiche, microbiologiche e analisi merceologica dei rifiuti. Regione Piemonte - Assessorato all'Ambiente Eds., 1992.

No Waste by 2010: leading the way

C. Horsey
ACT NOWaste, Department Urban Services,
Australian Capital Territory (ACT) Government, Australia

Abstract

The Government of the Australian Capital Territory (ACT) was the first in the world to establish a *No Waste by 2010* Goal in 1996 and has since established itself as an international leader in sustainable waste management practices. The No Waste Strategy is the result of extensive community consultation, which identified a strong community desire to achieve a waste-free society by 2010. Since its inception, the No Waste goal has had strong government and community backing and support that has resulted in significant changes in the way waste is viewed and handled in our community.

Implementation of the No Waste Strategy has already achieved a resource recovery rate of over 73%, with some 550,000 tonnes of resources being recovered and processed out of the 770,000 tonnes of waste generated in the Territory annually. This 73% resource recovery rate has been achieved despite a steady increase in total waste generation that is linked to increasing consumption patterns of an affluent western society.

The No Waste Strategy is based on the implementation of an Integrated Resource Recovery Approach that incorporates: the use of economic mechanisms; provision of appropriate infrastructure and services; strategic partnerships and alliances; the development of a resource recovery industry within the ACT; market development for products derived from wastes; strong community engagement programs; and recognition mechanisms for No Waste leaders in our community. Importantly, legislation and regulation mechanisms have not been utilised at this stage to drive resource recovery outcomes.

The model is based on four distinct corner stones:

- The recovery of standard recyclable materials from both the residential and business sectors;
- The recovery and processing of mixed building and demolition wastes and mixed commercial and industrial wastes;
- The recovery, separation and processing of organic materials, with separate processes for garden and food wastes; and
- Then dealing with the more complex composite residual waste products that make up about 5% of our waste stream.

The paper outlines the Integrated Resource Recovery Model and the approaches that have been adopted for the ACT context; model rationale and details on mechanisms utilised; a progress report; successes and challenges and our future directions.

Keywords: Australian Capital Territory; No Waste; sustainable waste management; Integrated Resource Recovery Approach; resource recovery industry; waste minimisation strategy.

WIT Transactions on Ecology and the Environment, Vol 92, © 2006 WIT Press
www.witpress.com, ISSN 1743-3541 (on-line)
doi:10.2495/WM060541

1   Introduction

Many jurisdictions around the world have established waste minimisation targets in varying forms. These targets will deliver a range of differing outcomes due to the policy directions that result from their implementation. The Australian Capital Territory's (ACT's) No Waste by 2010 Policy is significant in that it is guiding a philosophy of not only creating a resource recovery culture and developing alternatives to landfilling but is also striving to eliminate landfill activities.

In 1996 the ACT Government was the first in the world to establish the ambitious goal of the "No Waste by 2010" target and policy. This made it the first government to formalise a policy direction towards recovering all resources from the waste stream, with the aim of eliminating landfill. The impacts of establishing such a policy have been far reaching, and include:

- Residents, businesses and government agencies changing their waste practices;
- Waste service providers re-orientating from waste transport and disposal operations to sorting and recovering wastes;
- Development controls established for building design, demolition, construction and operational stages;
- Ongoing, broad scale, and strong political and community support for the No Waste initiatives; and most importantly
- A sense of community ownership and pride in our directions and success.

1.1 The Australian Capital Territory (ACT) Context

The ACT is seat of Australia's Federal Government and occupies a catchment area of 2,358 square kilometres with a population of 325,000 people and 125,000 households. The ACT gained self-government in 1989 as a Territory and is governed by a single tier of government that integrates both local and state government functions. This is a unique situation in the Australian context where all other States, and the Northern Territory, have a two tiered government system of separated local and state functions. This means that the ACT has greater control over its waste management legislation, polices and practices through ACT NOWaste (a section of the Department of Urban Services), thus enabling better coordination and delivery of waste management services and minimisation programs.

The ACT is positioned to be the first jurisdiction in the world to achieve a No Waste goal. There are a number of factors that will contribute to such an achievement including:

- Low levels of heavy industry, with the main industries in the ACT being government (Federal and ACT) and the service sector;
- A waste stream with a high percentage of readily recyclable materials;
- The Canberra community seeing itself as a "bush capital" with a clean and green image;

- A highly educated and environmentally / socially aware population; and
- Strong government and community support for the No Waste Strategy and sustainability in general.

1.1.1 Waste streams in the ACT

The ACT total waste to landfill stream is broken down into waste materials delivered from the commercial and industrial sector (48%), households (39%) (with 29% from domestic collections and 10% privately delivered) and the construction and demolition sector (13%).

The majority of the ACT waste stream is composed of readily recyclable materials, with waste composition data indicating that a total of 95% of the waste stream is recoverable as valuable resources (APrince [1]). The remaining 5% is made up of what have been termed complex composite or problematic materials. These wastes are composed of a range of material types that are complex to separate into recyclable materials or are made from materials that for some reason are not readily recyclable. Problematic wastes such as tyres, computers, and electrical items belong to this group.

It is important to distinguish that No Waste has not been interpreted to mean zero waste to landfill. Rather it is more about the action of not wasting resources and preventing valuable resources from being landfilled. In the ACT, with such a high percentage of the waste stream being readily recyclable, a benchmark of achieving a 95% resource recovery rate has been established, leaving 5% of residual waste materials requiring landfill disposal as at 2010. With a focus at a national level on establishing national industry agreements covering extended producer responsibility, product design and whole of lifecycle assessments, it is anticipated that this 5% residual waste will also be further reduced in the longer term.

1.1.2 Readily recyclable materials

Within the ACT there are recycling options that currently exist for the following materials:

- Paper
- Cardboard
- Glass
- Aluminium and steel
- Other metals
- Liquid paperboard
- PET, HDPE and mixed plastics
- Garden/vegetation
- Wood/timber
- Motor and cooking oils
- Soils/rocks
- Plasterboard
- Concrete, bricks, tiles
- Textiles (limited to clothing/rags)
- Reusable (including furniture, fixtures, fittings, electrical etc)

2   Progress to date

Over the past 10 years, the ACT has steadily increased its resource recovery rates with more and more materials being diverted from landfill into repair, reuse and recycling operations. Over 550,000 tonnes of material are currently being recycled per year in resource recovery operations and this industry continues to grow with many new businesses being established and over 300 jobs being created in this fledgling sector.

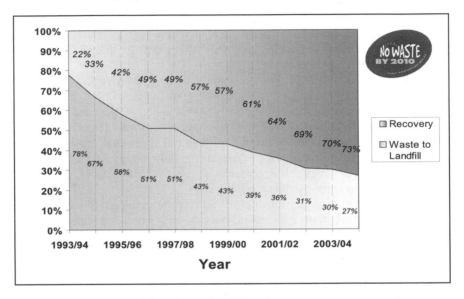

Figure 1:   ACT Waste to landfill and resource recovery trends.

The ACT is currently achieving a 73% resource recovery rate from the total waste stream in the Territory. That is, 550,000 tonnes of resources were recovered from the total waste stream of 750,000 tonnes during 2004-2005. This result has been achieved despite ongoing significant increases in total waste generation within the jurisdiction, as can be seen in Figure 2.

Figure 2 illustrates the significantly increased resource recovery rates that have been achieved in the ACT over the past decade. It also shows that while an initial rapid reduction in waste to landfill has occurred, this has stabilised over the past five years. The reason for this stabilisation trend is that while higher rates of resource recovery (73%) have been achieved, total waste generation levels have also increased and kept pace, resulting in a slowing of the reduction of waste to landfill. It is worth noting that if lower resource recovery rates had been achieved then waste to landfill would have significantly increased rather than coming down by almost 20%.

2.1 The consumption pattern problem

Sustainability has now become a global mission with all levels of government and the private sector searching for valid ways to meet this challenge. In the waste management arena, many have adopted a waste hierarchy of *Avoid, Reuse, Recycle and Disposal* as a set of priority actions in order of effectiveness in minimising waste.   The importance of managing waste once it has been generated is balanced by the need to avoid creating it in the first place.   However with a globally dominant paradigm of economic growth and a culture of consumption, it is worth questioning if Australian society is in conflict with waste minimisation goals, particularly waste avoidance.

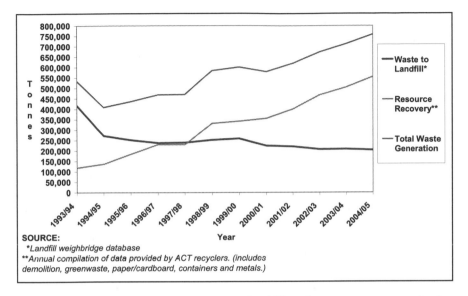

SOURCE:
*Landfill weighbridge database
**Annual compilation of data provided by ACT recyclers. (includes demolition, greenwaste, paper/cardboard, containers and metals.)

Figure 2:   Waste generation, waste to landfill and resource recovery in the ACT.

Fifty years of 'consumption' brainwashing need to be undone.   The 1950s gave birth to an 'out with the old and in with the new' philosophy that in the late 1900s and early 2000s has grown into full blown consumerism – buying for the sake of buying.   In the 1950s, marketing consultant, Victor Lebow pushed for an increased focus on consumption for the post war US: "Our enormously productive economy...demands that we make consumption a way of life, that we convert the buying and use of goods into rituals, that we seek our spiritual satisfaction... We need things consumed, burnt up, worn out, replaced and discarded at an ever growing rate..." (Seymour and Giradet [2]).

It is apparent that we have realised this vision.

An Australian think tank, The Australia Institute, recently undertook a study into "Wasteful Consumption in Australia 2005" which has some pertinent findings and reflections on waste generation and consumption patterns:

"Economic growth has become a dominant objective in itself, irrespective of the extent to which it contributes to improving social well-being." (Hamilton et al. [3])

For some, shopping and buying have become activities that give pleasure while actually consuming the goods bought is secondary and may not take place at all.

According to the results of a 2005 survey commissioned by the Australia Institute, 62% of Australians believe that they cannot afford to buy everything they really need. It is not just low income earners that feel this way, with 47% of respondents in the richest 20% of households believing their incomes are inadequate for their needs (Hamilton *et al.* [3]).

On average each Australian household wasted A$1226 on items purchased but unused in 2004. This amounts to over A$10.5 billion dollars annually spent on goods and services that are never, or hardly ever, used. Food accounts for the most wasteful consumption. Overall during 2004 Australians threw away A$2.9 billion of fresh food, $630 million of uneaten take-away food, A$876 million of leftovers, A$596 million of unfinished soft drinks and A$241 million of frozen foods, a total of A$5.3 billion on all forms of food. This represents more than 13 times the A$386 million donated by Australian households to overseas aid agencies in 2003 (ACFID [4]).

Throughout history our leaders have visioned the future and enacted policies and mechanisms to achieve the desired end results. It is clear that whilst we certainly have achieved our consumption and economic objectives, if we are to genuinely move towards a sustainable future we must create a new vision for our society that reconciles consumption patterns with our environmental, social and personal wellbeing needs.

We need to reflect on why our society has ritualised and spiritualised consumption and determine how we can re-engineer this trend in the western world.

3 The No Waste Strategy

The No Waste Strategy has been formulated around a number of social change mechanisms that form an available tool box for local and state authorities to use to facilitate waste minimisation. The social change tool box used within the No Waste Strategy development encompasses the following:

- Policy development and implementation;
- Economic and market force mechanisms;
- Infrastructure and service provision (both public and private sector based);
- Market development for waste derived products;
- Community engagement including education, communication, training, direct participation opportunities and consultation processes;
- Strategic partnerships and alliances;
- Research and development;
- Data management and reporting; and
- Awards and recognition of No Waste Leaders within the community.

3.1 Policy development and implementation

The impacts of establishing a policy direction such as the No Waste by 2010 policy should not be underestimated. With high profile education, communication and marketing campaigns and a range of community engagement programs, the concept of No Waste has been broadly disseminated and, to a large degree, accepted by the community as a means of moving towards sustainability. No Waste is seen as a major arm of sustainability, with energy, water and waste agencies often working side by side on social change programs.

Through mechanisms such as the Waste Service Providers Forum, the waste industry is being engaged and encouraged to respond to the government's policy objective. Other consultative mechanisms such as Waste Forums, have allowed peak industry, environmental and educational groups and associations, such as the Regional Chamber of Commerce and Industry, to become familiar with the No Waste Strategy and to also facilitate member input and feedback on their needs in relation to supporting and implementing waste minimisation.

3.2 Economic and market force mechanisms

3.2.1 Recognising the true cost of waste to landfill

In 2001 the ACT undertook a study – "The Actual Costs of Waste Disposal in The Act 2001" (RPM Pty Ltd *et al.* [5]) – to determine the true cost of waste to landfill. Historically only the administrative, contractual and operational costs were considered in setting prices for wastes. The study examined all costs associated with landfill including social, economic and environmental costs. These include: loss of amenity; surrounding land value issues; lost resources and lost opportunity costs; environmental monitoring and protection; remediation and rehabilitation; greenhouse; and the traditional costs of capital works, and operations and service contracts.

The study revealed that the true cost of waste to landfill in the ACT was A$105 per tonne. At that time the current waste to landfill charge was A$35 per tonne. The ACT subsequently developed a "Waste Pricing Strategy 2002" that recognised that waste generators were being heavily subsidised for disposal of their waste (ACT Government [6]). The pricing strategy set about adjusting the cost of waste to landfill charges in line with true costs over a five-year period to minimise impacts on waste generators, particularly the business sector.

Table 1:   Waste to Landfill charges in ACT.

Year	Waste to landfill charge
2001-02	A$33
2002-03	A$44
2003-04	A$55
2004-05	A$66
2005-06	A$77
2006-07	A$88 (proposed)

3.2.2 Market forces create the emergence of resource recovery businesses

In addition to reflecting the true cost of landfill, a significant driver in the pricing strategy was enabling the emergence of resource recovery alternatives. If waste-to-landfill remained the cheapest option for waste management then establishing more costly waste diversion alternatives would result in market failure. The pricing strategy has been successful, with the emergence of new private sector resource recovery businesses in the market-place. These businesses have realised that they can now process waste materials and market the waste-derived products for a cost that out-competes waste to landfill charges. Initially this development occurred in the case of separated construction and demolition materials, including concrete, bricks, tiles, wood waste and the like. However, it has since spread to waste collection companies who now are taking mixed commercial, construction and demolition wastes back to processing facilities and sorting the materials to extract resources.

Waste pricing has been a successful market driven approach to establishing a viable resource recovery industry within the private sector without the investment of public funds for infrastructure and service provision. There have been significant benefits in jobs creation that go beyond the traditional waste industry model of collection and transport to landfill disposal. It is estimated that the growing resource recovery industry has generated over 300 jobs within the ACT region over the past decade. It should be recognised that the extent of this job creation could not have been realised under a traditional waste industry model. Furthermore the salaries and wages derived from these jobs have a positive flow on impact within the local economy.

3.2.3 Waste generator user pays model

The pricing strategy also signalled a change in responsibility for covering waste disposal costs. Traditionally, the ACT Government derived rates revenue and commercial fees and charges to partially cover waste management costs with a significant portion of the costs being met from government finances. The pricing strategy changed this situation and transferred waste disposal costs back onto the waste generator in an effort to control waste generation rates. The concept is no different from charging strategies in the utility sectors of water or electricity, where a user pays approach requires consumers to pay for the products they use. The pricing strategy requires waste generators to pay for the wastes they generate if they are require it to be landfilled.

There is a clear incentive for waste generators to now take up cheaper and more sustainable alternatives than waste to landfill disposal. The key message from the government is that: "We don't want people to pay the higher waste disposal costs, we want them to take up recycling and resource recovery alternatives."

As waste disposal costs continue to rise over the next few years it is anticipated that further development of the resource recovery sector will continue with government encouragement and assistance.

3.2.4 Aligning charges for a future Alternative Waste Treatment Plant

It is important that the true cost of waste to landfill also aligns with the expected costs of treating waste in an alternative waste treatment (AWT) plant. This will ease the transition between landfill and AWT by minimising the price differential.

3.3 Infrastructure and service provision

The ACT Government is responsible for the provision of certain waste infrastructure and services to the ACT community. It provides a full range of household services, and also operates a regional combined landfill and resource recovery facility for both domestic and commercial waste. The landfill industry is regulated through a central regulation agency, the Environmental Protection Authority, who issues licences for landfilling and waste processing operations. At present there is only one operational landfill run by ACT NOWaste, which accommodates the remaining 204,000 tonnes of waste to landfill. A commercial waste service provider industry exists to provide both waste and recycling services to the business and industry sector of the ACT.

The ACT Government provides a range of household services including a weekly 140 litre wheelie bin garbage service and a fortnightly fully co-mingled 240 litre wheelie bin recycling service. Multi-unit complexes receive similar but shared services using hoppers. Household garbage continues to be disposed of to landfill at present, with the recyclables being processed at a state of the art Materials Recovery Facility (MRF).

3.3.1 Household hazardous waste collection

Residents also have access to an on-call household hazardous waste collection service where they can book in for collections of paints, solvents, pesticides, fungicides, petroleum products and other chemicals that divert toxic substances away from landfill for treatment. Permanent drop-off facilities exist for motor and cooking oils, paints, fire extinguishers, gas bottles and car batteries.

3.3.2 Re-usables

Residents are able to take re-usable materials from households to one of a number of free drop-off centres where goods are repaired, reconditioned and resold. One of these facilities generates over A$1million in revenue per year. Materials such as furniture, office equipment, electrical items, white goods, kitchen wares, toys, sporting items, tools, and building materials such as doors and window frames are all accepted at these facilities. On the weekends these facilities look like a bustling market place.

3.3.3 Regional Recycling Drop-off Centres

Several Regional Recycling Drop-off Centres are strategically located throughout the ACT to enable businesses to recycle a range of materials free of charge. These facilities accept standard recyclables including paper, cardboard, glass, aluminium, steel cans, plastics and liquid paperboard. By far the largest quantity of material recovered is cardboard from the retail sector. Residents with surplus recyclables generated at peak consumption times such as Christmas or when moving can also access these free facilities to recycle materials.

3.3.4 Garden waste

Historically in the ACT small 55 litre hand bins were used for household garbage collection and this meant that residents were not able to put their garden waste in the household collection system. Residents became accustomed to self-transporting garden waste to a number of free regional garden waste acceptance and processing facilities. This practice has largely continued since the introduction of 140 litre wheelie bins for garbage, with residents delivering garden waste in trailers, utility vehicles and car boots.

The ACT has three major garden waste acceptance and processing facilities that take garden waste materials free of charge from households, businesses and waste service providers. Two of these facilities are under contract to the ACT Government and a third exists as an independent commercial resource recovery operation.

The garden waste is turned into valuable mulches, composts and soil and potting mix products and marketed back into the ACT and interstate. A significant amount of effort has gone into the development of high quality products by the operators and sustainable markets have been developed for all products over the past decade. Over 190,000 tonnes of garden waste is recovered and the products find their way back into Australian soils, adding value in terms of moisture retention, carbon content, microbial activity and improving soil structures, all of which results in enhanced ecosystems either in the urban or rural environment. The value of this organic recycling cannot be underestimated for a country such as Australia that has extremely thin and poor top soils.

There continues to be an estimated 14,000 tonnes of garden waste going to landfill from households, another 2,500 tonnes from businesses and 1,200 tonnes from construction and demolition activities, a total of 17,700 tonnes. With mixed waste sorting beginning to take effect it is anticipated that much of this garden waste material will be separated and diverted into the existing free garden waste processing facilities.

3.3.5 Construction and demolition waste

Construction and demolition waste can be taken to two major recycling facilities at cost to the waste generator. One of these facilities operates from within the ACT's landfill facility under contract to the Territory, the other is a commercial resource recovery operator. Both of these facilities accept and process concrete, bricks, tiles, plaster board, wood waste, steel and soils. The facility that operates within the ACT landfill also processes mixed construction and demolition wastes using a range of processing technologies. The processing includes tromels to screen soil products, magnets and eddy currents to remove metals and aluminium, an excavator to separate bulk steel, wood waste, plaster board and plastic wrap and a hand-sort line to separate remaining materials with a small amount of residual waste going to landfill.

3.3.6 Private sector responding to resource recovery

There are a number of independent resource recovery operators within the ACT that are accepting and processing waste materials. These businesses have become possible under the waste pricing strategy that has allowed them to now charge reasonable rates to accept, sort, separate and process waste materials at a charge that is still cheaper than sending the material to landfill. Several waste collection and transport companies have seized the opportunity to change the nature of their business and align it with the Government's No Waste Policy. This has not only provided a market advantage to them but has also lowered the waste-to-landfill costs of their business which have been used as an offset to establish sorting and processing operations.

3.3.7 Government assisted Resource Recovery Estates

Resource Recovery Estates have also been developed by the ACT Government to assist the development of the resource recovery industry within the region. There are two estates, one that serves as an incubator model where businesses can trial or undertake establishment activities to enter the market place and another for larger scale higher infrastructure investment operations. In both estates there is a system of discounting market based licences or sub-leases based on the priority waste materials targeted and the quantity of materials diverted from landfill.

3.3.8 Responding to possible market failures

While there has been a strong take-up of recycling services within the business and industry sectors there is still significant room for further take-up. In discussions with peak associations, such as the Chamber of Commerce and Industry, it has become apparent that businesses see recycling as an additional cost. Through further investigation this attitude appears to stem from the fact that business and industry, when approaching waste service providers, generally purchase an additional recycling system that stands alongside of their existing waste systems, resulting in the recycling system being an additional cost impost. It is difficult to pinpoint what is occurring here. Are waste service providers marketing recycling services as a tack-on service to existing waste services to gain maximum profit? Or, have businesses failed to understand that they need to review their waste service arrangements? This would include reducing the size of bins and frequency of waste collection, to deliver a cost savings that in most cases can be off-set against recycling services.

To tackle this problem the government has undertaken a mass-media campaign encouraging business and industry to take up recycling and review their current waste services to downsize them and reduce the frequency of collection. The waste service provider industry has also been encouraged to market their recycling services and to change the range of the services they provide to recover the readily recyclable materials.

ACT NOWaste is currently considering establishing a differential charge of A$140 per tonne for loads of waste containing greater than 50% recyclables to landfill to further provide mechanisms to drive the take-up of recycling services. If the market, being both the waste service providers and the business and industry sectors, does not respond over the next 12-24 months then regulations will be introduced under the Waste Minimisation Act 2001 that will require business and industry to establish recycling and waste diversion systems and services.

3.4   Market development for waste derived products

As the resource recovery industry has grown, so too has the need to establish and expand existing markets for waste derived products.   This area has been challenging for the ACT due to its limited population and economy base. Further market analysis work is required to examine future directions including forecasting waste-derived material types and quantities, looking at existing market capacities, mapping market gaps and undertaking research and development into waste derived products that will meet prospective user needs.

3.4.1   Market threats

There are considerable market threats from other jurisdictions that have differing circumstances to the ACT and may interfere in established markets.   For example, Sydney (only 3 hours by road transport) has the capacity to flood markets that the ACT has established with heavily subsidised products on the basis that Sydney processors' revenue is often derived at the acceptance and processing stage, allowing it to forego sales revenue simply to ensure product movement.

3.4.2   Progressive market development focuses

The ACT's initial focus for market development activities was centred on garden waste derived products such as mulches, composts, soil products and potting mixes.   One of the keys to market development of this material is in ensuring the standard of the products is high and also in tailoring the waste derived product to end users requirements.   Approximately 45% of the ACT garden waste derived products are exported interstate to agricultural activities and also compost and potting mix suppliers.

The ACT NOWaste focus has now moved on to market development for construction and demolition derived products.   Markets for wood/timber, concrete, bricks, tiles and steel are well-established, however as the quantities of materials recovered increases there is uncertainty regarding the existing markets' capacities to absorb additional materials.

As the resource recovery industry adapts to cater for mixed waste sorting and separation there is a need to ensure that smaller scale operators have the capacity to find markets for smaller quantities of readily recyclable materials.   A significant threat to these smaller scale resource recovery businesses is the loss of a key revenue stream if they cannot realise the market value in the resources recovered within a commodities type market.

3.5   Community engagement

Extensive community engagement programs have ensured strong support within the region for the No Waste Strategy, with stakeholders and the broader community taking up the call for action on waste minimisation.   ACT NOWaste community engagement programs include:
- Business Waste Reduction Program
- Government Leadership Program

- Construction and Demolition Waste Reduction Program
- Schools Program
- Public Events Waste Minimisation Program
- Tertiary Students and Volunteers Program
- NOWaste Community Workshops
- Media Coordination Program
- Awards and Recognition Program
- Public Reporting of No Waste Progress

These programs are designed to raise awareness, knowledge and skills to engender positive attitudes and ultimately to change behaviour and practices. This is achieved using a range of social change mechanisms such as communication, education, training, provision of advisory and support services, provision of waste audits and developing waste management plans on a fee-for-service basis.

3.6 Strategic partnerships and alliances

The success of the No Waste Strategy requires that the broader community respond to the No Waste Policy and actively take up the challenge of minimising waste in their area of influence. ACT NOWaste has a role in ensuring that a range of strategic partnerships and alliances were established to assist facilitate this objective. Strategic partnerships and alliances have been established with a diverse range of stakeholders including: other relevant government agencies; peak groups and associations; other national and international waste minimisation organisations and agencies; waste industry players; education institutions; Volunteering ACT; and many other organisations.

The benefits of these strategic partnerships and alliances include gaining leverage on social change, increasing the number of No Waste advocates, recruiting specialist expertise and increasing waste knowledge and capacity in the ACT community to name a few.

3.7 Holding back the legislative and regulatory options

It is interesting to note that the ACT has achieved its resource recovery progress to-date without the use of any legislative or regulatory mechanisms. The ACT has a Waste Minimisation Act 2001 which has a provision for the creation of Industry Waste Reduction Plans that require selected industries to undertake a range of waste minimisation related actions. At this stage there are no plans to enact this provision and the current cooperative and collaborative approaches stimulated by market forces will continue to be implemented.

3.8 Integrated Resource Recovery Approach

A key consideration in the No Waste Strategy approach has been to avoid unnecessary costs imposed on the community. Waste minimisation strategies not only need to be achievable but should also recognise the most economical way to deal with a range of waste types. As No Waste initiatives evolved in the

ACT, it became clear that diverting different types of waste materials to different processing and treatment technologies would result in the most economical outcomes.

The introduction of the alternative waste management technology industry into Australia was seen by many as a panacea for waste minimisation. This 'black box' solution seemed to mean the simple formula of 'municipal waste in and waste minimisation achievement out' allowed the delivery on ambitious waste minimisation goals. The problem that exists with such a philosophy is that the community, or more specifically waste generators, are paying a premium for waste materials that would be more economically treated if they were separated and diverted to a range of more appropriate treatment options rather than a single facility. This concern for minimising the cost impost on the community has given rise to an Integrated Sustainable Waste Management Approach within the ACT.

The No Waste Strategy is not only setting out to divert waste from landfill but is also mindful of its supporting role within sustainability through defining an Integrated Resource Recovery Approach. This approach views the total waste stream generated from within a region by its sub-streams and selects the most economical treatment processes along with the highest order resource use for the products derived.

The Integrated Resource Recovery Approach is based on:
- Recovery of standard recyclables from both the domestic and commercial and industrial waste streams;
- Processing and recovery of resources from mixed waste from the building and demolition and commercial and industrial sectors;
- Separation, processing and recovery of organic material with separate processes for garden and organic materials;
- Dealing with the more complex composite residual waste materials.

In the ACT there are clear cost differentials that exist between the appropriate processes for the recovery of different waste materials. The table below illustrates typical recovery costs for the various waste streams. These exclude transport costs as it is assumed that this is an inherent cost and is independent of the process the material is delivered to, resource recovery or landfill.

Table 2: Cost of resource recovery for different waste materials.

Recovery of standard recyclables	A$40 – A$50 per tonne
Recovery of recyclables from mixed waste	A$50 – A$70 per tonne
Recovery of garden waste resources	A$5 per tonne
Recovery of recyclables and organics from mixed waste, including food waste	A$80 – A$110 per tonne

The No Waste Strategy has focused on driving materials out of the waste stream through the most economically cost efficient process, rather than rely on a 'one stop shop' AWT plant. This approach will minimise the cost impost on the ACT community however it also relies on the community playing its role to avoid unnecessary costs.

3.9 Bio bin trial and a third bin –results and observations

In 2000-2001 ACT NOWaste conducted a trial for household organic materials using a third bin system. Over 1000 participants were involved in a collection program for food and garden wastes to determine if a clean stream of organics could be derived. The objective was to determine if a cheaper and simpler technology could be relied upon to process a clean organics stream. The third bin system, if introduced would cost around A\$2.5 – 3 million per year.

The trial results included a weekly bin presentation result of only 60%, however approximately 90% of all residents participated in the trial. Contamination rates of the organic collection system were around 10%. However, the problem was that 30% of organic material was still being left in the garbage bin system (ACT Government [7]). This meant that while a relatively clean organics stream could be diverted to a cheaper treatment facility, the garbage stream containing 30% organics would still require a more complex and costly treatment process.

Given the fact that a more complex waste treatment technology would still be required to deal with the organic material in the garbage system it was decided not to pursue an organics collection system. Furthermore, given that garden waste materials were being recycled for under A\$5 per tonne, the value for money aspects of such a decision were highly questionable.

4   The challenges ahead

The No Waste Strategy has been steadily delivering increased resource recovery outcomes with over 73% of all waste generated in the ACT currently being recovered. Waste generation and consumption patterns continue to be a threat for the full success of the strategy. There is a clear need to break the link between the current economic paradigm and waste generation or, alternatively, for resource recovery to increase at a much more rapid rate to outstrip waste generation increases.

Further expansion of the resource recovery industry is required along with expanded market development activities to ensure the sustainability of resource recovery operations. Increased take-up of standard recycling services will further divert recyclables from landfill and minimise the cost impost on the community. To assist the development and expansion of the resource recovery industry, waste to landfill charges in the ACT will continue to rise until they reach the estimated true cost of waste-to-landfill charge of A\$105 per tonne.

As the ACT Government draws closer to its benchmark of 95% resource recovery, an AWT plant will be require to extract any remaining recyclables in the waste stream and to process the residual organics (mostly food wastes) in the domestic and commercial waste stream.

Whilst considerable gains have been made in achieving the No Waste goal, continued political and community support will be required to ensure progress and momentum is not lost in this long term social change agenda of treating wastes as resources.

References

[1]   APrince Consulting, *Canberra Residential Waste Audit*, 2004.
[2]   Seymour and Giradet, *Blueprint for a Green Planet*, Dorling and Kindersley, London, 1989
[3]   Hamilton, C., Denniss, R., & Baker, D., *Wasteful Consumption in Australia*, The Australia Institute Discussion Paper No. 77, 2005.
[4]   ACFID, Facts and Figures, Australian Council for International Development, Online http://www.acfid.asn.au/factsandfigures/factsandfigures.htm [accessed Dec, 2005].
[5]   RPM Pty Ltd, Kenny Lin & Assoc., and Energy Strategies Pty Ltd, *The Actual Costs of Waste Disposal in the ACT*, 2001.
[6]   ACT Government, *Waste Pricing Strategy for the ACT*, 2002.
[7]   ACT Government, *Household Organic Material Collection Trial: Chifley August 2000 – June 2001*, 2001.

Health risk assessment for sewage sludge applied to land in France

L. Déléry[1], G. Gay[1], S. Denys[1], H. Brunet[2], I. Déportes[3],
A. Cauchi[4] & M. Aupetitgendre[4]

[1]*National Institute for Industrial Environment and Risks, France*
[2]*Professional Association for agricultural recycling, France*
[3]*French Agency for the Environment and Energy Management, France*
[4]*Professional association for water distribution, France*

Abstract

Chemical and microbial health risks associated with the application of municipal and industrial sewage sludge to agricultural land were addressed in order to set guidelines within the French legislative context. The chemical part was a quantitative risk assessment study based on the simplified source-pathway-receptor event tree approach. The selected chemicals were those quoted in the French legislation. Several receptors were considered: population consuming vegetables and meat impacted by the sludge application, nearby populations also walking close to fields receiving sludge, farmers plowing such fields, and future residents consuming their own vegetables on former fields. The assumptions and the uncertainties of the approach were discussed. The first quantitative results led to acceptable risk levels, and allowed one to identify the major exposure pathways. The microbial part was a feasibility study based on a literature review. In spite of many uncertainties, a preliminary quantitative risk assessment case study for Enterovirus, *Salmonella*, *E. coli* O157 :H7 and *Cryptosporidium parvum* was developed in order to identify the critical points of the process and to judge its relevance within the national context. Uncertainties of the quantitative risk approach were discussed in detail and led to the proposal of qualitative guidelines based on an event tree, due to the current state of knowledge.
Keywords: sewage sludge, chemical and microbial health risk assessment, France.

WIT Transactions on Ecology and the Environment, Vol 92, © 2006 WIT Press
www.witpress.com, ISSN 1743-3541 (on-line)
doi:10.2495/WM060551

1 Introduction

In France, municipal and industrial sludges from sewage treatment works are regarded as waste within the existing regulation (French law n°75-633 of July 15, 1975). The fertilising properties of these wastes have been recognised for many years and their application on agricultural lands for 30 years in France is considered as an environmental and economical sustainable management [1]. Because of the presence of chemical contaminants and pathogens in sludge, precaution measures are taken with respect to the practice of spreading to guarantee food safety and the preservation of soils and other environmental compartments. They are framed by the French law n°92-3 of January 3, 1992 known as law on water and law n°76-663 of July 19, 1976 related to the classified installations for the environmental protection, and soon by a European directive on sewage sludge (currently under development). Municipal and industrial sewage treatment works subjected to French authorisation have to fulfil a study relative to the human health impact of sludge spreading. However no guideline to carry out such health risk assessment is available so far.

The objectives of the present study were the development of health risk assessment methodologies for both chemical and microbiological hazards. For the chemical part, the methodology was developed on two real sites. For the microbial part, as there are no limit values for pathogen concentrations (except for sludges that undergo treatments to reduce *Salmonella*, Enterovirus and helminth eggs content and for which there are no distance or period restrictions), a feasibility study based on a literature review was undertaken.

2 Chemical health risk assessment

The chemical health risk assessment methodology was developed with respect of the general risk assessment principles defined by US-EPA and already used in other French methodologies dedicated to other environmental media (contaminated soils for instance). In this specific case, the human health risk assessment was considered after 10 years of sludge application for a given area, this time frame being considered to be the duration over which sludge application is allowed by the authorisation procedure. Frequency of sludge spreading on a same plot is defined by the legislation and assumed to be, for a same plot, every three years.

2.1 A four steps methodology

The description of the methodology set up can be done according to four steps mentioned below:
-   determination of site characteristics,
-   evaluation of toxic effects for the selected substances,
-   exposure evaluation,
-   risk quantification.

2.1.1 Determination of site characteristics

The methodology developed here is based on a site-specific approach. The risk assessor has first to evaluate the site properties and technical methods of spreading which will lead to an accurate risk quantification. Among these characteristics, sludge composition (type of pollutant and concentrations), pedology (texture, pH, etc.), agricultural practices (cultures and spatial and temporal distribution, sludge application, etc.) have to be properly known to define which substances will be selected during the quantification step. It was defined that each substance mentioned in the legislative text should be selected to assess potential risks of sewage sludge spreading on agricultural lands, these substances currently being in the French legislation heavy metals (cadmium, chromium, copper, mercury, nickel, selenium, lead and zinc), polycyclic aromatic hydrocarbons (fluoranthen, benzo[b]fluoranthen, benzo[a]pyren) and polychloro-bi-phenyls. Critical issue of this first step is the definition of a conceptual scheme defining substances, exposure pathway and receptors that have to be considered for the risk quantification procedure.

2.1.2 Evaluation of toxic effects for the selected substances

Toxic effects have to be described for each effect of the studied substance (threshold or non-threshold effects). A thorough description of the different organs on which substances have an effect has to be done. When available, quantitative data about the effects of the substances on Humans will be selected among data available into the international toxicological databases. These toxicological reference values (TRV) will be used for risk quantification.

2.1.3 Exposure evaluation

The objective of this step is the evaluation of the dose of exposure (ED) for each receptor and each exposure pathway. This requires the estimation of the substance concentration in the soil on which sludge is spread (source of contamination), the values of parameters needed to estimate the transfer of substances to Humans (through either direct or indirect contact) and the values of parameters concerning daily food ingestion, space and time budget for each receptor.

Receptors and exposure pathways that have to be considered by the risk assessor are described below (table 1).

Table 1:   Receptors and exposure pathways.

	Farmers	Inhabitants	General Population	Inhabitants on former plots
Direct Contact	x	x		
Soil ingestion	x	x		x
Soil inhalation	x	x		x
Dermal contact with soil	x	x		x
Indirect Contact				x
Agricultural products consumption	x	x	x	x
Contaminated water contact	x	x	x	x

In predictive scenarios, concentration of each substance has to be calculated theoretically considering sludge dilution in a volume of agricultural soil, this volume depending on legislation, soil density and depth of spreading.

2.1.4 Risk quantification

This final step allows one to calculate, for threshold and non-threshold effects of each substance, the risk for each identified exposure pathway. For each route of exposure (ingestion, inhalation, dermal contact) addition of risk values (RV) has to be carried out for substances having effects on the same organs.

For the threshold effects, the risk value is obtained according to eqn (1):

$$RV_{\text{threshold effect}} = ED/TRV \tag{1}$$

whereas for the non-threshold effects, the risk value is obtained according to eqn (2):

$$RV_{\text{non-threshold effect}} = ED * TRV \tag{2}$$

Analysis of these values by the risk assessors has to be done concomitantly to the examination of uncertainties related to the methodology and phenomena taken into account when modelling the fate of pollutants from soils containing sludge to Man. It is usually considered that $RV_{\text{threshold effect}} < 1$ and $RV_{\text{non-threshold effet}} < 10^{-5}$ are acceptable risks values.

The methodology described before was applied to assess the chemical health risks on two French sites. One site was dedicated to the spreading of industrial sludge whereas urban sludge was amended on the other site. Exposure pathways were also different between the two cases: the first case had to consider the soil-(plant)-grazing animal-human pathway whereas this way of exposure was not accurate for the second scenario as lands were entirely dedicated to cereals cultivation.

For the two sites and whatever the receptors, risks values were below 1 for the threshold-effect and below 10^{-5} for the non-threshold effects showing that, according to our methodology and under the different assumptions made for the calculation, the spreading of sewage sludge on agricultural lands was an acceptable practice. For the two cases, the more significant risks were associated with the consumption of plants harvested on amended soils.

3 Microbial health risk assessment

Since the first French report on microbial health risks for sewage sludge [2], French research studies [3, 4] and thesis [5–13] were carried out. The first objective of this work was the analysis of the current knowledge on all the steps of the risk assessment approach. The second objective was the evaluation of the feasibility to implement a quantitative assessment of the risks of infection to humans.

Reports of this type have been published in particular in the United Kingdom [14] and the United States [15], so it appeared interesting to develop a case study based on data adapted to the French context. Taking into account the relevance to apply this exercise within the framework of national studies resulted in the setting of qualitative guidelines.

3.1 Literature review

This part of the study provided current available data on loads of pathogens in sludge, their sensitivity to sludge treatments and to storage, their survival in soil and their transport to environmental media (soil, water and air) after spreading, data on other sources of pathogens in the agricultural environment, epidemiological data (linked to direct exposure to sludge, wastewater and sludge composts as well as a synthesis on environmental waterborne outbreaks in France) and their health outcomes. This review also allowed the main gaps in knowledge to be highlighted.

3.1.1 Major findings

During 30 years of practice in France, no environmental epidemic outbreak linked with the spreading of sludge was detected. We found a few epidemiological studies conducted specifically for sludge exposure in our literature review [16]. The examination of the whole available epidemiological data in France indicated that ingestion of contaminated drinking water without any proven link with land spreading of sludge is to the present day the only way of exposure blamed in the waterborne outbreaks of environmental origin.

For ten years, French studies collected data on detection and quantification of certain pathogenic agents (helminth eggs, *Giardia spp.* cysts, *Salmonella*, Enterovirus, VHA, verotoxic *E. coli*, *Listeria spp.*) in sewage sludge. Helminth *Ascaris sp.* has practically disappeared in humans and pigs in metropolitan France because of hygienic rules and industrial breeding. *Taenia saginata* probably represents the last principal parasitic issue in human and animal health since it is practically found in all the sludges analyzed under moderate climates [17]. The review also shows that contamination of raw municipal sludge is quasi systematic. Some pathogens are little studied (*Cryptosporidium sp.*, other protozoa like *Balantidium* sp., *Campylobacter sp.*, *Yersinia spp.*, *Vibrio spp.*, shigatoxins producing *E coli, etc.*) or are not studied at all in sludges (Adenovirus, Calicivirus, Astrovirus, mycobacteria, microsporidia...). Only 2 studies on the contamination of raw French slaughterhouses sludge were found [18, 19].

In France, there are some data on the contamination of treated municipal sludge (Enterovirus, *Listeria sp.*, *Giardia* cysts, helminths eggs) but not on those of slaughterhouses or dairies sewage treatment works. All in all, the advanced treatments for the reduction of pathogens recommended by the European Community in 2001 [20] are: composting, thermal drying, thermophilic digestion, followed heat treatment followed by digestion and liming.

The French standardized methods for the analysis of pathogens in sludge are only available for helminth eggs, *Salmonella* and soon for the Enterovirus. The

European project Horizontal is intended to supplement currently available tools for sampling and analysis for the sludge matrix.

Many factors influence the survival of pathogens in the environment. Most of the time, survival is decreasing within 2 months because of the adverse conditions (dessication, temperature, natural competition) in the environmental media.

The pathogens of small size (virus, bacteria and some protozoa) can be found in bioaerosols [21] formed by the spreading with spray tanker or spray irrigation application of liquid sludge in windy conditions.

Water constitutes the main pathway of pathogens dissemination in the environment. However, in France, contamination of water is reduced by the respect of the regulation (distances of land spreading from sensitive uses). Moreover, taking into account the French regulated times of spreading, it is not very probable to find pathogens on plants grown on agricultural soils spread with sludge and the role of animals grazing on those sites in the transmission of pathogens with man is reduced.

3.1.2 Research needs

According to our findings, the principal needs of data concern: i) the determination of the species of the biological agents in sludge (salmonella, *campylobacter spp.*, *cryptosporidium sp., etc.*) as finer knowledge would make it possible to identify the human, animal, environmental origin of the pathogens ii) the presence, quantification and virulence of pathogens in raw and treated municipal sludge (endotoxins, *Staphylococcus aureus*, *Vibrio spp*, viruses - rotavirus, calicivirus, adenovirus, etc. –, pathogenic *E coli*, *Pseudomonas aeruginosa Cryptosporidium spp.*, mycobacteria, microsporidia, etc.) and in industrial sludge (*Pseudomonas spp., Listeria monocytogenes, Campylobacter spp., Salmonella*) iii) the determination of presence and ecology of *Legionella* and its methods of detection in sludge; iv) the construction of a data base on the survival and transport of pathogens (and in particular of viruses) in the environment (soil and plants); v) the human exposure to bioaerosols and composts of sludge at the time of spreading on land.

3.2 Preliminary risk assessment

In spite of many uncertainties linked to the state of knowledge, a preliminary quantitative risk assessment case study was developed for liquid raw sludge in order to identify the critical points of the process and to judge its relevance within the national context of health impact studies.

3.2.1 Rationale for selecting pathogens

From a list of principal pathogens of concern in sewage sludge established by ADEME [22] and EPA [23], Enterovirus, *Salmonella*, *E coli* O157 :H7 and *Cryptosporidium parvum* were selected on two criteria: existing data on load of pathogen in sludge or sewage water in France and on dose-response curves by ingestion. Table 1 presents a summary of those data for each selected pathogen.

Table 2:   Summary of data used for microbial assessment.

	Load of pathogens in sludge		Probability of infection (with N dose of pathogen)	
Enterovirus	280 IU/g DS	[9]	$P(N) = 1-(1+N/0,42)^{-0,26}$	[3]
Salmonella	10^4 CFU/g DS	[3]	$P(N) = 1-(1+N/51,45)^{-0,1324}$	[24]
E. coli O157:H7	0,6 CFU/g DS	*	$P(N) = 1-(1+N/1,001)^{-0,05}$	[25]
Cryptosporidium parvum	100 oocysts/g DS	*	$P(N) = 1-\exp(-N/238)$	[3]

* calculated from an event tree with an isolation percentage during surveillance at slaughterhouses of 1% for *E. coli* O157:H7 and from a load in sewage water of 30 oocysts/L for *Cryptosporidium*.

3.2.2 The conceptual model

The conceptual model was based on the simplified source-pathway-receptor event tree approach. Two receptors and three exposure pathways were considered: farmers plowing and ingesting dust from soil receiving sludge and nearby populations consuming their own vegetable products contaminated either by biological aerosols from the land receiving sludge either by erosion of amended soil. Many worst case assumptions were considered based on experts' judgment. The main assumptions are presented in figure 1.

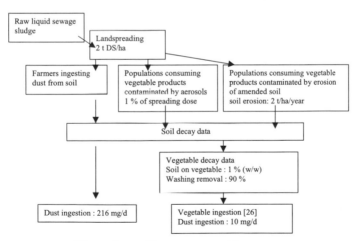

Figure 1:   Exposure event tree.

3.2.3 Results

Under main worst case assumptions and in the current state of knowledge, preliminary results for the liquid raw sludge spreading showed that: calculated exposures for both receptors are very weak (always lower than 1 pathogen/d); whatever pathogen is considered, the risks to have at least one infection over the exposure time is in decreasing order: population/biaerosol > farmer >

population/erosion; whatever receptor is considered, the risks to have at least one infection over the exposure time is in decreasing order: Enterovirus > *Salmonella* > *Cryptosporidium parvum* > *E coli* O 157: H7.

3.2.4 Uncertainty

This preliminary quantitative risk assessment was developed on a great number of working assumptions surrounded by uncertainty. Indeed, in the absence of sufficient data, values considered as realistic or worst case were used following a consensus of experts.

The main sources of uncertainty in the model are: i) the estimation of the pathogen loading in sludge that is directly related to the natural heterogeneity and to uncertainties of microbiological methods of analysis ii) the dose-responses relations that are rare in the literature, mostly available for ingestion and not built for the sludge matrix, in addition the existence of an acquired immunity or the genetic specificity of certain individuals are not taken into account because it is not easily quantifiable iii) the modelling of pathogens decay in soil environment and vegetables is very simplified and there is currently no experimental evidence to support the pathogens concentration results iv) the exposure scenarios used for both receptors correspond to the worst case situations (for example use of an open tractor for the farmer which increases significantly the quantity of dust ingested, and for the residents, use of uncertain modelling of bioaerosols deposition because of the lack of experimental data) which are not representative of the "normal" conditions of spreading in France vi) an important phenomenon (losses in the environment at the time of the rain) was not taken into account.

3.3 Recommended guidelines

The feasibility study enabled us to propose guidelines based on a qualitative analysis of the microbial risks and control measures. The analysis can be carried out according to the decisional tree that includes the data needed to describe sludge spread to land, the exposed populations and the preventive measures. Measures of risk reduction are based on solutions proposed by the French regulation: either a reduction of the sludge pathogens load (hygienized sludges) or a reduction of the human exposures (for other sludges).

Acknowledgements

This work was supported by funding from ADEME, SYPREA, SPDE and INERIS. We cordially address our thanks to all experts that contributed to it.

References

[1]   Agence de l'eau Rhin-Meuse, *Audit environnemental et économique des filières d'élimination des boues d'épuration*, cahier n°70, études des agences de l'eau, pp.124, 1999.

[2]   CSHPF, *Risques sanitaires liés aux boues d'épuration*, Technique & Documentation, pp. 1-18, 1998.

[3]   AGHTM, *Impact du futur projet européen sur la valorisation des boues en agriculture - campagne d'analyses sur 60 boues de STEP*, pp. 139, 2002.

[4]   NANCIE, *Micropolluants organiques et germes pathogènes dans les boues d'eaux résiduaires*, programme national, pp. 250, 2000.

[5]   Cardiergues, B., *Boues d'épuration et microorganismes pathogènes: influence de différents traitements et du stockage*, thèse de l'université Henri Poincaré, Nancy, pp. 240, 2000.

[6]   Garrec, N., *Détection et étude de la survie de Listeria monocytogenes dans les boues d'épuration destinées à l'épandage*, thèse de l'Ecole doctorale d'Angers, pp.169, 2003.

[7]   Gaspard, P., *Contamination parasitaire dans l'environnement: prospective pour un gestion des risques sanitaires,* thèse de l'université Henri Poincaré, Nancy, pp. 228, 1995.

[8]   Madeline, M., *Evaluation du risque sanitaire (parasitaire et virologique) des boues résiduaires urbaines en agriculture et des eaux épurées dans l'environnement*, thèse de l'université de Caen, pp. 223, 2003.

[9]   Monpoého, S., *Quantification génomique de 2 virus entériques (entérovirus et HAV) dans les boues de stations d'épuration. Estimation de l'impact sanitaire lié à leur valorisation agricole,* thèse de la Faculté de Pharmacie de Nantes, pp. 177, 2001.

[10] Moussavou-Boussougou, M.N., *Epandage des boues d'épuration urbaine et des lisiers sur les pâturages: risque parasitaire pour les ruminants*, thèse de Sciences, Tours, pp. 139, 2004.

[11] Paillard, D., *Prévalence et résistance aux antibiotiques de Listeria spp. dans les effluetns de stations d'épuration,* thèse de l'université de Pau, 2003.

[12] Thiriat, L., *Valorisation agricole des boues résiduaires: dénombrement des kystes de Giardia sp. et estimation de leur impact sur le risque sanitaire*, thèse de l'université Henri Poincaré de Nancy, pp. 231, 1998.

[13] Vansteellant, J., *Evaluation des risques de contamination microbiologique liés aux épandages de matières organiques sur prairie de montagne*, thèse de l'université de Savoie, pp. 172, 2004.

[14] UKWIR, *Pathogens in biosolids- Microbiological risk*, UK Water Industry Research, Environmental Agency, DEFRA, pp. 125, 2003.

[15] WERF, *A dynamic model to assess microbial health risks associated with beneficial uses of biosolids: phase 1*, Water Environment Research Foundation, 2003.

[16] NRC, (eds). *Biosolids applied to land: advancing standards and practices,* The national Academies Press: Washington DC, 2002.

[17] Cabaret, J., Geerts S. et al., The use of urban sewage sludge on pastures: the cysticersosis threat, *Vet. Res.*, **33**, pp. 575-597, 2002.

[18] INRA-ENVT, *Bactéries pathogènes dans les effluents d'abattoirs: aide à l'évaluation des risques pour la santé publique*, Toulouse, équipe environnement de l'UMR 960, pp. 72, 2002.

[19] Hydro-M, *Microbiologie et environnement: les effluents de la filière viande*, INTERBEV/ FNEAP - AELB/AESN, pp. 9, 2004.

[20] European Community, *Evaluation of sludge treatments for pathogen reduction*, pp. 44, 2001.

[21] Brooks, J., National Study on Residential Impact of Biological Aerosols from Land Application of Biosolids. *Journal of Applied Microbiology*, **99**, pp. 310-322, 2005.

[22] ADEME, *Les agents biologiques d'intérêt sanitaire des boues d'épuration urbaines*, Guides et cahiers techniques Connaître pour agir, pp. 180, 1999.

[23] USEPA, *Control of pathogens and vector attraction in sewage sludge*, Office of Research and Development, pp. 177, 1999.

[24] WHO/FAO, *Risk Assessments of Salmonella in Eggs and Broiler Chickens*, *Microbiological Risk Assessment Series-2*, pp. 50, 2002.

[25] Teunis P., Dose response for infection by *Escherichia coli* O157: H7 from outbreak data, *Risk analysis,* **24(2)**, pp. 401-407, 2004.

[26] ADEME, *Réhabilitation des sites et sols pollués Ciblex banque de données*, Cédérom - Co-édition: IRSN, Réf. 4773, 2003.

Towards an effective electronic waste management scheme in Attica, Greece

M. Menegaki & D. Kaliampakos
School of Mining and Metallurgical Engineering,
National Technical University of Athens, Greece

Abstract

Electronic wastes (e-wastes) pose a significant threat to human health and the ecosystem, due to both the volume of wastes produced and the hazardous materials contained. Discarded electronic equipment is one of the fastest growing waste streams worldwide, due to the increasing growth rate of the Information Technology (IT) market and the rapid obsolescence of the equipment.

The European Union has characterized the Waste from Electrical and Electronic Equipment (WEEE), in general, as one of the key waste streams and has already developed the framework for their management.

Although a lot of EU countries have already launched their electronic wastes policy, Greece started the efforts quite recently. Granted that the Greek IT market is expected to grow significantly in the near future, it is obvious that the need for immediate action is more urgent than ever, so as to be able to deal with the problems that will occur due to the accumulation of significant quantities of e-wastes.

The paper presents the process so far of a project, funded by European Regional Development Fund, which aims to develop an integrated management system for electronic wastes in the Attica Region.
Keywords: electronic waste management.

1   Introduction

The Waste from Electric and Electronic Equipment (WEEE) constitutes one of the most rapidly developing waste streams. The WEEE annual production in EU is estimated at 6 million tons, while the WEEE annual growth rate is about 3 to

5%. The significantly increasing rate is mainly the result of the rapid expansion of the industry of personal computers (PCs).

The manufacture of a personal computer demands the complex assembly of hundreds of materials. At the same time, an innovative product becomes obsolete in a short time period. Nowadays, the average life of PCs ranges between 2 and 5 years and is expected to become shorter in the near future, due to the continuously increasing needs of the end users.

European Union has characterized the WEEE, in general, as one of the key waste streams and has already developed the framework for their management (EU Directives: 2002/96/EC; 2002/95/EC; 2003/108/EC). EU legislation requires each Member State to draw up one or more waste management plans in accordance with relevant EU directives. As a result, a lot of the EU countries have already developed a plan for the management of electronic wastes.

In Greece, the need to design an electronic waste management plan became visible a few years ago, when there was an explosive entry of PCs in the marketplace. More specifically, the size of the Greek market of PCs and servers has been tripled from 1995 to 2000.

This situation has led, gradually, to the accumulation of large quantities of electronic wastes, which, in the most of the cases, are disposed of together with municipal wastes. The problem is expected to be more intense in the near future, due to the increasing quantities of wastes that will be produced, as a result of the continuing growth of the Greek information technology market.

On the ground of the above, a systematic effort has started, so as to draw the lines of a successful management scheme, focused on the Attica Region, since a large part of the Greek population (approximately 34%) is concentrated there and at the same time, Attica presents the biggest percentage of PC users among all other Greek Regions. Moreover the majority of the Greek IT market (approximately 68%) is located in the wider area of Attica.

This effort constitutes one of the main actions of the regional program "Information Society for the quality of life in the Region of Attica", funded by the European Regional Development Fund (ERDF).

The paper presents the methodological approach as well as the main results achieved until now.

2   Overview of the problem

Computer waste is a serious environmental concern, primarily because of its toxicity. According to studies, electronics account for only 1 percent of the content of landfills by volume, but they contribute up to 70 percent of their toxic content [1, 2]. More specifically:

-   The cathode ray tubes (CRTs) in most computer monitors have x-ray shields that contain significant quantity of lead, mostly embedded in glass. Discarded monitors together with televisions are believed to be the largest sources of lead in landfills [3].

- A PC's central processing unit (CPU) typically contains toxic heavy metals such as mercury (in switches), lead (in circuit boards), and cadmium (in batteries).
- Plastics used to house computer equipment and cover wire cables often contain polybrominated flame-retardants, a class of chemicals similar to PCBs.

Lead, mercury, cadmium and polybrominated flame-retardants are all persistent, bioaccumulative toxins (PBTs) that can contaminate groundwater and pose other environmental and health risks when computers are manufactured, incinerated or landfilled [4]. Even recycling of e-waste, if not properly treated, can cause additional environmental problems.

The quantities that are recycled remain very low, while the main volume is being landfilled, incinerated, or stored away in closets or warehouses. The latter is mainly due to the reluctance of the people to discard electronic equipment as trash. Also, in a lot of cases, consumers are unclear about how to dispose of or get rid of old computers and e-waste. At the same time, a lot of producers and retailers insist on ignoring the problem. If this situation continues for a long time, there would be a large quantity of "historical" wastes that will escalate the problem of electronic waste management.

3 The Greek IT market

The Greek IT sector, although lagging behind in comparison with the EU countries, is characterized by a strong rate of growth (Table 1). During the last years, the growth rate of the IT market in Greece was at least two times higher than the average growth rate of the respective market in the other EU countries (Table 2). This was the result of the development and modernization of the public and the private sector. However, it should be noted that the growth rate of the Greek market has started slowing down from the year 2000. According to analysts, this was the result of the great delay of the inflow of capitals from the 3rd Community Support Framework. Therefore, the most pessimistic projections state that in the near future the Greek IT market will present a continuously anodic drift, which will be at least the same as in the rest EU countries.

Table 1: Growth rate of the Greek IT market (including PCs and servers) from 1995 to 2003 [5].

Year	Market size (in items)	Growth rate	Year	Market size (in items)	Growth rate
1995	115.000	-	2000	402.000	34.0%
1996	130.000	13.0%	2001	460.000	14.4%
1997	160.000	23.1%	2002	445.000	-3.3%
1998	200.000	25.0%	2003*	458.000	2.9%
1999	300.000	50.0%			

*Estimation; Source: ICAP.

Table 2: Annual consumption per capita for IT products and services in Western Europe (in euro) [5].

Country	1999	2000	2001	2002
Austria	720	791	791	766
Belgium/Luxemburg	718	767	782	752
Denmark	1,224	1,281	1,231	1,197
Finland	866	935	927	931
France	789	819	847	842
Germany	788	828	833	802
Greece	154	178	179	172
Ireland	621	667	633	606
Italy	399	414	431	428
Netherlands	1,023	1,055	1,041	1,032
Portugal	230	248	262	249
Spain	271	283	292	291
United Kingdom	991	1,074	1,080	1,075
Sweden	1,267	1,296	1,310	1,298
Switzerland	1,573	1,635	1,621	1,562
Norway	1,256	1,345	1,337	1,293

Source: EITO 2003.

4 Current management practices

4.1 The recycling techniques

The available recycling techniques do not differ significantly. More specifically, the treatment procedure is either fully or partial automated and includes the following stages: dismantling, depollution, shredding and sorting of materials. After dismantling, the parts that can be reused for upgrading of old computers are forwarded to the refurbishment process. During the depollution stage, the dangerous components (switches, batteries, etc.) are removed. These components are treated at accredited processing plants.

The shredding equipment used is depending, mainly, on the required recovery rate. The crushed materials can be sorted using various separation techniques. The main products from the recycling process are ferrous metals, non-ferrous metals and plastics. The non-ferrous metals are transported to special fusion furnaces. The metals leave these furnaces in such a pure state that can be reused as raw materials for making cables and new electronic components. The recovered ferrous metals are used in the steel industry. The recovered plastics can be reused, after special treatment, in the car and furniture industries. However, because of their low quality, the recovered plastics are usually used as fuels in kilns, while the secondary plastics are incinerated.

An extensive research is currently carried out for the treatment of the CRT monitors. Following dismantling, the picture tubes can be processed according to two methods. In the first, the glass is crushed and then washed, so as to remove the fluorescent powder. In the second, the screen is split into two parts (front and rear glass) and the phosphorescent powder is sucked off. Then, the two parts of glass are crushed separately, since they contain different hazardous substances. The cleaned glass is used, mainly, in the manufacture of new picture tubes or in ceramics industry.

Special treatment is implemented, in several cases, for the Printed Circuit Boards (PCBs), which are led directly to smelters for precious metals recovery.

4.2 The management schemes in Europe

Europe presents the most integrated management schemes worldwide, although there are significant differences between them. The factors that affect the scheme operation include geography, population size/density, labour cost, product scope, industry participation and organisation, legal requirements and standards, and scheme maturity. Scheme performance is, also, largely dependent upon the prevailing national recycling culture and public willingness to engage.

The majority of schemes run by Producer Responsibility Organisations (PROs). The exception is Denmark, where municipal and regional authority collectives currently operate the WEEE schemes [6]. Many management schemes handle electronic together with electrical waste (e.g. El Kretsen in Sweden). However, there are some management schemes that work only on electronic waste (e.g. ICT Milieu in Netherlands).

The rate of separate collection of at least four kilograms of WEEE per inhabitant per year, imposed by the EU directive, has already been achieved in most of the European countries. Further, many countries try to increase the above rate as much as possible.

The collection of the waste is the most difficult part that a management scheme has to face. The problem is more intense in the CRT monitors, which are too heavy and fragile. There are three primary channels for collection of e-waste: municipal collection sites, in store retailer take-back, and producer take-back. The majority of schemes have adapted the municipal collection system. Some schemes, such as ICT Milieu in Netherlands, use this channel exclusively. Others, such as Recupel in Belgium and El-Retur in Norway, encourage retailer participation, but this does not exceed the 30% of total volume collected. All schemes outsource the majority of their transport and treatment activities to commercial suppliers, usually on the basis of 2–3 year competitively tendered contracts.

The level of administrative complexity reflects the way they raise revenues, as well as the financial and operational role that they play. In general, the financial structure is based upon a fee. This fee can be charged directly to the producer, according to the number and category of equipment sold in a certain period. This cost is passed onto the consumer at the time of purchase of a new product. Although EU directives mention that this extra cost should not be shown to the purchasers, some schemes have established a visible fee (e.g. Recupel). However, the majority of schemes prefer to charge the participants in the system according to their market share.

The free riders, which currently represent the 10 to 20% of the volume of products placed on the market, remain a problem for all the schemes. The total cost for the management of wastes in the various European management schemes ranges between 0.45 and 0.80 €/kg. The biggest proportion of this amount concerns the collection and transportation cost.

There is a wide variation in the level of expenditures for communication and marketing activities between schemes, and as a consequence, in the level of consumer awareness. Visible fee schemes, such as NVMP and Recupel invest up to 4% of budget in public relation activities, including television, print media and point of sale (POS) materials. They also conduct consumer research, indicating that they have achieved levels of consumer awareness of approximately 70%.

5   Building an efficient e-waste management system

Taking into account that there has been a total lack of management consideration in regard to e-waste, in Greece, the management system should be designed from scratch. The development of the methodology is crucial for the drawing of the policy that should be followed so as to end to a successful electronic waste management scheme. There are a lot of critical questions that have to be answered, with the most important being:
-   The volume and the annual rate of e-waste entering into the system.
-   The proportion of free-rider/orphan products entering into the system.
-   The collectable volume of historical waste.
-   The way that the producers will be charged for their waste.
-   The distribution of management cost for the historical waste and the orphan products.
-   The role of municipalities in the collection process.
-   The motives to the users so as to ensure their participation in the collection process.
-   The degree of the system automation (manual or automation dismantling, separation and recovery rate, etc.).
-   The training of the personnel.
-   The possible engagement of sensitive social groups (e.g. disabled).
-   The potential subcontracting for transportation and recycling.
-   The way of re-marketing systems and components that can be reused.
-   The capability of the market to absorb the recycling products or exporting some of them.
-   The way of disposal of non-recyclable components and residues from the recycling process.

The answer to the above questions will finally lead to the realization of an integrated business and action plan, which will ensure the technical feasibility and the economic viability of such an investment.

The main steps of the methodology are given in Figure 1.

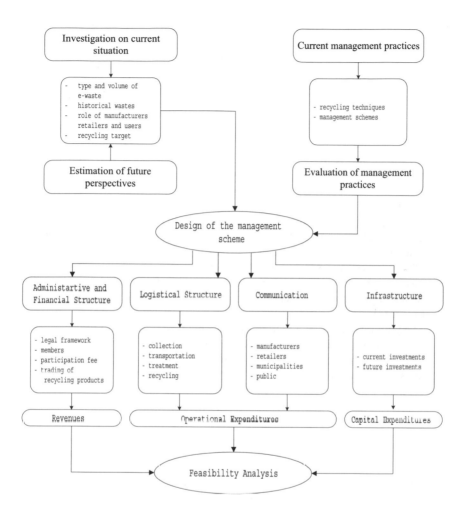

Figure 1:   Main steps of the methodology.

Up to now, the investigations on the e-waste management policies and techniques that are in use worldwide, as well as the examination of the current situation in Greece, have been completed.

6 Results - discussion

According to the experience of other European countries, the electronic wastes generated annually, do no exceed the 12.5% of the total WEEE. Based on the estimations that have been made from researchers [7], in regard to the annual WEEE amounts in the Region of Attica for the years 2003 to 2007, the respective amount of e-wastes was calculated as shown in Table 3.

Table 3:   Annual e-wastes amounts (in tons) for the years 2003–2007 in the Attica Region.

	2003	2004	2005	2006	2007
WEEE	59,344	59,718	60,094	60,472	60,853
e-wastes (12.5% of the WEEE)	7,418	7,465	7,512	7,559	7,607

The investigation of the current situation was based on data, which were collected by means of telephone and face-to-face surveys on PC users, manufacturers and retailers. From the data collected until now the following results are obtained:

- There is a significant percentage among the respondents, which ignores the existence of hazardous substances into the PCs.
- The majority of the domestic users do not throw away their old PC. Some of them prefer to store it, while others sell it or gift it, which practically means that a part of the old equipment is currently reused. Moreover, there are a lot of commercial users that store the obsolete equipment in warehouses. As a result, there are already significant quantities of historical wastes to deal with straightforward. Otherwise significant quantities of e-waste will end up to landfills.
- The public, although not well informed, seems to understand the necessity of the development of an e-waste management system.
- The most convenient take back system seems to be the collection from the house in regard to the domestic users. For that reason, the municipalities should play a major role in the system.
- The manufacturers and retailers are not willing, at least for the time being, to get involved in the management system. Therefore, a way to enforce them to take responsibility immediately should be found. It is obvious, that their participation will accelerate the whole processes.
- The manufacturers and retailers estimated that the market could not afford an extra fee more than 5 euro in the purchase value of a new PC. On the other hand, and this is most important, the domestic and commercial users seem to be willing to pay an average fee of about 31 euro and 20 euro respectively.
- As for the recycling products, plastics are the most difficult to deal with, due to the lack of incineration technologies in Greece. A possible solution for the second category plastics is their utilization as fuels in kilns of metallurgy or cement production.

References

[1]   McDonnell, E., Turning Wastestream Materials into Economic Opportunities, Demanufacturing Partnership Program Newsletter, New Jersey Institute of Technology, 1997.
[2]   U. S. Environmental Protection Agency. Computers, E-Waste, and Product Stewardship: Is California Ready for the Challenge?, Global Futures Foundation, 2001.

[3]   USA Sitting on Mountain of Obsolete PCs, USA Today, 22 June 1999.
[4]   Pare, M., Metech Seizes Opportunity in Computer Salvage, High Tech, 30 November 1998.
[5]   ICAP. Study on the Greek IT sector, Athens, 2004.
[6]   U. S. Environmental Protection Agency. Computers, E-Waste, and Product Stewardship: Is California Ready for the Challenge?, Global Futures Foundation, 2001.
[7]   Ecological Recycling Society. Economic impact study on the sustainable management of e-waste in Greece, LIFE ENV/GR/000688, Athens, 2003.

Section 16
Construction and demolition waste

C&D waste for road construction: long time performance of roads constructed using recycled aggregate for unbound pavement layers

F. Lancieri, A. Marradi & S. Mannucci
*Civil Engineering Department - Road and Transportation Division,
University of Pisa, Italy*

Abstract

This paper describes the results of the experimental measurements carried out during 2001 and 2005 on two secondary roads built in 1998 using C&D aggregates for subgrade and subbase layers. Field data included structural data in the form of nondestructive testing (NDT) performed with a Falling Weight Deflectometer and laser profilometer, visual condition survey, traffic measurements and construction and maintenance history data. The FWD deflection measurements were analysed to determine modulus values for the various pavement layers while profile data were used to calculate evenness index (IRI). Deflection tests performed in 2001 showed that in-situ pavement performance was better than expected. This performance improvement was explained by self-cementing properties. Field tests repeated in 2005 revealed a meaningful difference in the structural behavior of the road pavements examined: the first, subjected to heavy vehicles, showed a marked improvement in backcalculated moduli of the subbase pavement layer while the second, not subjected to heavy vehicles, maintained practically the same values of layer moduli. These field tests were integrated with laboratory tests performed in order to investigate the time evolution of mechanical characteristics of C&D materials and to evaluate the influence of compaction techniques on the improvement in resistance recorded with the gyratory compactor. The data and results presented confirm that road construction could offer a reliable use for C&D waste recycling.
Keywords: C&D , unbound layers, FWD, back analysis, gyratory compactor.

 WIT Transactions on Ecology and the Environment, Vol 92, © 2006 WIT Press
www.witpress.com, ISSN 1743-3541 (on-line)
doi:10.2495/WM060571

1 Introduction

During the last ten years, road works in Italy have on various occasions been carried out using recycled aggregate from construction and demolition waste, although codes for materials specifications have included such materials only since 2002.

The legislation currently in force in Italy requires Italian public authorities working in the field of building, road and environmental construction to cover at least 30% of their annual requirement of aggregate by using recycled aggregate. In the light of the provisions laid down in current legislation and the increasing difficulty in obtaining virgin materials, it is important to ascertain the type of mechanical performance that can truly be achieved by C&D materials once they have been set in place, and how their characteristics evolve over time as a function of their physical, mechanical and compositional acceptability requisites.

In order to improve and extend knowledge on these aspects, investigations were conducted at the Road Research Laboratory Civil Engineering Department of the University of Pisa, in order to assess the aggregate utilized in the construction of two road section in 1998. During subgrade and subbase construction, field tests were conducted to check that materials were correctly set in place (load plate test, dry density measurements). Subsequently, in 2001 and 2005, non-destructive Dynatest Falling Weight Deflectometer (FWD) testing was performed to monitor the evolution over time of the mechanical parameters of the pavement layers built with C&D materials. In 2005 the road profile of the two sites investigated was also assessed, by calculating the IRI (International Roughness Index) in order to evaluate surface evenness. For both infrastructures, whose pavement was not subjected to maintenance operations, two 200 m long sample sections were examined in the present study.

2 Description of the experimental roads

The sections subjected to investigation form part of two secondary roads, situated on the outskirts of the city of Pisa, opened to traffic in 1998. For both pavements the subbase layer was built with unbound recycled aggregate (a scheme of the pavement is shown in Figure 1). The two infrastructures are characterized by a different road section: the road indicated henceforth as Road 1 is constituted by a single carriageway with one lane in each direction, while the one indicated as Road 2 is constituted by a carriageway with two lanes in each direction.

In the 2001 and 2005 tests, vehicular traffic was also measured using magnetic induction sensors placed on the road pavement. The two roads were found to be subjected to appreciably different amounts and types of traffic: Road 1, during its eight years of use, had a constant TGM value of roughly 5000 v/d, with 6% of heavy vehicles, while Road 2 had a very modest TGM composed of passenger automobiles.

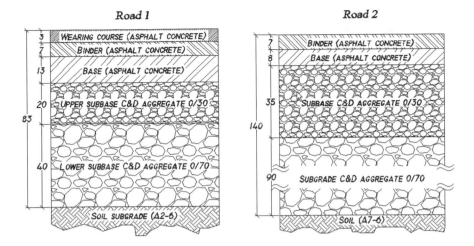

Figure 1: Scheme of pavements (measures in cm).

3 Tests at the end of construction works: year 1998

The characteristics of materials utilized for the asphalt concrete are shown in Table 1.

Table 1: Properties of asphalt concrete.

Experimental site	Road 1			Road 2	
Layer	wearing	binder	base	binder	base
Binder content (%)	5.6	4.8	4.5	4.9	4.3
Voids content (%)	5.5	8.3	10.0	9.1	10.6
Marshall stability (kN)	9.43	8.32	7.38	7.61	6.84

With regard to the C&D materials, in order to gain insight into the evolution of the mechanical characteristics over time, with reference for example to self-cementing properties, it is important also to have information on the composition of the mixtures utilized (Table 2).

Table 2: C&D aggregate composition (material retained 4 mm sieve).

Components (Weight %)		Concrete and Gravel	Brick and ceramic	Mortar	Glass	Asphalt	Clay lumps
Road 1	C&D 0/30	55.5	24.8	17.5	0.2	2.0	0.0
	C&D 0/70	54.7	25.6	4.7	0.4	14.6	0.0
Road 2	C&D 0/30	56.1	29.0	10.8	0.4	3.7	0.0
	C&D 0/70	52.0	23.4	3.2	0.4	12.3	8.7

The results of the characterization tests undertaken on the different types of C&D materials utilized in contructing the two roads are shown in Table 3, Marradi [1].

Table 3: Properties of recycled aggregates.

Experimental site	Road 1		Road 2	
Material	C&D 0/30	C&D 0/70	C&D 0/30	C&D 0/70
% Passing 2.5" ASTM sieve (63,5 mm)	100	100	100	100
% Passing 1.0" ASTM sieve (25,4 mm)	100	80	95	90
% Passing 10 ASTM sieve (2,00 mm)	39	22	46	62
% Passing 40 ASTM sieve (0,42 mm)	26	15	32	41
% Passing 200 ASTM sieve (0,075 mm)	13	8	11	19
Liquid limit w_L (%)	n.p.	n.p.	n.p.	30
Plastic limit w_P (%)	n.p.	n.p.	n.p.	21
Plasticity index PI (%)	n.p.	n.p.	n.p.	9
Los Angeles abrasion coefficient LA (%)[1]	41	43	39	42
AASHTO Maximum dry density γ_{dmax} (Mg/m³)	1.88	1.95	1.92	1.86
AASHTO optimum moisture content w_{opt}(%)	11.0	8.5	10.8	9.7
CBR Index (%)	105	105	115	71
CBR Index (soaked 4 day) (%)	104	98	73	35
CBR Index (soaked 8 day) (%)	81	78	-	-
CBR Index (soaked 16 day) (%)	113	85	-	-
Swelling of CBR samples (%)	<0.1	<0.1	0.2	0.7
[1] ASTM C 131 - grading A for C&D 0/70 and grading B for C&D 0/30				

After the various types of recycled aggregate had been set in place, three tests were carried out to determine the value of the in situ dry density "γ_{ds}" and load plate tests for determination of the modulus of deformation "M_d" in the pressure interval 0.15 – 0.25 MPa (Table 4).

Table 4: Mean values of "M_d" and dry density ratio "$\gamma_{ds}/\gamma_{dmax}$".

Experimental site	Road 1		Road 2	
Material	C&D 0/30	C&D 0/70	C&D 0/30	C&D 0/70
Load plate test – M_d (MPa)	85.63	36.14	94.35	34.18
Dry density ratio $\gamma_{ds}/\gamma_{dmax}$ (%)	98.3	-	99.6	-

4 Tests after 4 years of traffic: year 2001

After four years of traffic, FWD tests were carried out and the values of in-situ dry density "γ_{ds}" of the two subbases were determined. The FWD test practice and settings as well as mean "$\gamma_{ds}/\gamma_{dmax}$" values are shown in Table 5.

The deflection data acquired were analysed with the ELMOD 5 [2] program using the Method of Equivalent Thickness (MET), which makes it possible to backcalculate the values of the "E" elastic moduli of the pavement layers that most plausibly reproduce the stresses, strains and deflections measured under FWD loading (Ullidtz [3], Ullidtz and Zhang [4]). The values of the "E" moduli thereby obtained for the asphalt concrete layer, the subbase and the subgrade, are shown in Figs. 2 and 3.

Table 5:   FWD test practice and mean values of "$\gamma_{ds}/\gamma_{dmax}$" – year 2001.

Experimental site	Road 1	Road 2
Air Temperature (°C)	26.5	30.8
Asphalt Temperature (at asphalt mid-depth) (°C)	27.8	32.0
Loading plate type	Segmented – 0.3 m diameter	
Geophones distance from centreplate (m)	0, 0.2, 0.3, 0.45, 0.9, 1.20, 1.50	
Pressure levels (kPa)	600, 800, 1000	
Number of drops per station	9 (3 each pressure level)	
Step (m)	20	
Dry density ratio $\gamma_{ds}/\gamma_{dmax}$ (%)	101.3	100.4

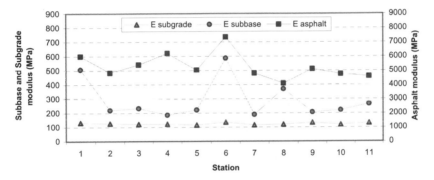

Figure 2:   Road 1 - year 2001.

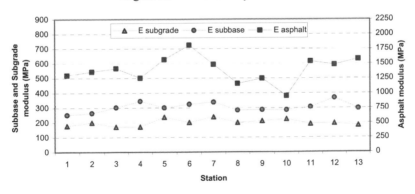

Figure 3:   Road 2 - year 2001.

5   Tests after 8 years of traffic: year 2005

In October 2005, after 8 years of traffic, both the FWD tests and the determinations of in-situ dry density "γ_{ds}" were repeated. The FWD tests practice and setting as well as mean "$\gamma_{ds}/\gamma_{dmax}$" values are shown in Table 6.

Table 6:   FWD test practice and mean values of "$\gamma_{ds}/\gamma_{dmax}$" – year 2005.

Experimental site	Road 1	Road 2
Air Temperature (°C)	21.4	22.5
Asphalt Temperature (at asphalt mid-depth) (°C)	22.0	24.1
Loading plate type	Segmented – 0.3 m diameter	
Geophones distance from centreplate (m)	0, 0.2, 0.3, 0.45, 0.6, 0.75, 0.9, 1.20, 1.50	
Pressure levels (kPa)	600, 800, 1000	
Number of drops per station	9 (3 each pressure level)	
Step (m)	20	
Compaction level $\gamma_{ds}/\gamma_{dmax}$ (%)	102.2	100.4

The values of the "E" moduli thereby obtained for the asphalt concrete layer, the subbase and the subgrade, are shown in Figs. 4 and 5.

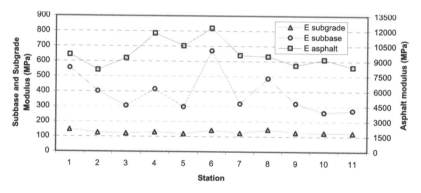

Figure 4:   Road 1 - year 2005.

Analysis of the data acquired by means of the Greenwood Laser Profilometer, with 25 mm sample distance, made it possible to determine the IRI in order to evaluate the longitudinal evenness of the two pavements after 8 years of traffic use. Mean IRI values obtained from three measurements for each of the road section analysed were 1.37 mm/m and 3.49 mm/m for Road 1 and Road 2 respectively.

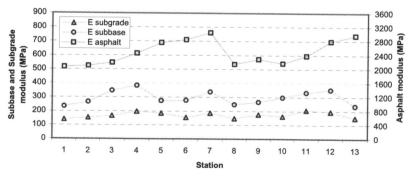

Figure 5:   Road 2 - year 2005.

6 Analysis of results

The values of the "E" moduli obtained for the layers constructed with recycled materials, as measured during the 2001 investigations, were found to be higher than had been expected on the basis of the values of the "M_d" deflection moduli measured by the load plate tests carried out in 1998. If the peak values of stations 1 and 6 of Road 1 are ignored, it can be observed that the mean value of the "E" moduli of the subbase are 235 MPa for Road 1 and 304 MPa for Road 2 (Table 7), roughly 3 times higher than the values obtained with the 1998 static tests. The increase in the moduli found at the end of the first four years of traffic could be attributed to the residual self-cementing properties of the mortar and concrete, contained in percentages of over 17% in the 0/30 C&D materials utilized for the subbase of Road 1 (Arm [5], Benedetto et al [6]). The mean values of the "E" moduli obtained by elaborations of the deflection data acquired in 2001 and 2005 are shown in Figs. 6 and 7 , while their mean values are summarized in Table 7.

Table 7: Mean values of the moduli backcalculated in 2001 and 2005.

Experimental site	Average moduli 2001 (MPa)			Average moduli 2005 (MPa)		
	Asphalt	Subbase	Subgrade	Asphalt	Subbase	Subgrade
Road 1	5296	235	122	5356[1]	379	127
Road 2	1397	304	200	1117[2]	297	170

[1] Moduli referring to the temperature of 27.8°C according to the relation proposed by AASHTO
[2] Moduli referring to the temperature of 32.0°C according to the relation proposed by AASHTO

It should be noted that although the materials utilized for the construction of the two roads were of the same type, they present different physical, mechanical and compositional characteristics. In addition, while Road 1 presents a closed type of wearing course ($V_v = 5.5\%$), which reduces rainwater infiltration into the underlying layers, the Road 2 pavement lacks the wearing course and the surface binder type asphalt layer is highly permeable as it is characterized by an elevated percentage of voids ($V_v > 9\%$).

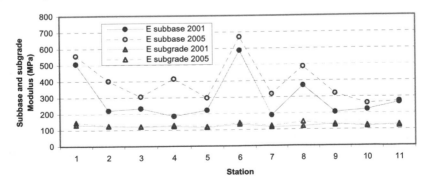

Figure 6: Road 1 – "E" modulus values for 2001 and 2005.

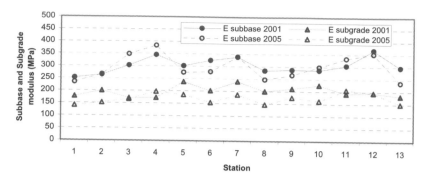

Figure 7:   Road 2 – "E" modulus values for 2001 and 2005.

For Road 2, the results obtained in 2005 show that the mean value of the "E" modulus of the subbase layer maintains practically the same value as that obtained in 2001, while the mean value of the "E" modulus of the subgrade is roughly 15% lower compared to that found in the 2001 investigations. This decrease is to be attributed to the high water table in the area where the road is situated (above all in periods of elevated rainfall); such a circumstance, together with rainwater infiltration, leads to a marked presence of moisture content in the subgrade, which, in laboratory tests, exhibited a non-negligible sensitivity to water (PI = 9).

For Road 1, while values of the "E" moduli of the soil subgrade remained almost unaltered from 2001 to 2005, values of the subbase built of C&D materials showed a mean increase greater than 34%. This increase cannot be explained by a different extent of the stresses and strains to which the subbase layer is subjected during the FWD tests, arising from a possibly greater stiffness of the overlying asphalt concrete. For as shown in Figure 8, the FWD tests conducted with 3 different levels of stress revealed a more or less linear behavior of the material examined: under the hypothesis that the modulus of the material varies as a function of the bulk stress (θ) according to the well-known K-Theta model (Hicks and Monismith [7]), for all the test stations the values of the exponential coefficient K_2 always ranged between -0.15 and 0.15, a value that was notably lower than values reported for natural materials in the literature (Domenichini and Di Mascio [8]).

The 34% increase in the mean value of the "E" modulus recorded between the fourth and eighth year of road use can only partially be explained by the previously mentioned self-cementing properties. It is more likely that this variation can be attributed to further increments in the dry density ratio of the subbase layer resulting from the passage of heavy vehicles (TGM = 320 hv/d,). It should be pointed out that this increase in the value of the "E" modulus is associated with a fairly moderate increase in the dry density ratio.

In order to gain further insight into this phenomenon, specific laboratory investigations were conducted both on the C&D material and also on an optimal subbase limestone with the same particles size. For this purpose the gyratory compactor was used. This is a viable test device to induce a state of stress in the gyratory sample that simulates the passing of a loaded vehicle (McRea [9], George [10]). The values of the dry density ratio that were thereby obtained, evaluated as the ratio between the value achieved with the gyratory compactor (γ_{dg}) and the maximum AASHTO (γ_{dmax}) and the corresponding values of the CBR index recorded at the end of Phase 2 of the compaction procedure, which simulates the effect of vehicular traffic, were compared with values recorded at the end of Phase 1, which simulates compaction at the end of road construction. The results shown in Table 8 demonstrate that for the C&D material analyzed, given a 0.7% increase in the dry density ratio at the end of Phase 2, the increase in the CBR index was roughly 60%. This indicates a higher sensitivity to the degree of compaction than was observed with the natural material. For the latter, given a three times higher increase (2.3%) in the dry density ratio, the increase in the CBR index was 16%.

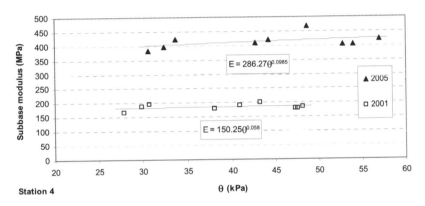

Station 4

Figure 8: Road 1 – Station 4 – Relationship between bulk stress and subbase modulus.

Table 8: Compaction with gyratory compactor.

	Phase 1		Phase 2	
Angle of gyration (°)	1.25±0.02		1.25±0.02	
Speed of gyration (rev/min)	30		30	
Number of gyrations and ram pressure	90 gyrations – P = 600kPa 90 gyrations – P = 300kPa		90 gyrations – P = 600kPa 90 gyrations – P = 300kPa 400 gyrations – P = 200kPa	
Aggregates for subbase layers	C&D 0/30	Virgin	C&D 0/30	Virgin
High specimen (mm)	117.6	124.7	116.7	121.8
Dry density ratio $\gamma_{dg}/\gamma_{dmax}$ (%)	101.85	100.26	102.53	102.51
CBR Index (%) (Mean of 3 values)	128	172	188	198

7 Conclusions

Analysis of the data acquired during the 2001 tests shows that for both of the roads examined, mean values of the "E" moduli of the respective subbases were roughly 3 times the "M_d" values obtained in 1998. On Road 1 the values of the moduli recorded in 2005 were, on average, 34% higher compared to those recorded in 2001. On Road 2 the values of the subbase obtained in 2005 virtually replicated the value recorded in 2001 (roughly 300 MPa) while the value of the modulus of the corresponding subgrade decreased by 15% (from 200 MPa to 170 MPa).

The increase in the moduli between the fourth and eighth year of traffic can be attributed to the residual self-cementing properties of the material, combined with the effect of traffic, which may have induced further compaction of the layer but without compromising the evenness of the pavement surface. This hypothesis was confirmed through purpose-designed laboratory tests with the gyratory compactor, which demonstrated that in this type of materials small variations in the dry density ratio can lead to significant increases in the CBR index (roughly 60%), unlike to findings for the optimal natural material with the same particles size used as term of comparison.

The reduction in the moduli of the subgrade of Road 2 can likely be attributed to the silty-clayey fractions contained in the C&D 0/70 and to the greater moisture content present in the subgrade itself.

However, for both of the roads analyzed, the layers constructed with materials deriving from recycled construction and demolition waste were shown to maintain over time a performance that was by no means inferior to that characterizing traditional materials. Furthermore, results both of in-situ and laboratory tests revealed that the load-bearing capacity of the material had elevated sensitivity to the dry density ratio.

We therefore conclude that in order to achieve a good performance with these materials it is advisable to carry out preliminary laboratory tests designed to establish a reliable and significant correlation between compaction and load-bearing capacity, so that appropriate knowledge can be acquired for correct utilization of the materials.

References

[1]   Marradi, A., Indagini sperimentali sui materiali inerti di riciclaggio, *Riciclare per l'ambiente*, ed. Grafica Pisana, Pisa, pp.72-89,1999.

[2]   Elmod 5 FWD data analysis software, Dynatest International A/S, Denmark, 2005.

[3]   Ullidtz, P., *Modelling Flexible Pavement Response and Performance*, Denmark, 1998.

[4]   Ullidtz, P. & Zhang, W., Back-calculation of pavement layer moduli and forward-calculation of stresses and strains. *ISAP 9th conference*, Copenhagen 2002.

[5]   Arm, M., Self-cementing properties of crushed demolished concrete in unbound layers: results from triaxial tests and field test, *Waste Management No. 21*, pp. 235-239, 2001.

[6]   Benedetto, A.,Benedetto, C. & De Blasiis, M. R., Le potenzialità del riciclato. *Recycling*, pp.81-91, July 2005.

[7]   Hicks, R.G. and Monismith, C.L., Factors Influencing the Resilient Properties of Granular Materials. *Transportation Research Record No. 345*, TRB, National Research Council, Washington DC, 1971, pp. 15-31.

[8]   Domenichini, L & Di Mascio, P., I materiali non legati impiegati nelle sovrastrutture stradali, pp. 32-48, 1990.

[9]   McRea, J. L., Gyratory compaction method for determining density requirements for subgrade and base of flexible pavements. *Miscellaneous Paper No. 4-494*, U.S. Army Engineering Waterways Experiment Station, Corps of Engineering, Vicksburg, Miss, 1962.

[10]   George, K.P., Resilient testing of soils using gyratory testing machine. *Transportation Research Record No. 1369*, TRB, National Research Council, Washington DC, pp.63-72, 1992.

An evaluation of kerbside recyclates collection as a means of enhancing waste recycling in Christchurch

C. Njue
School of Conservation Sciences, Bournemouth University, UK

Abstract

The fact that uncontrolled waste can lead to environmental and health risks has made it necessary to mitigate the degradation of water, soil and air. Several problems associated with traditional waste disposal methods such as lack of space for landfills and associated leachate, air pollution from incinerators and both the United Kingdom and European Union legislation on waste management has shifted attention to recycling as better option for waste disposal. The UK Government's Department of Environment, Transport and Rural Affairs (DEFRA) has given out guidelines on the latest performance indicator targets on recycling for individual local authorities. To enhance recycling and in order to meet the government's recycling targets, local authorities in the UK are currently using kerbside recycling programmes. To achieve the above targets, Christchurch Borough Council in the summer of 2003 introduced a new kerbside recycling scheme. Houses and businesses were provided with containers for separation of recyclable material. In June of 2003, a study was carried in Christchurch Borough, Dorset, England with the aim of evaluating kerbside recycling as a means of enhancing waste recycling. About 13,000 properties were surveyed. Findings from the research indicated a significant increase in the number of Christchurch residents participating in recycling. This increased participation was mainly attributed to the introduction of a new kerbside recycling scheme. Reasons were given for non-participation in the new kerbside recycling scheme. These form the basis of recommendations for an improved waste management framework. The substantial diversion of recyclables from landfills also proves why recycling and improved recycling schemes must play a major part in local, national and international waste management plans.
Keywords: kerbside recycling, waste recycling, performance indicator targets, evaluation, participation, improved waste management framework, mitigation, pollution, landfills, environment.

WIT Transactions on Ecology and the Environment, Vol 92, © 2006 WIT Press
www.witpress.com, ISSN 1743-3541 (on-line)
doi:10.2495/WM060581

1   Introduction

About 371 million tonnes of waste are generated in the UK per annum [9] and about 8 million tonnes of it is Municipal solid waste [1]. This is about 0.9 kg of Municipal solid waste per person per day [3]. Waste generation in the UK increased by 2.7% between 1999-2000 and 2000-2001. It is estimated that at this rate, by the year 2020, the amount of waste generated will be twice the current and at the same time increasing the disposal costs to about £1.6 billion (Recycling and Waste World, 2003). About 50-60% (by weight) of the household waste is recyclable material and yet the recycling rate for the same was only 5%-6% [6]. To protect the environment and public health, the UK Government set out a policy framework for sustainable waste management for the next 20 years. This policy framework was aimed at using material resources in a more efficient way with a view to reducing the bulk of generated wastes.

1.1   Waste recycling

Among the measures introduced in the policy framework were new statutory laws requiring waste collection authorities to formulate waste recycling plans [10]. Thus focus was shifted to waste recycling which comes third in the waste disposal hierarchy (i.e. reuse, waste exchange, recycling, composting and sanitary landfills). As at the year 1994, the average amount of household waste recycled nationally in UK was less than 5% with about 95% going to landfills. This contrasted with recycling figures of 15-20% in Europe and USA during the same period while in Japan, the high costs of landfill had encouraged recycling figures to 30% [9]. For local authorities, the set target by the year 2000 for waste recycling was 50% of the recyclates from domestic waste. The national recycling rates in UK were 11.2% and 13% by April 2001 and 2001/2 respectively (DEFRA 2003, cited in [11]). The set target for local authorities by the year 2003/4 was 22%. However despite some promising results from some of the waste collecting authorities, it was not possible to achieve the national target of 25% recycling and composting. The reason for this was attributed to lack of financial resources, lack of government guidelines and failure to fully implement the European Union Directive. This target therefore had to be extended up 2005 (DETR 1999, cited in [10]). This resulted in local authorities revising their waste strategies with a view to meeting legislative targets. New kerbside recycling schemes were introduced to enhance recycling rates (Waite 1995, cited in [10]).

1.2   Kerbside recycling

Kerbside recycling is a means of providing households and businesses with a container suitable for separating of recyclable material. Householders are supposed to separate recyclates at source, keep them and put them by the kerbside to be collected [9]. In some schemes, the collection of green waste and uncooked food is included and may involve the provision of additional bins, plastic sacks or boxes [10]. The permitted recyclates for containers in various areas are different and also depend on the prevailing markets for recyclates. After collection, the recyclates are transported to a Materials Recycling Facility (MRF). Here they are sorted out and taken to the buyers [10]. In their research [10] reported that recyclates collected from this type of separation are uncontaminated. Among the items separated in this way and collected by for example compartmentalized collection lorries include: paper, glass, metals and plastics. The kerbside waste recycling is used to recycle most of solid waste in many countries [2]. Kerbside recycling is said to be liked by the public as opposed to the recycling centers. This is because some people do not have means by which to transport the recyclates to the recycling centers which otherwise goes to landfills. The response to kerbside recycling is said to be high and incorporating it with general waste collection would result in high tonnages of recycling [9]. It is suggested that environment is put into consideration and strike a balance by not using for instance double the energy in the collecting and delivering recyclates than what it will take to recycle it. At the same time it is necessary that a majority of the kerbside schemes will have to be molded in order to meet the local needs in addition to the available facilities and markets [9]. To enhance solid waste recycling, Kerbside solid waste recycling is now seen to be common phenomena in most of the local authorities in UK and the set recycling target may not be possible in the absence of the kerbside recycling [9]. Barriers to kerbside recycling include; inconvenience, Inadequate facilities and storage handling. However there is an increase of the number of residents who claim to recycle with the introduction of a kerbside waste recycling and therefore overcoming such problems (Barton and Perrin 2001).

1.3 Kerbside waste recycling in Christchurch

Christchurch is located in Dorset, England and covers about 50 square Kilometres. The location of Christchurch is shown in the map below. The population of Christchurch is about 44,869 (Census population 1981,1991 and 2000, cited in [4]). The population density is about 891 people per square kilometre, the second highest in UK (Regional Trends ND, cited in [4]). As at the year 2002, about 87% of residents in Christchurch used recycling facilities (Christchurch Borough Council Panel Survey 2002, cited in [4]). This is an upward trend from

70% in 1999 and 1996 (Residents Survey of 1996 and 1999, cited in [4]). Some of the common recyclates then were paper and glass. The high proportion of residents using recycling facilities at the time could have been attributed to the availability of recycling facilities in that about 84% of residents are reported to have access to recycling facilities (Christchurch Borough Council Panel Survey 2002, cited in [4]). The other reason is the understanding by about 97% of residents of the need for recycling (Ibid ND, cited in [4]). In ranking order, some of the reasons for not recycling by residents as at the year 2002 were the unavailability of kerbside recycling, (63%), the distance to recycle banks (44%) and inability to get to the amenity sites (41%) (Ibid ND, cited [4]).

Figure 1:   Map of the UK [7].

2 Methodology

To achieve the objectives of this study both primary and secondary sources of data were explored. The author accompanied the kerbside rounds lorry crew on their five-week rounds which covered 24 days. To obtain primary data, a survey was undertaken which involved field observation and use of questionnaires [9]. A list of property numbers for households who put out a green boxes or other containers for storing recyclates for example plastic bags by the kerbside in the target streets was recorded in a field note book. The number of individual properties whose residents participated on the scheduled recyclates collection dates were counted and recorded. Some flats were not included in this study as it would not have been possible to establish participating and non-participating households because the recycling facilities were communal.

The nature of the survey also warranted the use of self and verbally administered questionnaires. To establish reasons for non participation in the kerbside scheme, a survey of households not participating in the scheme was also simultaneously undertaken by door knocking and verbally administering the questionnaire to residents on reasons for not recycling. Reasons given for non participartion in the kerbside recycling scheme were noted. Data was also collected from secondary sources such as organisational documents, minutes of meetings, journals, association records and state statistical records. Care was taken as data from secondary sources for a study can be intended for a different primary reason [12].

2.1 Sampling

A sample survey of households was therefore undertaken in June and July 2003. Samples were drawn from a total of 527 streets and intended to cover 13,576 properties within five weeks. This is as depicted table 1.

Table 1:   Recycling rounds and number of properties.

Recycling rounds	No. of streets	No.of households in the elect. roll
Round A Week 1	121	2470
Round A Week 2	104	3093
Round B Week 1	94	2896
Round B Week 2	107	2678
Round C Week 1	129	2439
Totals	**527**	**13576**

Households in five different residential areas and on 527 different streets were selected for the study which took 24 days to complete. The sampled residential areas displayed similar characteristics. The samples were randomly selected and were therefore representative of the total properties in Christchurch. It was therefore considered that sampling about half the number of households out of a total number of 22,000 households who were issued with recycling boxes was expected to give a true representative on the rate of participation in the new kerbside recycling scheme.

2.2 Scope and limitations of the study method

It was difficult interviewing non-participating residents with hearing disability. In some cases there were more properties counted on the ground than those listed in the electoral roll while in some cases it was vice versa. Some houses were not numbered. The author either guessed the number from neighbouring houses and where this was not possible "+1" was recorded under the names of respective streets. Changed collection schedule dates was another problem. For example it was not possible to collect data on 17th July 2003 scheduled for Round C Week 1. This was because the collection dates had been rescheduled. At the time of the data collection, it was in summer and some residents had gone on holidays. Time and financial limitation was another limitation because the study was not fully funded and was to cover data collection for only five weeks and not six weeks as originally planned.

3 Results and discussion

3.1 Problems experienced during the launch of the new kerbside scheme

A week prior to the launch of the kerbside scheme, recyclates boxes were delivered to households by a contracted company on behalf of the Council. Some of the problems encountered during this period were: mechanical problems with delivery vehicles, some properties received cards with the wrong collection dates and the lorry recyclates crew collection missed some green boxes and streets during their collection rounds. The overall impact of these problems adversely affected the overall tonnage of the recyclates collected during the first two weeks of the implementation of the scheme. This is as depicted in figure 2 and table 2.

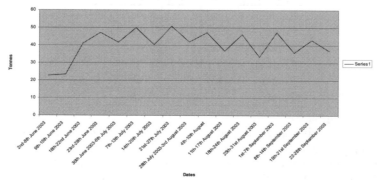

Figure 2:   Recyclates tonnage [4].

An average of about 40 tonnes of recyclates was collected per week regime during the 17 weeks in the 4-month period (June to September 2003). During the first two weeks of scheme launch, only about 46 overall tonnes of recyclates were collected as opposed to an average of about 80 tonnes of recyclates which would otherwise have been collected [4].

3.2 Level and participation percentage in the new kerbside scheme

Data gathered from 527 streets over a five-week period covering 24 days is illustrated in table 2.

Out of the total 12,974 properties surveyed, participants in the kerbside recycling scheme were about 8963. Those households who did not put out the green boxes during the recycling rounds were about 4011. To get the participation % for the sample area the formula below was used.

$$\text{Participation \%} = \frac{n}{m} \times 100$$

n stands for number of households using the recycling box at least once in a 5-week period and m for total number of households with recycling boxes in the same 5-week period. This gives a recycling participation rate of 69.04%. Table 2 also gives estimates on the total recyclates per week within the fortnight collection as well as the average weight of recyclates per household. It is worth noting that this percentage does not take into account flats in the 527 streets sampled who collectively shared recycling facilities. The reason for this was time and financial constraints which made it impossible to determine participants and non-participants in such cases as it would have meant interviewing individual households in these flats. This might also be the reason for the variance between the total properties recorded in the sampled area which were 12,974 and the total number of properties in the electoral roll which were listed as 13,576. The difference was 602 properties and was 4.6% above the actual recorded properties in the study survey from the sample area. The pilot kerbside scheme carried out in 1989 covered about 3,200 properties. This fortnightly collection was for mixed paper, magazines, plastic bottles and cans. About 24 tonnes of recyclates were collected monthly with a participation rate of 52%. In the new kerbside scheme, about 22,000 green boxes were given to households for glass bottles, mixed cans, newspaper and magazines. About 171 tonnes of recyclates are collected monthly and the participation rate is about 69.08%. Some streets reported a 100% participation. This is a remarkable improvement taking into account that the collection of plastics is not included in the new kerbside scheme.

Table 2:   Kerbside participation levels.

Rounds	Part.	Non part.	Total	% of part.	Tonnes of recyc.	Kgs/property
A Week 1	1642	718	2360	69.57	22.74	13.84
A Week 2	1613	726	2339	68.96	23.31	14.45
B Week 1	2128	915	3043	69.93	41.14	19.33
B Week 2	2108	963	3071	68.64	47.34	22.45
C Week 1	1472	689	2161	67.89	41.76	28.36
Total	**8963**	**4011**	**12974**	**69.08**	**176.29**	**19.66**

Source: Author's survey (2003)

3.3 Reasons for non participation

Some residents did not have boxes and newsletters. Other residents who were interviewed said that they are not willing to participate in the recycling project because they were not consulted. However these claims are reported not to be correct as the Council is said to have written in-house magazines that were distributed to every resident in Christchurch prior to the start of the kerbside scheme. Lack of space to store the green boxes in terms of space in the house and the house garages. This is common with households living in high-rise accommodation. This means that households with storage problems may not participate in kerbside recycling. Lack of adequate space to place recyclates boxes. Some residents were not using the recycle bins because sorting out recyclable materials and washing the tins is time consuming, cumbersome and inconveniencing. This was a response that came mostly from households having three or more young children and was therefore tied by the family cores. It is true that the Council requested residents (in the newsletters) to wash unbroken glass bottles and jars, food tins and drink cans. The reason for the cleaning was to make them clean for recycling. This is another barrier to waste recycling reported in the works of Vining and Ebreo (1990, cited in Tucker *et al.* 1998). Aylesford (1996, cited Turner, 1998) has indicated that recycling rates in households go below the nation average levels with the presence of young children. This is attributed to the fact that households are very busy when children are at young ages but as children become more enlightened, pressure is put on families to recycle. In a study, [2] reported that the average time taken by an American family is only 16 minutes a week to separate recyclable material. This translates to about 2 minutes a day. It would therefore be important to educate residents that recycling would not take much of their valuable time after all. The small size of the green box. Christchurch Borough Council has provided every household with a single green box. Any other extra box only available from the Council offices can be bought at a fee of £2.00. In several instances during the data collection on kerbside collection rounds, extra recyclable material was noticeable in plastic bags placed next to filled up green boxes. Some residents voiced concern that when it rains the recyclates in the paper bags get soaked. In practice, this means such recyclates might be rejected and therefore sent to landfills. This would defeat the intended purpose of recycling. It would be important for the Council to establish the percentage requiring extra boxes and if need be provide the same. Providing wheelie bins is another option the Council can try considering putting into account their viability as well as their acceptance by residents especially those lacking storage space. This would encourage more participation in the scheme. Disabilities. The green box is heavy especially when filled with such recyclates as bottles and residents with disabilities are therefore limited as to what can be put in. This forces some elderly residents to place their green boxes at the doorsteps of their houses and not by the kerbside. These boxes were not emptied because the recycling crew is limited to collect the boxes by the kerbside. However some disabled residents have improvised and have placed their green boxes on mobile trolleys. That disability was cited as one of the reasons for non-participation is an issue worth consideration. The demographic trend in Christchurch is skewed towards the elderly [4]. This is as depicted in figure 2.

If the council were able to identify the number and location of residents with disabilities (as a result of age, sickness or otherwise) and who are willing to participate in the kerbside scheme and hence consider providing such residents with wheel bins, then the participation rates would be rise. Forgetfulness of the collection dates. This was the case with a number of residents who had not put their boxes out for collection. However in some instances, it was clearly noticeable that some residents have stuck their kerbside collection date card schedules on top of their box lids.

Recylates collection schedules are attached to the communal containers for flats. This a good improvise as it clearly reminds them of the collection dates. However green boxes distributed to flats had the collection dates on cards stuck to them. The collection dates for the recyclates can also be posted on the Council's website.

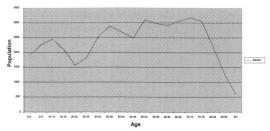

Figure 3:   Population of Christchurch-2001. Adopted from the Census of Population (2001, cited in [4]).

3.4 Environmental implications

That recycling saves resources and energy is not in doubt. About 685 tonnes of recyclates from kerbside were diverted from landfills within four months of the inception of the new kerbside recycling scheme (June-September 2003). This was a saving on raw materials, energy, landfill space and subsequent reduction in Global warming which can also impact on climate change. However there was a possibility of some errors during the weighing of recyclates from the collection lorries such as the weighing of the lorry crew members during the weighing of recyclates collection lorries in one of the MRF. This can affect the actual recyclates tonnage.

4   Conclusions and recommendations

4.1   Conclusions

Overcoming some of the problems experienced during the launch of the project was crucial to the success of the new kerbside scheme. The effect of this was seen in the rise in recyclates collection from the third week of the new kerbside launch. That the public showed support for the new kerbside scheme is in no doubt as the high recycling rates in Christchurch suggest that residents have accepted the new kerbside recycling scheme. They are also conscious of the correct recyclates as the contamination for recyclates transported to Sita recycling facility is only about 4%. That some additional boxes were being bought by residents from the council is even more encouraging. Christchurch Borough Council needs to focus its attention on reasons for non participation by some residents. Costs for the kerbside recycling programmes may be high and may not even be commensurate with financial returns at the moment but the long-term environmental benefits should be expected to be enormous. From the evaluation, a number of areas for improvements were identified. There was a significant shift from 52% kerbside recycling participation rate in the former kerbside scheme to about 70% participation rate in the new kerbside scheme. However despite this major improvement in the new kerbside participation rate, it still seems impossible for Christchurch Borough Council to achieve the government set targets. If the Council made improvements on critical areas, then there was the potential of achieving the 75% participation rate through the new kerbside recyclates scheme. This would further enhance the chances of Christchurch Borough Council nearing the government set targets of 22% by the end of 2003/04 financial year.

4.2 Recommendations

Among these are continuous consumer education, use of recycling symbols in packaging, inclusion of green waste in kerbside recycling, visits to best practices and further research on improvements to kerbside recycling.

References

[1]   Biffa, (1997). The environmental Balance sheet, An analysis of Britain's waste production and disposal account, with implications for industry and government. The Beacon press, United Kingdom.
[2]   Bulchholtz R. A. (1998). Principles of Environmental Management, The Greening of Business 2nd Ed. New Jersey Prentice-Hall Inc, USA.
[3]   Carra J.S. and Cossu R. (1990). International Perspectives on Municipal Solid Wastes and Sanitary Landfilling. Academic Press, Harcourt Brace Jovanovich, Publishers, London.
[4]   Christchurch Borough Council (2003). The state of Christchurch; A profile of Christchurch and its residents, Christchurch Borough Council.
[5]   Diaz L. F., Savage G. M., Eggerth. L. L. and Golueke C. G. (1993). Composting and Recycling. Lewis Publishers, Florida.
[6]   McHarry J. (1993). Reuse, Repair, Recycle; Gaia books Limited, London.
[7]   Ordinance Survey (2000). The map of United Kingdom, Ordinance Survey map. 'Reproduced with the permission of Ordinance Survey on behalf of The Controller of Her Majesty's Stationery Office © Crown Copyright' Licence No. ED 100018802 Bournemouth University, Talbot Campus, Poole, BH12 5BB.
[8]   Pentecost A. (1999). Analysing Environmental data. , Addison Wesley Longman Singapore (Pte) Ltd Singapore.
[9]   The Kindred Association (1994). A practical recycling handbook. Thomas Telford Services Ltd, London.
[10]   Woodard R; Harder M. K; Bench M. and Philip M. (2001). Evaluating The Performance of a fortnightly collection of household waste separated into compastibles, recyclates and reuse in the south of England. Resources, Conservation and Recycling, Volume 31, Issue 3, Pages 265-284. Available from http://www.sciencedirect.com/science. [Accessed on 16th September 2003]
[11]   Faithful M. (2003). MRW material recycling week. Councils create extra capacity through MRF reconstruction, Vol.181 issue 20. Headly Brothers, Kent.
[12]   Brunt P. (1997). Market research in travel and tourism. Butterworth-Heinman, Oxford.

Section 17
Costs and benefits
of waste management options

Welfare economic assessment of processing impregnated waste wood

V. Kjærbye[1], A. Larsen[1], B. Hasler[2], M. R. Schrøder[1] & J. Cramer[3]
[1]Akf, Institute of Local Government Studies, Denmark
[2]National Environmental Research Institute, Denmark
[3]Force Technology

Abstract

Waterproof waste wood contains a series of chemicals, especially chromium, copper and arsenic, which can be hazardous to human health and the natural environment in concentrated quantities. In this welfare economic analysis the economic and environmental consequences of four methods of processing impregnated waste wood are considered and compared: deposition, incineration, gasification and an extraction process. The quantity of impregnated waste wood is not a limiting factor for the individual method. The analysis includes both the socio-economic and the environmental consequences of applying these methods. The results of the analysis show that incineration and gasification are the cheapest wood processing methods in a welfare economic perspective. The reason is that both methods produce heat and thereby avoid the use of other more polluting fuels. Deposition is quite expensive, and it neither recycles nor uses the energy in the wood. If one only looks at the direct costs of the processes, and does not estimate and include the value of the environmental consequences, the differences between the methods are smaller. The basis for the article is a report by the authors for The Danish Environmental Protection Agency. The authors would like to express their gratitude to the agency for funding.
Keywords: impregnated waste wood, welfare economic assessment.

1 Introduction

This analysis is launched to provide answers as to how society should obtain a welfare economic optimal treatment of impregnated waste wood. The welfare economic analysis includes both the direct costs of processing the wood as well as the environmental consequences.

Impregnated waste wood is most problematic when it comes to the removal, as the waste from the impregnation of the wood is minor. Some of the heavy metals and organic solvents are washed out to the ground and to groundwater during the lifespan of the impregnated wood, but most of the heavy metals will remain in the wood until the wood is removed. By the removal the contents of chromium, copper and arsenic are especially problematic. These contents are not removed by ordinary combustion, but will remain in the residual products, in the slag and the ashes.

As a consequence of the Danish Waste Policy (Waste 21) all impregnated wood, except the creosote impregnated, should be assigned to a deposit facility at present, and this has been the situation since 2001. The creosote impregnated wood can be treated in special combustion plants.

The wood impregnated with creosote does not result in these residual products and waste after combustion, however [3], and therefore the treatment of creosote wood is not subject to the present welfare economic assessment.

The current practice is embedded in the so-called 'hierarchy of waste'. This hierarchy is built so that recycling is weighted higher than incineration, which is weighted higher than deposition. It is an ongoing discussion as to what extent this hierarchy should be used, and the discussion has been intensified in Denmark, among other reasons because the pollution with chemicals and other compounds. The effects of chromium, copper and arsenic from impregnated wood in the environment depend on the concentration to which humans and nature is exposed, but they stretch from locally irritating to poisonous and cancer-causing. There are some positive environmental effects from processing impregnated waste wood, and that is recycling or the displacing of other more polluting fuels.

This paper examines the economic as well as the environmental consequences of four methods of processing. These four methods are assessed: deposition, incineration, gasification (by processes at the plant 'Kommunekemi') and an extraction process (RGS90 Watech). When the wood is deposited the energy is not used nor is it recycled. When incinerating, the energy is utilised, but there is no recycling. The methods of both Kommunekemi and RGS90 Watech utilise the energy and have an element of recycling; hence they are both ranked higher in the hierarchy of waste.

2 Data and scenarios

2.1 Data

The input data to the analysis comprise data on potential amounts of impregnated wood, the composition of the wood and budget data from the processing plants. The content of the budget data is partly confidential and therefore not described further here.

Impregnated wood is mainly in poles, sleepers and waste wood from public waste collections. The total amount of impregnated wood is estimated to 50,000 tons in 2004, and is projected to be about doubled in 2010 [1]. In the waste strategy launched by the Danish Government in 2003 [2] it is assumed that approximately 4 million tons impregnated wood is accumulated since the 1960s and that this waste should be processed in 40 years' time. According to this strategy the energy and raw materials should be utilised.

Impregnated wood can be divided into two categories; the creosote treated and the non-creosote treated. The creosote treated wood is burned in incineration plants in a process that does not impact the environment. The non-creosote treated wood, however, burdens the environment and this must be dealt with when processing it. Therefore, this analysis focuses solely on the non-creosote treated wood when comparing the methods. This wood is mostly found in poles and waste wood from public collections hence only wood from these sources is included.

Data from 5 tons of impregnated wood collected by Kommunekemi are used to estimate the composition of the wood [12].

2.2 Scenarios

The baseline scenario represents the current treatment of the waste wood, where creosote treated wood is burned and the salt impregnated wood is deposited.

Two alternative scenarios are assessed; a maximum and a minimum scenario. They are divided so that the maximum scenario provides the best economy for the plants. For deposition this means a higher specific weight (kg/m^3) in the maximum scenario. The difference for the other three processing methods is that the calorific value is higher in the maximum scenario as compared to the minimum. For the maximum and minimum scenarios the content of Cu, Cr and As in 'clean' impregnated wood is assumed to be close to constant, as the leaching of the compounds mainly happens in the first years of the lifetime of the wood.

The costs of the alternative processing methods are assessed as compared to the baseline scenario, i.e. the current treatment. In the baseline the treatment is paid by the taxpayers at municipal level.

3 Method

The four methods of processing impregnated waste wood: deposition, incineration, gasification and extraction, are compared with respect to the costs and the benefits of each, using a welfare economic assessment method.

3.1 The welfare economic assessment

The basic idea behind a welfare economic analysis of benefits and costs is to determine the total effect of the project on the welfare of society as a whole. In the present analysis the 'project' is the processing of impregnated waste wood. The welfare economic assessment therefore implies estimation and prediction of the changes in consumption possibilities of the members of the society. Seen from a welfare economic viewpoint, the state or other governmental levels should implement policies where the social benefits derived from the policy exceed the economic or social costs associated with the implementation.

The assessment of the welfare economic costs and benefits takes its point of departure in a description of the consequences. The measurement of the welfare economic costs and benefits is, in other words, based on expectations of the consequences of the alternatives for processing impregnated wood. For the measurement of the relative marginal welfare economic consequences, so-called calculation prices are used. The so-called accounting price method is used for the estimations of the welfare-economic costs, and the principles behind this method are presented in more detail in Møller et al. [4] and Birr-Pedersen [5].

The welfare economic evaluation determines benefits and costs from the point of view of the economy of the society as a whole, and includes two sub-analyses:

1. The analysis of the *financial consequences,* upon which the distribution effects between the different actors can be measured, e.g. the effects for the processing plants, the municipalities etc. The prices used are the market prices either paid on the market for inputs in the form of producer or consumer goods, or obtained on the market from selling outputs, including all non-refundable taxes and subsidies. In the case of consumer goods all taxes are non-refundable and should therefore be included in the price [5].

2. The *welfare economic analysis*, where economic costs and benefits to the economy as a whole are measured with market prices adjusted to reflect the true economic costs and benefits to society. The external effects, i.e. the non-marketed effects of the environmental changes are as far as possible included in the welfare economic analysis.

The financial analysis that is performed upon market prices is the basis for the welfare-economic analysis. The inputs to this part of the analysis are delivered by the processing plants. In the economic analysis in this study, intersectoral transfers between sector and sub-economies are not counted, e.g. between the state and the processing plants.

In order to determine the correct price for the inputs used in the production process in cases where the project's output is producer goods, the prices are adjusted to reflect the consumer's willingness to pay:

- For *consumer goods* produced the price is simply the prevailing market price, gross of taxes and subsidies. No adjustments need to be made on these prices.

- For *producer goods* the market prices reported by the processing plants need to be adjusted, because consumers are also willing to pay sales tax and maybe other product specific taxes levied on the good during the production process. Møller et al. [4] suggest increasing producer prices of domestically traded goods and services (net of refundable taxes) with a so-called 'net-tax-factor' of 1.17, and for internationally traded goods a factor of 1.25 is suggested (see [5]).

- *Labour input:* Similar to the use of other resources in the production process the use of labour should be reflected by the market price (net tax factor 1.17).
- *Capital goods:* The costs associated with the inputs of capital goods (i.e. the equipment, the machines and the buildings) normally enter a financial cost-benefit analysis in the form of the assumed annual loss of value associated with the usage of the capital goods in the production process. In the welfare economic analysis the total investment amount, plus the net-tax factor, is divided equally over the assumed period of operation with a capital recovery factor. Investing in capital goods in one project gives rise to opportunity costs – foregone returns or foregone consumption. In financial calculations these foregone returns are reflected in the discount rate chosen to derive the net present value of the annual cash flows. To account for the opportunity costs of investments the annual investment amount is increased with a 'return on investment factor for capital'. The term most often used for this factor is 'shadow price on capital'.
- This return on investment factor is equal to the present value of one DKK invested in the second-best project alternative and is calculated using an (economic) investment rate and the social time preference rate for discounting. Møller [6] recommended an economic investment rate of 6% which is in accordance with the recommendations for socioeconomic cost-benefit analysis of the Danish Ministry of Finance [7].

After calculating the annual investment amount by using a capital recovery factor and taking into consideration the opportunity costs from foregone investments by multiplying the annual amount with the return on investment factor, the resulting amount is then increased with the net-tax factor for either domestically or internationally traded goods.

Present and expected market prices are used for the measurement of the changed use of marketed goods, but the largest problem is the valuation of the non-marketed goods, i.e. the environmental effects.

3.2 Quantification of the environmental effects

Last but not least the environmental consequences are assessed quantitatively in physical terms as well as in monetary terms, whenever possible. The environmental consequences comprise air emissions, but also emissions to soil and water, comprising emissions of SO_2, NO_x, particles, CO_2, but also heavy metals, dioxin and chemicals. The emissions to air mostly stem from the combustion process. The emissions to soil and water mainly stem from the deposit of waste and slag, which is a residual product from the combustion. Ground- and surface water can also be damaged because of emissions of percolation from the deposits. The magnitude of transport influences the air emissions.

These emissions cause damage. On the positive side, there are benefits when heat and energy are produced as residual products from the combustion process, and some of the metals can be reutilised as well. Of the greatest difference between the financial and the welfare economic assessment is that the welfare economic assessment comprises the environmental effects stemming from each of the processes, but as mentioned it is necessary to quantify the environmental effects as far as possible in monetary terms, to include them in the welfare economic assessment.

Because these environmental effects are not traded on a market, these goods have no price. Exceptions are heat and metals, however. Accounting prices for environmental goods can be used as prices for the non-marketed goods, but should be interpreted with caution because of the uncertainty. These accounting prices are so far as possible built on existing knowledge about the revealed or stated preferences in the population. Many methods exist to reveal the preferences through willingness to pay-assessments, and both revealed and stated methods are commonly used to elicit the social value of non-marketed public goods. These elicitations make the willingness to pay for these goods comparable to the willingness to pay for other, traded goods. Alternatively, accounting prices for the environmental goods can be elicited by estimating damage costs, or by using cost-based prices for a 'statistical life', where the probability of death and illness as a consequence of the emissions is used in connection to the price for the statistical life (cf. [8, 9].

Table 1:   Accounting prices for environmental effects and emissions.

	Accounting price		Reference
CO_2	DKK/kg	0.02	Andersen and Strange 2003
Methane	DKK/kg	X	No account exist
Particles (PM2,5)**	DKK/kg	1308	Andersen et al. 2004/p. 8-9
Particles (PM10)*	DKK/kg	783	Andersen et al. 2004/p. 8-10
VOC	DKK/kg		No account
NO_x	DKK/kg	83	Andersen et al. 2004/p. 8-9
SO_2	DKK/kg	583	Andersen et al. 2004/p. 8-9
CO	DKK/kg	0	Andersen, pers. comm..
HCl	DKK/kg	X	No account exist
Cd	DKK/kg	X	No account exist
Lead and other heavy metals	DKK/kg	13142	Andersen, pers. comm. Spadaro and Rabl, 2003**
Arsenic	DKK/kg	5358	Andersen, pers. comm. Spadaro and Rabl, 2003**
Dioxin	DKK/kg	9,000,00 0	Andersen et al. 2004

* Calculated from PM2.5 by division by 1.67 (Andersen, pers. comm.)
** Calculated as an average for cities between 100,000 and 500,000 inhabitants.

Based on accounting prices from the EU-funded BeTa.system (benefit tables) [10] and the ExternE [11] Andersen et al. [9] have elicited accounting prices for air emissions, and the prices are used in the welfare economic assessments in the present study. The prices are apparent from table 1. In connection with waste disposal Andersen and Strange [8] pinpoint that the most important damage effect is that from lost amenity value. No Danish studies are performed until now to qualify if this holds under Danish conditions, and therefore the potential amenity loss is not included in the present analysis.

The accounting prices used are apparent from table 1.

The emissions of methane, HCl, VOC and Cd have not been possible to estimate because of lack of data. One of the assumptions behind the estimations of the accounting prices is that the effects are accounted for citizens in 'average-sized' cities between 100,000 and 500,000 inhabitants. The accounting prices change for SO_2 and particles if the basic assumptions for larger cities above 500,000 inhabitants are used, but the other prices are not changed, and therefore, the prices estimated for average cities between 100,000 and 500,000 inhabitants are used for the estimations.

In addition to these direct environmental consequences there is also production of heat and energy from the combustion process, and this is positive as conventional energy sources are substituted. Hereby the emissions and damage costs from the production and use of these energy sources are avoided. These substitution effects are included in the assessments. The processing of copper and chromium for market purposes likewise displaces the emissions and waste. Energy, heat, copper and chromium are priced in welfare economic prices, and the value of the avoided environmental consequences is included in the assessments.

Finally, only domestic consequences are considered in this analysis, in other words potential consequences outside Denmark are not considered.

Furthermore, in this welfare economic analysis we have assumed that the project implies a choice of technique, where there will be no sunk costs. This assumption is chosen to avoid that choices are dominated by investments and decisions already made, and hereby avoid decisions that will be inefficient in the long run. The time horizon for the analysis follows the lifetime of the plants and technologies; from 10 to 30 years. In this analysis the scenario covers the period from 2000 to 2030.

4 Results and conclusions

The analysis indicates that the collection potential is not a limiting factor for the plants and the processing of the wood, but there is uncertainty about the quantity of the wood collected by the public schemes. Using samples of the quantity collected by these schemes suggests a minimum and maximum scenario.

In the financial economic analysis illustrated in Figure 1, the difference between methods of processing can be seen. These figures are estimated without the environmental effects. Minimum and maximum scenarios are indicated by 'min' and 'maks'.

Making a welfare economic analysis, including the environmental consequences the differences between the methods are clearer, cf. figure 2. The figure shows the welfare economic costs of the four methods, where the costs are calculated as the costs in DKK of processing one ton of impregnated waste wood. The figure also contains some sensitivity analyses in order to provide more robust results.

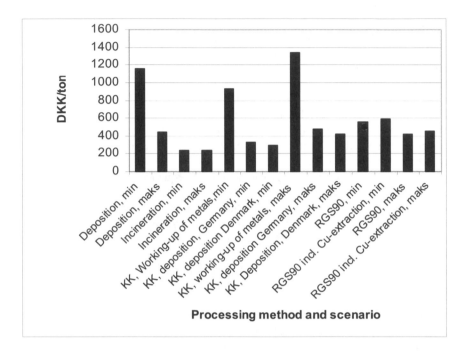

Figure 1:   Economic costs (DKK/ton processed impregnated waste wood) for both min. and max. scenario (2004/2005-prices) [13].

Of the four methods it can be seen that incineration and Kommunekemi (with deposition of metals) are the cheapest. The reason is partly that the heat of the processes is used, and thereby other more polluting fuels are displaced. In the figure it is also worth noticing that deposition is quite expensive if the use of heat does not displace fuels. Ordinary deposition neither utilises the wood by recycling nor the energy in the wood. If one only looks at the direct costs of the processes, and does not try to estimate the value of the environmental conse-quences, the differences in methods are much smaller, as can be seen above, re figure 1.

The environmental consequences are hard to estimate the value of. The potential long-term effects of leaching from deposition and the effects of heavy metals in slag and discharge water are not included in the report, but CO_2, NO_x, SO_2, arsenic, dioxins, lead and other heavy metals are. To make the results as robust as possible several sensitivity analyses have been made, e.g. by changing the prices on the environmental effects, changing the rate of interest and assuming different transport costs. The results of these analyses are unambiguous in their ranking of the methods, but they obviously change to which degree the methods differ.

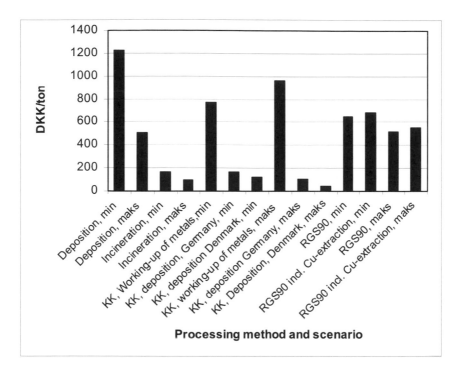

Figure 2:   Welfare economic costs (DKK/ton processed impregnated waste wood) for both min. and max. scenario (2004/2005-prices) [13].

The difference between the results in figure 1 and 2 mainly comes from the fact that the environmental consequences are only taken into account in the welfare analysis in figure 2. The gasification and the incineration processes improve their position by displacing other more polluting fuels. If it is assumed that the wood stemming from RGS90 Watech's process is used to produce electricity or heat it is not unrealistic that RGS90 Watech's method would be competitive to incineration and gasification.

Concluding on the results does not give occasion for believing that the hierarchy of waste should be ignored when choosing the optimal processing of impregnated waste wood. Deposition, however, seams to be expensive given the method neither recycles nor utilises the energy.

References

[1]   The Danish EPA, *Waste 21*, Ministry of the Environment, 1999.
[2]   The Danish Government, *Waste Strategy 2005-08* (Affaldsstrategi 2005-08), Copenhagen, (2003).
[3]   Centre for knowledge about waste (in Danish), Imprægneret træ. Deponi. Forbrænding. http://www.affaldsinfo.dk/ 2004.

[4]   Møller, F., Andersen, S.P., Grau, P., Hussom, H., Madsen, T., Nielsen, J. & Strandmark, L., (in Danish), *Samfundsøkonomisk vurdering af miljø-projekter*, Danmarks Miljøundersøgelser, Miljøstyrelsen og Skov- og Naturstyrelsen, 2000.
[5]   Birr-Pedersen, K., Welfare Economic Cost-Benefit Analysis, Working Paper from the project *"ARLAS"*, National Environmental Research Institute, 2001.
[6]   Møller, F. (in Danish), Forrentningsfaktoren og diskontering (supplement to) *Samfundsøkonomisk vurdering af projekter*, Danmarks Miljøunder-søgelser, 2001.
[7]   Ministry of Finance (in Danish), *Vejledning i udarbejdelse af samfunds-økonomiske konsekvensvurderinger*, Copenhagen, 1999.
[8]   Andersen, M.S. & Strange, N. (in Danish with an English summary), *Miljøøkonomiske beregningspriser. Forprojekt. Danmarks Miljøunder-søgelser*. Technical report, National Environmental Research Institute, no. 459, 2003.
[9]   Andersen, M.S., Frohn, L.M., Jensen, S.S., Nielsen, J.S., Sørensen, P.B., Hertel, O., Brandt, J. & Christensen, J. (in Danish with an English sum-mary), *Sundhedseffekter af luftforurening – beregningspriser*, Technical report from NERI, no. 507 2004. www.dmu.dk
[10]   Spadaro J.V. & Rabl, A., *Pathway Analysis for Population – Total Health Impacts of Toxic Metal Emissions*, 2003.
[11]   European Commission (EC), *ExternE, Externalities of Energy* (Vol. 1), Summary, 1995.
[12]   Kristensen, O., *Note on the composition of impregnated waste wood* (in Danish), 2003.
[13]   Hansen, V., Cramer, J., Hasler, B., Larsen, A. & Bruun Poulsen, P. (in Danish with an English summary) *Miljø- og samfundsøkonomisk analyse af indsamling og behandling af imprægneret affaldstræ*, Miljøstyrelsen, (forthcoming).

Efficiency of solid waste collection in Spain

I. M. García Sánchez
Department of Administration and Business,
University of Salamanca, Spain

Abstract

One of the major environmental problems for society is the great quantity of solid waste generated. The management of urban solid waste is one of the most important services and for this reason town councils have to maintain the cities in the proper hygienic and aesthetic conditions for their inhabitants as well as for tourists or visitors.

The loss of credibility of the Spanish municipal public sector as a manager of this service is accompanied by manifestations demanding and forcing the sector to act by applying the principle of efficiency. These new demands require the development and application of control techniques that provide relevant information for decision-making.

In the present work we examine the waste collection scheme in Spain using the Data Envelopment Analysis (DEA) methodology in terms of calculation of the efficiency, showing the vast information that may be provide by this technique.

The results of this study showed that the average technical efficiency of waste collection is situated at 56.94%. Out of the 34 towns examined, the 73.53% were found inefficient. The analysis of slacks reveals a resource excess of about 9% above the optimal collection activity.
Keywords: data envelopment analysis, solid waste collection, public sector, municipalities.

1 Introduction

One of the major environmental problems for society is the great quantity of solid waste generated. The management of urban solid waste is one of the most important services and for this reason town councils have to maintain the cities

WIT Transactions on Ecology and the Environment, Vol 92, © 2006 WIT Press
www.witpress.com, ISSN 1743-3541 (on-line)
doi:10.2495/WM060601

in the proper hygienic and aesthetic conditions for their inhabitants as well as for tourists or visitors.

The loss of credibility of the Spanish municipal public sector as a manager of this service is accompanied by manifestations demanding and forcing the sector to act by applying the principle of efficiency. These new demands require the development and application of control techniques that provide relevant information for decision-making about the relation between the quantity of resources used in the production and development of the appropriate quantity and quality of goods or services in a suitable time. Owing to the scarce significance of the value of the public output, its measurement is defined as the estimation of technical efficiency by inputs and outputs expressed in physical terms.

The present work focuses on the study of the efficiency of urban solid waste collection. Hence, the technical efficiency of the service is calculated with the aim of detecting potential savings in the use of physical resources, which lead to an increase in productivity.

According with the evidence of the large number of studies carried out in various countries – e.g. Portugal [3], the United Kingdom [4], Australia [7], France [2], Switzerland [1], Finland [5] and America [6]- Data Envelopment Analysis (DEA) was utilized to estimate the efficiency.

2 Data Envelopment Analysis

DEA yields a *piecewise linear production surface* that, in economic terms represents the best practice production frontier. By projecting each unit onto the frontier, it is possible to determine the level of inefficiency by comparison to a single reference unit or a convex combination of other reference units. The projection refers to a hypothetical DMU which is a convex combination of one or more efficient DMUs and not an actual DMU.

The basic DEA model, named CCR is expressed as follows (1):
Given a set of J DMUs, the model determines for each DMU_0 the optimal set of input weights and output weights that maximizes its efficiency score δ_0. A score less than one means that a linear combination of other units from the sample could produce the vector of outputs using a smaller vector of inputs. Mathematically, a DMU is termed *efficient* if its efficiency rating δ_0 obtained from the DEA model is equal to one. Otherwise, the DMU is considered inefficient.

3 Aims and selection of variables

3.1 Sample

In order to estimate the efficiency, the population selected comprised the 113 towns of over 50,000 inhabitants that exist in Spain in 1999. Specifically, we obtained information from 34 towns that make up 30.09% of the population. The technique used for obtaining information was the questionnaire, which, forwarded to the population, guarantees randomness in the data obtained.

3.2 Inputs and outputs

In this section the indicators that represent the functions of this service will be selected, grouped in indicators of input and output.

3.2.1 Inputs
The production process of municipal solid waste collection is highly contingent upon the supply of capital and human resources. The latter is represented, in general, by *total staff* and our analysis will include the indicator expressed in terms of total workers. In relation to the capital goods, the basic element is the vehicles measured in physical units.

3.2.2 Outputs
The variables tonnage (*Output1*) and collection points (*Output2*) correspond to the most used indicators for identifying the final product of the activity of waste collection. They represent the volume of solid waste generated and the number of places where it is collected, respectively.

4 Empirical analysis

4.1 Efficiency index

In this stage, we estimate the technical inefficiencies of the Spanish towns for solid waste collection and street cleaning. The results are given in table 1.

According to the results, of the 34 units examined 25 are inefficient, that is, approximately 73.5% of the total of local authorities evaluated. Only nine towns were considered efficient: Guadalajara, Barcelona, Castellón, Hospitalet, Bilbao, Mieres, Ciudad Real, Tenerife and Madrid.

The technical inefficiency measured is situated at 56.94%, there being a significant slack of potential improvement, which is expressed in a possible average reduction of the inputs of around 9%. The greatest slack, figure 1, is basically correlated with the variable tonnage, collection points being carried out more optimally.

4.2 Best-practice in solid waste collection and inefficient units

DEA displays performance information relating to inefficient towns and shows the difference between their performance and the "best practice" (100% efficient) municipalities to which they have been compared.

Potential improvements indicate by how much and in what areas an inefficient town needs to improve in order to be efficient. For example, Figure 2 shows what percentage Salamanca needs to either decrease its inputs or increase its outputs in order to become 100% efficient: it needs to reduce its number of staff by nearly 40% or its vehicles around 7%, while maintaining the same level of outputs.

Table 1: Technical Efficiency Index.

MUNICIPALITIES	Technical Efficiency Index	MUNICIPALITIES	Technical Efficiency Index
GUADALAJARA	100	SEVILLA	38.60
BARCELONA	100	TORREJON	36.62
CASTELLON	100	MANRESA	35.29
HOSPITALET	100	MALAGA	35.12
BILBAO	100	OVIEDO	32.21
CIUDAD REAL	100	SANTA COLOMA	31.30
MIERES	100	FUENLABRADA	29.36
TENERIFE	100	VITORIA	29.34
MADRID	100	EL FERROL	27.60
SALAMANCA	92.14	SAGUNTO	27.17
LLEIDA	84.33	LORCA	23.99
GRANADA	80.95	MELILLA	22.22
GIJÓN	70.49	CORDOBA	21.37
PAMPLONA	67.74	ALCALA DE GUADAI	19.73
A CORUÑA	48.12	ALGECIRAS	12.34
PALMA DE MALLOR	45.16		
MARBELLA	42.81	**AVERAGE EFFICIENCY**	**56.94**
VILANOVA	42.76	**EFFICIENCT UNITS**	**9 (26.47%)**
VALENCIA	39.04	**INEFFICIENT UNITS**	**25 (73.53%)**

If the assessment of a town as inefficient is felt to be justified then the information provided can be used as a basis for setting targets for the municipality. As a first step insetting targets, the inefficient unit should be compared with the towns in its reference set.

The reference set is the set of efficient municipalities to which the unit has been most directly compared when calculating its efficiency rating, table 2. Salamanca has efficient towns Barcelona, Ciudad Real and Tenerife.

However the reference set towns do not all contribute equally to the target-values for an inefficient municipality. Some reference set town are more important than others. Figure 3 shows all the towns in Salamanca's reference set and how much in percentage terms they have each contributed to forming the target values for each of Salamanca's inputs and outputs.

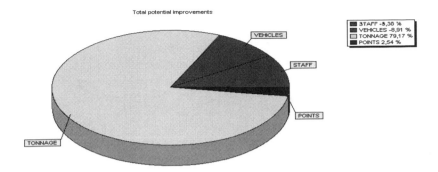

Figure 1:   Potential average savings and increases.

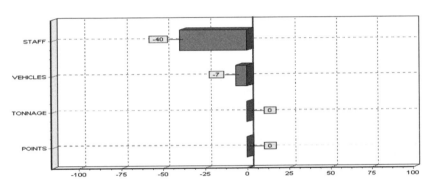

Figure 2:   Salamanca: potential improvements.

Table 2:   Group of references.

INEFFICIENT MUNICIPALITIES	EFFICIENT MUNICIPALITIES								
	Guadalajara	Barcelona	Hospitalet	Castellon	Bilbao	Ciudad Real	Mieres	Tenerife	Madrid
SALAMANCA		X				X		X	
LLEIDA	X		X	X	X				
GRANADA	X	X				X		X	
GIJÓN		X				X			
PAMPLONA		X	X	X	X				
A CORUÑA		X		X		X		X	
PALMA DE MALLOR	X	X	X	X	X				
MARBELLA		X	X						
VILANOVA				X			X	X	
VALENCIA								X	X
SEVILLA	X	X				X		X	
TORREJON	X			X		X	X		
MANRESA	X				X		X		
MALAGA							X		
OVIEDO	X	X	X	X	X				
SANTA COLOMA	X	X	X						
FUENLABRADA	X	X	X						
VITORIA	X		X		X				
EL FERROL	X			X		X	X		
SAGUNTO	X				X		X		
LORCA	X				X		X		
MELILLA	X								
CORDOBA		X	X	X	X				
ALCALA DE GUADAI	X	X	X	X					
ALGECIRAS		X							
FREQUENCY	15	14	10	10	9	7	7	6	1
Percentage	60%	56%	40%	40%	36%	28%	28%	24%	4%

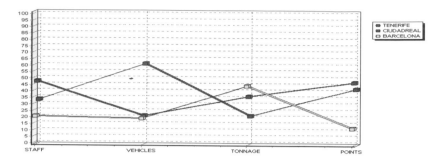

Figure 3:   Salamanca: reference contributions.

5   Conclusions

A methodology of evaluation of the efficiency of waste collection has been established in this work. It provides not only an efficiency score for each town but also indicates by how much and what areas an inefficient municipality needs to improve in order to be inefficient. As regards the results reveals that:
- Of the 34 towns examined, 25 are inefficient, 73.53%.
- The average technical efficiency is relatively short, 56.94.
- The analysis of slacks reveals surplus resources of around 9%.
- The collection activity being carried out more optimally.

$$\text{Max} \quad \delta_0 = \frac{\sum_i u_i \ y_{io}}{\sum_j v_j \ x_{jo}} \tag{1}$$

Subject to:

$$\frac{\sum_i u_i \ y_{ik}}{\sum_j v_j \ x_{jk}} \leq 1 \qquad \text{for all DMUs K} = 1, ..., n$$

$$u_i, v_j \geq 0$$

where,

δ_0 = the efficiency score of the DMU 0 under analysis; n = number of DMUs under analysis;
i = number of outputs;   j = number of inputs;
Y_k = ($y_{1k}, y_{2k},...,y_{ik},...,y_{lk}$) is the vector of outputs for DMU k with y_{ik} being the value of output i for DMU k;
X_k = ($x_{1k}, x_{2k},...,x_{jk},...,y_{lk}$) is the vector of inputs for DMU k with x_{ik} being the value of input j for DMU k;
μ and v the vector on multipliers respectively set on Y_k and X_k where μ_i, v_j = the respective weights for output i and for input j;

References

[1] Burgat, P. and Jeanrenaud, C., *Measure de l'efficacité productive*, Working Paper IRER, Universitat de Neuchatel, 1990.

[2] Distexhe, V., .L'Efficacité Productive des Services D'Enlévement des Immondices en Wallonie. *Cahiers Economiques de Bruxelles* 137, p.p. 119-138, 1993.

[3] Gaiola, A.J.F., Efficiency Evaluation in the Urban Solid Waste Systems of Portugal Using Data Envelopment Analysis, *Symposium at a Glance*, Moscow, 2002.

[4] Haas, D.A., Murphy, F.H., and Lancioni, R.A., Managing Reverse Logistics Channels with Data Envelopment Analysis. *Transportation Journal* 42(3), p.p. 105-113, 2003.

[5] Jenkins, L. and Anderson, M., A Multivariate Statistical Approach to Reducing the Number Variables in Data Envelopment Analysis. *European Journal of Operational Research* 147, p.p. 51-61, 2003.

[6] Segal, G.F., Moore, A.T. and Nolan, J., *California Competitive Cities: A Report Card on Efficiency in Service Delivery in California's Largest Cities*, Reason Public Policy Institute, California, 2002.

[7] Whortington, A.C. and Dollery, B.E., Measuring Efficiency in Local Government: An Analysis of New South Wales Municipalities Domestic Waste Management Function. *Policy Studies Journal* 29(2), p.p. 4-24, 2001.

Automated truck design for glass-waste collection with optimised compactor for glass recycling

M. Zordan & S. Morassut
Regola Team Srl, Porcia (PN), Italy

Abstract

Waste recycling begins from the collection phase. The ordinary activity of emptying bins in trucks for subsequent stocking, material separation and recycling treatment has great impact over the whole process efficiency. With optimised garbage collection it is possible not only to achieve a better recycling rate in terms of quantity and material quality, but also to deeply cut the costs of material selection and treatment in the recycling process. This especially applies to collected glass where the recycling steps require multiple treatments for sorting, processing and glass re-use either with "new" material, to obtain glass-based items, or mixed to other substances, even to produce completely different composites. This paper highlights some "design guidelines" of a truck aimed at urban glass-waste collection. Two features give unique capabilities to the vehicle: the automation of both loading and compacting processes and the compactor performance with regard to the size of the treated material, cut down in pieces suitable for optimal recycling efficiency.

Keywords: compactor, glass, recycling, truck, vehicle, fieldbus.

1 Introduction

1.1 Waste collection

Garbage collection involves many interactions among technical and organisational aspects. Logistic problems concentrate in the management of a fleet of vehicles (trucks, loaders, etc.) and in the coordination of the related personnel. The decision where to set collection points (waste bins, containers, etc.) is a trade-off between user commodity, time spent for collection and

WIT Transactions on Ecology and the Environment, Vol 92, © 2006 WIT Press
www.witpress.com, ISSN 1743-3541 (on-line)
doi:10.2495/WM060611

foreseen waste quantity. Typical parameters to be evaluated in such decisions (such as habitants/area, bins/road-length, habitants/blocks) must be weighted by correction factors taking into account the distribution of residential and industrial zones in the same neighbourhood. Even if urban garbage collection is mainly related to domestic waste, special collections must also be considered for demolition waste and heavy/cumbersome materials especially concentrated around commercial premises.

Technical matters relate to waste separation, stocking and disposal (dump, landfill). Material treatment is then an issue when recycling processes are to be carried on from collected waste. Cities usually differentiate various materials: aluminium, glass, plastic, paper, organic material. Many of them undergo recycling treatment while the remaining is disposed in landfill sites.

1.2 Glass-waste collection

Glass-waste, together with paper-waste is the first material that has been considered for recycling and the related technology is nowadays mature. Glass-waste collection is usually carried out using different types of containers: "bell bins" (containers with openings on their sides to insert the disposed glass, which are emptied through their bottom, once they are lifted over a collecting truck) and "wheelie bins" (relatively small plastic containers with limited capacity, 120-240 litres). "Bell bins" usually serve wide areas (blocks, streets, etc., corresponding to about 600 m in size) while "wheelie bins" are suited for domestic needs (they could be placed every ten/twenty residential units, or about 50-100 m from each other). It is easier to collect from wheelie bins because:

- reduced waste quantity requires small vehicles to be used for collection;
- small bin dimension facilitates the loading phase (which is comparable to the collection of "general" waste);
- safety issues are reduced due to small quantity/weight of material being raised and emptied on collecting vehicles;
- the loading phase is much more suitable to be performed automatically;
- less time required per loading allows a smoother collection phase to be performed in busy road traffic conditions.

Therefore the trend in glass-waste collection is towards allocation of many collection spots ("wheelie bins"), especially in the city centre, instead of concentrating the collection in "bell bins" (usually kept in the suburbs).

The truck whose design guidelines are presented in this work is intended for glass-waste collection using "wheelie bins". The vehicle must perform the collection together with the compacting action aimed at reducing the glass in fragments suitable to be recycled afterwards (about 5 cm in size). If glass fragments exceed that dimension, an additional step in the recycling process must be performed, aimed at crumbling the material to make it suitable for the following treatments. On the other hand if glass is broken down into too small pieces, it becomes difficult to separate from other materials and to manage in the waste disposal area. The compacting action avoids to perform the crumbling process once the glass is stocked and it also makes easier the separation of different materials (plastic bags, aluminium tops, etc.) sometimes collected with the glass: the compactor doesn't cut in small fragments non-glass material.

2 Compactor mechanical design

2.1 The container

The truck hosts a container to store the collected glass and to hold the material during the compacting process: this is necessary in order to fragment the glass in parts whose dimensions are suited for later treatment.

The container is made of a single iron sheet metal, and it has the inside covered with anti resound paint, in order to prevent both loud acoustic noise during loading operations and exposure of the metal structure to corrosive climatic agents. The supporting structure is an iron frame at the compactor edges: no other strengthening elements have been considered, in order to obtain a light but strong metal structure. The container section is rounded towards the outside of the structure: this evenly distributes the containing forces and helps the fragmented glass to compensate the pressing forces by "going around" the compactor mechanical parts (details in section 2.3). The truck weight is actually a key-element of the design as the lighter is the structure, the larger becomes the container capability.

Various unloading alternatives have been studied. The most effective one (considered in the truck) requires the back of the container to open while the container itself rises on the front part: the collected glass therefore drops down by gravity and accumulates behind the truck (which can drive straight on once the unloading has been completed). This is achieved with the back of the container hinged at its upper side while the container rises thanks to an oleodynamic cylinder lifting its front (figure 1). As the back of the container acts also as contrasting structure to the compactor pressure (details in section 2.2), great care has been spent on the fastening blocks. A two-movement lock has been designed: with a single movement, the back of the container at first rises up unlocking the container and then rotates to open position (figure 2).

2.2 "Slider" and "Shovel"

The compacting force presses the collected glass against the rear part of the container. This is achieved using two moving mechanical parts: a "slider" and a "shovel". Their joint movements also ensure the glass remains confined for enhanced vehicle stability.
1) The slider is a horizontal support that moves from the front to the back of the container. It holds the shovel and it is driven by oleodinamic cylinders.
2) The shovel (figure 3) rotates around hinges fixed to the slider and it presses the collected glass. It is driven by oleodinamic cylinders.

To obtain even grinding of the collected glass, the mechanical parts have been designed appropriately: an optimised choice of their joint movements, with regulated pressure against the material, yields the desired result.

WIT Transactions on Ecology and the Environment, Vol 92, © 2006 WIT Press
www.witpress.com, ISSN 1743-3541 (on-line)

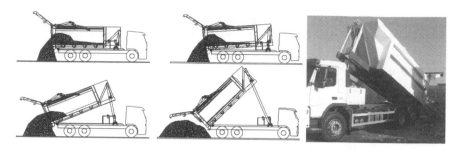

Figure 1:   Truck unloading.

Figure 2:   Container locking elements.

Figure 3:   Shovel seen from the back of the container.

2.3 Compactor movements

Extensive simulation analysis has been carried on, considering the movements of the slider and the shovel as separate actions. The fist one (slider moving with the shovel in vertical position) presses the glass against the back of the container while the second one (shovel rotation) presses the glass against the bottom of the container. To ensure uniform crumbling of the whole collected glass, a sliding movement, with uniform pressure of the shovel against the glass, has to be performed (figure 4). The slider moves from the front of the container holding the shovel in vertical position. This collects the glass which has just been loaded (if any) to the rear part of the container. As soon as the shovel starts compressing the glass with a force exceeding a chosen value (sensed on the oleodynamic pipe), its driving cylinders are released making the shovel stop pressing the glass. As the slider keeps moving, the shovel will soon compress the glass with greater pressure than allowed, turning on again the regulating effect. This results in a continuous sliding movement with uniform pressure against the glass. To smooth the pressure of the compactor and to make it more uniform on the whole collected glass, the shovel has been designed with reduced dimensions with respect to the inside of the container (figure 3). This makes the glass move through escaping paths around the lateral edges of the shovel if uneven pressure is applied. The regulation effect of this solution results in very effective, uniform glass crumbling, better than acting only with precise regulation of the oleodynamic forces on both the slider and the shovel. The tuning of the oil-pressure power circuitry is still necessary for a fine glass treatment. Practical tests on a prototype confirmed the quality of the compacting action and the uniform fragmentation of the collected glass. Force adjustments can also be made by setting the pressure value at which the shovel force is released.

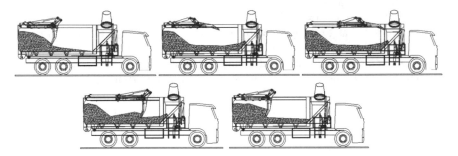

Figure 4:   Compactor movements.

Figure 5:   Loading operation.

2.4 The loading arm

A side loader has been considered to empty the wheelie bins in the container. Its position allows fast and prompt collection of the bins on the pavements. The loading arm is bent in the middle so that it remains in folded position when it is retreated on the vehicle. The outer end is provided with two holders that take the bin at its sides (after the arm is extended with horizontal movement); the arm then lifts it up over the container making the glass fall down on the inside (figure 5). This movement also opens the bin cover and set it to closed position once the bin is put back in its original place. For intervention on standard pavements the loading arm is required to extend to about 3 m. To minimise the space occupied by the loading arm in closed potion, one of its grabbing ends is folded, and it is automatically extended during operation.

3 Vehicle operating specifications

The main characteristics of the glass collection vehicle are listed hereafter and they result from a trade-off between easiness of use, compliance of actual regulations (garbage treatment), safety at work and effectiveness of glass pre-treatment (installed compactor).
1) The vehicle has to be operated by a single person, even if two operators are expected to speed up the collection.
2) The loading phase must be independent from the compacting phase.
3) The operator must control every appliance from the driving cabin.
4) Dangerous operations (directly on the compactor and/or on the loading arm, outside the driving cabin) must be executed with "two hands" commands.

4 Automation system

The truck has been provided with sensors and actuators for automatic operation of all mechanical systems. The centralised control unit, in the cabin, allows the operator to control the bin loading arm and the glass compactor with simple commands on a comfortable user interface (graphic display with touch screen). Once the bin is aligned with the loader, the operator has to simply press a button and the glass is collected without any further human intervention.

Such a control system has been designed using traditional computational devices (industrial PC), running executable code programmed using high level language in order to exploit at best the computational resources. It can be easily adapted to other trucks operating garbage collection and, in general, to vehicles fitted with tanks and containers to carry and treat specific materials (liquid sewage, mud and water discharge transport, etc.). In similar situations the devices dedicated to specific tasks are mounted onto the truck structure; if necessary, the vehicle provides the mechanical power (through the drive shaft connected to the engine) and the electrical power supply (batteries and energy generation system).

4.1 The fieldbus

The system uses a fieldbus interconnection to provide an efficient interface to sensors and actuators. Moreover, the fieldbus allows optimal cabling in terms of used space and intervention time required for both installation and maintenance. With regard to the devices mounted on the truck, they are:

- sensors (such as mechanical transducers, for both angular and linear movements, proximity, level, temperature, etc.)
- actuators (such as oleodynamic and pneumatic electric valves, mechanical releasers, relais commands, signalling lamps, etc.)

    Employed sensors are:

- inductive proximity transducers have been positioned at the ends of the moving zones of every mechanical component;
- encoders sense the absolute position of the loading arm in its two joints;
- ultrasonic meters provide the distance of the bin from the truck.

Every sensor has its signal output: proximity transducers give on/off status with binary digital signal output, ultrasonic distance meters treat continuous information and thus they interface through analogue output voltage, complex devices require the information to be supplied using multiple signals and/or to be digitized and transferred through particular protocols.

Actuators act on the devices by delivering electric and pneumatic commands. In the truck the whole oleodynamic system is governed by electric valves and therefore digital (on/off) and analogue (continuous over the whole output range) power outputs are required; the grabbing ends of the loading arm are extended through pneumatic commands.

Among various fieldbus analysed, the Actuator/Sensor Interface (AS-I) has been considered due to its simple device management, flexible supported network structure, easy cable interconnection and certified safety capability integrated in the transmission protocol. The datarate supported by this fieldbus standard, well fits the requirements for interconnection to the devices mounted on the truck, where mainly small packets of information have to be delivered between the sensors / actuators and the central processing unit (industrial PC interfaced through AS-I master). Moreover this multi-vendor open standard makes available on the market a large amount of compliant devices, from generic interfaces to specific components (figure 6). Among them, absolute encoders with direct AS-I interface have been used and easily snapped to the fieldbus cable without any additional interface.

The device power supply is also facilitated by the two-wires "data and power" yellow cable which holds up to 8 amps. If more power is required by consuming devices (such as some electric valves employed in the truck) an additional black cable delivers the requested energy without any further burden affecting data transmission. The advantage of the AS-I solution in case of failures is enhanced by the integrated fault reporting feature implemented in the protocol: corrective actions have been programmed to promptly face the faults by driving the system to a safe shutdown or to safe positions. Once the faulty device is located, its simple replacement restores the full system efficiency being the fieldbus protocol in charge of re-configuring the new device with the same parameters of the previous one. This feature has been greatly appreciated by maintenance teams which can operate with shorter intervention times (that have been cut down to an average 60% with respect to conventional systems).

The AS-I standard includes the implementation of a safety protocol ("Safety at Work") which has been approved for applications up to Category 4 according to EN 954-1 and it also meets the requirements of Safety Integrity Level 3 (SIL3) of IEC 61508 on safe transmission.

Figure 6:   AS-I sensor / actuator interface devices and encoder

4.2 Monitoring and controlling software

The monitoring and controlling procedures (parameterised for full configurability) are structured in modules (libraries) and performed by software executed on standard operating systems. This helps the operator intervention, being also facilitated by the provided touch screen (tested for glove wearing users, as happens during waste collection) for direct interaction with the displayed interface elements, buttons and controls (figure 7).

Accurate software programming allows the implementation of concurrent control processes (multithread programming) helping the execution of self-contained routines, independent from one another: for example it is possible to verify the device status and to drive actuators meanwhile the control algorithm manages the system. Three main modules have been considered: implementation of the operating procedures for control and system configuration, fieldbus management, user interface management.

4.2.1 Module implementing the operating control procedures and the system configuration

It detects sensors and actuators, it recognises and matches their capabilities to the operating functionalities of the glass-waste collection truck. It resolves every addressing issue related to the communication through the fieldbus and it identifies every device task (for example it relates a movement limit to its corresponding sensor, an electric valve to an actuator, etc.). If faults are detected, not allowing safe execution of every system functionality (such as the lack of detection in the reach of a mechanical limit), the software enters a "user warning mode" thus allowing problem identification and solving.

The module also implements the following procedures for the control of the loading arm and the glass compactor.

- Automatic loading: the operator aligns the truck to the bin and he starts the collection. Automatically the arm extends to reach the bin, it grabs it and empties it into the container. Afterwards the arm puts down the bin in its original place and it returns on its rest position.
- Manual loading: the operator switches the system to use a joystick to control the loading arm (figure 7). This is useful if waste bins are in wrong positions, inaccessible by the standard automatic procedure, but at the same time it avoids the operators to go out of the driving cabin and manually move the bin in a correct position.
- Semi-automatic loading: the operator can grab manually the bin and the automatic procedure completes the loading operation.
- Compaction: the operator can start a compacting action. It is even possible to set the compactor to act automatically after every collection.
- Unloading: the operator can manage the container unloading procedure according to applicable safety regulations.

Figure 7:   Operator interface (touch-screen display and joystick).

4.2.2 Module implementing the fieldbus management

Data acquisition and command delivering is performed through the AS-I master device connected to the fieldbus: the interconnection routines have been optimised for speed without compromising error and data consistency checks. Remedial and/or recovering actions can be triggered in case of fault detection.

The possibility to access the fieldbus (retrieving data and/or sending commands) is granted independently from the corresponding function calls in the control software. For example this allows more than one control process to access the fieldbus through the same AS-I master device, provided that the different processes act on different sets of slave devices, otherwise contention errors may occur. This capability has been used in the independent control of the loading arm and the compactor.

4.2.3 Module implementing the user interface

It manages the user interaction and commands execution through the industrial PC display. Showing the system status can be effective if data are organised in graphic form, thus allowing a fast and easy way to recognise the configuration of the installed devices and the instantaneous position of every mechanical part of the loading arm and the glass compactor (figure 7). This software module also performs the initial setup of the control program, the insertion of the operating parameters and the checking of the executed actions (e.g. using graphical views, log files, etc.). If anomalous running conditions are detected, or unaccepted system configurations are set, it also activates warning alarms.

The complete monitoring of the controlling software (also saving data in a "log file") is essential in order to keep track of the use of every vehicle (waste collectors, compactors, etc.) which is part of the fleet dedicated to garbage collection. Thanks to the module structure of the software, additional improving functions could be easily implemented: for example remote data collection can be considered by using medium-long wireless interconnection, with values stored in a comprehensive database for later evaluation. Remote monitoring could address vehicle routing and locating issues (logistic), loading optimisation by evaluating the weight of collected glass in real time, etc.

The automated system allows a single operator to drive the truck and, at the same time, to manage the waste collection. In practice, also in order to support other activities required during the collection, two operators are employed. Nevertheless it has been confirmed that the operators' burden during glass-waste collection is greatly reduced together with the risks involved by this task.

5 Conclusions

This paper presented some design guidelines related to a glass-waste collection truck with integrated compactor. The collection is achieved using a side-loading arm controlled by an industrial-PC based system that includes a fieldbus for fast command delivering and system monitoring. A suitable, customised software manages the whole system, in which the two parts, compactor and loading arm, are actuated either synchronously or independently from each other.

Few operators are required: the complete collection route can be entirely completed by one person only, the driver of the vehicle. In practice two operators are sufficient and manual intervention is reduced to a minimum.

The optimisation achieved in the compactor design gives optimal glass fragmentation for recycling: this results from the mechanical design of the compactor shovel and the tuning of its oil-pressure power circuitry.

Acknowledgments

The authors gratefully acknowledge Muzzin Srl for the precious technical advices during the design and prototyping stages of the work, and Pepperl+Fuchs Italy for the AS-I fieldbus support.

Author Index

Environmental Economics and Investment Assessment

Edited by: K. ARAVOSSIS, University of Thessaly, Greece, C. A. BREBBIA, Wessex Institute of Technology, UK, E. KAKARAS, National Technical University of Athens, Greece, A. G. KUNGOLOS, University of Thessaly, Greece

The current emphasis on sustainable development is a consequence of the general awareness of the need to solve numerous environmental problems resulting from our modern society. This book addresses the topic of Investment Assessment and Environmental Economics in an integrated way; in accordance with the principles of sustainability; considering social and environmental impacts of new investments. Bringing together papers from the First International Conference on Environmental Economics and Investment Assessment, papers encompass topic areas such as: Economy and the Environment; Investment Planning and Assessment; Environmental Economics and Entrepreneurship; Environmental Investment Planning; Sustainable Environmental Management; Environmental Impact Assessments and Investments; Environmental Performance Indicators; Environmental Management Systems; Legislation and Law Enforcement; Cost Benefits Analysis; Natural Resources Management; Social Issues and Environmental Policies; Risk Management in Environmental Investment; Location Optimization.

WIT Transactions on Ecology and the Environment, Volume 98
**ISBN: 1-84564-046-2 2006 apx 350pp
apx £129.00/US$233.00/€193.50**

Brownfields III

Prevention, Assessment, Rehabilitation and Development of Brownfield Sites

Edited by: C. A. BREBBIA, Wessex Institute of Technology, UK, U. MANDER, University of Tartu, Estonia

This book focuses upon the problems facing the public and private sectors and the engineering and scientific communities, in terms of the decrease of available new land for development purposes. The volume looks at long term plans for the productive re-use of properties that have been abandoned or lie idle, in order to satisfy current needs without compromising the ability of future generations to meet their own requirements. Featuring papers from the Third International Conference on Prevention, Assessment, Rehabilitation and Development of Brownfield Sites, the text will be vital to practitioners and those businessmen in industry and commerce as well as those in research organisations interested in the problems facing the prevention, assessment, rehabilitation and development of brownfields. Topics featured include: Environmental Assessment; Risk Assessment and Management; Monitoring of Contaminated Sites; Cleanup Methodologies; Lessons from the Field; Case Studies; Development Issues; Community and Public Involvement; Financial Aspects; Multimedia Modelling and Assessment; Hydrological Aspects; Differences in Environmental Laws.

WIT Transactions on Ecology and the Environment, Volume 94
**ISBN: 1-84564-041-1 2006 apx 350pp
apx £129.00/US$225.00/€193.50**

WITPRESS

Development and Application of Computer Techniques to Environmental Studies X

Editors: G. LATINI and G. PASSERINI, Universita Politecnica delle Marche, Italy and C.A. BREBBIA, Wessex Institute of Technology, UK

This volume features the proceedings of the Tenth International Conference on the Development and Application of Computer Techniques to Environmental Studies (ENVIROSOFT), and will be of interest to many specialists such as environmental scientists, planners and administrators, program developers, regulators and industrialists.

The papers included cover a wide range of issues and are divided under the broad headings: Environmental Management and Decision Analysis; Environmental Impact Assessment; and Air, Water and Soil Pollution, while several contributions confirm the necessity of specialised software for local, small scale issues.

Series: Environmental Studies, Vol 11
ISBN: 1-85312-718-3 2004 208pp
£90.00/US$144.00/€135.00

WIT Press is a major publisher of engineering research. The company prides itself on producing books by leading researchers and scientists at the cutting edge of their specialities, thus enabling readers to remain at the forefront of scientific developments. Our list presently includes monographs, edited volumes, books on disk, and software in areas such as: Acoustics, Advanced Computing, Architecture and Structures, Biomedicine, Boundary Elements, Earthquake Engineering, Environmental Engineering, Fluid Mechanics, Fracture Mechanics, Heat Transfer, Marine and Offshore Engineering and Transport Engineering.

Brownfield Sites II
Assessment, Rehabilitation and Development

Editors: A. DONATI and C. ROSSI, University of Siena, Italy and C.A. BREBBIA, Wessex Institute of Technology, UK

Bringing together information, experience and research from many countries in order to give readers the ability to help revitalize their communities, this volume features papers from the Second International Brownfields Conference.

Invaluable to all specialists involved in this challenging subject, the book covers: Environmental Assessment; Risk Assessment and Management; Monitoring of Contaminated Sites; Cleanup Methodologies; Lessons from the Field; Case Studies; Development Issues; Community and Public Involvement; and Financial Aspects.

ISBN: 1-85312-719-1 2004 336pp
£134.00/US$214.00/€201.00

All prices correct at time of going to press but subject to change.
WIT Press books are available through your bookseller or direct from the publisher.

WIT eLibrary

Home of the Transactions of the Wessex Institute, the WIT electronic-library provides the international scientific community with immediate and permanent access to individual papers presented at WIT conferences. Visitors to the WIT eLibrary can freely browse and search abstracts of all papers in the collection before progressing to download their full text.

Visit the WIT eLibrary at
http://library.witpress.com

Brownfields

Multimedia Modelling and Assessment

Editor: G. WHELAN, Pacific Northwest National Laboratory, USA
Multimedia modelling and assessment compartmentalizes the real-world environment into its respective components or processes and describes the linkages and interactions between these in order to facilitate more informed decision-making. This book contains invited contributions first presented at the Second International Brownfields Conference. Covering some of the most mature and widely used multimedia software technology products and approaches designed to support brownfields and hazardous waste site decision makers, the volume describes software tools and methods, and illustrates applications.

ISBN: 1-85312-755-8 2004 152pp
£68.00/US$109.00/€102.00

Waste Management and the Environment II

Editors: V. POPOV, Wessex Institute of Technology, UK, H. ITOH, University of Nagoya, Japan, C.A. BREBBIA, Wessex Institute of Technology, UK and A. KUNGOLOS, University of Thessaly, Greece
Highlighting present challenges and opportunities for progress, this book contains over 65 papers from the Second International Conference on Waste Management and the Environment. The topics discussed will be of interest to a wide readership including government officials, waste disposal experts, research scientists specialising in this area, and environmental engineers.

The contributions are divided under the following broad subject headings: Advanced Waste Treatment Technology; Hazardous Waste Management; Disposal of Hazardous Waste in Underground Mines; Biological Treatment of Waste; Biosolids, Composting and Agricultural Issues; Environmental Effects and Remediation; Waste Reduction and Recycling; Landfills, Design, Construction and Monitoring; Waste Management, Strategies and Planning; Waste Management in Greece; Waste and Wastewater Treatment; and Methodologies and Practices.

ISBN: 1-85312-738-8 2004 696pp
£245.00/US$392.00/€367.50

Waste Management in Japan

Editor: H. ITOH, University of Nagoya, Japan
This book contains contributions first presented in the special session Advanced Waste Treatment and Management in Japan at the Second International Conference on Waste Management and the Environment. Topics discussed include the novel utilization of wasted materials for glass-ceramics or pollutant absorbers, hazardous waste detoxification and extraction techniques, the recycling of organic or agricultural wastes, and advanced incineration technology for thermal recycling.

The papers featured will help the international waste management community gain an appreciation of current issues in Japan together with their technical solutions, and increase the potential for international cooperation.

ISBN: 1-84564-000-4 2004 192pp
£90.00/US$144.00/€135.00